MICROBIOLOGY
OF
WOUNDS

MICROBIOLOGY
OF
WOUNDS

Edited by
Steven Percival and Keith Cutting

CRC Press
Taylor & Francis Group
Boca Raton   London   New York

CRC Press is an imprint of the
Taylor & Francis Group, an **informa** business

CRC Press
Taylor & Francis Group
6000 Broken Sound Parkway NW, Suite 300
Boca Raton, FL 33487-2742

Library of Congress Cataloging-in-Publication Data

Microbiology of wounds / editors, Steven Percival, Keith Cutting.
 p. ; cm.
 Includes bibliographical references and index.
 ISBN 978-1-4200-7993-7 (hardcover : alk. paper)
 1. Wounds and injuries--Infections. 2. Wounds and injuries--Microbiology. I. Percival, Steven L. II. Cutting, Keith F. III. Title.
 [DNLM: 1. Wound Infection--microbiology. 2. Wound Infection--prevention & control. 3. Biofilms. 4. Chronic Disease. 5. Wound Healing--physiology. WC 255 M626 2010]

RD95.M53 2010
617.1--dc22   2010005449

Visit the Taylor & Francis Web site at
http://www.taylorandfrancis.com

and the CRC Press Web site at
http://www.crcpress.com

Steven L. Percival would like to dedicate this book to Carol, Alex, Tom, Mum, Dad, Nigel, and Emma.

Keith F. Cutting would like to dedicate this book to Maud, Amelia, and Henry, and to all the patients who have allowed me to learn from them.

Contents

Preface...ix
Editors...xi
Contributors xiii

Chapter 1   An Introduction to the World of Microbiology and Biofilmology 1
Steven L. Percival, John G. Thomas, and David Williams

Chapter 2   Human Skin and Microbial Flora 59
Rose A. Cooper and Steven L. Percival

Chapter 3   An Introduction to Wounds.. 83
Michel H.E. Hermans and Terry Treadwell

Chapter 4   Burn Wound Management 135
Michel H.E. Hermans

Chapter 5   Cell Biology of Normal and Impaired Healing................................. 151
Keith Moore

Chapter 6   The Microbiology of Wounds 187
Steven L. Percival and Scot E. Dowd

Chapter 7   Types of Wounds and Infections 219
Randall D. Wolcott, Keith F. Cutting, Scot E. Dowd, and
Steven L. Percival

Chapter 8   Biofilms and Significance to Wound Healing.................................. 233
Keith F. Cutting, Randall D. Wolcott, Scot E. Dowd, and
Steven L. Percival

Chapter 9   Wounds, Enzymes, and Proteases.. 249
Steven L. Percival and Christine A. Cochrane

Chapter 10 Wound Healing Immunology and Biofilms 271

Emma J. Woods, Paul Davis, John Barnett, and Steven L. Percival

Chapter 11 Antimicrobial Interventions for Wounds 293

Steven L. Percival, Rose A. Cooper, and Benjamin A. Lipsky

Chapter 12 Wound Dressings and Other Topical Treatment Modalities in
Bioburden Control 329

Richard White

Chapter 13 Factors Affecting the Healing of Chronic Wounds:
An Iconoclastic View 345

Marissa J. Carter and Caroline E. Fife

Index 373

Preface

Chronic wounds are a serious public health issue. The incidence and prevalence of the different types of chronic wounds are largely unknown worldwide, but 13 years ago George[1] estimated the worldwide burden of wounds to be:

- Surgical wounds, 40 to 50 million
- Leg ulcers, 8 to 10 million
- Pressure ulcers, 7 to 8 million
- Burns, 7 to 10 million

In the United States alone, the estimated number of chronic wounds includes 1 to 2 million diabetic foot ulcers, 1 to 2 million venous leg ulcers, 3 to 5 million pressure ulcers, and 1% surgical site infections. One of the underlying pathologies known to increase the prevalence of chronic wounds is diabetes mellitus. Diabetes mellitus in the Western world is growing continuously at a double-digit rate. However, this figure is not truly representative of the extent of the problem. Figures from the Centers for Disease Control and Prevention (CDC) state that there are approximately 24 million patients with diabetes mellitus (24 million diabetics). Cutaneous wounds in the United States alone cost society over $25 billion annually.

The management of infected wounds remains an area of confusion and hence great debate. No definition or authoritative clinical guidelines of what constitutes an infected wound exists. Terminology in wound care such as colonization, critical colonization, biofilm, and other descriptions of bacterial behavior on the surface of the wound are not clearly defined. Even the term *infection* requires redefining in light of recently generated insight into the prevalence and behavior of the biofilm phenotype. In addition, many of the concepts concerning wound infections are not backed up with meaningful scientific support. Consequently many terms used in wound care have led to confusion and unnecessary or inappropriate management of chronic wounds.

It is well established that wound healing is dynamic, infinitely complex, non-linear, and prodigiously individualized to the context of the patient. Understanding the intricacies of chronic wounds becomes even more complex when one considers the myriad of host variables that contribute to the disease state.

The plausible common barrier that may impair many of these wounds from healing is chronic infection as a result of biofilm infection. Chronic biofilm-based infections constitute 80% of all human infection. Accordingly, acute infections remain as the minority census of all infectious disease. The definition of acute infection is based on clinical characteristics of rapid onset and aggressive bacterial behavior, which responds rapidly and completely to antibiotics or the host immune response. Chronic infections are persistent and recalcitrant. It is interesting to note that acute and chronic infections have not been clearly differentiated on a molecular level and may be explained by bacteria pursuing widely divergent survival strategies only now becoming elucidated through research.

Bacteria producing chronic infections employ a biofilm phenotype for their infectious strategy. In this type of survival strategy, the bacteria attach to the host and subvert a number of host systems. First, the bacterium rapidly encases itself inside an extracellular polymeric substance (EPS), which protects the biofilm members from the host's immunity. Bacteria within the biofilm secrete communication molecules termed *quorum-sensing molecules,* which direct the activity of many bacteria within the community. Subsequently, the bacteria are collectively under partial regulatory control by the community. This highly organized and competent communal structure is biofilm.

This biofilm, now attached and centrally regulated, strives to reproduce itself and obtain sustainable nutrition from host sources. Through multiple modes, the biofilm exploits host inflammatory pathways. By commandeering other host pathways, the biofilm is able to prevent apoptosis in the cells that constitute the wound bed. Consequently, a senescent wound bed is created which may provide a stable base of attachment. Further, the biofilm prevents host neutrophils from lysing the surface of the wound bed to remove the attached biofilm. The biofilm also downregulates bacterial protease activity, and alternatively, stimulates the host inflammatory response to produce increased host protease activity, thereby generating plasma transudate as a constant source of fluid and nutrition for the biofilm community. As previously stated, through quorum sensing the biofilm can regulate the size and activity of the entire community in its pursuit of a parasitic strategy.

The issue of addressing whether or not chronic wounds are "infected" is controversial, and that controversy seems to rely on the traditional diagnostic markers of infection (acute or chronic). Chronic wounds exhibit high proinflammatory cytokines, high host protease activity, and excessive neutrophil infiltration, which is predictably typified by most other tissues affected by the persistent biofilm infection. Also, the secondary signs of infection most common in wounds—such as excessive exudate, a soft degraded and senescent wound bed that fails to progress—are all consistent with a host response to a biofilm.

Acknowledgment of the presence of chronic infection and biofilm in most chronic wounds as an important barrier to healing allows a single, unified perspective for the approach and treatment of chronic cutaneous wounds. Indubitably, patient comorbidities such as neuropathy, immobility, poor perfusion, impaired immunity, malnutrition, and systemic diseases must be aggressively managed as a parallel strategy to optimize the treatment regimen, but if we ignore the contribution of biofilms to infection and fail to manage appropriately, then our care will be suboptimal.

Steven L. Percival
Keith F. Cutting

REFERENCE

1. George, G. (1996) *Wound Management.* Richmond Surrey, U.K., PJB Publications.

Editors

Steven Percival, PhD, qualified from the Department of Microbiology, University of Leeds, United Kingdom, with a PhD in medical microbiology and biofilms. He holds a BSc in applied biological sciences, postgraduate certificate in education, an MSc in epidemiology and health sciences, and an MSc in medical and molecular microbiology. He is also a Fellow of the Institute of Biomedical Sciences. Following 6 years as a senior university lecturer in medical microbiology, Dr. Percival was awarded a prestigious senior clinical research fellowship to investigate biofilm control and catheter-related bloodstream infections at the Centers for Disease Control and Prevention, Atlanta, Georgia, and Leeds Teaching Hospitals Trust. In addition to intravascular catheter research, Dr. Percival has undertaken, provided consultancy, and directed research for major companies and government departments worldwide on antimicrobials, waterborne diseases, urinary catheter-related infections, ventilator-associated pneumonia, infection control, and hospital acquired infections. He has written over 150 scientific peer-reviewed journal papers, book chapters, and conference abstracts on microbiology and biofilms and has authored and edited four biofilm and microbiology textbooks. He has also presented extensively worldwide on biofilms, infection control, and medical microbiology. From 2003 to 2009 he held a senior position in research and development and was responsible for medical microbiology and anti-infectives research and product innovation at Bristol Meyers and Squibb (ConvaTec Ltd., UK). In 2009 Dr. Percival joined Advanced Medical Solutions, Ltd., UK to head up innovation, research, product support, and new technology development. He also holds the position of adjunct Professor of Medical Microbiology at the Medical School, University of West Virginia. He presently has a number of patents filed on areas related to anti-biofilm/antimicrobial compositions and diagnostics for use in wound care.

Keith F. Cutting, PhD, received his master's degree from the College of Medicine, University of Wales, Cardiff, United Kingdom. He holds a diploma in nursing from London University and has a certificate in education (FE) from University of Wales, Cardiff. Dr. Cutting has been involved in tissue viability for over 20 years and is currently a visiting professor at Buckinghamshire New University, United Kingdom. In addition to lecturing, he has maintained clinical and research roles. He has published widely and written over 100 journal papers, book chapters, and conference abstracts and is editor of *Trends in Wound Care*, volumes IV and V. Dr. Cutting has a particular interest in diagnosis and management of wound infection and was consultant editor and contributor to the European Wound Management Association (EWMA) position document, Criteria for Wound Infection (September 2005), and is editor and contributing author to *Advancing Your Practice: Understanding Wound Infection and the Role of Biofilms* (Association for the Advancement of Wound Care, Malvern, Pennsylvania, 2008). He has presented his work nationally and internationally at conferences including the World Union of Wound Healing Societies, European Wound

Management Association, the European Tissue Repair Society, and the Symposium on Advances in Wound Care. Cutting is clinical editor for *Wounds*, United Kingdom journal, a member of a number of wound healing societies, a Fellow of the Higher Education Academy, and a Regional Fellow of the Royal Society of Medicine. He works with a number of international corporations and medical agencies as an independent consultant.

Contributors

John Barnett
Department of Microbiology,
    Immunology, and Cell Biology
West Virginia University School of
    Medicine
Morgantown, West Virginia

Marissa Carter
Strategic Solutions, Inc.
Cody, Wyoming

Christine A. Cochrane
Leahurst School of Veterinary Science
University of Liverpool
Liverpool, United Kingdom

Rose A. Cooper
Cardiff Institute of Higher Education
Cardiff, Wales, United Kingdom

Keith Cutting
Buckinghamshire New University
High Wycombe, Buckinghamshire,
    United Kingdom

Paul Davis
Department of Chemistry
University of Warwick
Coventry, United Kingdom
Mologic Ltd.
Sharnbrook, Bedford, United Kingdom

Scot E. Dowd
Research and Testing Laboratories of
    the Southwest
Lubbock, Texas

Caroline Fife
Department of Medicine, Division of
    Cardiology
University of Texas Health Science
    Center
Houston, Texas
Memorial Hermann Center for Wound
    Healing
Humble, Texas

Michel H.E. Hermans
Hermans Consulting Inc.
Newtown, Pennsylvania

Benjamin A. Lipsky
Primary and Specialty Medical Care
Veterans Affairs Puget Sound Health
    Care System
Division of General Internal Medicine,
    Department of Medicine
University of Washington School of
    Medicine
Seattle, Washington

Keith Moore
WoundSci
Somerton, Somerset, United Kingdom

Steven L. Percival
West Virginia University School of
    Medicine
Robert C. Byrd Health Sciences
    Center–North
Morgantown, West Virginia
Advanced Medical Solutions
Winsford, Cheshire, United Kingdom

John G. Thomas
West Virginia University School of
   Medicine
Robert C. Byrd Health Sciences Center
North Morgantown, West Virginia

Terry Treadwell
Institute for Advanced Wound Care
Montgomery, Alabama

Richard White
University of Worcester
Worcester, United Kingdom

David Williams
School of Oral Microbiology
University of Cardiff
Cardiff, Wales, United Kingdom

Randall D. Wolcott
Southwest Regional Wound Care
   Center
Lubbock, Texas

Emma J. Woods
Manchester, United Kingdom

1 An Introduction to the World of Microbiology and Biofilmology

Steven L. Percival, John G. Thomas, and David Williams

CONTENTS

Introduction 3
A Brief History of Microbiology 3
The Microbial World 5
Bacteria 5
    Shapes 5
    Size 5
    Structure of Bacteria 6
        Cell Wall 6
        Structure of Cell Wall in Bacteria 7
        Capsule/Slime Layer 7
        Ribosomes 8
        Nucleoid 8
        Plasmids 8
        Pili and Fimbriae 8
        Flagella 8
    Chemotaxis 8
Growth of Bacteria 9
    Lag Phase 9
    Exponential Phase 10
    Stationary Phase 10
    Death Phase 10
Factors That Affect Growth 11
    pH 11
    Temperature 11
    Oxygen 11
Control of Microorganisms 12
    Terms Used to Describe Microbial Control 13
    Death of Microorganisms following Exposure to Antimicrobial Agents 13
    Mechanism of Action of Antimicrobials 14

Effectiveness of the Antibiotic or Antimicrobial... 14
Mechanism of Bacterial Resistance 15
Transmission of Antibiotic Resistance 16
Epidemiology 16
Sources of Pathogens 17
Acquisition of Pathogens 17
Control of Disease and Infection.. 17
Control of Nosocomial Infections... 17
Historical Aspects of Biofilms 18
Occurrence of Biofilms.. 20
Stages in the Formation of Biofilms 22
Development of the Conditioning Film and Substratum Effects 22
Events That Bring Microorganisms into Close Proximity with the Surface........... 24
Microbial Adhesion... 26
Role of Pili and Fimbriae in Adhesion 27
Growth and Division of the Microorganisms at the Colonized Surface 29
Microcolony and Biofilm Formation.. 29
Extracellular Polymeric Substances 30
Gene Transfer 31
Biofilm Structure 31
Factors That Govern the Development of Biofilms 32
Quorum Sensing 33
Detachment and Dispersal of the Biofilm 33
Public and Medical Health Consequences of Biofilms... 36
Drinking Water.. 36
Hospital and Domestic Water.. 37
Dental Water Units 37
Kidney Stones 37
Endocarditis... 38
Cystic Fibrosis... 38
Otitis Media... 38
Osteomyelitis... 38
Prostatitis... 39
Intra-amniotic Infection 39
Indwelling and Medical Devices... 39
Urinary Tract Infections 39
Central Venous Catheters 40
Endotracheal Tubes (ETTs)... 40
Rhinosinusitis.. 40
Ophthalmic Infections 42
Oral Infections 42
Chronic Wounds 43
Biofilm Resistance 43
Binding/Failure of the Antimicrobial to Penetrate the Biofilm.......................... 44
Slow Growth and the Stress Response 44

Heterogeneity 45
Induction of a Biofilm Phenotype 45
Conclusion 45
References.. 46

INTRODUCTION

Microorganisms can simply be defined as organisms that cannot be viewed without the aid of a microscope. Included in this category are viruses, bacteria, certain species of algae and fungi, and protozoa. Microorganisms are very important in the ecology of the planet and the existence of man, playing essential roles in the carbon, nitrogen, and sulfur cycles. They have historically been exploited within human society for the synthesis of foodstuffs such as cheese, beer, bread, wine, and vinegar. More recently, with the advent of genetic engineering it has been possible to extend this list to the manufacture of specific enzymes, antibiotics, vitamins, and human hormones such as insulin.

The majority of microorganisms are generally beneficial, but to the "non-microbiologist," microorganisms are all too often considered harmful and a cause of human disease and infections. The notoriety associated with pathogenic bacteria is a result of the devastating effect that widespread and virulent bacterial disease can wreak on individuals and communities. For example, the involvement of pathogenic bacteria in the Black Death (bubonic plague) in 1346–1352 resulted in the death of approximately 25 million people, one-third of the population of Europe. The microorganism responsible for this disease, a rod-shaped bacterium, was only identified in the latter part of the 19th century and is now known as *Yersinia pestis*.

A BRIEF HISTORY OF MICROBIOLOGY

Lucretius (98–55 B.C.), a Roman philosopher and poet, suspected the existence of microorganisms and their involvement in causing disease. Later, Girolamo Fracastroro (1478–1553), an Italian physician, echoed these sentiments and suggested that diseases were caused by invisible living creatures. Despite these historical accounts, it was not until the work of Antonie van Leeuwenhoek (1632–1723) in 1673 that microorganisms were first visualized and described using his rudimentary microscope. Leeuwenhoek used the term *animalcules* to describe his observations of microorganisms that he acquired from dental plaque samples.

A number of individuals became associated with the growth in understanding of microorganisms and their impact on society. For example, in 1748 John Needham observed the result of the growth of microorganisms following the addition of organic matter to water and seeds. Louis Pasteur (1822–1895) provided evidence about the growth of microorganisms; in particular, how to protect solutions from microbial contamination (i.e., pasteurization, which uses heating to inhibit microbial growth).

As mentioned previously, the role microorganisms played in human disease processes was first acknowledged by Girolamo Fracastroro. However, at the time, many other physicians and scientists believed that human diseases were related to forces that were unnatural and occurred through poisonous vapors. The Greek physician

Galen suggested that it was imbalances in the humors (humorism) of the human body that caused disease in the body. These humors were referred to as blood, phlegm, yellow bile, and black bile.

However, the role of microorganisms in disease, the *germ theory,* was not acknowledged until about 1835 by Agostino Bassi, who demonstrated that a disease (white muscadine) in the silkworm was caused by a fungus, now known as *Beauveria bassiana.* In 1845, Berkeley showed that potato blight was also caused by a fungus (*Phytophthora infestans*).

The results of these pioneering studies and concerns over the silk industry led Pasteur to investigate silkworm disease further. Pasteur actually discovered a separate silkworm disease (pébrine) which was due to a protozoan (*Nosema bombycis*) infection. By recognizing that diseased worms were spreading infection to uninfected silk worms, he devised simple strategies to keep uninfected silkworms healthy, which ultimately saved the silk industry in France at the time.

Evidence for the involvement and prevention of infections in wounds was reported by studies undertaken by Joseph Lister (1827–1912). Lister applied phenol to surgical dressings and heat sterilized many instruments being used in wound treatment. These approaches were shown to have a significant positive impact on the outcome of wound infections and diseases and provided clear evidence on the role of phenol as an agent that could kill microorganisms.

Evidence that bacteria could cause disease transpired following the work of Robert Koch (1843–1910), who focused mainly on the role of the bacterium *Bacillus anthracis* and its ability to cause anthrax (1876). As a result of these studies, Koch stipulated that a number of criteria needed to be met for a microorganism to be identified as a disease-causing agent. These criteria are today referred to as Koch's postulates and are summarized as follows:

1. The microorganism must be evident in all episodes of the disease.
2. The microorganism has to be isolated and grown as a pure culture.
3. When the microorganism is inoculated into a healthy host, it must cause the same disease.
4. The microorganism must be isolated from the host that is diseased.

Koch continued with his work and produced numerous studies, many of which were confirmed or supported by eminent scientists in the field. The prolific output from Koch gave rise to the use of agar as the base for microbiological culture media by one of Koch's assistants, Walther Hesse. Another of Koch's assistants, Richard Petri, then developed the Petri dish. The combination of the Petri dish and agar greatly aided Koch in further developing culture media for the growth of bacteria recovered from humans, which allowed further advances in the study of human disease. The fact that the Petri dish and agar remain the mainstay of cultural microbiology today highlights the significance of this breakthrough.

As time progressed, further work on microorganisms led to the discovery of viruses and their role in disease. In addition, the first protective role of vaccines (1796) had been shown by Edward Jenner (1749–1823) in his study of smallpox, and work by Pasteur and Charles Chamberland (1851–1908) extended the development of vaccination to produce live attenuated vaccines. Only 3 years after Koch's work on *B. anthracis,* Pasteur successfully generated a protective vaccine, and this was followed with equally successful vaccine against rabies (1885).

It is through the work of these pioneer microbiologists that our current understanding of disease and infection stems, and it is notable that these historical concepts are still practiced even today in "modern" microbiology.

THE MICROBIAL WORLD

There are two types of cells found within the human body; namely eukaryotic and prokaryotic cells. Human cells are examples of eukaryotic cells which are outnumbered greatly by colonizing prokaryotic cells that are typified by bacteria. Prokaryotic cells are structurally more basic than eukaryotic cells, and although the former contain DNA, there is no nucleus and an absence of membrane-bound organelles.

BACTERIA

SHAPES

Cocci are spherically-shaped bacteria and may appear under the microscope in pairs (diplococci) (e.g., *Neisseria*), in chains (e.g., *Streptococcus* and *Enterococcus*), in clusters (e.g., *Staphylococcus*), or as tetrads (e.g., *Micrococcus*). Bacillus or rod-shaped bacteria are frequently encountered in the environment and mammalian infections (e.g., *Pseudomonas* and *Salmonella*). Curved bacilli are also evident and collectively referred to as vibrios, and cocco-bacilli are short but wide-shaped bacteria (e.g., *Acinetobacter* sp). Many bacteria occur as long twisted rods called spirilla if rigid, or spirochaetes if flexible. Even though bacteria are found in many different shapes and sizes, they should not be stereotyped on this basis, as numerous variables are known to influence the shape of bacteria. Bacteria are referred to as pleomorphic when they can exist in a variety of different shapes and sizes (e.g., *Corynebacterium* and *Mycoplasma*).

SIZE

Bacteria exist in many different sizes. For example, nanobacteria range in size from about 0.2 μm to less than 0.05 μm. *Nanobacterium sanguineum* was proposed in 1998 as a cause of pathological calcification (apatite in kidney stones). *Bdellovibrio* are comma-shaped motile rods about 0.3–0.5 × 0.5–1.4 μm in size. This particular genus of bacteria can parasitize Gram-negative bacteria by entering into their periplasmic space and feeding on the proteins and nucleic acids of their hosts. Mycoplasma can be as small as 0.3 μm in diameter. The largest currently known bacterium (*Thiomargarita namibiensis*) has a diameter of 750 μm. However, most commonly, bacteria tend to be of the size order of 1.1–1.5 μm wide and 2.0–6.0 μm long.

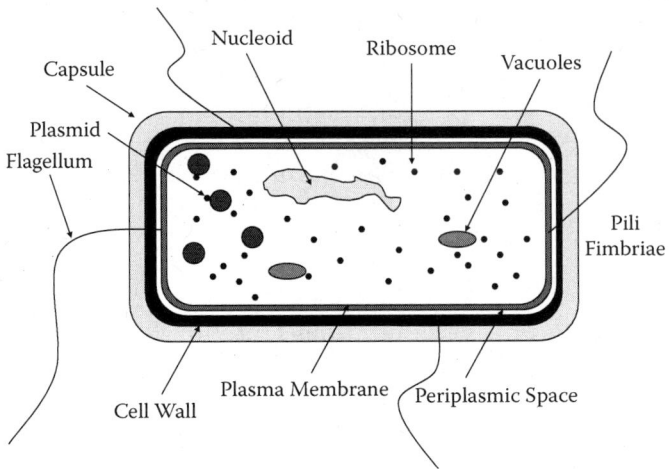

FIGURE 1.1   Representative structure of a bacterium.

TABLE 1.1
Components of a Bacterium

Structure	Function
Capsule/slime layer	Resist phagocytosis; important for adhesion
Cell wall	Produces bacterial shape/protection
Periplasmic space	Contains enzymes, binds proteins (nutrients)
Plasma membrane	Permeable (selective), metabolic processes, important for chemotaxis (detection)
Vacuole (gas)	Buoyancy
Ribosomes	Synthesis of proteins
Nucleoid	DNA storage area (genetic material)
Flagella	Movement
Plasmid	Circular double-stranded DNA (independent of chromosome)
Pili/fimbriae	Adhesion to surfaces

STRUCTURE OF BACTERIA

A representative structure of a bacterium can be seen in Figure 1.1. Table 1.1 shows the structures that make up the bacterium.

Cell Wall

In 1884, the Danish scientist Christian Gram discovered that bacteria could be divided into two types—namely, Gram positive or Gram negative, based on their appearance after a particular staining process (Gram stain). This occurs due to the fundamental differences in the cell wall structure between the two cell types. After

Gram staining, Gram-positive bacteria stain purple and Gram-negative bacteria appear pink or red.

The Gram stain involves adding crystal violet to a slide of heat-fixed bacteria. All bacteria will take up the crystal violet, but due to a thick (20–80 nm) peptido-glycan layer within the Gram-positive cell wall, it is only this group of bacteria that retains the crystal violet after subsequent staining and washing steps. The next process is the addition of Gram's iodine (iodine and potassium iodide) which complexes with the crystal violet, creating a larger molecule to aid retention in the cell. Ethanol is added, which dehydrates the peptidoglycan, effectively locking the crystal violet complex within Gram-positive cells. In Gram-negative bacteria, the peptidoglycan layer is much thinner (2–7 nm) with the result that the crystal violet and iodine complex is not retained and the cells appear pink or red after the final addition of a counter stain, typically safranin.

Compared with Gram-positive bacteria, the cell wall of Gram-negative bacteria is a much more complex structure. The Gram-negative cell wall actually consists of an inner and outer membrane separated by a large periplasmic space. Although a periplasmic space occurs in Gram-positive bacteria, it is much thinner or often absent. The periplasmic space in Gram-negative bacteria has been reported to contain many hydrolytic enzymes.

Structure of Cell Wall in Bacteria

The chemical content of the Gram-positive bacterial cell wall is composed principally of peptidoglycan with large amounts of teichoic acids (polymers of glycerol or ribitol). Attached to the glycerol and ribitol are sugars such as glucose or the amino acid D-alanine.

In Gram-negative bacteria, outside the thin peptidoglycan layer is the outer membrane. The outer layer contains lipopolysaccharide (LPS) that consists of lipids and carbohydrates that constitutes lipid A. Lipid A is the anchor for LPS in the cell membrane, to which an O side chain is attached via a core polysaccharide. LPS in general provides the bacterium with a negative charge. The lipid A component of the LPS is known to be toxic and is referred to as an endotoxin.

Some bacteria, called L forms, are known to lack a cell wall and can occur in both Gram-positive and Gram-negative bacteria.

The cell wall is very important as it determines the shape of the bacterium and aids to protect the bacterium from toxic agents as well as antimicrobial agents, pH changes, and ionic and osmotic effects.

Capsule/Slime Layer

A bacterial capsule is a layer found on the outside of a bacterium. The capsule cannot easily be removed by washing. Contrary to this is the slime layer, which consists of a more diffuse material that can be easily removed from a bacterium by washing. The material on the outside of the bacterial cell wall is often called the glycocalyx. The glycocalyx is composed of polysaccharides and other components that have included glycoproteins and lipids. The glycocalyx is considered to have a significant role to play in mediating bacterial attachment to surfaces. The slime layer and the capsule are considered by some to be synonymous with the glycocalyx.

Not all bacterial species possess a capsule, but for those that do, it appears to provide protection to free-floating or planktonic state cells against phagocytosis by white blood cells. Capsules are also known to protect bacteria from desiccation.

Ribosomes

The location of bacterial protein synthesis is at ribosomes, which are often loosely attached to the inner surface of the bacterial plasma membrane. The ribosomes are complex and composed of both protein and ribonucleic acid (RNA). The ribosome has a mass of about 2.5 MDa, with RNA accounting for two-thirds of its mass. Electron microscopy shows that the ribosome consists of two subunits denoted 30S (small subunit; S, Svedberg unit) and 50S (large subunit). When these subunits are joined, the ribosome has a sedimentation coefficient of 70S as opposed to 80S due to its tertiary structure.

Nucleoid

Deoxyribonucleic acid (DNA) is found in the nucleoid. The DNA is not surrounded by a nuclear membrane as would be the case in eukaryotic cells.

Plasmids

Plasmids are circular double-stranded DNA that exist and replicate independently of the chromosome (nucleoid). They are known to carry genes that may confer resistance to antibiotics and heavy metals, or genes that can produce enzymes necessary to break down certain new metabolites.

Pili and Fimbriae

Many Gram-negative bacteria have short and fine projections that are not involved in motility. These are referred to as fimbriae but are sometimes referred to as pili. Some bacteria are associated with having in excess of several 1000 pili. Pili have been shown to be very important in mediating adhesion of bacteria to a surface and appear to be significant during the formation of a biofilm. Bacteria also contain sex pili, which are generally larger than fimbriae and are important in bacterial reproduction.

Flagella

Flagella are important for motility in bacteria and are seen as long appendages protruding from the outer surface of the bacterial cell. Flagella are on average 20 nm in width and 15–20 μm in length. The distribution of flagella on a motile bacterium varies depending on their distribution on the outer surface of the cell and may be located at the polar regions of the cell (polar flagella) or uniformly distributed around the entire cell (peritrichous flagella).

Chemotaxis

Bacteria are attracted to many agents that are important for their survival. The movement of a bacterium toward or away from an agent is referred to as chemotaxis.

Bacteria detect chemicals by chemoreceptors that are found in the periplasmic space or the plasma membrane and are able to bind to specific chemicals.

GROWTH OF BACTERIA

In a liquid, microorganisms follow a specific growth curve as seen in Figure 1.2, which typifies the growth of bacteria in a closed batch system. A batch system is one where there is only one medium source that does not get replenished. Consequently, during batch culture, waste will accumulate over time and nutrients become depleted. The growth or mean generation times of bacteria are shown in Table 1.2.

LAG PHASE

When bacteria are first added to a new environment (i.e., fresh culture medium), they do not immediately increase in number. Essentially, they become static. This period is referred to as the lag phase of microbial growth. During this period of the bacterial growth cycle, the cell begins to synthesize new components required for growth. Essentially the bacteria are assessing their new environment. If the media

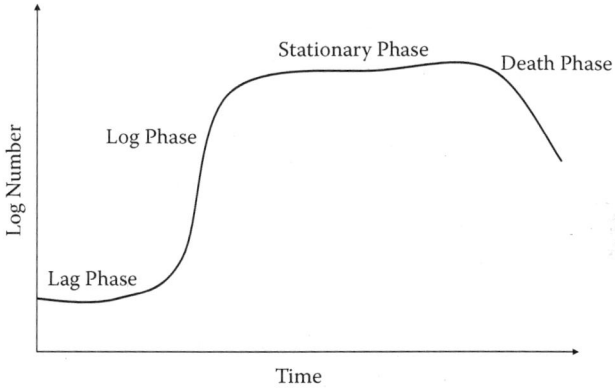

FIGURE 1.2 Growth of planktonic bacteria in liquid.

TABLE 1.2
Mean Generation Times for Microorganisms

Microorganism	Mean Generation Times (h)
Bacteria	
Escherichia coli	0.35
Bacillus subtilis	0.43
Yeast	
Saccharomyces cerevisiae	2

in which the bacteria are being grown is highly nutritious, this may shock the bacterium and delay or prevent them from growing. In addition to this, the media used to grow bacteria may be different to the nutrients it has been used to in the original environment it had been removed from. Because of this, new enzymes would need to be synthesized to break down these new nutrients the bacteria have detected. Also the bacteria may have been injured (e.g., heat shock) and would require some form of resuscitation. The lag phase of growth can be prolonged depending on the type of bacteria and the conditions to which it is being put into.

EXPONENTIAL PHASE

During the exponential phase of growth, the bacterial population has assessed its new environment and begins to multiply at its optimal rate, and often this is in the form of *planktonic* or free-floating dividing cells. However, the actual speed of growth will be dependent on the nature of the growth medium and environmental parameters such as incubation temperature, pH, and oxygen level of the growth medium. It is important to recognize that even in an exponential growth phase, individual bacteria may differ with respect to their growth rates.

STATIONARY PHASE

When a population of microorganisms ceases to grow, the growth curves take on a different shape and a plateau is reached. This *stationary phase* is achieved at approximately 10^9 cells per milliliter and essentially represents a balance between cell division and cell death or indicates that the population has become metabolically inactive.

The reason a microbial population reaches this stage in its life cycle is largely because of nutrient constraints (i.e., essential nutrients for growth have become depleted, and an accumulation of waste products in the medium has occurred). For aerobic bacteria, the environment of a closed system may have started to become anaerobic and therefore only at the surface of the liquid culture will oxygen levels be at a level suitable for growth. Space also becomes a concern as bacteria require a certain space around them to allow nutrients to diffuse into the cell.

DEATH PHASE

Nutrient depletion and toxic waste accumulation lead to a decrease in the number of microorganisms within a batch culture, and as such the degree of bacterial multiplication may fall below the rate of cell death with the result that the bacterial numbers decrease significantly. However, to truly determine if a cell has died and not become dormant, it is important to reinoculate it into a fresh medium and see if it grows. With the development of molecular techniques, this has enabled us to move away from culture media to determine if bacteria are alive or dead, by analyzing ribosome accumulation. The death of microorganisms decreases after the initial recorded death increase.

FACTORS THAT AFFECT GROWTH

Growth of bacteria is affected by an array of physical and chemical agents. It is very important to understand the factors known to affect the growth of bacteria as this can have important implications for their control.

pH

Hydrogen ion activity in a solution is measured by pH. Bacteria that exhibit optimum growth at pH 0 to 5.5 are called acidophiles, neutrophiles are bacteria that grow optimally at pH 5.5 to 8.0, and alkalophiles are bacteria that have an optimum growth at pH 8.5 to 11.5. Bacteria known to grow above pH 11.5 are referred to as extreme alkalophiles. Most bacteria are neutrophilic, but fungi prefer slightly acidic conditions (pH 4 to 6). In most environments, a wide variation in pH can occur, and microorganisms adapt to such changes by possessing specific mechanisms to maintain a neutral internal pH. In addition, microorganisms have the ability to alter their local environment so that optimum growth can be maintained in adverse conditions.

TEMPERATURE

Microorganisms are very susceptible to temperature changes. Enzymes and their activity are affected significantly by temperature. Temperature increases from a low base will generally increase growth rates. For every 10°C rise in temperature, it is thought that there is an equivalent doubling in the rate of enzyme reactions. However, when a certain temperature limit is reached, the growth rate decreases because of denaturation of enzymes, transport mechanisms, and proteins in general.

Bacteria able to grow below 0°C, with an optimum growth temperature of −10 to 20°C are called psychrophiles. Bacteria that can grow at 0°C but have an optimum of 20 to 30°C are called facultative psychrophiles. Mesophiles have an optimum growth at around 20 to 45°C with a maximum growth temperature of 45°C. Most bacteria fall into the category of mesophiles. However, some bacteria are able to grow at temperatures in excess of 55°C and these are termed *thermophiles*. Hyperthermophiles are capable of growth above 90°C.

OXYGEN

Bacteria able to grow in oxygen are called aerobes (Figure 1.3), and those that grow in the absence of oxygen are called anaerobes. The bacteria that are completely dependent on oxygen for growth are called obligate aerobes. Facultative anaerobes do not require oxygen to grow, but they do grow better in the presence of oxygen. Aerotolerant anaerobes are able to grow with or without oxygen. However, strict or obligate anaerobes are not able to tolerate oxygen and die when it is present. Other bacteria called microaerophiles are affected by atmospheric levels of oxygen and therefore require oxygen levels around 2% to 10% (e.g., *Helicobacter pylori*).

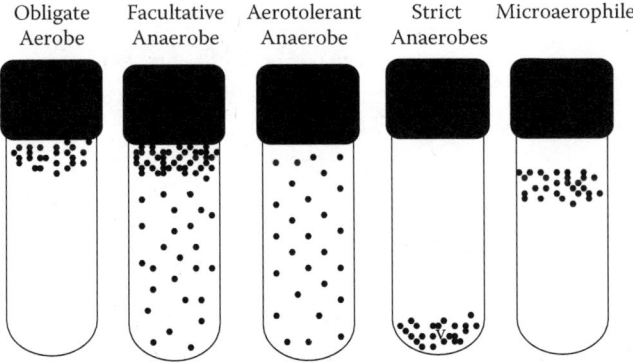

| Obligate Aerobe | Facultative Anaerobe | Aerotolerant Anaerobe | Strict Anaerobes | Microaerophile |

FIGURE 1.3   Typical location of microbial growth in a batch culture test tube.

CONTROL OF MICROORGANISMS

Controlling the levels of microorganisms is very important for infection control and therefore the reduction in the incidence of infections and disease. The development of control processes for microorganisms by chemical means began many years ago. It was the work of Paul Ehrlich (1854–1915) which showed that some chemicals were able to selectively kill some bacteria without killing human cells. He discovered a dye, trypan red, that could be selectively used to kill a microorganism known as a trypanosome, which was responsible for African sleeping sickness. He and his colleagues also established that arsenic compounds could kill the bacterium that caused syphilis in rabbits. It was not until the works of Ernest Duchesne in 1896 and later in 1928 by Alexander Flemming, that an agent called penicillin was reported. This agent was accidentally discovered following the contamination of staphylococci agar plates by *Penicillium notatum*. The discovery of the antibiotic penicillin started the quest to find other antibiotics that could be used for human disease. For example, in 1944 Selman Waksman discovered the antibiotic streptomycin, and later in 1953 other antibiotics including tetracycline, terramycin, and neomycin were isolated from microorganisms.

Antibiotics have selective toxicity, which kills or inhibits microorganisms but they generally do not damage the host cell. However, the toxicity of antibiotics is affected by many factors including dose. The level of toxicity can be measured by a ratio called the therapeutic index. Antibiotics are referred to as having a narrow spectrum of activity or a broad spectrum of activity based on the different microorganisms the agent can kill. Many of these antibiotics can also be classed as either cidal (kill) or static (stops the growth) in their effects.

The effectiveness of an antibiotic is measured by its minimal inhibitory concentration (MIC). This is referred to as the lowest concentration of an antibiotic that stops or prevents the growth of a specific bacteria. Other measures of antibiotic efficacy have included the minimum bactericidal concentration (MBC), which is the lowest concentration of an antibiotic that kills the specific bacteria. The MBC is generally two or four times higher than the MIC. The term MBEC or minimal biofilm eradication concentration is the lowest concentration of an antibiotic or antimicrobial able to kill bacteria within a biofilm or in an attached state. The MIC can be up to 1000 times higher than the MBC. The MBEC is a term that relates more specifically to the true efficacy of an antimicrobial agent and its effectiveness *in vivo*.

TERMS USED TO DESCRIBE MICROBIAL CONTROL

Sterilization is the process by which all living cells including microorganisms are destroyed. That means the *sterilized* product is free of all microorganisms and spores. When sterilization is undertaken by chemical agent, the agent is called a sterilant. Disinfection is the killing or inhibition of microorganisms that are able to cause disease and infection. Such agents are generally used only on inanimate objects. Disinfectants will not necessarily sterilize an object.

Sanitization involves reducing the microbial population to levels that are safe for public health.

Antiseptics prevent infection or sepsis. They are chemicals that are applied to tissue and prevent infection by killing or inhibiting pathogens. *Germicides* kill pathogens but not endospores. *Bactericides, fungicides, viricides*, and *algicides* are agents that kill bacteria, fungi, viruses, and algae, respectively.

DEATH OF MICROORGANISMS FOLLOWING EXPOSURE TO ANTIMICROBIAL AGENTS

The death of microorganisms when exposed to an antimicrobial is generally exponential or logarithmic. If death is rapid, the rate of cell death may slow with the development of resistant strains of microorganisms within the population.

The effectiveness of an antimicrobial agent is affected by an array of different factors that include:

1. *Size of the population.* Essentially, the bigger the population of microorganisms is, the longer it will take to kill them.
2. *Composition of population.* The efficacy of an antimicrobial agent varies and depends on the nature of the microorganism being treated. Older and more mature bacteria are generally more difficult to kill than younger cells. Many microorganisms have an inherent and enhanced tolerance to antimicrobials.
3. *Concentration and intensity of the antimicrobial agent.* The more concentrated an antimicrobial is, the quicker it kills the microorganisms. However, for some agents a small increase in concentration can lead to an enhanced efficacy in comparison to what the benchmark is expected to be. In addition to this, some antimicrobials are more effective at lower concentrations. For example, ethanol is more efficacious at 70% compared to 95%.
4. *Duration of exposure.* The longer a microbe is exposed to an antimicrobial agent, the greater the number of microorganisms that will be killed.
5. *Temperature.* By increasing temperature, the activity of the antimicrobial will also increase.

6. *The environment*. Numerous environmental factors affect antimicrobial efficacy—particularly, pH.

MECHANISM OF ACTION OF ANTIMICROBIALS

The mode of action of a number of antimicrobials will be discussed in more detail in further chapters. However, in brief, the effects of various antibiotics can be seen in Table 1.3.

EFFECTIVENESS OF THE ANTIBIOTIC OR ANTIMICROBIAL

The effectiveness of an antimicrobial is fundamental to the eradication of an infection or a disease. To be effective, the antimicrobial must be administered at the correct concentration and be able to reach the site of infection. For example, antibiotics administered by the oral route must be able to withstand the acidity of the stomach

TABLE 1.3
Mode of Action of Antibiotics

Antibiotic	Mode of Action
Cell Wall Synthesis Inhibition	
Penicillin	Inhibit the transpeptidation enzyme
Ampicillin	involved in peptiodoglycan
Meticillin	production
Vancomycin	
Bacitracin	
Protein Synthesis Inhibition	
Streptomycin	Binds to ribosomes and inhibits
Gentamicin	protein synthesis; interferes with
Chloramphenicol	t-RNA and block translocation
Tetracycline	
Fusidic Acid	
Inhibition of Nucleic Acid Synthesis	
Rifampin	Inhibits DNA gyrase, blocks RNA
Ciprofloxacin	synthesis
Cell Membrane Disruption	
Polymyxin B	Disrupts the structure and permeability
	of the plasma membrane
Metabolic Disruption	
Isoniazid	Inhibits folic acid synthesis, interferes
Sulfonamides	with folic acid synthesis or inhibits
Trimethoprim	the synthesis of mycolic acid

before they can be absorbed through the intestinal wall. Because of this, a number of antibiotics that are acid labile must be administered by injection, intravenously or intramuscularly, and the antibiotic will be transported via the blood or lymphatic system to the site of action.

The pathogen's susceptibility to an antimicrobial agent being administered topically or systemically is significant. Of particular concern is whether the bacteria have an acquired or inherent resistance to an antibiotic or, as is the case with penicillin, are active only when the microorganism is both growing and dividing. In addition, if an antibiotic works on the cell wall it will not be effective against L form bacteria devoid of the cell wall target. It is also important that the level of antibiotic administered reaches the pathogen at a level above the bacteria's MIC. In the case of a biofilm, it must be remembered that the MBEC can be 1000 higher than the MIC. Therefore the effectiveness of an antibiotic will be affected by the concentration of antibiotic administered and the maintenance of the correct level of the antibiotic at the site of action for a period that is long enough to kill the bacteria and thereby reduce the infection. However, with the use of higher levels of antibiotic comes the added concern of toxicity to the host.

It is accepted that the antibacterial activity and its efficacy are proportional to the concentration of an antibiotic, but in 1948 Eagle and Musselman noticed a paradoxical effect of penicillin against streptococci and *Staphylococcus aureus*. That is to say, an increasing antibiotic concentration to one higher than the accepted MBC actually resulted in a reduced antimicrobial activity. To date, this "Eagles Effect" has been noted in an array of different microorganisms when exposed to different antibiotics.

MECHANISM OF BACTERIAL RESISTANCE

The mechanisms by which bacteria become resistant to antibiotics are wide and varied. Some bacteria have natural or inherent resistance to many antibiotics. In addition, bacteria can undergo spontaneous mutation that may result in the expression of a protein conferring a resistant phenotype to the cell. This resistant mutant will then be selected for in a population when challenged by an antibiotic.

Bacteria commonly become resistant to antibiotics because they are able to prevent the entry of the antibiotic into the cell, so the antibiotic is unable to get to its site of action. For example, many bacteria have a cell envelope that prevents antibiotic entry. For penicillin to be effective, it must be able to bind to penicillin-binding proteins on the outer surface of the bacteria. If the bacteria change the structure of these binding sites, penicillin may not be able to exert its effects on that specific bacteria.

Some bacteria are able to actively pump an antibiotic out of the cell once it has entered. These pumping systems are generally not specific, so such bacteria can expel many antibiotics. Such a pumping system is referred to as a multidrug resistance pump. These have been identified in many bacteria, including *Staphylococcus aureus* and *Pseudomonas aeruginosa*.

In addition to the above, a number of bacteria are able to inactivate antibiotics and antimicrobial agents by modifying the active agent. One of the most widely reported resistance mechanisms has been reported with penicillin. Some bacteria are able to hydrolyse the beta lactam ring of penicillin using enzymes called penicillinases.

Other bacteria are able to phosphorylate or acetylate antibiotics. This has been known to occur in aminoglycosides and chloramphenicol resistance.

As antibiotics act on a specific target, if this target becomes modified in any way the typically highly-specific antibiotic may become ineffective against that bacteria. For example, many bacteria are able to alter or bypass a metabolic pathway that is inhibited by a specific antibiotic.

TRANSMISSION OF ANTIBIOTIC RESISTANCE

Resistance to antibiotics is related to the bacterial genes that are present in both the bacterial chromosome and in plasmids. Resistance to antibiotics can be due to spontaneous mutations that may occur in the bacterial chromosome. However, such strains of bacteria are initially referred to as mutants. In patients exposed to antibiotics, this has been shown to result in the selection of mutant-resistant pathogens that will proliferate and be selected for in this environment.

Some bacteria develop resistance because they have acquired a plasmid from other bacteria that have resistant genes. These are called R plasmids, or resistance plasmids. Many genes are able to exist on a plasmid that then carries resistance to an array of different antibiotics. These genes are capable of producing new enzymes that may help to degrade antibiotics (e.g., hydrolysis of penicillin). R plasmids can be transmitted to other bacteria by a process known as conjugation, transduction, or transformation.

The extensive and indiscrimate use of antibiotics may aid in the development and spread of antibiotic-resistant bacteria. The reason for this is due to the fact that bacteria susceptible to antibiotics are killed, but the bacteria that are resistant—which occur in many bacterial populations—will be selected for. The antibiotic- or antimicrobial-resistant pathogens that are produced are then able to spread from patient to patient.

Efforts continue worldwide to reduce and try to prevent the spread and escalation of antibiotic resistance of pathogens of public health significance. The way to do this is to reduce the usage of antibiotics and to use them only when it is necessary. Appropriate and responsible use of antibiotics is necessary in medicine and despite evidence of ongoing worldwide education on this matter, indiscriminate use remains a problem.

EPIDEMIOLOGY

Epidemiology is concerned with the occurrence, determinants, distribution, and control of both health and disease in humans and animals. The origin of the word "disease" comes from old French and means "lack of ease." It is essentially the impairment of the normal state that also hinders a function. The study of epidemiology is very important to infections and diseases and helps in the development of control measures that can be used in the present and the future.

The pathogen and its link to infectious disease and infections in general relates to a cycle of events. To help in controlling any disease or infection, this cycle has to be broken.

As with any disease or infection, the pathogen causing this condition has to be discovered and established as a risk to the population. This is the basis of Koch's postulates. To determine whether a microorganism fits under the umbrella of a pathogen, the bacteria must be isolated and proven to cause a disease.

If a pathogen can be passed on from one host to another, it is capable of causing a communicable disease. The potential for a pathogen to cause a disease is called its *pathogenicity,* and this potential is determined by the virulence of that pathogen.

SOURCES OF PATHOGENS

Pathogens can be found in an array of different environments and their source is very important to understand, particularly when trying to develop some form of control. If this reservoir can be controlled or reduced, this will help to minimize the spread, or transmittability, of the pathogen.

The source of a pathogen can include inanimate areas (such as water, soil, and food) and animate sources (such as animals and humans). Many animals and humans are classed as carriers. That is, they are able to transport the pathogen but do not show any signs or symptoms related to this carriage. Carriers can be defined as casual, acute, or transient. If pathogens are carried for a lifetime, the carrier is referred to as a chronic carrier. If pathogens can be transmitted from an animal to a human, they are called zoonotic pathogens and the disease is referred to as a zoonoses.

ACQUISITION OF PATHOGENS

Pathogens can be acquired by indirect or direct contact. Airborne transmission is a common route for acquiring a pathogen, as pathogens can reside in droplets of fluid or dust and spread easily. When individuals cough, sneeze, or talk, small droplets of fluid may be aerosoled and easily passed on to a receiver. Contact or direct transmission occurs when a pathogen is passed on by touching (e.g., person-to-person transmission). Indirect transmission occurs from inanimate objects such as eating utensils, cups, and bed linen. There are also vehicles of transmission that serve to spread the pathogen (e.g., animals), as well as vector-borne transmission, where pathogens are transmitted by animals; in particular, domestic pets.

CONTROL OF DISEASE AND INFECTION

There are numerous ways of controlling disease and infection. These include the isolation of the patient, control of the pathogens at the source (e.g., their reservoir), treatment of the source (e.g., chlorination of water), and of course, treatment of the infected patient with appropriate antibiotics.

CONTROL OF NOSOCOMIAL INFECTIONS

Nosocomial is derived from the Greek word "nosokomeion;" "nosos" meaning disease and "komeo," meaning to take care of. Nosocomial infections occur in hospitals and are caused by specific pathogens. These pathogens can affect patients and also hospital staff. Nosocomial infections can occur within the hospital or can develop after leaving the hospital, having acquired the pathogen in the hospital environment. If a patient is harboring a pathogen prior to admittance to a hospital and that pathogen does not cause a disease until the patient is in the hospital, then the infection is referred to as community acquired.

The sources of all pathogens in nosocomial diseases are endogenously or exogenously acquired. Endogenously means the individual's own microbiota or flora are responsible. These microorganisms are brought into the hospital by the patient. Exogenous pathogens are those acquired external to the patient (e.g., water, contact from hospital staff or other patients, medical equipment, etc.).

Controlling the transmission of microorganisms, particularly pathogens, is very important. Control measures that hospitals have can include isolation of patients, the use of aseptic techniques, improved handling of fecal matter, and the use of dressings in wound care. These measures have been shown to reduce the risk of cross contamination to other patients and hospital personnel. Hospitals have a permanent epidemiologist who monitors the sources, spread, and control of these infections and forms part of an infection control team. Of particular concern to hospitals is the control of biofilms.

HISTORICAL ASPECTS OF BIOFILMS

Historically, the majority of microorganisms, particularly pathogens, microbiology studies have involved the use of single-species microorganisms cultured in liquid media. As mentioned previously in this book, microorganisms cultured in this manner are described as being in a "free-living," or planktonic, state. However, it is now evident that in most natural and clinical environments, microorganisms grow within communities, often attached to a solid surface and embedded within an extracellular polymeric matrix produced by the attached "sessile" microorganisms. Such an existence is now referred to as a biofilm.[1]

Donlan and Costerton[2] stated that a biofilm is "a microbially derived, sessile community, characterized by cells that are irreversibly attached to a substratum or interface, or to each other, embedded in a matrix of extracellular polymeric substances (EPS) that they have produced, and exhibit an altered phenotype with respect to growth rate and hence transcription." However, many definitions exist, and for this book we propose that biofilms are a community of microorganisms, either evident as monospecies or mixed species of microorganisms, attached to a surface (abiotic or biotic) or each other, encased within a matrix of extracellular polymeric substances (EPS) and internally regulated by the inherent population. Biofilms are ubiquitous, and their presence in many areas associated with healthcare in particular is generally regarded as unfavorable and harmful, either because of infection risks presented to patients and healthcare workers, or due to the detrimental effect they can have on medical device function. It is partly because of these issues that the study of biofilms, referred to as "biofilmology," has become an area of active research worldwide.

An important question that is often asked is, "why do bacteria form biofilms?" Many years ago this question was addressed and answered anthropomorphically, and research is ongoing in this area. Essentially, bacteria are thought to form biofilms for the following reasons:

1. Ahesion—Protection, security, and availability of nutrients
2. Enhanced metabolic synergies between cohabiting bacteria
3. To gain cooperative populations of bacteria
4. Defense—To have resistance to physical forces (e.g., water flow, blood or urine, saliva), pH changes, temperature changes, enhanced resistance to antimicrobial agents, and the body's immune response

Although the accumulation of biofilms has benefits for the inherent bacterial population, their development in the wrong environments comes with concerns for the host.

To fully appreciate and control harmful biofilms, research in this area has principally focused on:

- How microorganisms attach to surfaces
- How a biofilm develops, proliferates, and sustains itself in adverse conditions and ultimately disperses the inherent microbial population
- The regulatory processes involved in biofilm development
- How to control and inhibit biofilms

The results being generated from these studies are helping to direct future antibiofilm technologies applicable to the vast array of environments that biofilms are associated with.

Antonie van Leeuwenhoek, in the 17th century, was the first to observe biofilm-originating organisms, which were his "animalcules," which we now know were dental plaque bacteria. Despite this "early biofilm analysis," it was not until 1940 that the concept of "biofilm" was introduced into science when observations of a "bottle effect" (a concept known today as microbial regrowth) confirmed evidence of its existence. This initial biofilm work effectively demonstrated enhanced growth of marine bacteria when attached to solid surfaces.[3]

Further evidence for the existence of biofilms occurred in 1943, when Zobell[4] proposed that the adhesion of bacteria to a surface was a two-step sequence of reversible and irreversible binding. Twenty years passed until the first recognition of beneficial biofilms was acknowledged in trickling water filters of wastewater treatment plants,[5] a feature that is still routinely exploited today as a method to improve the quality of water.

Pioneering research into biofilms commenced in the early 1970s, when Characklis[6] established the detrimental effects biofilms inflicted on industrial water pipeline systems. Research conducted by Marshall and colleagues[7] observed that bacteria utilized "very fine extracellular polymer fibrils" to mediate attachment to a surface. Today these structures are considered to be organelles and are referred to as *Caserna,* which are cabling systems that enable bacteria to communicate between different compartments within the biofilm.

Early findings reported by Costerton and colleagues[8] from studies in aquatic ecosystems reported that bacteria existed within polysaccharide matrices that promoted bacterial attachment to surfaces. Further research has demonstrated that bacteria found in the planktonic state differed profoundly to their attached or sessile counterparts. In 1987 more pioneering biofilm work, again from Bill Costerton and co-workers, established that biofilms consisted of microcolonies of microbial cells embedded in a hydrated exopolymeric anionic matrix.[9]

As the years progressed, advances in our understanding of biofilms developed, and in particular the recognition that differential gene expression occurred during microbial adhesion. In 1998 further advances in biofilm physiology were made possible by the utilization of enhanced scientific tools; in particular, techniques that made use of molecular technologies and confocal laser scanning microscopy (CLSM). CLSM helped researchers visualize three-dimensional biofilms in real time. The exploitation of such methods enabled researchers to significantly enhance the understanding of the ecological activity, ecological interaction, architecture, genetics, and mechanical properties of bacteria within biofilms.

Later research in the 1990s highlighted that biofilms were the underlying cause of many industrial pipeline problems, and in particular, microbially induced corrosion (MIC). An example of MIC is the situation within oil pipelines where biofilms are the primary cause of "pepper pot corrosion," costing industry billions of dollars per annum in lost revenue. Eradication of these problematic biofilms with traditional biocides, such as chlorine and antibiotics, is often found to be ineffective. This evidence of recalcitrance of biofilms to antimicrobials highlighted the inherent resistance feature of biofilms. This phenomenon is recognized today in many recalcitrant and chronic diseases in humans and animals such as chronic wound infections and catheter-related bloodstream infections. The involvement of biofilms within the clinical setting has thus been a major focus for recent research.[10] The importance of biofilms in causing hospital-acquired infections and inducing the failure of medical devices is now recognized as a major healthcare problem exacerbated by the difficulty in their removal once they have become established.

Bacteria within a biofilm are considered to exist in a community and exhibit altruistic behavior. This concept is considered to contradict the theory of evolution that relates to "survival of the fittest," as Darwin's theory applies specifically to nonaltruistic behavior. This area is conceptually interesting and provides food for thought to the reader and the anthropomorphic studies applied to microbial behavior.

OCCURRENCE OF BIOFILMS

For the purpose of this chapter, particular emphasis and reference to biofilms will be placed on clinical areas, and comparisons to environmental and industrial biofilms will be addressed where warranted. One fundamental statement that the reader needs to appreciate is the fact that microorganisms generally develop biofilms in the same fundamental ways, irrespective of the ecosystem from where they are derived (see Table 1.4 and Figure 1.4).

Despite the obvious clinical problems associated with biofilms, it must be emphasized that not all biofilms are detrimental to health. Biofilms are found indigenously in certain regions of the human body and have a protective role against infections and diseases, referred to as colonization resistance. For example, biofilms located in the gastrointestinal tract,[11] the skin,[12] oral cavity,[13] and the female urogenital area[14] serve as barriers to invading pathogens capable of causing disease. Despite evidence

TABLE 1.4
Examples of Where Biofilms Occur

Environmental

Water treatment	Pulp and paper
Algal blooms	Concrete
Drinking water	Oil industry
Food industry	

Dental and Medical

Dental caries	Continuous ambulatory peritoneal dialysis (CAPD)
Periodontitis	Schleral buckles
Otitis media	Urinary catheter cystitis
Musculoskeletal infections	Intrauterine device (IUD) infection
Necrotizing fasciitis	Endotracheal tubes
Biliary tract infection	Hickman catheters
Osteomyelitis	Central venous catheters
Bacterial prostatitis	Mechanical heart valves
Native valve endocarditis	Vascular grafts
Cystic fibrosis pneumonia	Biliary stent blockage
Meloidosis	Gallstones
Nosocomial infections	Orthopedic devices
ICU	Penile prostheses
Sutures	Chronic wounds
Arteriovenous shunts	Nonhealing surgical site infections

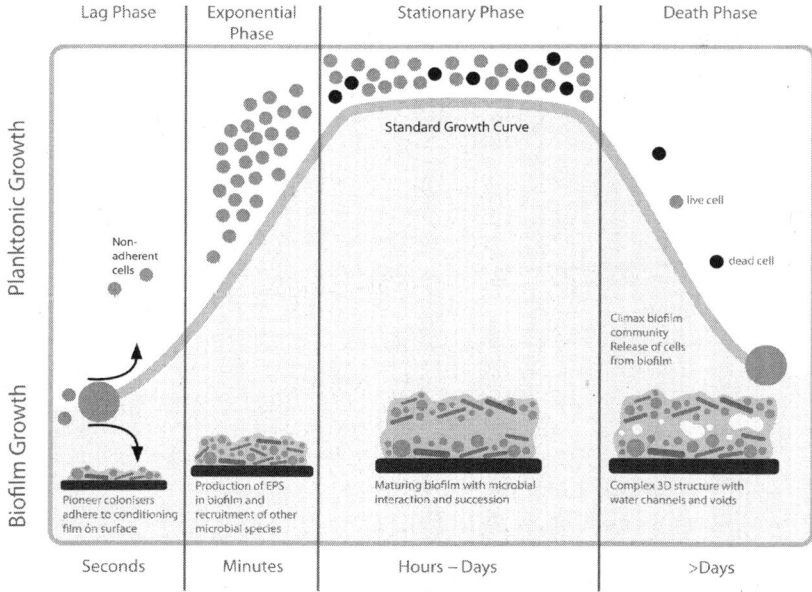

FIGURE 1.4 An overview of the microbiology life cycle. (Adapted from Thomas, J.G. and Nakaishi, L.A. (2006) Managing the complexity of a dynamic biofilm. *JADA* 137: 105–155.)

of these "beneficial" biofilms, situations do arise when normally harmless biofilms can become pathogenic. Such clinically significant biofilms have been implicated in the rejection of a vast array of biomaterial and medical devices used in the human body, which have been used to help improve and enhance the quality of human life. Examples of where "harmful" biofilms have caused concern have included the colonization of artificial hips, heart valves, artificial voice boxes, catheters, intrauterine devices, and dental and an array of other medical prostheses.

STAGES IN THE FORMATION OF BIOFILMS

The development of a biofilm is a complex and dynamic process. There are five stages in the formation of a biofilm which have been recognized,[9,15] including the following (see Figures 1.5 through 1.7):

1. Development of a surface conditioning film (appropriate for medical devices, drinking water pipes, food-processing equipment, to name a few). Conditioning agents have included water, fats and lipids, proteins (albumin), glycoproteins, amines, and generally any bathing fluid within which the surface is found.
2. Events that bring the microorganisms into close proximity of the surface (e.g., liquid flow observed in blood vessels [laminar flow], pipe systems, and rivers [turbulent and laminar flow]).
3. Adhesion (reversible and irreversible adhesion of microbes to the conditioned or unconditioned surface) aided by microbial adhesins, polysaccharides, and physical/chemical interactions.
4. Growth of the microorganisms at the colonized surface, extracellular polymeric substance production, microcolony formation and biofilm "maturation" (climax community), phenotypic and genotypic changes, quorum sensing, and microbial interaction.
5. Detachment/sloughing/dispersal of the biofilm.

Each of the above stages will be considered in turn, and their relationships on a broader scale will be linked to medically related conditions where appropriate.

DEVELOPMENT OF THE CONDITIONING FILM AND SUBSTRATUM EFFECTS

It is generally appreciated that a "naked" surface, biotic or abiotic, is quickly coated, within milliseconds, with an organic conditioning film. Consequently, in most environments bacteria do not in fact attach directly to a surface but to the conditioning film. The adsorption of this conditioning film to a surface is considered to precede the attachment of microorganisms, and will alter the surface chemistry of the substratum.

Within the medical setting, the conditioning film can be derived from the constituents of blood, tears, urine, saliva, intravascular fluid, and respiratory secretions. More specifically, constituents of the conditioning film have included components such as fibrinogen, collagen, lipids, water, extracellular polymers, and serum albumin.[16] The

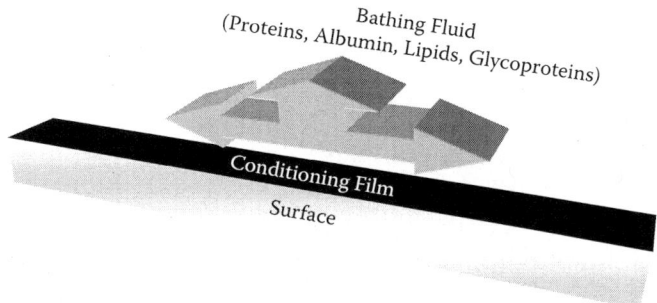

FIGURE 1.5   Conditioning film.

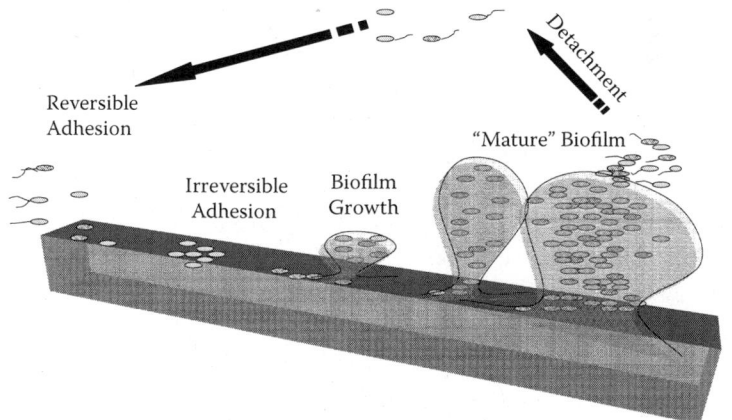

FIGURE 1.6   Development of a biofilm.

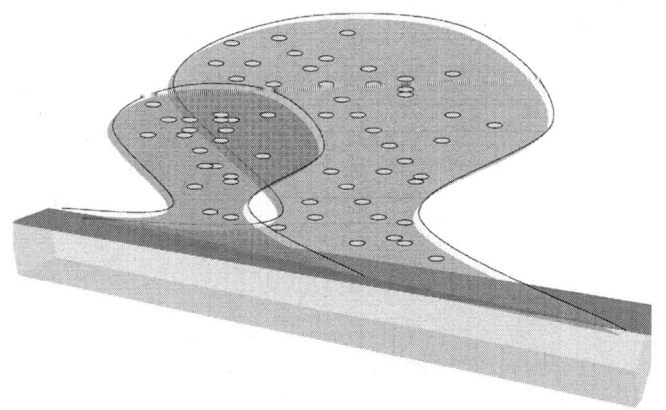

FIGURE 1.7   Mature biofilm.

formation of a conditioning film has been reported particularly on urinary catheters, orthopedic devices, contact lenses, and stents.

Whether a conditioning film is a prerequisite for microbial attachment remains debatable. However, research has shown that the presence or absence of a conditioning film has a major impact on microbial adhesion and therefore biofilm development.[17]

On a tooth surface, for example, a *pellicle* forms immediately after cleaning of the enamel surface. This pellicle is composed of salivary and dietary proteins and glyco-proteins and can be detected within 30 seconds of placing a clean enamel surface into saliva.[18] The thickness of the salivary pellicle has been shown to be between 100 and 1000 nm.[19,20] This conditioning pertinacious film has been shown to act as a receptor for bacterial attachment.[21] Consequently, bacteria that are known to adhere to teeth do not bind directly to the tooth surface but to a predominately proteina-ceous pellicle.[22]

The chemical and physical properties of a surface to which a microbial cell attaches is fundamental to biofilm formation and therefore to the subsequent pro-gression of a biofilm through to maturation. For example, in the case of surfaces found in the food and water industries, and in medical prosthetics, the chemical com-position, surface topography, and particularly surface roughness all seem to have a role to play in aiding and enhancing bacterial adhesion.[23–27]

An important factor shown to encourage biofilm development is the roughness of the surface to which the microorganisms attach. As a result of extensive research, many theories have been proposed as to why surface roughness aids microbial colo-nization. It has been suggested that an appropriate size of roughness possibly affords shelter for attached microbes from the effects of shear forces caused by fluids flowing over the surface. Such situations occur in environments where there is turbulent water flow, or where there is constant movement of body fluids across a surface. Examples of where this occurs include urinary catheters, prosthetic heart valves, and dentures. Surface roughness has been shown to affect bacterial adhesion in the food-process-ing industry,[28,29] on denture materials,[30,31] and in pipes within the water industry.[32] Overall, many research groups have observed that with a high surface roughness, adhesion increases. However, a number of researchers have observed no correlation.

In addition to surface roughness, the physicochemical properties of a surface are also known to affect microbial adhesion. For example, microorganisms have been shown to attach more rapidly to hydrophobic, nonpolar surfaces such as polytetra-fluoroethylene (PTFE; Teflon®) and other plastics, compared with hydrophilic mate-rials such as glass or metals.[33,34] However, such findings need to be considered with an open mind, as many studies in this area have proven contradictory.

EVENTS THAT BRING MICROORGANISMS INTO CLOSE PROXIMITY WITH THE SURFACE

Nonmotile bacteria can be brought close to a surface by random chance or Brownian motion, guided by water flow. Contrary to this, motile bacteria can search for a surface using processes aided by attraction or evasion to chemical (chemotaxis), air quality (aerotactics), and movement toward light (phototaxis).

The transport of microbial cells and nutrients to a surface, particularly in the presence of laminar flow, is generally achieved by a number of well-established fluid dynamic processes. Such processes have included, among others, diffusion, sedimentation, convection mass transport, and thermal and gravitational effects.[35] Within water pipes, two main flow conditions exist, namely laminar flow (also observed in blood vessels and in the movement of urine through the urethra) and turbulent flow (observed in drinking water pipes and rivers). Laminar flow is characterized by the evidence of parallel smooth flow patterns with limited lateral mixing. In laminar flow, the fastest flow occurs in the center.[36,37] During laminar flow, microorganisms and nutrients maintain a straight path and remain in a stabilized position dictated by the flow rate.[37] In contrast, flow that is considered turbulent has a random and chaotic motion (Figure 1.8). From a microbiological perspective, turbulent flow will aid mixing of bacteria and nutrients[38,39] and will increase organic material and microbial cell attachment to a surface.[2] Within turbulent flow, small eddying currents (random and unpredictable flow) develop, which cause vertical sweep forces that help propel bacteria to within short distances of a surface, enhancing the chances for individual cells or microbial aggregates to attach.[40] This helps to avoid the effects of Gibbs free energy that enables bacteria to adhere to a surface. Gibbs free energy is equal to the sum of the van der Waals forces with electrostatic interactions (often negatively charged) between the bacteria and the surface.

Another mechanism known to aid in the microbial colonization of a surface is gravitational cell sedimentation. This phenomenon has been observed principally in flowing systems where coaggregation of bacteria is evident.[41] Coaggregation of

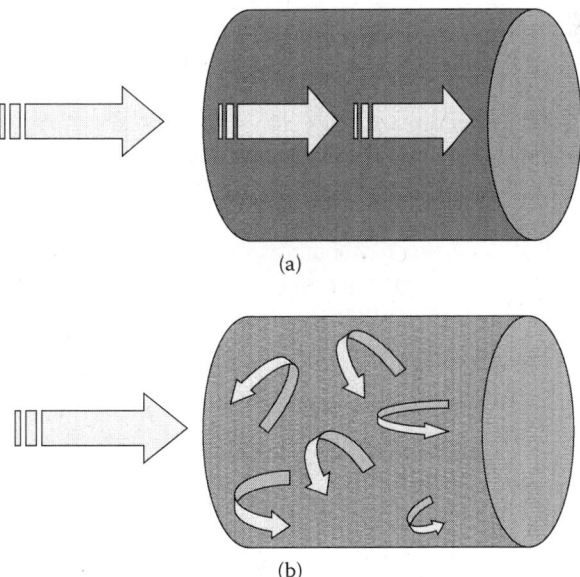

(a)

(b)

FIGURE 1.8 Transport mechanisms: (a) laminar flow and (b) turbulent flow.

bacteria is considered an important factor during the formation of a biofilm and is now being considered particularly relevant to skin and wound care, dentistry, and water.

MICROBIAL ADHESION

Microbial adhesion to a surface is a multifactorial process. Planktonic bacteria are considered to randomly make contact with a surface. However, as mentioned earlier, this can also be driven by chemotactic responses and bacterial motility. Early research into bacterial adhesion has shown that the rates of adhesion are determined by many factors. In particular, the physical and chemical nature of the surface including hydrophobicity, the genera and species of bacteria, the proteins and carbohydrate components on the bacterial surfaces,[42] the availability of adhered nutrients and bathing fluid, hydrodynamics, and cellular communication (between prokaryotic and prokaryotic, and prokaryotic and eukaryotic cells), together with the associated regulatory processes.[43,44]

Bacterial adhesion to a surface was first proposed by Zobell in 1943.[4] Zobell postulated that adherence of bacteria to a surface consisted of a two-step sequence of events constituting both a reversible and irreversible phase. Reversible adhesion is considered to be an initial weak attachment of microbial cells to a surface.[45] Conversely, irreversible adhesion is a permanent bonding of the microorganisms to the surface. The microorganism, or more specifically the characteristics of the microorganism in question, determines the effectiveness of this process. For irreversible adhesion to occur, EPS becomes significant,[46] as do the specific interactions that occur between the substratum and the microorganism.[47] For example, on an abiotic surface, the attachment of microbes is mediated by nonspecific interactions that occur through van der Waals or hydrophobic forces, and electrostatic interactions. Conversely, on biotic surfaces, microbial adhesion is considered much more complex, involving specific molecular mechanisms mediated by adhesins and lectins.[48]

A large amount of research has investigated bacterial adhesion to abiotic surfaces as opposed to biotic surfaces. Conclusions drawn from these studies have shown that bacterial adhesion is often related to the distance between the bacterium and the surface to which it will adhere.[49] Busscher and Weerkamp[49] first proposed three theories regarding bacterial adhesion. Theory one was that if a bacterium was farther than a distance of 50 nm from a surface, then weak and long-range nonspecific interactions mediated by van der Waals forces would exist that would help drive the bacterium to the surface. Theory two proposed that if a bacterium was at a distance of 10 to 20 nm away from a surface, then both van der Waals and electrostatic interactions would occur. Theory three suggested that when a bacterium was less than 1.5 nm from a surface, van der Waals, electrostatic and specific interactions, including hydrogen bonding, covalent bonding, and hydrophobic forces, could occur between bacteria and surface. All the forces above are considered significant in bacterial adhesion and exert both a positive and negative influence on the bacteria, serving to both enhance and significantly retard successful adhesion. The longer-range forces observed between a bacterium and a surface are found to be nonspecific. These forces are described by the Derjaguin, Landau, Verwey, and Overbeek (DLVO) theory.[50,51] The DLVO theory proposes that the stability of a colloidal system is a result of the sum of attractive Lifshitz–van der Waals and electrical double-layer repulsive forces that exist between particles as they approach each other due to Brownian motion. It follows, therefore, that for successful adhesion, a bacterium would initially have to overcome the electrostatic repulsion between the surface and the bacterium. Once achieved, a net attraction between the bacterium and the surface would then occur.

The bacterial cell has a major role to play in adhesion to a substratum. Surface hydrophobicity (interactions tend to increase when one or both of the surfaces involved are nonpolar), fimbriae and flagella, and particularly the extent and composition of the bacterial EPSs and proteins,[52,53] all influence the rate and extent of microbial cell attachment to both abiotic and biotic surfaces. Antibiotics have an effect on the cell surface hydrophobicity of bacteria and are known to affect bacterial adherence. During treatment with antibiotics, the cell surface charge of streptococci has been found to significantly alter the result that a decrease in cell surface hydrophobicity causes. Under these conditions, bacterial adherence to human buccal epithelial cells is reduced.[54] Research has also shown that once at a surface, redistribution of sessile bacteria can occur. This is found to be aided by the motility of the bacteria, a phenomenon known as twitching motility.[55,56]

The majority of bacterial cells have a net-negative charge at neutral pH,[57] which is reduced as the pH is lowered and varies depending upon the conditions in which the bacteria are cultured, the culture age, and the ionic strength.[58] The negative charge observed at pH 5 to 7 is due to the presence of excessive amounts of carboxyl and phosphate groups on the outside of the bacteria when compared to the amino groups. This negative charge will reduce bacterial attachment to a negatively charged surface. In aqueous environments, most surfaces have been reported to be negatively charged.[59] In biological systems, the distribution of ions is more complex and in a state of flux.

There is a constant battle between the host's immune system and bacteria to attach to a surface. Bacteria have many strategies to evade the host's immune system. If the host is compromised, it becomes colonized by an array of different "commensal" microorganisms that form biofilms. Once bacteria have secured a habitable environment in the body, they possess many mechanisms to ensure they remain attached. Bacteria possess proteins on their surface to bind to the host's extracellular matrix, such as fibronectin, fibrinogen, vitronectin, and elastin. These are referred to as MSCRAMMs (microbial surface component recognizing adhesive matrix molecules) and are known to be significant to microbial adhesion to the host.[60]

Role of Pili and Fimbriae in Adhesion

Many interactions between bacteria and surfaces have been observed to be stereospecific. The bacterial surface-associated molecules that aid stereospecific adhesion of bacteria are referred to as adhesins. Adhesins are located on bacterial appendages such as flagella, capsules, fimbriae, and outer membranes.[61] These adhesins promote bacterial adhesion to a surface by bridging the gap between the bacterium and the repulsive electrostatic forces, which aid to repel bacteria.

O'Toole and Kolter[55] demonstrated that mutants of *Pseudomonas aeruginosa* required type IV pili-mediated twitching motility to enable aggregation at a surface. Twitching motility is thought to aid the migration of bacteria across a surface, enabling coaggregation, and therefore enhance coadhesion and enhance survival.[62]

Fimbriae play a role in cell surface hydrophobicity and attachment of bacteria to a surface.[63] It is probable that fimbriae are able to overcome initial electrostatic repulsion that exists between the bacterium and the substratum.[64,65]

Bacterial mutants that lack flagella have been shown to have reduced adhesive abilities.[44,66] Bacteria that have reduced motility, possibly as a result of culture age, have a reduced adhesion rate compared to bacteria that have a high degree of motility.[67] Korber and colleagues[68] investigated the attachment of motile and nonmotile strains of *Pseudomonas fluorescens*. The authors reported that motile bacterial cells attached in greater numbers and against resistant fluid flow more rapidly than nonmotile bacterial strains. Nonmotile strains were found to be less able to recolonize or seed vacant areas on a substratum compared with motile strains. Flagella can also play an important role in the early stages of bacterial attachment. This is again thought to be mediated by overcoming the repulsive forces associated with the substratum and the bacteria.

In addition to flagella, other bacterial cell surface structures including lipopolysaccharide (LPS) and EPS can play important roles in the attachment process. Cell surface polymers with nonpolar sites such as fimbriae, other proteins, and components of certain Gram-positive bacteria (mycolic acids) appear critical factors in attachment to hydrophobic substrata, and EPS and LPS are more important in attachment to hydrophilic materials. Following treatment of adsorbed cells with proteolytic enzymes, a marked detachment of bacteria has been noted.

Once a bacterial cell attaches to a surface, vast arrays of phenotypic changes are known to occur. These include changes in the rate of oxygen uptake, respiration rate, synthesis of extracellular polymers, substrate uptake, rate of substrate breakdown, heat production, and changes in growth rate.

Molecular changes have been demonstrated to occur when a bacterium attaches to a surface. The up- and down-regulation of many genes during adhesion of bacteria to a surface suggests that the mechanisms for bacterial adhesion are genetically complex. For example, as early as 1993, Davies and colleagues[69] demonstrated a fivefold up-regulation of the gene *algC* in *P. aeruginosa* within minutes of attaching to a surface. In addition, Whiteley and colleagues[70] reported that 70 genes of *P. aeruginosa* underwent alteration in expression during adhesion and biofilm maturation. Prigent-Combaret et al.[71] also showed altered transcription levels for 38% of *Escherichia coli* genes in the biofilm state. Becker and colleagues[72] showed that genes of *Staphylococcus aureus* were also up-regulated in the developed biofilm.

Swarming motility is also a process by which bacteria can move along a surface, but for this to occur, flagella are required. Such a process has been observed in both Gram-negative and Gram-positive bacteria and is considered a significant process required by bacteria to colonize a tissue. Together with swarming and biofilm formation, this aids to increase the survival of bacteria in hostile environments.

GROWTH AND DIVISION OF THE MICROORGANISMS AT THE COLONIZED SURFACE

When a bacterium has adhered to a surface, it becomes more firmly attached. This process is aided by EPS. At a surface, the sessile microbial cells begin to proliferate and form cell clusters or microcolonies. The observation of microcolonies on a surface is considered to constitute a structural marker of biofilms, particularly within infections found in individuals with cystic fibrosis and chronic wounds.

Microcolony and Biofilm Formation

Following the irreversible adhesion of pioneering bacteria to a surface (observed principally in oral bacteria and often referred to as coaggregation), the adhered bacteria begin to grow and extracellular polymers are produced which accumulate, with the result that the bacteria are eventually embedded in a complex three-dimensional matrix of hydrated polymeric substances. Once attached, motile bacteria are able to move from where the first few cell divisions have taken place to different regions of the developing biofilm.[73,74] Once biofilm bacteria become immobilized in EPS, they then become dependent upon substrate flux from the liquid phase or exchange of nutrients with their neighbors in the biofilm in order to proliferate and flourish.

An important feature of the biofilm environment is that the microorganisms are immobilized in relatively close proximity to one another. As the biofilm begins to develop further, microbial succession of bacteria has been observed, where pioneer colonizing bacteria help to modulate the environment to one conducive to the adhesion and proliferation of secondary, tertiary, and quaternary bacteria. This serves to develop a complex ecosystem containing a community of interacting microorganisms.

During the initial development of the biofilm, many different species of bacteria will immigrate and migrate from the biofilm. Many transient microorganisms will be introduced into the biofilm, but they are often unable to permanently establish themselves.

Within the biofilm community, specific functional types of organisms may, through their activities, create conditions that favor other complementary functional groups. This continual flux will lead to the establishment of spatially separated, but interactive, functional groups of bacteria. Within the biofilm, the exchange of metabolites at group boundaries occurs, which leads to a physiological cooperation between sessile bacteria.[75]

As microbial communities tend to be complex both taxonomically and functionally, there is considerable potential for synergistic interactions to occur among constituent organisms with a biofilm. This helps in the development of a potentially homeostatic environment aiding to protect the encased biofilm bacteria from outside perturbations. This is extremely important in natural and medical communities where bacteria are exposed to constant nutritional and pH fluxes.

Within biofilms, there is evidence of a high level of cellular interaction and competitive behavior.[76,77] As a result of competition strategies by specific species of bacteria, the biofilm system is under constant flux during its early development.[76,78–80] As the biofilm "matures," heterogeneity increases and chemical and physical microgradients develop including those of pH, oxygen, and nutrients. This heterogeneity is often aided by both the sessile microbes and the "extracellular soup" of the biofilm or EPS.

Extracellular Polymeric Substances

If microorganisms reside at a surface for long enough, irreversible adhesion occurs, mediated by EPS. EPS is a cementing substance and is an extracellular microbial product. The extracellular material associated with the cell has been referred to as a glycocalyx, a slime layer, a capsule, or a sheath.[8,81]

Colonization of microorganisms involves the synthesis of large amounts of EPS, multiplication of the attached organisms, or attachment of other bacteria to the already adhered cells (coadhesion). It has been shown for *S. aureus* that a polysaccharide intracellular adhesin (ica) is also synthesized. This adhesin aids in binding the cells together and thereby promoting aggregation and biofilm formation.[81,82] After this stage, further growth of attached organisms continues, resulting in the formation of dense bacterial aggregates characteristic of a maturing or climaxing biofilm.

Mature biofilms are composed primarily of microbial cells and EPS, which may account for 50% to 90% of the total organic carbon of the biofilm.[83] EPS can vary both chemically and physically, but it is primarily composed of polysaccharides, which are neutral, or polyanionic in the case of the EPS of Gram-negative bacteria. Uronic acids (such as D-glucuronic, D-galacturonic, and mannuronic acids) or ketal-linked pyruvates are known to constitute part of the EPS matrix, as well as proteins and nucleic acids. These molecules give anionic properties to the biofilm, allowing cross-linking of divalent cations such as calcium and magnesium.[83–85] The composition of biofilms, particularly those that contain Gram-positive bacteria, has an effect on the chemical composition of the EPS causing it to be primarily cationic.

The EPS of *P. aeruginosa* has been extensively studied. It has been shown to contain alginic acid that is controlled by the *algACD* gene cassette of *P. aeruginosa*.[69] Within the biofilm state, the gene *algC* is expressed at a level 19 times higher than evident in the planktonic state.

The polysaccharide content of EPS has a marked effect on the biofilm,[85] as the composition and structure of the polysaccharides determine the primary conformation of the EPS. Bacterial EPS primarily possess backbone structures containing 1,3- or 1,4-β-linked hexose residues that are rigid and generally poorly soluble or insoluble, whereas some EPS molecules are more readily soluble in water. The EPS of biofilms is not generally uniform, and it seems that the amount of EPS increases with age of the biofilm.[88]

EPS provides many benefits to a biofilm; it not only offers biofilm cohesive forces, but also absorbs organic and inorganic materials that act as nutrients for the proliferating bacteria. EPS also adsorbs microbial products and other microbes, sequesters heavy metals, and provides protection to the immobilized cells from rapid external changes. EPS facilitates intercellular communication in the biofilm and has been shown to enhance intercellular transfer of genetic material.

The polysaccharides associated with EPS are known to help anchor bacteria to the substratum. This is brought about by high numbers of polyhydroxyl groups present in the polysaccharide backbone. Extending lengths of polymers attached to the microorganisms interact with vacant bonding sites on the surface by polymer bridging that helps secure the microorganisms near the surface. Several mechanisms for polymer bridging within biofilms have been suggested but are not fully understood.

Additional chemicals found within the biofilm organic matrix include glycoproteins, proteins, and nucleic acids.[89]

EPS influences the physical properties of the biofilm, including diffusivity, thermal conductivity, and rheological properties. EPS, irrespective of charge density or ionic state, has some of the properties of diffusion barriers, molecular sieves, and adsorbents. EPS can influence physiochemical processes that include diffusion and fluid frictional resistance.

As EPS is predominantly polyanionic, it can act as an ion exchange matrix, serving to increase local concentrations of ionic species such as heavy metals, ammonium, and potassium, to name but a few, while having the opposite effect on anionic groups. Bacteria are thought to concentrate and use cationic nutrients such as amines. This suggests that EPS can serve as a nutrient trap to enhance bacterial proliferation and biofilm growth.[10]

It is thought that the penetration of charged molecules, including biocides and antibiotics, may be partly restricted by EPS.

Gene Transfer

Microorganisms need to disseminate genetic information to promote species proliferation. For this to happen, bacteria contain plasmids that are transportable from one bacteria to another, generally within the same species or genera, although cross-genera transfer of plasmids occurs. Plasmids carry an array of genes involved in antibiotic resistance, virulence, and toxin and enzyme production.

The biofilm EPS matrix with its water channels or voids acts as a gene pool that allows for genetic acquisition and exchange to take place. This helps to enhance the complexity of the biofilm, aiding in its survival and sustainability in adverse conditions. Bacteria are able to acquire genes for antibiotic resistance by horizontal transfer (plasmids, transposons, or integrons) or through specific gene mutations. The ability of bacteria to acquire genes is likely to occur in biofilms by a process known as conjugation. This gene transfer has been shown in *Streptococcus* species.[90] Few gene transfer studies have been reported and demonstrated in biofilms, but because of the high microbial population density evident in the biofilm, gene transfer is likely to occur.

Biofilm Structure

The structure of a biofilm is influenced by an array of factors including EPS composition and production, microbial growth rate, quorum sensing between bacteria, twitching motility of bacteria, and external pertubations. Young or immature biofilms generally contain few species of microorganisms. This reflects the low diversity of pioneering microbial populations.[91] This diversity increases, culminating in the formation of a stable *climax community*. The microbial richness of the climax community is often underestimated due to the selectivity and inadequacy of pure-culture isolation techniques.[92] It has been estimated that a climaxed or mature biofilm contains only 10% or less of its dry weight, in the form of microbial cells.[93]

The structure of a biofilm has been reported extensively within the literature and observed in both mixed and pure culture systems in many different environments.

Most of these systems demonstrate the existence of structural heterogeneity, often evident in a patchy configuration.

Many researchers have established that biofilms consist of cell clusters, which are discrete aggregates of cells located within the EPS matrix. These clusters have been shown to vary in shape, often ranging from cylinders to filaments and forming "mushroom" structures.[94] Within these systems, due to the evidence of water channels or patches of biofilms, a number of biofilm arrangements have been noted, including aggregates, cell clusters, streamers, and stacks.[72,95–100] The open channels evident in biofilms are referred to as voids and pores. Evidence of these structures highlights the high degree of spatial and temporal complexity found within biofilms. Mushroom-shaped colonies are often observed in *in vitro* biofilms grown in laminar flow. In turbulent flow, the biofilm structure becomes more filamentous with evidence of streamers a common feature.

Depending on the site of biofilm formation, biofilms are complex, with many different genera, species, and numbers of microorganisms. Collectively, therefore, the biofilm is a microbial community consisting of an array of ecosystems and niches.[101,102]

Bacteria that grow in a biofilm can create their own environment, which in turn influences the physiology of its component cells. In medical biofilms, biofilm structure is substantially influenced by the interaction of particles of the nonmicrobial origin from the host or environment. In the human body, biofilms on native heart valves provide a clear example of this type of interaction in which bacterial microcolonies in the biofilm develop in a matrix of platelets, fibrin, and EPS.[103] A fibrin capsule that develops will protect the organisms in these biofilms from the white blood cells of the host.

Biofilms on urinary catheters have been found to contain organisms that have the ability to hydrolyze urea in the urine to form free ammonia through the action of the enzyme urease. The produced ammonia then increases the pH at the biofilm–liquid interface. This has been found to result in the precipitation of minerals such as calcium phosphate (hydroxyapatite) and magnesium ammonium phosphate (struvite).[104] These minerals become entrapped in the biofilm and cause encrustation and catheter blockage with subsequent promotion of urinary tract infection.

In mature biofilms of *P. aeruginosa,* there are substantially different protein profiles when compared to planktonic bacteria, and in one particular study[105] more than 300 proteins were detectable in mature biofilm samples that were not present in planktonic bacterial cultures.

FACTORS THAT GOVERN THE DEVELOPMENT OF BIOFILMS

Overall, the development of a biofilm is generally governed by a number of parameters.[106] These include ambient and system temperatures, which are related to season, day length, climate, and wind velocity; hydrodynamic conditions (shear forces, friction drag, and mass transfer); nutrient availability (concentration, reactivity, antimicrobial properties); roughness, hydrophobicity, and electrochemical characteristics of the surface; pH (an approximately neutral pH of the water is optimal for the growth of most biofilm-forming bacteria); the presence of particulate matter (this can become entrapped in the developing biofilm and provide additional attachment sites); and effectiveness of biofilm control measures.

From the list above, it is clearly evident that many parameters play a role in affecting and also determining the structure of a biofilm. Generally four major factors influence biofilm structure[107] and include the surface or interface properties, hydrodynamics, nutrients, and biofilm consortia. This list is by no means exhaustive, but does reflect the number of factors that can affect the developing biofilm. Hydrodynamic forces that are known to operate within flowing conditions are considered by some researchers to affect biofilm structure. It is however now well established that biofilms, which are exposed to high turbulent flow, experience and develop a phenomenon known as *streaming*. The significance of this is still under investigation.

Quorum Sensing

Many bacteria communicate to the same species or to different species using chemical signals that are synthesized by bacteria and then secreted. The release of communicating molecules is related specifically to the microbial cell density (quorum), but the molecules are also known to be produced at different stages in their growth cycle and are not just restricted to release on a cell density basis. Quorum sensing was first documented in *Vibrio fischeri*. However, quorum sensing has since been documented in many bacteria, including *Pseudomonas*, *Staphylococcus*, *Streptococcus*, and *Escherichia coli*.

DETACHMENT AND DISPERSAL OF THE BIOFILM

Biofilm detachment generally refers to the release of cells (either individually or in clumps) from a biofilm or a surface, and is often considered to be an active process and therefore physiologically regulated.[108] The local loss of detached cells in the human body leads to a temporary loss of colonization resistance of a surface. If this occurs within or upon the host, susceptibility to infection and disease increases. This phenomenon becomes more significant when we consider infections and diseases that occur in the gastrointestinal tract and on the surface of the skin.

Detachment of a biofilm has been defined as erosion (single cells or small component parts of the biofilm) or sloughing (clusters of cells or large aggregates of cells,[109,110] abrasion[111]). Detachment can also occur because of human intervention or predator grazing. Detachment is a physical process and occurs predominately in aqueous systems and in the food industry.[112]

Biofilm erosion has been categorized as mechanical, abrasion, or physiologically mediated erosion.[113] The extent of mechanical erosion is determined by the rate of liquid flow over the biofilm. For physiological erosion, the biofilm components must be destabilized for erosion to occur. As a rule, erosion increases with increasing biofilm thickness. In newly formed, immature biofilms, erosion is not often evident. [114]

Sloughed material that has detached from a biofilm is thought to physiologically mimic that of a biofilm and is believed to provide protection for the component populations of bacteria. As sloughed material contains large populations of bacteria, dissemination of these organisms to other virgin surfaces occurs, and as such provides a mechanism that can aid adhesion and colonization of new sites.

Sloughing possibly occurs due to mechanical instability of the biofilm in relation to shear and induction of certain enzymes. Detachment caused by physical forces has been studied in great detail, and it would seem that the rate of erosion of cells from the biofilm is enhanced with increased biofilm thickness and occurrence of fluid shear at the biofilm–bulk liquid interface. Sloughing is more random than erosion and is thought to result from nutrient or oxygen depletion within the biofilm structure. Sloughing is more commonly observed with thicker biofilms that have developed in nutrient-rich environments.

Emerging scientific evidence has shown that microorganisms disseminating from biofilms may overcome the host immune system and cause infection.[115] Shiau and Wu[116] have shown that the EPS matrix produced by *S. epidermidis* can interfere with macrophage phagocytic activity. In addition, Meluleni and colleagues[117] found that opsonic antibodies in patients with chronic cystic fibrosis were unable to mediate phagocytosis and eliminate bacterial cells growing in biofilm microcolonies. Yasuda et al.[118] showed that resuspended biofilm cells of *E. coli* were less sensitive to the killing activity of human polymorphonuclear leukocytes (PMNLs) *in vitro*. The study reported that this was due to resistance of the biofilm organisms to the active oxygen species produced by the PMNL. This indicated that cells detaching from biofilms in indwelling medical devices may have the ability to survive the PMNL phagocytic activity in the bloodstream to initiate a bloodstream infection.

The mode of dispersal is reported to affect the phenotypic characteristics of the organisms within the biofilm. As mentioned above, eroded or sloughed aggregates from the biofilm are likely to retain certain biofilm characteristics in particular antimicrobial resistance properties, whereas cells that have been shed as a result of growth may revert quickly to the planktonic phenotype. Biofilm dispersal/detachment has a very important implication in public health medicine. Raad et al.[119] determined a direct relationship between biofilm formation and catheter-related septicemia. The detachment of biofilm clumps in medical situations also has implications in infective endocarditis, when biofilms are detached from native heart valves. These clumps of cells from biofilms may also contain platelets or erythrocytes that lead to the production of emboli, which may cause serious complications to the host.

Biofilms in hospital water systems containing potentially pathogenic organisms might also detach as aggregates. Those microorganisms with a low infective dose may be consumed and may therefore result in infection. This may be one cause of the increasingly observed incidence of nosocomial or hospital associated infection.

The dispersal of a biofilm is considered to be a dynamic, highly regulated process, controlled by an uncharacterized hierarchical set of genes and triggered by an array of medical or environmental trigger mechanisms.[120]

Different parameters are known to affect biofilm dispersal and detachment and have been studied in detail using *in vitro* model systems. These have included changing the pH, temperature, and presence of organic macromolecules, either absorbed on the substratum or dissolved in the liquid phase.[121] All these conditions and many more have been shown to affect bacterial detachment and are often not species specific.

Surface roughness of the substratum is thought to play a significant role in biofilm detachment, with early events in biofilm formation being controlled by hydrodynamic forces.[122] As detachment increases with increasing fluid shear stress at the substratum surface, macro- and microroughness may significantly influence detachment rates of the biofilm due to a sheltering effect from hydrodynamic shear. The detached cells may be transported close to the surface (in viscous sublayers), resulting in a greater number of surface collisions and thus providing more opportunity for reattachment.

Despite a lack of research in the area of dispersal and detachment of biofilms from a surface (abiotic or biotic) into surrounding environments, detachment and dispersal of bacteria have significant implications. This, as mentioned previously, has been documented within the manufacturing, medical, and public arenas and remains a public health and infection control concern. In microbiological terms, detachment from surfaces may at first seem to be disadvantageous to the biofilm, but in fact, biofilms with greater detachment rates will have larger fractions of active bacteria and consequently may be the most problematic.

Detachment and dispersal have been documented to occur as a result of low nutrient conditions, indicating that it can also function as a survival mechanism for bacteria. In fact, this process has been shown to be genetically determined in some species of bacteria. Consequently, the detachment and dispersal of the biofilm seem not only to be important for promoting genetic diversity within a biofilm, but also for bacteria to escape unfavorable habitats.

Bacteria evident within biofilms found specifically in aquatic systems are often preyed upon by free-living protozoa[123] and bacteriophages.[124] Similarly, within the human body, biofilms are preyed upon by PMNLs and other cells that constitute components of both the innate and acquired immune system.

As biofilms develop, they will become thicker and less uniform. In addition, different susceptibilities to the forces of dispersion will become evident. As the biofilm complexity increases, regions will exist where the dispersion force is greater than the adhesion and cohesion forces. The resulting feature of this will be the development of localized detachment of microbial cells.

The dispersal of bacteria from a biofilm was described by Davies[125] as a physiologically regulated process that occurred naturally during biofilm development and maturity. Dispersion is a coordinated disaggregation resulting in the release of bacteria.[126] It has been reported that when dispersion occurs, cell clusters become hollow in the middle[105,125,127–129] and therefore appear in donut shape,[130] an effect defined as *hollowing*.

Davies[125] suggested that biofilm dispersion is modulated through the accumulation of signaling molecules. Davies also proposed that once these molecules reach a threshold, this triggers the release of enzymes leading to the disaggregation of cell clusters.

Other dispersal theories have suggested that dispersal occurs due to an accumulation of metabolic products and depletion of metabolic substrates.[131,132]

Dispersion has been referred to as detachment, dissolution, disaggregation, and as a starvation and nutrient-induced response. It is possible that an increase in concentration of an inducer molecule may be responsible for the release of matrix polymer–degrading enzymes, which result from detachment from the biofilm. This has been found for *Streptococcus mutans* that produces a surface protein-releasing enzyme (SPRE) that mediates the release of cells from biofilms.[133] A number of investigators have shown that homoserine lactones may play a role in detachment.[134]

Biofilm cells may also be dispersed either by the shedding of daughter cells from actively growing cells, detachment as a result of nutrient levels or quorum sensing, or shearing of biofilm aggregates (continuous removal of small portions of the biofilm) because of flow effects.

Interestingly, a number of bacteria have been shown to be able to change the components within their outer cell membrane and wall which in turn has been shown to alter the hydrophobicity of the cell. This change has been shown to aid in microbial release from a surface.[134] Given the above influences on the dynamics of a biofilm, it is clear that the microbial community of biofilms exhibits constant flux. This flux would seem to be more significant during the early development of a biofilm where any implications may significantly alter the microbial community of the biofilm. The significance of this will be discussed below.

PUBLIC AND MEDICAL HEALTH CONSEQUENCES OF BIOFILMS

Nosocomial infections have been reported to be the "fourth leading cause of death in the United States," which has resulted in a cost of $5 billion per annum.[135] Approximately 70% of nosocomial infections are associated with medical devices, and within the United States alone, some 5 million medical devices are implanted per annum.

It has been estimated that biofilms are associated with 65% of nosocomial infections. [136] In addition, biofilm-related infections are now recognized as a major cause of morbidity and mortality accounting for 80% of all known infections. The major incidences of biofilm-related effects are particularly associated with individuals who have compromised immune systems and those who have implantable medical devices.

Antibiotic resistance and the treatment of infections caused by antibiotic-resistant microorganisms is a difficult problem to resolve. In patients with implantable devices, the only appropriate course of action if infections of this kind do occur is to remove the device. This, however, is of great inconvenience to the patient and is also very costly to health services. For example, the estimated cost of a hip replacement is presently £3500 in the United Kingdom, but when the costs of a biofilm-related infection in relation to the hip replacement are included, the figure can be as high as £30,000.[137] The treatment of biofilm-based infections is estimated to cost >$1 billion annually in the United States alone.[138–140]

DRINKING WATER

Drinking water is known to be contaminated with an array of different microorganisms. A number of these microorganisms are considered significant to public health and have often resulted in waterborne diseases in both developed and developing countries.[141] Of public health significance to drinking water is the ability of bacteria to grow on the inside of water distribution pipes where they form biofilms.[142,143] These biofilms are thought to harbor pathogens and allow for the resuscitation of "injured" pathogens. Possible survival and proliferation of potential pathogens in biofilms constitutes a concern when these biofilms detach from the surface into flowing water.[144–146] Particular pathogens that have been known to survive in drinking water biofilms and are now being studied in detail include *Mycobacterium avium* and *Helicobacter pylori*.[147,148]

HOSPITAL AND DOMESTIC WATER

Hospital water supplies are continually monitored and treated to help reduce the risk of nosocomial infections in patients. As with drinking water supply pipes, the plumbing materials used to transport both cold and hot water are known to be contaminated with biofilms.[149–151] Areas where biofilms are known to develop in the hospital environment include shower heads, dialysis water systems, humidifiers, air-conditioning systems, hot water tanks, taps, storage tanks, hydrotherapy pools, and baths. Patients at the highest risk from nosocomial infections are those who are immunocompromised.

Within hospital water supplies, numerous microorganisms have been found to exist within a biofilm state. These have included *Mycobacterium avium*, *Legionella pneumophila*, *Aeromonas hydrophila*, *Aspergillus* spp., *P. aeruginosa*, *Staphylococcus aureus*, and *Acinetobacter* sp.[152,153] Of particular concern in hospitals is the aerosolization and ingestion of water droplets containing *Legionella* spp. and *Mycobacetrium* spp.[154,155] Control of biofilms in these system is considered important.[156]

DENTAL WATER UNITS

Biofilm growth in dental water units has been considered by many to be a potential public health concern. The reason for the concern is that patients can be exposed to both strict and opportunistic pathogens from the biofilms found on the walls of the water lines of the dental water unit, and although a rare occurrence, a number of infections related to exposure to water from the dental water units have been documented. Consequently, cross contamination is viewed as a concern when using these devices.[157–159] Because of these concerns, the use of biocides in these units is considered necessary.[160–162]

KIDNEY STONES

Struvite [Mg (NH$_4$)PO$_4$.6(H$_2$0)] kidney stones are formed due to interactions between urease producing bacteria, including *Pseudomonas*, *Klebsiella*, *Providencia*, and *Proteus* spp.[163] and substrates that are found in urine resulting in the formation of biofilms. Bacteria embedded in a biofilm were first indicated in 1938 as having a role to play in kidney stone development.[164] Further evidence of biofilms within kidney stones was confirmed by Nickel and colleagues.[165] Markers within this study which were used to confirm evidence of biofilms included the presence of microcolonies, which were surrounded by a matrix of polysaccharide, and struvite crystals. The formation of kidney stones leads to urine flow obstruction, inflammation, and in some circumstances, kidney failure. Despite the use of antibiotics for patients with this condition, infections are often reoccurring, a common event associated with a biofilm-related infection.

ENDOCARDITIS

Lesions that become infected on cardiac heart values have been reported to be composed of both the components of the host and microorganisms. Within these lesions, biofilms are found containing bacteria, extracellular material produced by the bacteria, platelets, and fibrin.[166]

CYSTIC FIBROSIS

With the genetic disease cystic fibrosis, individuals are susceptible to infections. These infections are divided into two stages. First is the formation of a bronchitis-like intermittent infection due to an array of different bacteria including *H. influenzae, P. aeruginosa,* and *S. aureus.*[167] Second is the development of a chronic and permanent infection that is documented to continue throughout the life of the patient. Patients with CF have reoccurring infections that carry on for many years eventually leading to failure in the respiratory processes.[168] Both mucoid and non-mucoid *P. aeruginosa* exist in these patients and the strains are avid biofilm formers. When analyzing the sputum from patients with CF, these samples have been shown to be composed of microcolonies of bacteria encased within a matrix of extracellular polymeric substances.[169] As with all biofilm-related infections, when antibiotics are administered to patients with CF, antibiotic resistance is significantly increased when compared to *in vitro* antibiotic-resistant profiles observed in planktonic bacteria.[170]

OTITIS MEDIA

With this condition, fluid is known to build up in the middle ear. Within this liquid and on the mucosa bacteria have been identified. Repeated infections are noted particularly in acute otitis media (OM) and also OM with effusion. When antibiotics are administered for these conditions, only short-term gains to the patient are often achieved.[171]

In animal models, biofilms composed of bacteria have been visualized on the mucosa of the middle ear.[172] Because of this and numerous research findings, biofilms are now considered to be involved in acute OM and OM with effusion.[173] It is thought that evidence of biofilms may explain why OM reoccurs frequently.[174] Particularly in children with OM with effusion, 50% of these patients will have preceding infection of acute OM.[175]

However, despite numerous findings, there is no conclusive evidence that biofilms cause OM with effusion, but studies continue in this area.[176]

OSTEOMYELITIS

Osteomyelitis is an infection in a bone; in particular the long bones found in the legs. However, each bone of the body is susceptible to infection. The different types of osteomyelitis are related to the source of infection (hematogenous or contiguous) and the patient's vascular condition. Contiguous focus osteomyelitis is associated with patients with diabetes mellitus involving polymicrobial or biofilm infections of the feet. Hematogenous osteomyelitis is common in children. During the initial phases of osteomyelitis, conditions become favorable for the growth of biofilms. Evidence of biofilms on infected bones has been documented.[177] *Staphylococcus aureus* is found to be a common cause of infection, and this is mediated by biofilm growth.[178]

PROSTATITIS

Microscopic analysis of biopsy samples removed from patients with chronic prostatitis has shown the samples to be colonized with microcolonies of bacteria.[179,180]

INTRA-AMNIOTIC INFECTION

Invasion of the amniotic cavity by bacteria has been reported.[181] In order to determine infection, amniotic fluid samples are often taken. The presence of "amniotic fluid sludge" has been found to be associated with infections in the amniotic cavity.[182] In addition to this, sludge has been associated with the presence of bacteria and also inflammation. This sludge has been documented as a biofilm, leading to its link as a cause of infection and inflammation of amniotic cavity.[183]

INDWELLING AND MEDICAL DEVICES

Biofilm-related infections have been reported to be associated with the use of central venous catheters, artificial hearts, contact lenses, intraocular devices, prosthetic and orthopedic devices, cardiac pacemakers, shunts, and contraceptive devices. Biofilms have been observed and associated with infections in an array of indwelling devices and catheters. Dialysis patients are susceptible to biofilm-related infections, particularly from Gram-positive bacteria including *Staphylococcus aureus* or coagulase-negative staphylococci (CNS) and Gram-negative bacteria and yeasts. Infections on medical devices are difficult to eradicate; therefore, their existence can cause increased mortality and morbidity.[184]

Staphylococcus epidermidis is a frequent cause of infections of implanted medical devices.[185]

URINARY TRACT INFECTIONS

The catheterized urinary tract has been documented to provide an ideal environment for microbial proliferation. A number of bacterial species are known to colonize indwelling catheters as biofilms.

The growth of biofilms on urinary catheters is known to increase the patient's risk of further complications. Crystalline biofilms have been documented to occlude the catheter lumen,[186,187] which is known to cause pyelonephritis and septicemia. The formation of crystalline biofilms is due to urease-producing bacteria, particularly *Proteus mirabilis*. Clinical prevention strategies for this condition are urgently required, particularly as bacteria found within these crystalline biofilms are reported to be resistant to antibiotics.[188] The ability of bacteria to persist and grow in a biofilm seems to be one of the important factors in both the resistance to antibiotics and the severity of urinary tract inflammation.[189] Many methods have been employed to remove biofilms from catheters, including tricolsan, citric acid, and ethylenediamine tetracetic acid (EDTA).[186,190]

CENTRAL VENOUS CATHETERS

In the United States alone, some 5 million central venous catheters are inserted per year.[191] However, catheter-related bloodstream infections (CRBSIs) are associated with the use of intravascular catheters.[192] CRBSIs are the third most common cause of nosocomial infections in the intensive care unit (ICU).[193] Biofilm-embedded bacteria have been documented to form on catheters within the first 24 h after insertion. Because biofilms are responsible for infections in these catheters, a number of technologies have been developed to help prevent and control biofilms, with some positive results.[193,194–200]

ENDOTRACHEAL TUBES (ETTs)

Ventilator-associated pneumonia (VAP) is considered one of the most frequent nosocomial infections in intensive care units and is thought to be second only to urinary tract infection among hospital-acquired infections. The actual true incidence of VAP is unclear as no "gold standard" diagnostic test exists, but it is believed to develop in between 8% and 28% of susceptible patients, with mortality rates ranging between 24% and 76%.[201] Intubation and mechanical ventilation serve to bypass normal host defense mechanisms and the presence of an endotracheal tube (ETT) is strongly associated with the subsequent occurrence of pneumonia. Biofilm formation within the ETT occurs in 80% of patients[202] and provides a potential source of respiratory pathogens that can infect the lung field, causing VAP. Interestingly, it has been found that in mechanically ventilated patients, dental plaque biofilm is also modified by the additional presence of potential respiratory pathogens such as *Staphylococcus aureus* and Gram-negative bacteria including *Pseudomonas aeruginosa* and *Enterobacteriacea*.[201,203] There is mounting evidence that such oropharyngeal colonization is a prerequisite for the development of VAP[204] and might be the original source of organisms that "seed" the ETT biofilm. Figures 1.9 and 1.10 show evidence of biofilms on ETTs.

RHINOSINUSITIS

With increased interest in the role biofilms play in many acute and chronic conditions, more interest is growing concerning the involvement of biofilms in upper airway infections and in particular, rhinosinusitis.[205] However, more conclusive research is required in this area to determine the role biofilms play in rhinosinusitis.

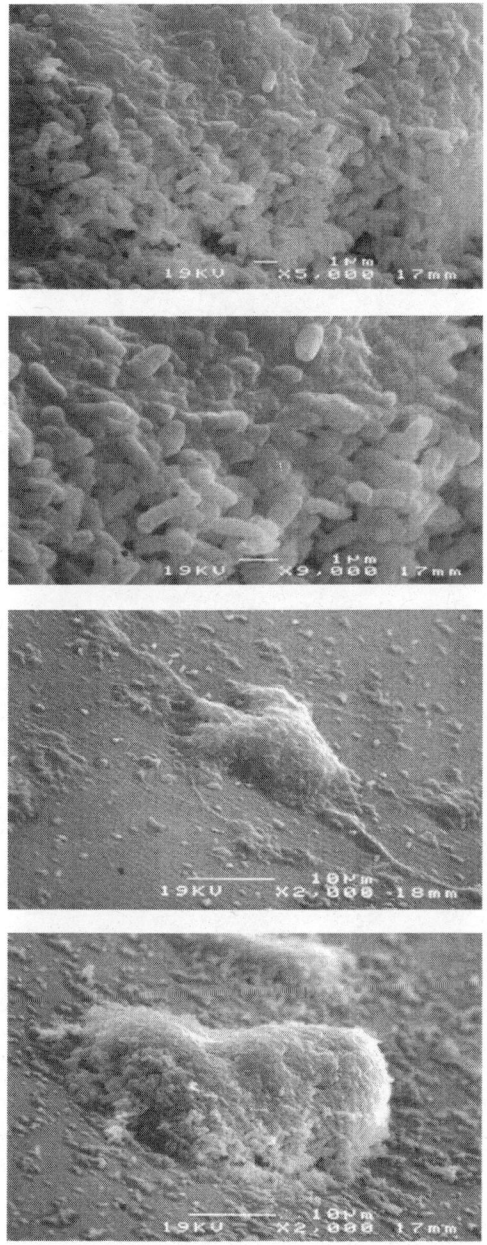

FIGURE 1.9 Evidence of biofilms on the lumen of endotracheal tubes.

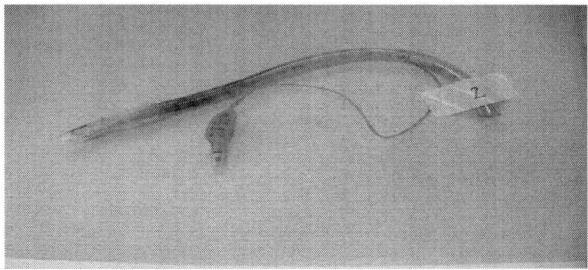

FIGURE 1.10   Biofilm formed on the inner lumen of an extubated endotracheal tubes.

Ophthalmic Infections

Evidence is increasing that biofilms have a role to play in ocular infections. These infections are generally associated with the use of abiotic prosthetic materials including contact lenses, intraocular lenses, and scleral buckles.[206] Bacterial biofilms have been reported on contact lenses, scleral buckles, suture material, and intraocular lenses. Approximately 56% of all corneal ulcers in the United States alone have been documented to be associated with individuals who wear contact lenses.[207]

Oral Infections

The oral cavity contains a rich diversity of microbial species, which undoubtedly reflects the wide range of habitats that exist within the mouth.[208] Sites where microorganisms can be recovered include all of the oral mucosa surfaces, the teeth, gingival crevice, and any inserted orthodontic appliance or denture. The types of microbial species recovered at each site can be distinct and relate not only to the nature of the colonized surface but also to the availability of oxygen and nutrient supply. Historically, the first biofilms analyzed were those of dental plaque, which were viewed by Antonie van Leeuwenhoek using his pioneering microscope in 1676.

From a clinical perspective, oral biofilms are responsible for two of the most prevalent infectious diseases of humans, dental caries and periodontal disease.[209] Although both infections are of biofilm origin, the specific causes are distinct. Dental caries is the dissolution of enamel by acids generated from the microbial fermentation of sugars. It has been shown in gnotobiotic studies that a range of bacteria can cause dental caries including *Lactobacillus*, *Actimycosis*, and *Streptococcus* species. However, the principal species involved is believed to be *Streptococcus mutans*. Importantly, not only is the species highly acidogenic and acidophilic, it also very adept at producing biofilms. A key strategy employed by *S. mutans* is the rapid generation of a "sticky" extracellular polysaccharide, referred to as glucan. This polymer is generated along with lactic acid from sucrose and serves to retain and recruit plaque bacteria at the tooth surface.

Periodontal diseases are collectively defined as the microbial mediated destruction of the tissues (gingival mucosa, periodontal ligament, and bone) that support the teeth. The exact organisms involved are not known, but associations between these infections and the presence of *Porphyromonas gingivalis*, *Tannerella forsythia*, *Treponema denticola,* and *Aggregatibacter actinomycetemcomitans* have been made.[210] Such bacteria are found in subgingival plaque, and being strictly anaerobic species, readily survive in the anaerobic environment of the gingival crevice. These plaque species are asaccharolytic, gaining their carbon sources largely from the proteolytic breakdown of proteins. This activity is believed to be partly responsible for periodontal disease, although the ensuing host-mediated inflammatory reaction also contributes.

In addition to bacterial-mediated diseases, oral disease relating to fungal biofilms is also relatively frequently encountered. The most prevalent fungal infections in the oral cavity are those caused by *Candida* species, particularly *C. albicans.* The form of oral candidosis most directly associated with *Candida* biofilms is chronic erythematous candidosis. In this infection, denture colonization by *Candida* biofilm is a key feature. In chronic erythematous candidosis, the fitting surface of the denture is a source of infectious organisms, which given their close proximity to the palatal mucosal cause local infection of that tissue. Normally, the mucosal surface is an effective barrier to infection, however, an ill-fitting denture can induce frictional irritation of mucosa, thus facilitating invasion of *Candida* into the superficial layers of the epithelium.

The wearing of a denture is a predisposing factor to other forms of oral candidosis and infection, most notably chronic hyperplastic candidosis and angular cheilitis. It is thought that the elevation in *Candida* numbers in the oral cavity arising from the denture biofilm favors a shift from normal commensal existence of *Candida* to an opportunistic pathogenic one.

CHRONIC WOUNDS

Following many years of speculation,[211–216] biofilms have been observed and reported in chronic wounds.[215–225] Because of this evidence, the management of biofilms is being considered significant to the healing of chronic wound infections and chronic wounds that have not healed for decades.[224–228]

BIOFILM RESISTANCE

Cells within biofilms have been reported to be 10 to 100 times more resistant to antibiotics[229,230] when compared to their planktonic counterparts. A number of detailed reviews have been published on this area.[231–234] Many factors are thought to contribute to the biofilm's ability to withstand high concentrations of antimicrobials.[235]

These include the following:

1. The binding of the antimicrobial to the extracellular matrix of the biofilm.
2. The inactivation of the antimicrobial by enzymes trapped in the biofilm matrix.
3. The reduced growth rate of bacteria in biofilms renders them less susceptible to the antimicrobial agent.
4. The altered microenvironment within the biofilms (e.g., pH, oxygen content) can reduce the activity of the agent.
5. Altered gene expression by organisms within the biofilm can result in a phenotype with reduced susceptibility to the antimicrobial agent.

Characteristics of the biofilm that provide it with antimicrobial resistance will be discussed in turn.

BINDING/FAILURE OF THE ANTIMICROBIAL TO PENETRATE THE BIOFILM

The EPS of the biofilm has been suggested as a mechanism to prevent ingress of the antibiotics. This involves either a reaction of the antimicrobial compound with, or sorption to, the components of the biofilm. This has been documented for some antimicrobials to limit the transport of antimicrobial agents to the microorganisms within the biofilm. A difference between thick and thin biofilms and their resistance to antibiotics has been observed leading to the conclusion that the biofilm acts as a barrier to antimicrobial agents.[236,237] For certain compounds, however, the exopolymers within the biofilm matrix have been shown to form an initial barrier hindering penetration of the antimicrobial agent. Despite this and other research papers published on this matter, many studies have established that the exopolymer matrix does not form an impenetrable barrier to the diffusion of antimicrobial agents. Consequently, other mechanisms must be in place to promote microbial cell survival in biofilms when exposed to antimicrobials.

SLOW GROWTH AND THE STRESS RESPONSE

It is now evident that within a biofilm there are regions where the bacterial cells become starved of a particular nutrient. The consequence of this will be to slow the growth of the effected bacteria. A decrease in growth has been shown to be synonymous with an increase in resistance to antibiotics.[238,239] It has been suggested that this physiological change can account for resistance of biofilms to antimicrobial agents.

Growth-rate-related effects under controlled growth conditions for planktonic cultures and biofilms of *P. aeruginosa*, *Escherichia coli*, and *S. epidermidis*[240,241] have been studied. Findings have shown that the sensitivities of both the planktonic and biofilm cells to either tobramycin or ciprofloxacin increased with increasing growth rate. This would therefore support the suggestion that the slow growth rate of biofilm cells protects the cells from antimicrobial action.[242] They found that resistance increased as the planktonic cultures and the biofilm cells approached stationary phase. These results and others suggest that some determinant other than growth rate is responsible for a certain level of resistance, and slow growth adds additional protection.

The slow growth rate of some cells within the biofilm is not due to nutrient limitation per se, but to a general stress response initiated by growth within a biofilm.[243] The stress response results in physiological changes that act to protect the cell from various environmental stresses. The central regulator of this response is the alternate factor, RNA polymerase, sigma S (sigma 38) factor (RpoS). Recent studies suggest that RpoS is induced by high cell density and that cells growing at these high densities seem to have undergone the general stress response.[244] *E. coli* cells that lack RpoS are unable to form normal biofilms, whereas planktonic cells are apparently unaffected by the absence of this factor.[245] There is some evidence to suggest that

RpoS have a role in biofilm resistance to oxidative biocides; it is clear, however, that other factors must contribute to this resistance.

HETEROGENEITY

Cells within the biofilm will experience a slightly different environment compared with other cells within the same biofilm and thus will be growing at a different rate. Gradients of nutrients, waste products, and signaling factors form to allow for this heterogeneity within the biofilm. There is evidence for gradients of physiological activity in response to antimicrobial treatment.[246,247] These studies reveal that the response to antimicrobial agents can greatly vary, depending on the location of a particular cell within a biofilm community.

INDUCTION OF A BIOFILM PHENOTYPE

An emerging concept is that a biofilm-specific phenotype is induced in a subpopulation of the community which results in the expression of active mechanisms to combat the detrimental effects of antimicrobial agents.[248–250]

When cells attach to a surface, they will express a general biofilm phenotype, and work has begun to try to identify genes that are activated or repressed in biofilms compared with planktonic cells.[251] Furthermore, it is possible that all or just a subset of these biofilm cells could express increased resistance to antimicrobial agents. This resistant phenotype might be induced by nutrient limitation, certain types of stress, high cell density, or a combination of these phenomena. Recent work has focused on the identification of genes that could contribute to this increased-resistance phenotype.

Multidrug efflux pumps can extrude chemically unrelated antimicrobial agents from the cell. In *E. coli*, upregulation of the *mar* operon results in a multidrug-resistant phenotype. The efflux pump thought to be responsible for this resistance is AcrAB.

There are three known multidrug-efflux pumps in *P. aeruginosa,* and there are several other putative pumps that have been identified. Another resistance mechanism that can be induced in biofilm cells is the alteration of the membrane-protein composition in response to antimicrobial agents. This change could result in decreased permeability of the cell to these compounds. Mutations in *ompB* (a regulator of the genes encoding the outer membrane porin proteins OmpF and OmpC) and in *ompF* increased the resistance of *E. coli* to a small beta-lactam antibiotic.[252] The environmental conditions within the biofilm can lead to alterations within the cell envelope that protect the bacteria from the detrimental affects of antimicrobial agents.

CONCLUSION

Biofilms are ubiquitous and associated with many environments. Their clinical significance has been documented widely when associated with the use of medical devices. Recognition of the role biofilms play in other infections in the human body has only recently gained momentum despite the fact that evidence was noted as far back as 1977.[227,253]

Biofilms are both beneficial and harmful to life, particularly with reference to clinically related diseases and infections. Many factors are known to affect the development of a biofilm and include the nature of the surface to which bacteria attach (e.g., its chemical and physical properties, nutrient availability, hydrodynamic forces, and communication systems between microorganisms). Biofilms are inherently resistant to antimicrobial agents, as well as environmental stresses and the innate immune host defense processes. As biofilms are associated with 80% of all known infections and 65% of nosocomial infections, it is clear that the mechanisms relating to resistance to antibiotics and antimicrobials in general is an area of intensive current and undoubted future research.

REFERENCES

1. Costerton, J.W., and Stewart, P.S. (2001) Battling biofilms. *Sci Am* 285:75–81.
2. Donlan, R.M., and Costerton, J.W. (2002) Biofilms: Survival mechanisms of clinically relevant microorganisms. *Clin Microbiol Rev* 15:167–193.
3. Heukelekian, H., and Heller, A. (1940) Relation between food concentration and surface for bacterial growth. *J Bacteriol* 40:547–558.
4. Zobell, C.E. (1943) The effect of solid surfaces upon bacterial activity. *J Bacteriol* 46:39–56.
5. Jones, H.C., Roth, I.L., and Saunders, W.M. III. (1969) Electron microscopic study of a slime layer. *J Bacteriol* 99:316–325.
6. Characklis, W.G. (1973) Attached microbial growths-II. Frictional resistance due to microbial slimes. *Water Res* 7:1249–1258.
7. Marshall, K.C., Stout, R., and Mitchell, R. (1971) Mechanism of the initial events in the sorption of marine bacteria to surfaces. *J Gen Microbiol* 68:337–348.
8. Costerton, J.W., Geesey, G.G., and Cheng, K.J. (1978) How bacteria stick? *Sci Am* 238:86–95.
9. Costerton, J.W., Cheng, K.J., Geesey, G.G., Ladd, T.I.M, Nickel, J.C., Dasgupta M., and Marie, T.J. (1987) Bacterial biofilms in nature and disease. *Annu Rev Microbiol* 41:435–464.
10. Costerton, J.W., Irvin, R.T., and Cheng, K.J. (1981) The bacterial glycocalyx in nature and disease. *Annu Rev Microbiol* 35:299–324.
11. Macfarlane, S. (2008) Microbial biofilm communities in the gastrointestinal tract. *J Clin Gastroenterol* 42:S142–143.
12. Coenye, T., Honraet, K., Rossel, B., and Nelis, H.J. (2008) Biofilms in skin infections: *Propionibacterium acnes* and acne vulgaris. *Infect Disord Drug Targets* 8(3):156–159.
13. Gera, I. (2008) The bacterial biofilm and the possibilities of chemical plaque control. *Fogorv Sz* 101(3):91–99.
14. Fraga, M., Perelmuter, K., Delucchi, L., Cidade, E., and Zunino, P. (2008) Vaginal lactic acid bacteria in the mare: Evaluation of the probiotic potential of native *Lactobacillus* spp. and *Enterococcus* spp. strains. *Antonie van Leeuwenhoek* 93(1–2):71–78.
15. Palmer, R. Jr., and White, D.C. (1997) Developmental biology of biofilms: Implications for treatment and control. *Trends Microbiol* 5:435–440.
16. Dune, W.M. (2002) Bacterial adhesion: Seen any good biofilms lately? *Clin Microbiol Rev* 15:155–166.
17. Mittelman, M.W. (1996) Adhesion to biomaterials. In: Fletcher, M., editor. *Bacterial adhesion: Molecular and ecological diversity.* New York: Wiley-Liss, pp. 89–127.

18. Vacca Smith, A.M., and Bowen, W.H. (2000) *In situ* studies of pellicle formation on hydroxyapatite discs. *Arch Oral Biol* 45(4):277–291.
19. Amaechi, B.T., Higham, S.M., Edgar, W.M., and Milosevic, A. (1999) Thickness of acquired salivary pellicle as a determinant of the sites of dental erosion. *J Dent Res* 78(12):1821–1828.
20. Hannig, M. (1999) Ultrastructural investigation of pellicle morphogenesis at two different intraoral sites during a 24-h period. *Clin Oral Investig* 3(2):88–95.
21. Shimotoyodome, A., Kobayashi, H., Tokimitsu, I., Hase, T., Inoue, T., Matsukubo, T., and Takaesu, Y. (2007) Saliva-promoted adhesion of *Streptococcus mutans* MT8148 associates with dental plaque and caries experience. *Caries Res* 41(3):212–218.
22. Marsh, P.D. (2004) Dental plaque as a microbial biofilm. *Caries Res* 38(3):204–211.
23. Characklis, W.G., McFeters, G.A., and Marshall, K.C. (1990) Physiological ecology in biofilm systems. In: Characklis. W.G., and Marshall, K.C., editors. *Biofilms.* New York: John Wiley & Sons, pp. 341–394.
24. Percival, S.L. (1999) The effect of molybdenum on biofilm development. *J Industrial Microbiol Biotech* 23: 112–117.
25. Litzler, P.Y., Benard, L., Barbier-Frebourg, N., Vilain, S., Jouenne, T., Beucher, E., Bunel, C., Lemeland, J.F., and Bessou, J.P. (2007) Biofilm formation on pyrolytic carbon heart valves: Influence of surface free energy, roughness, and bacterial species. *J Thorac Cardiovasc Surg* 134(4):1025–1032.
26. Díaz, C., Schilardi, P., and de Mele, M.F. (2008) Influence of surface sub-micropattern on the adhesion of pioneer bacteria on metals. *Artif Organs* 32(4):292–298.
27. Rodriguez, A., Autio, W.R., and McLandsborough, L.A. (2008) Effect of surface roughness and stainless steel finish on *Listeria monocytogenes* attachment and biofilm formation. *J Food Prot* 71(1):170–175.
28. Korber, D.R., Choi, A., Wolfaardt, G.M., Ingham, S.C., and Caldwell, D.E. (1997) Substratum topography influences susceptibility of *Salmonella enteritidis* biofilms to trisodium phosphate. *Appl Environ Microbiol* 63(9):3352–3358.
29. Schlegelová, J., and Karpísková, S. (2007) Microbial biofilms in the food industry. *Epidemiol Mikrobiol Imunol* 56(1):14–19.
30. Charman, K.M., Fernandez, P., Loewy, Z., and Middleton, A.M. (2009) Attachment of *Streptococcus oralis* on acrylic substrates of varying roughness. *Lett Appl Microbiol* 48(4):472–477.
31. Teughels, W., Van Assche, N., Sliepen, I., and Quirynen, M. (2006) Effect of material characteristics and/or surface topography on biofilm development. *Clin Oral Implants Res* 17:68–81.
32. Percival, S.L., Knapp, J.S., Wales, D.S., and Edyvean, R.G.J. (1998) Biofilm development in mains water. *Water Research* 32(1):243–253.
33. Pringle, J.H., and Fletcher, M. (1983) Influence of substratum wettability on attachment of freshwater bacteria to solid surfaces. *Appl Environ Microbiol* 45:811–817.
34. Westman, E.H., Ek, M., Enarsson, L.E., and Wågberg, L. (2009) Assessment of antibacterial properties of polyvinylamine (PVAm) with different charge densities and hydrophobic modifications. *Biomacromolecules* 10:1478–1483.
35. Characklis, W.G. (1981) Fouling biofilm development: A process analysis. *Biotechnol Bioeng* 23:1923–1960.
36. Fletcher, M., and Marshall, K.C. (1982) Are solid surfaces of ecological significance to aquatic bacteria? *Adv Microbial Ecol* 12:199–236.
37. Boyle, J.D., and Lappin-Scott, H. (2007) The effect of flow direction and magnitude on the initial distribution of *Pseudomonas aeruginosa* cells attached to glass. *Biofouling* 23(3–4):139–150.

38. Characklis, W.G., McFeters, G.A., and Marshall, K.C. (1990). Physiological ecology of biofilm systems. In: Characklis, W.G., and Marshall, K.C., editors. *Biofilms*. New York: Wiley, pp. 344–393.

39. Characklis, W.G., Turakhia, M.H., and Zelver, N. (1990) Transfer and interfacial transport phenomena. In: Characklis, W.G., and Marshall, K.C., editors. *Biofilms*. New York: Wiley, pp. 265–240.

40. Lister, D.H. (1979) Corrosion products in power generating systems. In: Somerscales, E.F.S., and Knuden, J.G., editors. *Fouling of Heat Transfer Equipment*. Washington, DC: Hemisphere, pp. 135–200.

41. Walt, D.R., Smulow, J.B., Turesky, S.S., and Hill, R.G. (1985) The effect of gravity on initial microbial adhesion. *J Colloid and Interface Sci* 107:334–336.

42. Fletcher, M., and Loeb, G.I. (1979) The influence of substratum characteristics on the attachment of a marine *Pseudomonas* to solid surfaces. *Appl Environ Microbiol* 37:67–72.

43. Davey, M.E., and O'Toole, G.A. (2000) Microbial biofilms: From ecology to molecular genetics. *Microbiol Mol Biol Rev* 64(4):847–867.

44. Shirliff, M.E., Mader, J.T., and Camper, A.K. (2002) Molecular interactions in biofilms. *Chem Biol* 9:859–871.

45. Rittman, B.E. (1989) The effect of shear stress on biofilm loss rate. *Biotechnol Bioeng* 24:501–506.

46. Jain, A., Gupta, Y., Agrawal, R., Khare, P., and Jain, S.K. (2007) Biofilms—A microbial life perspective: A critical review. *Crit Rev Ther Drug Carrier Syst* 24(5):393–443.

47. Whittaker, C.J., Klier, C.M., and Kolenbrander, P.E. (1996) Mechanisms of adhesion by oral bacteria. *Ann Rev Microbiol* 50:513–552.

48. Perez Vidakovics, M.L., and Riesbeck, K. (2009) Virulence mechanisms of *Moraxella* in the pathogenesis of infection. *Curr Opin Infect Dis* 22(3):279–285.

49. Busscher, H.J., and Weerkamp, A. (1987) Specific and non-specific interactions: Role in bacterial adhesion to solid substrata. *FEMS Microbiol Rev* 46:165–173.

50. Derjaguin, B.V., and Landau, L. (1941) Theory of the stability of strongly charged lyophobic sols and of adhesion of strongly charged particles in solution of electrolytes. *Acta Physiochimica URSS* 14:633–662.

51. Hermansson, M. (1999) The DLVO theory in microbial adhesion. *J Colloid Interface Sci* 14:105–119.

52. Bashan, Y., and Levanony, H. (1988) Active attachment of *Azospirillum brasilense* Cd to quartz sand and to a light-textured soil by protein bridging. *J Gen Microbiol* 134:2269–2279.

53. Danielsson, A., Norkrans, B., and Bjornsson, A. (1977) On bacterial adhesion—The effect of certain enzymes on adhered cells in a marine *Pseudomonas* sp. *Bot Marina* 20:13–17.

54. Araújo, A.M., Oliveira, I.C., Mattos, M.C., and Benchetrit, L.C. (2008) Cell surface hydrophobicity and adherence of a strain of group B streptococci during the postantibiotic effect of penicillin. *Rev Inst Med Trop Sao Paulo* 50(4):203–207.

55. O'Toole, G.A., and Kolter, R. 1998. Flagellar and twitching motility are necessary for *Pseudomonas aeruginosa* biofilm development. *Mol Microbiol* 30:295–304.

56. Shirtliff, M.E., Mader, J.T., and Camper, A.K. (2002) Molecular interactions in biofilms. *Chem Biol* 9(8):859–871.

57. Rijnaarts, H.H.M., Norde, W., Lyklema, J., and Zehnder, A.B. (1999) DVLO and steric contributions to bacterial deposition in media of different ionic strengths. *J Colloid Interface Sci* 14:179–195.

58. James, A.M. (1991) Charge properties of microbial cell surfaces. In: Mozes, N., Handely, P.S., Busscher, H.J., and Rouxhet, P.G., editors. *Microbial Cell Surface Analysis*. Weinheim, Germany: VCH Weinheim, pp. 221–263.

59. Hancock, I.C. (1991) Microbial cell surface architecture. In: Mozes, N., Handely, P.S., Busscher, H.J., and Rouxhet, P.G., editors. *Microbial Cellsurface Analysis*. Weinheim, Germany: VCH Weinheim, pp. 21–59.

60. Patti, J.M., Allen, B.L., McGavin, M.J., and Hook, M. (1994) MSCRAMM—A mediated adherence of microorganisms to host tissues. *Annu Rev Microbiol* 48:585–617.

61. Coutte, L., Willery, E., Antoine, R., Drobecq, H., Locht, C., and Jacob-Dubuisson, F. (2003) Surface anchoring of bacterial subtilisin is important for maturation function. *Mol Microbiol* 49(2):529–539.

62. Varga, J.J., Nguyen, V., O'Brien, D.K., Rodgers, K., Walker, R.A., and Melville, S.B. (2006) Type IV pili-dependent gliding motility in the Gram-positive pathogen *Clostridium perfringens* and other *Clostridia*. *Mol Microbiol* 62(3):680–694.

63. Gaddy, J.A., and Actis, L.A. (2009) Regulation of *Acinetobacter baumannii* biofilm formation. *Future Microbiol* 4:273–278.

64. Corpe, W.A. (1980) Microbial surface components involved in adsorption of microorganisms onto surfaces. In: Bitton, G., and Marshall, K.C., editors. *Adsorption of microorganisms to surfaces*. New York: John Wiley & Sons, pp. 105–144.

65. Bullitt, R., and Makowski, L. (1995) Structural polymorphism of bacterial adhesion pili. *Nature* 373:164–167.

66. Tomich, M., Herfst, C.A., Golden, J.W., and Mohr, C.D. (2002) Role of flagella in host cell invasion by *Burkholderia cepacia*. *Infect Immun* 70(4):1799–1806.

67. Fletcher, M. (1977) The effects of culture concentration and age, time, and temperature on bacterial attachment to polystyrene. *Can J Microbiol* 23:1–6.

68. Korber, D.R., Lawrence, J.R., Sutton, B., and Caldwell, D.E. (1989) Effect of laminar flow velocity on the kinetics of surface recolonization by Mot$^+$ and Mot$^-$ *Pseudomonas fluorescens*. *Microb Ecol* 18:1–19.

69. Davies, D.G., Chakrabarty, A.K., and Geesey, G.G. (1993) Exopolysaccharide production in biofilms: Substratum activation of alginate gene expression by *Pseudomonas aeruginosa*. *Appl Environ Microbiol* 59:1181–1186.

70. Whiteley, M., Bangera, M.G., Bumgarner, R.E., Parsek, M.R., Teitzel, G.M., Lory, S., and Greenberg, E.P., et al. (2001) Gene expression in *Pseudomonas aeruginosa* biofilms. *Nature* 413:860–864.

71. Prigent-Combaret, C., Vidal, O., Dorel, C., and Lejeune P. (1999) Abiotic surface sensing and biofilm-dependent regulation of gene expression in *Escherichia coli*. *J Bacteriol* 181(19):5993–6002.

72. Becker, P., Hufnagle, W., Peters, G., and Herrmann, M. (2001) Detection of different gene expression in biofilm-forming versus planktonic populations of *Staphylococcus aureus* using micro-representational difference analysis. *Appl Environ Microbiol* 67:2958–2965.

73. Caldwell, D.E., Korber, D.R., and Lawrence, J.R. (1992) Confocal laser microscopy and digital image analysis in microbial ecology. *Advances in Microbial Ecology* 12:1–67.

74. Korber, D.R., Lawrence, J.R., Lappin-Scott, H.M., and Costerton, J.W. (1995) The formation of microcolonies and functional consortia within biofilms. In: *Bacterial Biofilms*, pp. 15–45.

75. Blenkinsopp, S.A., and Costerton, J.W. (1991) Understanding bacterial biofilms. *Trends Biotechnol* 9:138–143.

76. Connell, J.H., and Slatyer, R.O. (1977) Mechanisms of succession in natural communities and their role in community stability and organization. *Am Naturalist* 111:1119–1144.

77. Fredrickson, A.G. (1977) Behaviour of mixed cultures of microorganisms. *Annu Rev Microbiol* 33:63–87.

78. Lear, G., Anderson, M.J., Smith, J.P., Boxen, K., and Lewis, G.D. (2008) Spatial and temporal heterogeneity of the bacterial communities in stream epilithic biofilms. *FEMS Microbiol Ecol* 65(3):463–473.

79. Baier, R.E. (1984) Initial events in microbial film formation. In: Costlow, J.D., and Tipper, R.C., editors. *Marine Biodetermination: An Interdisciplinary Approach.* London: E and F.N. Spon, pp. 57–62.
80. Wahl, M. (1989) Marine epibiosis. 1. Fouling and antifouling: Some basic aspects. *Marine Ecol Progress Ser* 58:175–189.
81. Sousa, C., Teixeira, P., and Oliveira, R. The role of extracellular polymers on *Staphylococcus epidermidis* biofilm biomass and metabolic activity. *J Basic Microbiol* 49:363–370.
82. Crampton, S.E., Gerke, C., Schnell, N.F., Nichols, W.W., and Gotz, F. (1999) The intracellular adhesion (ica) locus is present in *Staphylococcus aureus* and is required for biofilm formation. *Infect Imun* 67:5427–5433.
83. Flemming, H.-C., Wingender, J., Griegbe, C., and Mayer, C. (2000) Physico-chemical properties of biofilms. In: Evans, L.V., editor. *Biofilms: Recent Advances in Their Study and Control.* Amsterdam: Harwood Academic, pp. 19–34.
84. Banin, E., Brady, K.M., and Greenberg, E.P. (2006) Chelator-induced dispersal and killing of *Pseudomonas aeruginosa* cells in a biofilm. *Appl Environ Microbiol* 72(3):2064–2069.
85. Sutherland, I.W. (2001) Biofilm exopolysaccharides: A strong and sticky framework. *Microbiology* 147:3–9.
86. Samrakandi, M.M., Roques, C., and Michel, G. (1997) Influence of trophic conditions on exopolysaccharide production: Bacterial biofilm susceptibility to chlorine and monochloramine. *Can J Microbiol* 43(8):751–758.
87. Ceyhan, N., and Ozdemir, G. (2008) Extracellular polysaccharides produced by cooling water tower biofilm bacteria and their possible degradation. *Biofouling* 24(2):129–135.
88. Leriche, V., Sibille, P., and Carpentier, B. (2000) Use of an enzyme-linked lectinsorbent assay to monitor the shift in polysaccharide composition in bacterial biofilms. *Appl Environ Microbiol* 66:1851–1856.
89. Humphrey, B.A., Dickson, M.R., and Marshall, K.C. (1979) Physiological and *in-situ* observations on adhesion of gliding bacteria to surfaces. *Arch Microbiol* 120:231–238.
90. Roberts, A.P., Cheah, G., Ready, D., Pratten, J., Wilson, M., and Mullany, P. (2001) Transfer of Tn916-like elements in microcosm dental plaques. *Antimicro Agents Chemother* 42:2943–2946.
91. Atlas, R.M. (1984) Diversity of microbial communities. *Adv Microbial Ecol* 7:1–47.
92. Brozel, V.S., and Cloete, T.E. (1990) Evaluation of agar plating methods for the enumeration of viable aerobic heterotrophs in cooling water. *6th Biennial Congress of the South African Society for Microbiology*, Abstracts 22.13.
93. Hamilton, W.A. (1985) Sulphate-reducing bacteria and anaerobic corrosion. *Annu Rev Microbiol* 39:195–217.
94. Lewandowski, Z., Stoodley, P., and Roe, F. (1995) Internal mass transport in heterogeneous biofilms. *Recent Advances in Corrosion/95,* paper no. 222, NACE International, Houston, TX.
95. Ma, L., Conover, M., Lu, H., Parsek, M.R., Bayles, K., and Wozniak, D.J. (2009) Assembly and development of the *Pseudomonas aeruginosa* biofilm matrix. *PLoS Pathog* 5(3):e1000354.
96. Costerton, J.W., Lewandowski, Z., de Beer, D., Calwell, D., Korber, D., and James, G. (1994) Biofilms, the customised microniches. *J Bacteriol* 176:2137–2142.
97. de Beer, D., Stoodley, P., Roe, F., and Lewandowski, Z. (1994) Effects of biofilm structures on oxygen distribution and mass transport. *Biotechnol Bioeng* 43(11):1131–1138.
98. Keevil, C.W., Dowsett, A.B., and Rogers, J. (1993) *Legionella* biofilms and their control. Society for Applied Bacteriology Technical Series: Microbiofilms, pp. 203–215.
99. Lewandowski, Z., and Stoodley, P. (1995) Flow induced vibrations, drag force, and pressure drop in conduits covered with biofilm. *Wat Sci Tech* 32(8):19–26.

100. Percival, S.L., Knapp, J.S., Wales, D.S., and Edyvean, R.G.J. (1999) The effect of flow and surface roughness on biofilm formation. *J Microbiol Biotechnol* 22:152–159.
101. Tolker-Nielsen, T., and Molin, S. (2000) Spatial organization of microbial biofilm communities. *Microb Ecol* 40:75–84.
102. Lewandowski, Z. (2000). Structure and function of biofilms. In: Evans, L.V., editor. *Biofilms: Recent Advances in Their Study and Control.* Harwood Academic.
103. Durack, D.T. (1975) Experimental bacterial endocarditis. IV Structure and evolution of very early lesions. *J Pathol* 115:81–89.
104. Tunney, M.M., Jones, D.S., and Gorman, S.P. (1999) Biofilm and biofilm-related encrustations of urinary tract devices. In: Doyle, R.J., editor. *Methods in Enzymology, vol. 310. Biofilms.* San Diego: Academic Press, pp. 558–566.
105. Sauer, K., Camper, A.K., Ehrlich, G.D., Costerton, J.W., and Davies, D.G. (2002) *Pseudomonas aeruginosa* displays multiple phenotypes during development as a biofilm. *J Bacteriol* 184:1140–1154.
106. Palmer R. (2009) Oral bacterial biofilms—History in progress. *Microbiology* May 21.
107. Stoodley, P., Boyle, J.D., Dodds, I., and Lappin-Scott, H.M. (1997) Consensus model of biofilm structure. In *Biofilms: Community Interactions and Control.* Wimpenny, J.W.T., Handley, P.S., Gilbert, P., Lappin-Scott, H.M., and Jones, M., editors. Third Meeting of the British Biofilm Club, Gregynog Hall, Powys, 26–28 September, 1997, pp. 1–9.
108. Stoodley, P., Wilson, S., Hall-Stoodley, L., Boyle, J.D., Lappin-Scott, H.M., and Costerton, J.W. (2001) Growth and detachment of cell clusters from mature mixed species biofilms. *Appl Environ Microbiol* 67:5608–5613.
109. Characklis, W.G., and Cooksey, K.E. (1983) Biofilms and microbial fouling. *Adv Appl Microbiol* 29:93–138.
110. Applegate, D.H., and Bryers, J.D. (1991) Effects of carbon and oxygen limitations and calcium concentrations on biofilm removal processes. *Biotechnol Bioeng* 37:17–25.
111. Rochex, A., Massé, A., Escudié, R., Godon, J.J., and Bernet, N. (2009) Influence of abrasion on biofilm detachment: Evidence for stratification of the biofilm. *J Ind Microbiol Biotechnol* 36(3):467–470.
112. Telgmann, U., Horn, H., and Morgenroth, E. (2004) Influence of growth history on sloughing and erosion from biofilms, *Water Res* 38(17):3671–3684.
113. McBain, A.J., Allison, D., and Gilbert, P. (2000) Emerging strategies for the chemical treatment of microbial biofilms. *Biotechnol Genet Eng Rev* 17:267–279.
114. Chang, H.T., and Rittman, B.E. (1988) Comparative study of biofilm shear loss on different adsorptive media. *J Water Pollut Control Fed* 60:362–368.
115. Ward, K.H., Olson, M.E., Lam, K., and Costerton, J.W. (1992) Mechanisms of persistent infection associated with peritoneal implants. *J Med Microbiol* 36:406–403.
116. Shiau, A.L., and Wu, C.L. (1998) The inhibitory effect of *Staphylococcus epidermidis* slime on the phagocytosis of murine peritoneal macrophages is interferon independent. *Microbiol Immunol* 42:33–40.
117. Meluleni, G.J., Grout, M., Evans, D.J., and Pier, G.B. (1995) Mucoid *Pseudomonas aeruginosa* growing in a biofilm in vitro are killed by opsonic antibodies to the mucoid exopolysaccharide capsule but not by antibodies produced during chronic lung infection in cystic fibrosis patients. *J Immunol* 155:2029–2038.
118. Yasuda, H., Ajiki, Y., Aoyama, J., and Yokota, T. (1994) Interaction between human polymorphonuclear leucocytes and bacteria released from in vitro bacterial biofilm models. *J Med Microbiol* 41:359–367.
119. Raad, I.I., Sabbagh, M.F., and Rand, K.H., and Sherertz, R.J. (1992) Quantitative tip culture methods and the diagnosis of central venous catheter-related infections. *Microbiol Infect Dis* 15:13–20.
120. Goller, C.C., and Romeo, T. (2008) Environmental influences on biofilm development. *Curr Top Microbiol Immunol* 322:37–66.

121. McEldowney, S., and Fletcher, M. (1988) Effect of pH, temperature, and growth conditions on the adhesion of a gliding bacterium and three nongliding bacteria to polystyrene. *Microb Ecol* 16:183–195.
122. Powell, M.S., and Slater, N.K.H. (1982) Removal rate of bacterial cells from glass surfaces by fluid shear. *Biotechnol Bioeng* 24:2527–2537.
123. Murga, R., Forster, T.S., Brown, E., Pruckler, J.M., Fields, B.S., and Donlan, R.M. (2001) The role of biofilms in the survival of *Legionella pneumophila* in a model potable water system. *Microbiology* 147:3121–3126.
124. Donlan, R.M. (2009) Preventing biofilms of clinically relevant organisms using bacteriophage. *Trends Microbiol* 17(2):66–72.
125. Davies, D.G. (1999) Regulation of matrix polymer in biofilm formation and dispersion. In: Wingender, J., Neu, T., and Flemming, H.-C., editors. *Microbial Extracellular Polymeric Substances*. Berlin: Springer-Verlag, pp. 93–117.
126. Sauer, K., Camper, A.K., Ehrlich, G.D., Costerton, J.W., and Davies, D.G. (2002) *Pseudomonas aeruginosa* displays multiple phenotypes during development as a biofilm. *J Bacteriol* 184(4):1140–1154.
127. Stewart, P.S., Rani, S.A., Gjersing, E., Codd, S.L., Zheng, Z., and Pitts, B. (2007) Observations of cell cluster hollowing in *Staphyloccus epidermidis* biofilms. *Lett Appl Microbiol* 44(4):454–457.
128. Kirov, S.M., Webb, J.S., O'May, C.Y., Reid, D.W., Woo, J.K., Rice, S.A., and Kjelleberg, S. (2007) Biofilm differentiation and dispersal in mucoid *Pseudomonas aeruginosa* isolates from patients with cystic fibrosis. *Microbiology* 153(10):3264–3274.
129. Purevdorj-Gage, B., Costerton, W.J., and Stoodley, P. (2005) Otypic differentiation and seeding dispersal in non-mucoid and mucoid *Pseudomonas aeruginosa* biofilms. *Microbiology* 151(Pt 5):1569–1576.
130. Stapper, A.P., Narasimhan, G., Ohman, D.E., Barakat, J., Hentzer, M., Molin, S., Kharazmi, A., Høiby, N., and Mathee, K. (2004) Alginate production affects *Pseudomonas aeruginosa* biofilm development and architecture, but is not essential for biofilm formation. *J Med Microbiol* 53:679–690.
131. Schleheck, D., Barraud, N., Klebensberger, J., Webb, J.S., McDougald, D., Rice, S.A., and Kjelleberg, S. (2009) *Pseudomonas aeruginosa* PAO1 preferentially grows as aggregates in liquid batch cultures and disperses upon starvation. *PLoS ONE* 4(5):e5513.
132. Hunt, S.M., Werner, E.M., Huang, B., Hamilton, M.A., and Stewart, P.S. (2004) Hypothesis for the role of nutrient starvation in biofilm detachment. *Appl Environ Microbiol* 70(12):7418–7425.
133. Lee, S.F., Li, Y.H., and Bowden, G.H. (1996) Detachment of *Stretococcus mutans* biofilm cells by an endogenous enzymatic activity. *Infec Immun* 64:1035–1038.
134. Neu, T.R. (1996) Significance of bacterial surface-active compounds in interactions of bacteria with interfaces. *Microbiol Rev* 60:151–166.
135. Wenzel, R.P. (2007) Health care-associated infections: Major issues in the early years of the 21st century. *Clin Infect Dis* 45:S85–S88.
136. Allison, D.G., Ruiz, B., SanJose, C., Jaspe, A., and Gilbert, P. (1998) Extracellular products as mediators of the formation and detachment of *Pseudomonas flourescens* biofilms. *FEMS Microbiol Lett* 167:179–184.
137. Licking, E. (1999). Getting a grip on bacterial slime. *Business Week* 13 September, pp. 98–100.
138. Bayston, R. (2000) Biofilms and prosthetic devices. In: Allison, D.G., Gilbert, P., Lappin-Scott, H.M., and Wilson, M., editors. *Community Structure and Co-operation in Biofilms*. Cambridge: Cambridge University Press, pp. 295–307.
139. Costerton, J.W., Lewandowski, Z., Caldwell, D.E., Korber, D.R., and Lappin-Scott, H.M. (1995) Microbial biofilms. *Annu Rev Microbiol* 49:711–745.

140. Archibald, L.K., and Gaynes, R.P. (1997) Hospital acquired infections in the United States: The importance of interhospital comparisons. *Nosocomial Infect* 11: 245–255.
141. Percival, S.L., Hunter, P.R., Chalmers, R., Sellwood, J., and Levy, D. (2004) *Microbiology of Waterborne Diseases: Public Health Significance.* New York: Academic Press.
142. Percival, S.L., Knapp, J.S., Wales, D.S., and Edyvean, R.G. J. (1999) Biofilm development in potable quality water. *Biofouling* 13(4):259–277.
143. Flemming, H.C., Percival, S.L., and Walker, J.T. (2002) Contamination potential of biofilms in water distribution systems. *Water Sci Technol: Water Supply* 2(1):271–280.
144. Percival, S.L., and Walker, J.T. (1999) Biofilms and public health significance. *Biofouling* 14:99–115.
145. Percival, S.L., Walker, J.T., and Hunter, P. (2000) *Microbiological aspects of biofilms and drinking water.* Boca Raton, FL: CRC/Lewis Press.
146. Percival, S.L., Knapp, J.S., Wales, D.S., and Edyvean, R.G.J. (1998) Biofilm development in mains water. *Water Res* 32(1):243–253.
147. Fazakerley, C., Pryor, M., and Percival, S.L. (2001) The isolation of *Mycobacterium* spp from groundwater, chlorinated distribution systems and biofilms. In: Gilbert, P.G., Allison, D., Walker, J.T., and Brading, M., editors. *Biofilm Community Interactions: Chance or Necessity?* Cardiff: Bioline, pp. 53–57.
148. Percival, S.L., and Thomas, J.G. (2009) Transmission of *Helicobacter pylori* and the role of water and biofilms. *J Water Health* 7(3):469–477.
149. Percival, S.L., Knapp, J.S., Wales, D.S., Edyvean, R.G.J., Beech, I.B., and Videla, H.A. (1996) Biofilm development on 304 and 316 stainless steels in a potable water system. Second NACE Latin American Region Corrosion Congress, September. NACE, Houston. Paper LA 96204.
150. Percival, S.L., Beech, I.B., Knapp, J.S., Wales, D.S., and Edyvean, R.G.J. (1997) Biofilm development on stainless steel in a potable water system. *J Inst Water and Environ Manage* 11(4):289–294.
151. Percival, S.L. (1998) Biofilm development in engineered systems. *Br Corrosion J* 33:130–137.
152. Exner, M., Kramer, A., Lajoie, L., Gebel, J., Engelhart, S., and Hartemann, P. (2005) Prevention and control of health care-associated waterborne infections in health care facilities. *Am J Infect Control* 33:S26–S40.
153. Exner, M., Kramer, A., Kistemann, T., Gebel, J., and Engelhart, S. (2007) Water as a reservoir for nosocomial infections in health care facilities, prevention and control. *Bundesgesundheitsblatt Gesundheitsforschung Gesundheitsschutz* 50(3):302–311.
154. Ozerol, I.H., Bayraktar, M., Cizmeci, Z., et al. (2006) Legionnaire's disease: A nosocomial outbreak in Turkey. *J Hosp Infect* 62:50–57.
155. Lau, H.Y., and Ashbolt, N.J. (2009) The role of biofilms and protozoa in Legionella pathogenesis: Implications for drinking water. *J Appl Microbiol* 10:156.
156. Walker, J.T., and Percival, S.L. (2001) Control of biofouling in drinking water systems. In: Walker, J.T., Surman, S., and Jess, J., editors. *Industrial Biofouling: Detection, Prevention and Control* (Vol. 1). New York: John Wiley & Sons, pp. 55–76.
157. Szymańska, J. (2005) Electron microscopic examination of dental unit waterlines biofilm. *Ann Agric Environ Med* 12(2):295–298.
158. Walker, J.T., Bradshaw, D.J., Finney, M., Fulford, M.R., Frandsen, E., ØStergaard, E., Ten Cate, J.M., Moorer, W.R., Schel, A.J., Mavridou, A., Kamma, J.J., Mandilara, G., Stösser, L., Kneist, S., Araujo, R., Contreras, N., Goroncy-Bermes, P., O'Mullane, D., Burke, F., Forde, A., O'Sullivan, M., and Marsh, P.D. (2004) Microbiological evaluation of dental unit water systems in general dental practice in Europe. *Eur J Oral Sci* 112(5):412–418.
159. Liaqat, I., and Sabri, A.N. (2009) Isolation and characterization of biocides resistant bacteria from dental unit water line biofilms. *J Basic Microbiol* 49:275–284.

160. Liaqat, I., and Sabri, A.N. (2008) Effect of biocides on biofilm bacteria from dental unit water lines. *Curr Microbiol* 56(6):619–624.

161. Shorman, H., Nabaa, L.A., Coulter, W.A., Pankhurst, C.L., and Lynch, E. (2002) Management of dental unit water lines. *Dent Update* 29(6):292–298.

162. Walker, J.T., and Marsh, P.D. (2007) Microbial biofilm formation in DUWS and their control using disinfectants. *J Dent* 35(9):721–730.

163. Griffith, D.P., and Klein, A.S. (1983) Infection-induced urinary stones. In: *Stones: Clinical Management of Urolithiasis.* Roth, R.A., and Finlayson, B., editors. New York: Lippincott Williams & Wilkins, pp. 210–226.

164. Hellstrom, J. (1938) The significance of staphylococci in the development and treatment of renal and urethral stones. *Br J Urol* 10:348–372.

165. Nickel, J.C., Reid, G., Bruce, A.W., and Costerton, J.W. (1986) Ultrastructural microbiology of infected urinary stone. *Urology* 28:512–515.

166. Freedman, L.R., and Valone, J. Jr. (1979) Experimental infective endocarditis. *Prog Cardiovasc Dis* 22:169–180.

167. Rajan, S., and Saiman, L. (2002) Pulmonary infections in patients with cystic fibrosis. *Semin Respir Infect* 17:47–56.

168. Wagner, V.E., and Iglewski, H. (2008) *P. aeruginosa* biofilms in CF Infection. *Clinic Rev Allerg Immunol* 35:124–134.

169. Singh, P.K., Schaefer, A.L., Parsek, M.R., Moninger, T.O., Welsh, M.J., and Greenberg, E.P. (2000) Quorum-sensing signals indicate that cystic fibrosis lungs are infected with bacterial biofilms. *Nature* 407:762–764.

170. Hoiby, N., Krogh, J.H., Moser, C., Song, Z., Ciofu, O., and Kharazmi, A. (2001) *Pseudomonas aeruginosa* and the *in vitro* and *in vivo* biofilm mode of growth. *Microbes Infect* 3:23–35.

171. Rosenfield, R.M., and Post, J.C. (1992) Meta-analysis of antibiotics for the treatment of otitis media with effusion. *Otolaryngol Head Neck Surgery* 106:378–386.

172. Post, J.C. (2001) Direct evidence of bacterial biofilms in otitis media. *Laryngoscope* 111:2083–2094.

173. Ehrlich, G.D., Veeh, R., Wang, X., Costerton, J.W., Hayes, J.D., et al. (2002) Mucosal biofilm formation on middle-ear mucosa in the chinchilla model of otitis media. *JAMA* 287:1710–1715.

174. Bakaletz, L.O. (2007) Bacterial biofilms in otitis media: Evidence and relevance. *Pediatr Infect Dis J* 26:S17–S19.

175. Ogra, P.L., Giesuik, G.S., and Bareukamp, S.J. (1989) Microbiological, immunology and vaccination. *Ann Otol Rhinol Laryngol* 139:S29–S49.

176. Fergie, N., Bayston, R., Pearson, J.P., and Birchall, J.P. (2004) Is otitis media with effusion a biofilm infection? *Clin Otolaryngol* 29:38–46.

177. Marrie, T.J., and Costerton, J.W. (1985) Mode of growth of bacterial pathogens in chronic polymicrobial human osteomyelitis. *J Clin Microbiol* 22:924–933.

178. Brady, R.A., Gadd, J.G., Calhoun, J.H., Costerton, J.W., and Shirtliff, M.E. (2008) Osteomyelitis and the role of biofilms in chronic infection. *FEMS Immunol Med Microbiol* 52:13–22.

179. Nickel, J.C., and Costerton, J.W. (1993) Bacterial localization in antibiotic-refractory chronic bacterial prostatitis. *Prostate* 23:107–114.

180. Hua, V.N., Williams, D.H., and Schaeffer, A.J. (2005) Role of bacteria in chronic prostatitis/chronic pelvic pain syndrome. *Curr Urol Rep* 6(4):300–306.

181. Nguyen, D.P., Gerber, S., Hohifeld, P., Sandrine, G., and Witkin, S.S. (2004) Mycoplasma hominis in mid-trimester amniotic fluid: Relation to pregnancy outcome. *J Perinat Med* 32:323–326.

182. Espinoza, J., Goncalves, L.F., Romero, R., et al. (2005) The prevalence and clinical significance of amniotic fluid "sludge" in patients with preterm labor and intact membrane. *Ultrasound Obstet Gynecol* 25:346–352.

183. Romero, R., Schaudinn, C., Kusanovic, J.P., et al. (2008) Detection of a microbial biofilm in intraamniotic infection. *Am J of Obstetrics* 135:e1–e5.

184. Dasgupta, M.K. (2002) Biofilms and infection in dialysis patients. *Semin Dialysis* 15:338–346.

185. McCann, M.T., Gilmore, B.F., and Gorman, S.P. (2008) *Staphylococcus epidermidis* device-related infections: Pathogenesis and clinical management. *J Pharm Pharmacol* 60:1551–1571.

186. Hatt, J.K., and Rather, P.N. (2008) Role of bacterial biofilms in urinary tract infections. *Curr Top Microbiol Immunol* 322:163–192.

187. Holá, V., and Růzicka, F. (2008) Urinary catheter biofilm infections. *Epidemiol Mikrobiol Imunol* 57(2):47–52.

188. Stickler, D.J. (2008) Bacterial biofilms in patients with indwelling urinary catheters. *Nat Clin Pract Urol* 5(11):598–608.

189. Salo, J., Sevander, J.J., Tapiainen, T., Ikäheimo, I., Pokka, T., Koskela, M., and Uhari, M. (2009) Biofilm formation by *Escherichia coli* isolated from patients with urinary tract infections. *Clin Nephrol* 71(5):501–507.

190. Percival, S.L., Kite, P., and Stickler, D. (2009) The effect of EDTA instillations on the rate of development of encrustations and biofilms in Foley catheters. *Urological Res* 37:205–209.

191. Darouiche, R. (2001) Device-associated infections: A macroproblem that starts with microadherence. *Clin Infect Dis* 33:1567–1572.

192. Bacuzzi, A., Cecchin, A., Del Bosco, A., Cantone, G., and Cuffari, S. (2006) Recommendations and reports about central venous catheter-related infection. *Surg Infect (Larchmt)* 7: S65–S67.

193. Richards, M., Edwards, J., Culver, D., and Gaynes, R. (1999) Nosocomial infections in medical intensive care units in the United States. National Nosocomial Infections Surveillance System. *Crit Care Med* 27:887–892.

194. Matchett, A.A., Nickson, P.B., Wainwright, M., and Percival, S.L. (2001) Control of biofilms by photoactivation of exogenous phenothiaziniums. In: Gilbert, P.G., Allison, D., Walker, J.T., and Brading, M., editors. *Biofilm Community Interactions: Chance or Necessity?* Cardiff: Bioline, pp. 313–323.

195. Kite, P., Eastwood, K., Sugden, S., and Percival, S.L. (2004) Use of *in-vivo* generated biofilms from haemodialysis catheters to test the efficacy of a novel antimicrobial catheter lock for biofilm eradication *in-vitro*. *J Clin Microbiol* 42:3073–3076.

196. Kite, P., Eastwood, K., and Percival, S.L. (2005) Catheters and control of infection. Biofilm Club, September 6–8, Bioline, Greynog, Cardiff.

197. Percival, S.L., Kite, P., and Donlan, R. (2005) Assessing the effectiveness of tetrasodium ethylenediaminetetraacetic acid as a novel central venous catheter (CVC) lock solution against biofilms using a laboratory model system. *Infect Control Hosp Epidemiol* 26(6):515–519.

198. Eastwood, K., Kite, P., and Percival, S.L. (2005) The effectiveness of TEDTA on biofilm eradication. Biofilm Club, September 6–8, Bioline, Gregynog, Cardiff.

199. Percival, S.L., and Kite, P. (2007) Catheters and infection control. *J Vascular Access* 2:69–80.

200. Kite, P., Eastwood, K., Sugden, S., and Percival, S.L. (2004) Use of *in-vivo* generated biofilms from haemodialysis catheters to test the efficacy of a novel antimicrobial catheter lock for biofilm eradication *in-vitro*. *J Clin Microbiol* 42:3073–3076.

201. Chastre, J., and Fagon, J.-Y. (2002) Ventilator associated pneumonia. *Am J Respir Crit Care Med* 165:867–903.

202. Sottile, F.D., Marrie, T.J., Prough, D.S., et al. (1986) Nosocomial pulmonary infection: Possible etiologic significance of bacterial adhesion to endotracheal tubes. *Crit Care Med* 14:265–270.

203. Ramirez, P., Ferrer, M., and Torres, A. (2007) Prevention measures for ventilator-associated pneumonia: A new focus on the endotracheal tube. *Curr Opin Infect Dis* 20(2):190–197.

204. Sumi, Y., Miura, H., Michiwaki, Y., Nagaosa, S., and Nagaya, M. (2007) Colonization of dental plaque by respiratory pathogens in dependent elderly. *Arch Gerontol Geriatr* 44(2):119–124.

205. Kilty, S.J., and Desrosiers, M.Y. (2008) The role of bacterial biofilms and the pathophysiology of chronic rhinosinusitis. *Curr Allergy Asthma Rep* 8:227–233.

206. Lau, I., and Gilmore, M.S. (2008) Microbial biofilms in ophthalmology and infectious disease. *Arch Ophthalmol* 126:1572–1581.

207. Zegans, M.E., Becker, H.I., Budzik, J., and O'Toole, G. (2002) The role of bacterial biofilms in ocular infections. *DNA Cell Biol* 21(5–6):415–420.

208. Kerr, W.J.S., and Geddes, D.A.M. (1991) The areas of various surfaces in the human mouth from nine years to adulthood. *J Dent Res* 70:1528–1530.

209. Marsh, P.D., and Nyvad, B. (2008) The oral microflora and biofilms on teeth. In: Fejerskov, O., and Kidd, E.A.M., editors. *Dental Caries. The Disease and Its Clinical Management* (2nd ed.). Oxford: Blackwell, pp. 163–187.

210. Socransky, S.S., and Haffajee, A.D. (2005) Periodontal microbial ecology. *Periodontology* 38:137–187.

211. Percival, S.L., and Bowler, P. (2004) Biofilms and their potential role in wound healing. *Wounds* 16(7):234–240.

212. Percival, S.L. (2004) Understanding the effects of bacterial communities and biofilms on wound healing, *World Wide Wounds,* July.

213. Percival, S.L., and Rogers, A.A. (2005) The significance and role of biofilms in chronic wounds. In: McBain, A., Allison, D., Pratten, J., Spratt, D., Upton, M., and Verran, J., editors. *Biofilms, Persistence and Ubiquity* Cardiff: Bioline.

214. Thomas, J., and Percival, S.L. (2007) Synergy and enhanced virulence in biofilms. Biofilm Club, September 6–8, Bioline, Gregynog, Cardiff.

215. Walker, M., and Percival, S.L. (2008) Advances in wound healing: A topical approach. In: Hadgraft, R.M.J., Roberts, M.S., and Lane M.E., editors. *Modified Release Drug Delivery* (2nd ed., Chap. 52). Informa Healthcare, 325–338.

216. Clutterbridge, A., Cochrane, C.A., Woods, E., and Percival, S.L. (2007) Biofilms and their relevance to veterinary medicine. *Vet Microbiol* 121(1–2):1–17.

217. Serralta, V.W., Harrison-Balestra, C., and Cazzaniga, A.L. (2001) Lifestyles of bacteria in wounds: Presence of biofilms? *Wounds* 13:29–34.

218. Bello, Y.M., Falabella, A.F., Cazzaniga, A.L., Harrison-Balestra, C., and Mertz, P.M. (2001) Are biofilms present in human chronic wounds? Presented at the Symposium on Advanced Wound Care and Medical Research Forum on Wound Repair, Las Vegas, NV. April, 2001.

219. Wolcott, R.D., Rhoads, D.D., and Dowd, S.E. (2008) Biofilms and chronic wound inflammation. *J Wound Care* 17:333–341.

220. Davis, S.C., Martinez, L., and Kirsner, R. (2006) The diabetic foot: The importance of biofilms and wound bed preparation. *Current Diabetes Reports* 6:439–445.

221. Ngo, Q., Vickery, K., and Deva, A.K. (2007) Pr21 role of bacterial biofilms in chronic wounds. *ANZ J Surg* 77(1):A66.

222. Kennedy, P., Brammah, S., and Wills, E. (2009) Burns, biofilm and a new appraisal of burn wound sepsis. *Burns* June 10.

223. Cochrane, C., Woods, E., and Percival, S.L. (2009) DGGE analysis and biofilm formation of bacteria isolated from horse wounds. *Can J Microbiol* 55(2):197–202.
224. Wolcott, R., Cutting, K.F., Dowd, S., and Percival, S.L. (2008) Surgical-site infections— Biofilms, dehiscence, and delayed healing. *US Dermatol* 3:56–59.
225. Malic, S., Hill, K.E., Hayes, A., Thomas, D.W., Percival, S.L., and Williams, D.W. (2009) Detection and identification of specific bacteria in wound biofilms using peptide nucleic acid (PNA) fluorescent *in situ* hybridisation (FISH). *Microbiology*, 155:2603–2611.
226. Cutting, K.F., and Percival, S.L. (2009) Biofilm Management, *Nurs Stand* 23(32):64, 66, 68 passim.
227. Rhoades, D., Walcott, R., Cutting, K., and Percival, S.L. (2007) Biofilms and management in chronic infected wounds. Biofilm Club, September 6–8, Bioline, Gregynog, Cardiff.
228. Rhoades, D., Walcott, R., and Percival, S.L. (2008) Management of biofilms in chronic wounds. *J Wound Care* 17(11):502–508.
229. Prosser, B.L. et al. (1987) Method of evaluating effects of antibiotics on bacterial biofilm. *Antimicrob Agents Chemother* 31:1502–1506.
230. Nickel, J.C. et al. (1985) Tobramycin resistance of *Pseudomonas aeruginosa* cells growing as a biofilm on urinary tract catheter. *Antimicrob Agents Chemother* 27:619–624.
231. Ito, A., Taniuchi, A., May, T., Kawata, K., and Okabe, S. (2009) Increased antibiotic resistance of *Escherichia coli* in mature biofilms. *Appl Environ Microbiol* 75(12):4093–4100.
232. Schierholz, J.M., Beuth, J., Konig, D., Nurnberger, A., and Pulverer, G. (1999). Antimicrobial substances and effects on sessile bacteria. *Zentralbl Bakteriol* 289:165–177.
233. Xu, K.D., McFeters, G.A., and Stewart, P.S. (2000) Biofilm resistance to antimicrobial agents. *Microbiology* 146:547–549.
234. Allison, D.G., McBain, A.J., and Gilbert, P. (2000) Biofilms: Problems of control. In: Allison, D.G., Gilbert, P., Lappin-Scott, H.M., and Wilson, M., editors. *Community Structure and Co-operation in Biofilms*. Cambridge: Cambridge University Press, pp. 309–327.
235. Wilson, M. (2001) Bacterial biofilms and human disease. *Sci Prog* 84:235–254.
236. Stewart, P.S. (2002) Mechanisms of antibiotic resistance in bacterial biofilms. *Int J Med Microbiol* 292(2):107–113.
237. Cochran, W.L., McFeters, G.A., and Stewart, P.S. (2000) Reduced susceptibility of thin *Pseudomonas aeruginosa* biofilms to hydrogen peroxide and monochloramine. *J Appl Microbiol* 88:22–30.
238. Tuomanen, E. et al. (1986) The rate of killing of *Escherichia coli* by β-lactam antibiotics is strictly proportional to the rate of bacterial growth. *J Gen Microbiol* 132:1297–1304.
239. Tuomanen, E. et al. (1986) Antibiotic tolerance among clinical isolates of bacteria. *Antimicrob Agents Chemother* 30:521–527.
240. Evans, D.J. et al. (1991) Susceptibility of *Pseudomonas aeruginosa* and *Escherichia coli* biofilms towards ciprofloxacin: Effect of specific growth rate. *J Antimicrob Chemother* 27:177–184.
241. Duguid, I.G. et al. (1992) Growth-rate-independent killing by ciprofloxacin of biofilm-derived *Staphylococcus epidermidis*; evidence for cell-cycle dependency. *J Antimicrob Chemother* 30:791–802.
242. Desai, M. et al. (1998). Increasing resistance of planktonic and biofilm cultures of *Burkholderia cepacia* to ciprofloxacin and ceftazidime during exponential growth. *J Antimicrob Chemother* 42:153–160.
243. Brown, M.R., and Barker, J. (1999) Unexplored reservoirs of pathogenic bacteria: Protozoa and biofilms. *Trends Microbiol* 7:46–50.
244. Foley, I. et al. (1999) General stress response master regulator *RpoS* is expressed in human infection: A possible role in chronicity. *J Antimicrob Chemo* 43:164–165.
245. Adams, J.L., and McLean, R.J. (1999) Impact of RpoS deletion on *Escherichia coli* biofilms. *Appl Environ Microbiol* 65:4285–4287.

246. Huang, C.-T. et al. (1995) Nonuniform spatial patterns of respiratory activity within biofilms during disinfection. *Appl Environ Microbiol* 61:2252–2256.

247. Korber, D.R. et al. (1994) Evaluation of fleroxàcin activity against established *Pseudomonas fluorescens* biofilms. *Appl Environ Microbiol* 60:1663–1669.

248. Cochran, W.L. et al. (2000). Reduced susceptibility of thin *Pseudomonas aeruginosa* biofilms to hydrogen peroxide and monochloramine. *J Appl Microbiol* 88:22–30.

249. Gilbert, P. et al. (1997) Biofilms susceptibility to antimicrobials. *Adv Dent Res* 11:160–167.

250. Maira-Litran, T. et al. (2000) An evaluation of the potential of the multiple antibiotic resistance operon (*mar*) and the multidrug efflux pump *acrAB* to moderate resistance towards ciprofloxacin in *Escherichia coli* biofilms. *J Antimicrob Chemother* 45:789–795.

251. Kuchma, S.L., and O'Toole, G.A. (2000) Surface-induced and biofilm-induced changes in gene expression. *Curr Opin Biotechnol* 11:429–433.

252. Jaffe, A. et al. (1982) Role of porin proteins OmpF and OmpC in the permeation of small beta-lactams. *Antimicrob Agents Chemother* 22:942–948.

253. Savage, D.C. (1977) Microbial ecology of the gastrointestinal tract. *Annu Rev Microbiol* 31:107–133.

2 Human Skin and Microbial Flora

Rose A. Cooper and Steven L. Percival

CONTENTS

Introduction..59
The Anatomy and Characteristics of Human Skin..60
The Protective and Defensive Mechanisms of Human Skin...................................63
   The Distribution of Microbial Flora on the Skin 63
      Inhibitory Factors 64
      Other Factors 64
   Investigations into the Normal Flora of the Skin of Healthy Adults..................65
The Distribution of Indigenous Microbiota 65
   Molecular Approaches to the Investigation of Skin Microbiota 68
   Normal Residents of Skin in Health and Infection 69
      Corynebacterium 69
      Micrococcaceae..69
      Staphylococcus epidermidis 70
      Staphylococcus aureus 71
      Propionibacterium spp. 72
      Group A Streptococcus (*Streptococcus pyogenes*)..72
      Pseudomonas aeruginosa...72
      Acinetobacter species...72
   Fungal species...73
      Bacterial Interactions on the Skin 73
   Skin Flora and Infection...74
   Skin Microflora and Human Immunity...76
Conclusion 77
References..77

INTRODUCTION

Human skin has inherent properties that are important in preventing infection and promoting healing in wounds. The structure and function of skin is not uniform, and specific adaptations are found at different anatomical sites. Human skin is a multifunctional organ that provides sensation, thermoregulation, biochemical, metabolic, immune functions, and physical protection.[1] This protection is afforded by the mechanical rigidity of the stratum corneum, low moisture content, lipids, lysosyme acidity, and defensins.[2] Such factors help to create specific ecological niches on the skin surface aiding colonization by "specialized" microorganisms.[3]

The colonization of skin with microorganisms occurs almost immediately after birth via contact transfer with microbial reservoirs that include the birth canal, clothing, skin, and the local environment.[4] The population that establishes on the skin is subsequently varied due to the types of organisms transferred and their preferred anatomical location. Changes in host-dependent factors such as age, hormones, and health status continue to affect the population type and characteristics throughout the life of the host.[5]

The aim of this chapter is to provide an overview of the anatomy and physiology of human skin, to summarize the common microbial flora, and to consider the role these microbial species may play in infection. Additionally, the benefits of the skin microflora will be outlined. The normal microflora of the skin includes fungi, bacteria, and viruses, and this chapter will focus on bacteria.

THE ANATOMY AND CHARACTERISTICS OF HUMAN SKIN

Human skin is considered to be one of the largest organs of the body, with a total surface area ranging from 1.5 to 2.3 m^2, and a weight varying between 5 and 10 kg in an adult. Skin inherently has a protective role, by providing a tough but pliable barrier between the body's interior and the external environment. Skin actively regulates water loss and is permeable to both oxygen and carbon dioxide. Skin is also able to influence and regulate temperature. Skin is composed of a number of different tissue types (epidermal, nervous, muscular, and connective). Each of these tissue types contributes to wound prevention.

Anatomically, the skin is divided into two layers, namely the dermis and outermost region, the epidermis, together with the subcutaneous or adipose tissue layers (see Figure 2.1).

The epidermis is formed from many layers of closely packed cells and arises from a single layer of basal cells, which produce keratinocytes that differentiate as they approach the skin surface. As this process continues, these cells elongate and flatten to produce the stratum corneum (SC), the outermost layer. This layer is approximately 15 cell layers thick (15 μm). The SC acts as the interface between the internal body components and the external environment. The epidermis contains no blood vessels, and thickness varies. On the eyelids, the epidermis has been found to be less than 0.1 mm and 1 mm on the palms and soles. The five layers that make up the epidermis are shown in Table 2.1. Cells that make up the epidermis include keratinocytes that are tightly held together by spot desmosomes and interconnected by keratin filaments. These cells are constantly being produced by the stratum basale causing displacement of older cells by newer ones; during the maturation process, they undergo keratinization before death and keratin filaments progressively accumulate. Once dead, the keratinized cores of these cells (often referred to as *squames*) become the outermost layer of the epidermis or stratum corneum. The extensive amounts of keratin provide protection to the underlying cells from many potentially adverse events, such as microorganisms, heat, water, and chemicals. Lipids are deposited between these cells during the differentiation of keratinocytes and provide a continuous permeability layer. Keratinized cells are constantly being sloughed off the outermost

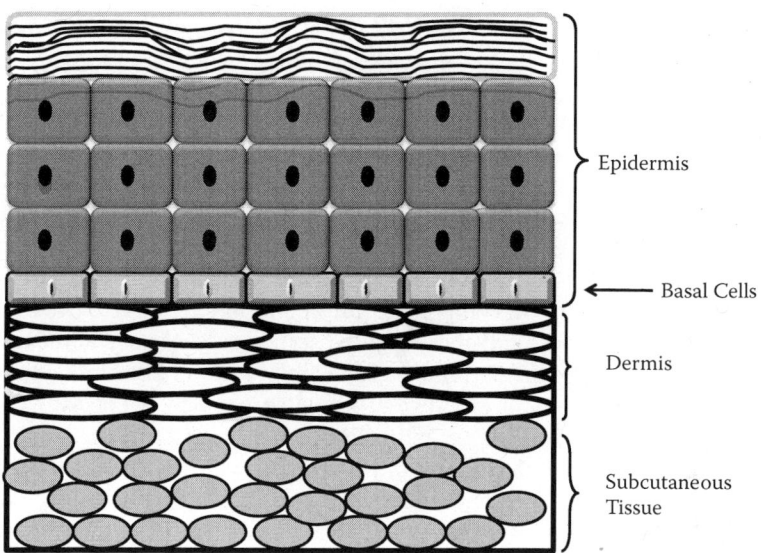

FIGURE 2.1   Overall structure of the skin.

TABLE 2.1
Components of the Epidermis of the Skin

Components of the Epidermis	Characteristics
Stratum basal	Composed of keratinocytes (single layer), Langerhans cells, melanocytes, Merkel cells
Stratum spinosum	Keratinocytes (multilayered), Langerhans cells, melanocytes; lipids are secreted
Stratum granulosum	Keratinocytes (multilayered and flattened); lipids are secreted
Stratum lucidum	Dead cells (on palms and soles of feet)
Stratum corneum	Dead cells (multilayered), lipids

layer of the epidermis, a process known as desquamation. It has been estimated that a typical adult accommodates approximately 2×10^9 squames. On average, it takes 4 weeks for cells generated in the basal region of the skin to enter the outermost layer of the skin. Consequently, the stratum corneum is replaced every 15 days.

Other cells found in the epidermis include the melanocytes. These cells are responsible for melanin (pigmented black or brown) production. Melanin is transferred to keratinocytes when cells are exposed to ultraviolet (UV) light. Melanin absorbs the UV light and therefore aids to protect the skin from UV light–induced damage such as premature aging, carcinogenesis, and so forth.

In addition to melanocytes, the epidermis is composed of Langerhans cells and Merkel cells. Langerhans cells are important during the immune response and therefore are significant to the first line of defense during microbial colonization. The

Merkel cells are involved in touch as they are associated with sensory neurons on the skin surface.

Below the epidermis is the dermis. The whole of the epidermis adheres to the dermis partly by the interlocking of its downward projections (epidermal ridges) with upward projections of the dermis (dermal papillae). The dermis is composed of a complex array of components that include connective tissue (collagen and elastin), papillae (contain nerve endings that are sensitive to heat and pain and project into the epidermis), hair follicles, erector pili muscles (to control the movement of hair), nerves, adipose tissue, capillaries and veins, sebaceous glands, and sudoriferous glands. Not all of these components are found in skin at all of the different anatomical sites of the body, and consequently, certain regions vary in the thickness of the dermis. Essentially, the dermis is a substantial layer of connective tissue that contains many structural fibers, blood vessels, and nerve fibers with elastin aiding to confer elasticity to connective tissue and collagen providing strength.

The presence of hair follicles that are associated with glands that open to the external environment provide a possible entry point for bacteria. Hair consists of dead keratinized cells with the shaft part of the hair protruding from the skin but firmly attached via a root at the base of the follicle. Around each hair follicle are touch-sensitive cells. Most hair follicles have one or two sebaceous glands associated with them. The sebaceous gland secretes sebum, an oily fluid, which aids to help prevent hair from drying out by preventing excessive moisture loss and helps to prevent bacterial colonization. Some sebaceous glands also open directly onto the skin surface and are not associated with hair follicles; these lubricate the skin, helping to waterproof it and prevent cracking. The concentration of sebaceous glands varies with anatomical site, for example, in some places such as the face, forehead, and scalp approximately 800 glands/cm² can be found. None are found on the palms of the hands or the soles and dorsa of feet.

As well as sebaceous glands, the dermis contains sudoriferous, or sweat-producing glands. Two types of sudoriferous glands exist—the apocrine and eccrine. The apocrine open directly into the follicular canal, whereas the eccrine enter directly onto the surface of skin by a sweat pore. Apocrine cells become significantly more active during puberty and are found in abundance in the axillae and perineum, where they produce viscous and odoriferous material. The secretions, which are initially odorless, are converted to odorous products by surface bacteria and are thought to be of pheromonal significance. On the contrary, eccrine glands are abundant throughout the rest of the body but especially on the axillae, palms of hands, and forehead. These glands produce sweat, a fluid containing water, salt, urea, and lactate. Sweat production helps regulate body temperature, remove waste, and distribute agents generated from the acquired and innate immune response.

At birth the skin is sterile, but microbial species are acquired during passage through the birth canal. These include staphylococci and corynebacteria, with lesser numbers of coliforms and sometimes streptococci.[6] With time, a number of structural and functional changes are induced in skin; some are readily recognized visually, and others are less obvious. One notable change occurs with puberty, when sebum production changes and increased levels of lipids on the skin result. In older adults, sebum production diminishes; wrinkling results from loss of elasticity of collagen and elastin fibers following cross-linking promoted by exposure to the ultraviolet irradiation in sunlight. Hair loss, graying, and increased areas of skin pigmentation and decreased skin depth are also linked to aging. Impaired rates of wound healing are associated with advanced age.

THE PROTECTIVE AND DEFENSIVE MECHANISMS OF HUMAN SKIN

Healthy human skin is the first line of defense against invading microorganisms.[7-10] As a mechanical barrier, the skin inherently has an armory of mechanisms at its disposal which prevents microbial penetration. Normal skin has its own lymphoid tissue that produces Langerhans cells or antigen-presenting cells. These cells, including dendritic cells, constitute a very small component of the epidermal cells. Lymphoid cells are important for initiating localized immune responses. Immunoglobulins A and G which are produced by the eccrine cells are located on the surface of the skin and assist in reducing and preventing microbial attachment. In addition to this, endothelial cells, keratinocytes, and lymphocytes produce cytokines that enhance the immune defenses. In conjunction with this, keratinocytes produce an array of peptides including human beta-defensins and adrenomedullin that further enhance the defenses of the skin. Lysozyme produced by keratinocytes is known to cleave glycosidic bonds in the peptidoglycan layer of the bacterial cell wall. The characteristics of the skin which mitigate against microbial colonization are shown in Table 2.2.

THE DISTRIBUTION OF MICROBIAL FLORA ON THE SKIN

Temperature, and humidity in particular, influence both the microbial density and coverage over the surface of the skin.[11-14] Extremes in both temperature and humidity are known to enhance both the survival and proliferation of many bacteria including *Staphylococcus* spp. Although the internal temperature of the human body is 37°C, the surface of the skin has been reported to be significantly lower. The regions of the body documented to have the highest localized temperatures have included the areas around the groin and axillae. In general, both these regions

TABLE 2.2
Defensive Mechanisms of Skin

Characteristics	Benefits
Moisture content	Usually low, which reduces colonization and enzymatic activity of bacteria
pH (acidic)	An acidic pH will restrict the colonization of certain bacteria
Squamous cell shedding	Cell shedding will enhance the sloughing of colonizing bacteria
Ionic conditions (high salt content)	A high salt level will restrict colonization of certain bacteria
Peptides (antimicrobial)	These will restrict colonization and kill contaminating bacteria
Stratum corneum	An integral stratum corneum will prevent bacterial penetration
Fatty acids and lipids	Reduces bacterial adhesion of contaminating bacteria
Immunoglobulins	Aid in the killing of contaminating bacteria

have been found to support the highest numbers of microorganisms when compared to cooler regions of the body. It is here that higher levels of sweat production are evident. High levels of moisture help to enhance both the numbers and diversity of microorganisms.

The pH of the skin at different regions around the body can vary. The reasons for this relate to many factors, including types of microorganisms, available nutrients, and age. In particular, acidic skin can be due to an accumulation of substances such as lactic acid and amino acids, fatty acids, and degradation of proteins.

Nutrient availability and distribution on the skin surface can significantly affect the colonization and proliferation of microorganisms. For growth, the majority of skin bacteria require carbon and nitrogen sources, minerals, and water. Sweat contains many of these components, including urea, proteins, and amino acids. Skin has also been found to contain a number of trace elements including selenium, molybdenum, and chromium. These elements can be utilized by bacteria as cofactors for a number of important enzymes. The sebaceous glands are found throughout the body surface. They are particularly abundant on the forehead, scalp, and face. The chemical composition of sebum varies, and this is dependent on the different regions of the body where it is found. It is composed of fatty acids, esters, squalene, glycerides, and vitamin E. Lipids found on the skin are broken down by the indigenous microbiota, causing the release of fatty acids and glycerol.

Inhibitory Factors

When sweat and water evaporate from the skin surface, a high concentration of salt is left behind. This creates a high osmolarity reducing and restricting microbial growth. Sweat also has been documented to contain dermcidin (an antimicrobial peptide) that is active against an array of bacteria such as *Staphylococcus aureus* and *Escherichia coli*. This peptide is able to function at low pH and in high salt concentrations.

Skin also has a constant airflow over its surface. This reduces the ability of airborne pathogens to adhere. This complements the constant shedding of dead keratinized cells from the skin surface previously mentioned which severely compromises microbial adherence.

Many of the fatty acids found on the skin surface have been shown to inhibit the growth of many bacteria. For example, sphingosine, which is commonly found on healthy skin, has been shown to be effective at killing *Staphylococcus aureus*. Lauric acid and other saturated fatty acids have been found to suppress the growth of *Propionibacterium acnes*, *Streptococococcus* sp, and *Candida* spp., to name a few. Some other interesting agents associated with the skin have included reactive nitrogen intermediates (RNIs), which are often released from the epidermis. These RNIs are formed when bacteria break down the nitrates, which are found in sweat, into agents such as dinitrogen tetroxide, perooxynitrite, and nitrous oxide. Many of these agents are inhibitory to different types of bacteria. During exposure of the skin to sunlight, vitamin D_3 is synthesized, and this has been shown to have some antimicrobial effects.

Other Factors

Other factors that affect the diversity of skin microbiota have included body location, hospitalization, illness, medications, sex, race, occupation, and the use of topical preparations such as soaps, cosmetics, and disinfectants. In general, the low moisture content, the acidic environment (pH of 5.5), the presence of antimicrobial peptides, high salt content, lipids and fatty acids, immunoglobulins, and lysozyme, together with the continual shedding of squames from the surface of the body, create an environment in skin that is not conducive to extensive microbial proliferation.

Investigations into the Normal Flora of the Skin of Healthy Adults

Price[15] reported that microorganisms found on the skin can be divided into resident flora, which are irreversibly attached to the skin or transient flora that do not grow on skin and usually remain dormant, die, or detach after a short period of time. The resident flora of the skin are referred to as the indigenous microbiota considered to exist as a biofilm and has largely been viewed as harmless being composed of commensals that rarely damage the host. The transient flora reflects the host's level of personal hygiene, lifestyle, and personal activities and level of environmental contamination. Transient organisms are generally not attached to skin and do not persist and are considered to be more associated with exposed areas of the skin. A third, more occasional category of skin flora has been described as temporary or nomadic flora. A normadic flora represents microorganisms that attach to skin and are able to multiply but only persist for relatively short periods.[16]

Much of the research into the indigenous microbiota of human skin was undertaken during the past 50 years. The data and information generated were based on the use of conventional, culturing techniques to both recover and identify species of bacteria[17–20] that probably led to a gross underestimation of the diversity of the skin's indigenous microbiota.[21–23]

Estimates of microbial population sizes on skin depend on the sampling methods employed,[24] and reports from numerous skin studies have not yielded consistent estimates. Sampling methods have included swabs, contact plates, stripping of surface layers with tape, biopsies, rinsing and scrubbing techniques, as well as air sampling to recover shedding squames. Counts ranging from 4×10^2 to 1.9×10^4 per cm^2 on the forearm and from 6.3×10^3 to 1.67×10^7 per 3.8 cm^2 on axilla skin have been reported.[17]

THE DISTRIBUTION OF INDIGENOUS MICROBIOTA

Skin flora can be classed as commensal or parasitic. It supports the protective function of the skin via factors such as bacterocin production, toxic metabolites, depletion of essential nutrients, prevention of adherence of competing bacteria, and degradation of bacteria. Microorganisms predominately found on the skin surface are Gram-positive bacteria; they include *Staphylococcus, Micrococcus, Acinteobacter, Corynebacterium, Propionibacterium, Malassezia, Dermabacter,* and *Brevibacterium*[25,26] (see Table 2.3). Although other bacteria have been isolated from skin, they are often found in small numbers and considered harmless.[27] The resident flora of normal skin has been found to form microcolonies on and within the stratum corneum,[28,29] whereas lipophilic organisms associate with the sebaceous glands.[30]

To reiterate, the microbial communities found on adult skin differ in composition according to anatomical site and the individual being sampled and any external or

TABLE 2.3
Bacterial Skin Residents and Their Associated Dermatoses

Bacterium	Location	Distinguishing Features	Pathology
Gram Positive			
Staphylococcus			
S. epidermidis	Upper trunk	Production of slime	Prosthetic joints and catheter infection
S. hominis	Glabrous skin		
S. haemolyticus			Endocarditis, septicemia, and joint infections
S. capitis	Head		
S. midis			
S. warneri			Associated with endocarditis and joint infections
S. saprophyticus	Perineum		Urinary tract infections
S. cohnii			
S. xylosus			
S. simulans			
S. saccharolyticus	Forehead/antecubital	Anaerobic	
Micrococcus			Rarely associated with skin/wound infections
M. luteus			
M. varians			
M. lylae	In children/cold temperature		
M. kristinae	In children		
M. nishinomiyaensis			
M. roseus			
M. sedentarius			Pitted keratolysis
M. agieis			
Corynebacterium			
C. minutissimum	Intertriginous	Lipophilic/porphyrin	Erythrasma
C. tenuis	Intertriginous	Lipophilic	Trichomycosis
C. xerosis	Conjunctiva	Lipophilic	Conjunctivitis
C. jeikeium	Intertrginous	Lipophilic/antibiotic resistance	
Rhodococcus	Lipophilic		Granuloma in HIV
Propionibacterium			
P. acnes	Sebaceous gland	Lipophilic/anaerobic	Acne
P. granulosum	Sebaceous gland	Lipophilic/anaerobic	Severe acne
P. avidum	Axilla	Lipophilic/anaerobic	

TABLE 2.3 (continued)
Bacterial Skin Residents and Their Associated Dermatoses

Bacterium	Location	Distinguishing Features	Pathology
		Gram Positive	
Brevibacterium	Toe webs	Nonlipophilic	Foot odor, white piedra
Dermabacter	Nonlipophilic		Pitted keratolysis
		Gram Negative	
Acinetobacter	Dry areas		Burn wounds

environmental perturbations.[31] From an anatomical point of view, the skin is divided into three distinctive regions (exposed, moist, and oily). Exposed areas of the body are the face, neck, and hands; these are more likely to host higher numbers of transient organisms than other areas of the skin. Regions with variations in microbial density and diversity may be found on the hands. For example, the area beneath the fingernail (an occluded region) is densely colonized by anaerobic bacteria, fungi, and Gram-negative bacteria.

The Gram-negative bacteria often found colonizing healthy adult skin include *Acinetobacter* spp. and *Pseudomonas* spp. which constitute about 25% of the adult skin microflora.[32] Investigation by Leyden et al.[33] has shown that the toe-webs and axillae are specifically and predominantly colonized by coryneforms and bacteria belonging to the micrococcaceae (Gram-positive) group. In the perineum region, for example, there are large proportions of micrococcaceae together with a large number of Gram-negative rods, possibly of fecal origin.[34]

The oily areas include the forehead and the dry areas, such as the forearm. The forehead has a high acidic pH, a variable temperature range, and a high density of sebaceous and eccrine glands. The main bacteria dominating the forehead, as with the scalp, include propionibacteria specifically *P. acnes*, staphylococci, specifically *S. capitis, S. hominis, S. epidermidis, Micrococci, Malassezia,* and low levels of coryneforms.[34–37] A large number of sebaceous glands are located on the scalp, together with hair coverage enhancing the moisture content. Propionibacteria are common inhabitants of hair follicles and sebaceous glands and are prevalent anaerobes of the normal flora of skin. The most predominant species found on the skin of the back, forehead, and scalp is *P. acnes*. *P. acnes* is particularly relevant at puberty, when densities start to peak. Other bacteria associated with follicles include staphylococci, specifically *S. capitis*.

Yeasts are recovered in higher numbers in an older adult population. This is possibly due to a decrease in sweat production that occurs in the older adult population.[38] Within the dry skin areas of the adult, staphylococci represent over 90% of the total population.[39] Skin flora tends to be more varied in children, and they support higher population densities of micrococci, coryneforms, and Gram-negative bacteria than older people.[40]

There have been numerous reports of the significant difference between hospital-ized patients and healthy people, in that the skin microflora of hospitalized patients has been shown to shift from predominantly Gram-positive organisms to a more Gram-negative microflora.[41–44]

One study showed that hospitalized patients acquired pathogenic and antibiotic-resistant strains as antibiotic-sensitive strains were lost.[45]

MOLECULAR APPROACHES TO THE INVESTIGATION OF SKIN MICROBIOTA

Many studies have identified an array of different microorganism on human skin, and with the development of genotyping and molecular technologies, the species diversity found on skin is continuing to increase.[46–48] These are based on the characterization of a conserved region of the 16S rRNA-encoding gene. It is becoming increasingly clear that conventional techniques may have revealed only 1% of the members of the normal skin microbiota.

A notable study that investigated the microbiology of the foreheads of five volun-teers also highlighted a high prevalence of *Staphylococcus* and *Propionibacterium* spp. and *Methylophilus* spp.[47] In addition to 10 previously characterized bacteria, 13 novel "phylotypes" (not-yet cultured microorganisms) were indicated and 9 spe-cies not previously associated with skin were recognized.[47]

A study carried out by Barton and colleagues[49] using denaturing gradient gel electrophoresis-polymerase chain reaction (DGGE-PCR) has shown that the ecol-ogy of the inner forearm skin of six volunteers has a microbiology composed of staphylococci, *Moraxella osloensis,* and micrococci. Other microorganisms cultured included *Streptomyces, Dermacoccus,* and *Lactobacillus helveticus.*

A comparison between the microbiota of facial skin of 13 patients with atopic dermatitis (AD) and that of 10 healthy individuals using several 16S rRNA analy-sis systems revealed 18 different organisms.[50] *Stenotrophomonas maltophilia* was detected significantly more frequently in AD patients compared to controls (5/13 and 0/10, respectively), and *Dietzia maris* was more common in controls. Unexpectedly, because *Streptococcus* species are not normally found in uninfected skin, these bac-teria were observed in 7 of the AD patients and 8 controls.[50]

Another study using molecular analysis of the flora in superficial forearm skin samples of six healthy subjects highlighted the fact that the skin microbiota was composed of hundreds of species of bacteria that included *Propionibacterium, Corynebacterium* spp., *Staphylococcus,* and *Streptococcus.*[51] From 1221 clones that were analyzed from the 6 humans sampled, 182 species belonging to 8 phyla were identified. Most of the clones fell into 3 phyla, and 86.5% of them represented known species; 30 were unknown, though. The study confirmed the presence of strepto-cocci as a constituent of normal skin flora and demonstrated substantial variation between 6 subjects.

The cutaneous *Malassezia* microbiota of 770 healthy Japanese using real-time PCR with a TaqMan probe was used to investigate the effects of age and gender on the *Malassezia* population. *Malassezia* levels increased in males up to 16 to 18 years of age and in females to 10 to 12 years old but decreased gradually in both genders until senescence. The total colonization of *Malassezia* was found to change with age and gender.[52]

A recent study employing molecular techniques to characterize microbiota of human skin of the inner elbow area compared the diversity of flora recovered from three different depths using three sampling techniques.[48] Surface flora was recovered by skin swabs, superficial flora from slightly lower levels were collected by skin scrapes, and subcutaneous flora was sampled by punch biopsy. The left and right arms of each of 5 subjects were tested and 113 phylotypes were detected. Analysis indicated the presence of similar species to a previous study[51] but surprisingly *Pseudomonas* accounted for 60% of the 16S rRNA genes and *Janthinobacterium* for 20%. Both of these Gram-negative bacteria are common inhabitants of soil and water, rather than skin. Whereas culturing studies have indicated *S. epidermidis* to be the predominant aerobic inhabitant of human skin which a high percentage are able to form biofilms,[53] the molecular approach used here demonstrated that *S. epidermidis* and *P. acnes* accounted for less than 5% of the flora.[48] Estimations of bacterial density suggested that swabs, skin scrapes, and biopsies, respectively, collected approximately 10,000, 50,000, and 1,000,000 bacteria per cm^2. Uniformity between samples collected from the same patient was found and one further observation was the similarity between the skin microbiota of humans and mice.[48]

NORMAL RESIDENTS OF SKIN IN HEALTH AND INFECTION

Corynebacterium

These bacteria are often referred to as the diphtheroids or coryneforms. Essentially, they are Gram-positive rods, often pleomorphic, nonsporing, nonbranching, and non-acid fast. Their tendency to form palisades of cells has led to their microscopic appearance being described as "Chinese letters." They are catalase-positive, non-motile, aerobes or facultative anaerobes. They have a G+C content of 46 to 74 mol%, evidence of mycolic acids (C22 to C36), and meso-diaminopimelic acid, a cell wall containing arabinogalactan. Cell wall analysis has been used to discriminate between classic *Corynebacterium* spp. and *Brevibacterium* spp.

The main carbon source for *Corynebacterium* spp. includes carbohydrates and amino acids, but *Corynebacterium* species have been found to adhere easily to lipids, particularly those found in sebum, and they have often been described as lipophilic. *Corynebacterium striatum*, *Corynebacterium amycolatum*, and *Dermabacter hominis* are thought to be a dominant part of the normal skin flora, but their ability to act as opportunist pathogens in immunocompromised patients has also been recognized.[54]

Micrococcaceae

The bacteria most frequently isolated from human skin are the Gram-positive, catalase-positive cocci of the Micrococcaceae family. These bacteria are divided into their respective genera on the basis of their fermentative capacity: members of the genus Micrococcus are strictly aerobic, whereas the members of Staphylococcus are facultative anaerobes. Staphylococci are subdivided into two groups on the basis of their ability to coagulate fibrinogen to fibrin; S. *aureus* is coagulase positive, and the remaining species are all coagulase negative.

The most abundant inhabitants of skin are coagulase-negative staphylococci (CNS), constituting 18 different species, and are located in high concentrations, particularly near hair follicles.[55,56] Fifty percent of the staphylococci found colonizing the skin are *Staphylococcus epidermidis*. S. *aureus* has been isolated frequently from adult skin, the anterior nares of humans,[57,58] and the perineum.[59] *Micrococcus luteus* are also found on the skin of the head, legs, and arms.[60] Species of *Micrococcus* are less frequently isolated from skin than staphylococci.

Staphylococcus epidermidis

Staphylococcus epidermidis is a common member of the indigenous microflora of the skin. These bacteria have been reported to inhabit the skin and mucosal membranes constituting over 80% of the indigenous microbiota. S. *epidermidis* is coagulase negative, white in color when grown on blood agar plates, and sensitive to desferrioxamine. Even though long considered as contaminants during the routine microbiological analysis of clinical specimens, it is now accepted that the coagulase-negative staphylococci are responsible for many nosocomial infections in compromised patients.[61] They have been responsible for infections in patients with indwelling catheters, where the formation of biofilms[62] aids in their survival and enhances resistance to antimicrobial agents. If S. *epidermidis* enters the body, they have been reported to cause sepsis and have the ability to cause endocarditis. They have been reported to be part of abscesses and cause cellulitis. However, it seems to be generally accepted that S. *epidermidis* on the skin surface is relatively benign, but their role in causing infection is related to the host predisposition to infection.

S. *epidermidis* is able to adhere to many devices using an autolysin protein AtlE and an array of adhesion factors that belong to the microbial surface component recognizing adhesive matrix molecules (MSCRAMM) group of proteins. Other proteins involved in attachment have included accumulation-associated proteins (AAP) Aas1, Aas2, and SdrF. They are capable of producing antibiotics (lanthionine-containing peptides) as well as bacteriocins. The most significant ones they produce include epidermin, epilancin K7, epilancin 15X, staphylococcin 1580, and Pep5 together with peptides that have been shown to kill other bacteria such as S. *aureus* and group A streptococci (S. *pyogenes*). S. *epidermidis* has been shown to rarely affect keratinocytes in the epidermis. Consequently, S. *epidermidis* is considered to provide protection to the host against certain bacteria. They are also thought to influence the innate immune response of keratinocytes using Toll-like receptors. These Toll receptors recognize molecules produced from pathogens, and it is thought that presence of S. *epidermidis* enhances the keratinocyte response to pathogens. Consequently, the presence of S. *epidermidis* on the skin surface is considered beneficial to the host, as it contain peptides that inhibit the colonization of harmful pathogens and could be faithfully classed as a mutual organism to the host when present on the skin surface.

Staphylococcus aureus

S. aureus is often considered a transient colonizer on human skin, but many consider this bacterium an indigenous resident of the nose. Approximately 87 million people are considered to be colonized with this bacteria, but this estimate varies widely. Evidence of this bacterium on the human body is not synonymous with infection. *S. aureus* are Gram-positive, coagulase-positive spherically shaped bacteria that are yellow or golden colonies and cause beta hemolysis on sheep blood agar. *S. aureus* is a major pathogen and has been associated with many forms of skin infections that range from minor to life-threatening infections. The main infections of the skin they cause include impetigo, folliculitis, subcutaneous abscesses, and furuncles. They have also been reported to cause meningitis, septicemia, pneumonia, arthritis, and osteomyelitis. Like *S. epidermidis, S. aureus* is a frequent cause of catheter-related infections.

S. aureus is known to produce many different virulence factors that are significant to disease and infections. They are able to bind onto neutrophils, therefore, affecting their efficacy the protein is called a chemotaxis inhibitory protein of staphylococci (CHIPS). CHIPS bind to both C5a and peptide receptors on neutrophils. They produce an array of toxins that are either superantigens such as toxic shock syndrome-1, A-E enterotoxins, ETA, B and D) as well as hemolysins and exotoxins, leucocidins E–D and, in particular, the Panton-Valentine leucocidin (PVL). *S. aureus* also produces enzymes that often result in tissue damage. These include lipases, proteases, hyaluronidase, and collagenase. They are adept at evading killing by phagocytosis due to possession of a polysaccharide capsule, and production of clumping factors and its yellow pigment has been shown to aid against neutrophil killing.

Of particular concern is evidence of antibiotic-resistant *S. aureus* such as methicillin-resistant *S. aureus* (MRSA). This rise has been predominately within hospitals and within the community (caMRSA). In addition to this, *S. aureus* are also gaining resistance to the potent, last-defense antibiotic such as vancomycin; as such, strains of *S. aureus* have been termed vancomycin-intermediate (VISA) and vancomycin-resistant *S. aureus* (VRSA).

MRSA occurs when *S. aureus* acquires a transferable DNA element called staphylococcal cassette chromosome *mec or* SCCmec, a cassette that carries the mecA gene Type I to V, encoding a penicillin-binding protein (PBP) 2a.[63] The DNA becomes inserted into the genome of a susceptible host bacterium. In antibiotic-susceptible strains, beta lactam antibiotics normally bind to PBP sites within the cell wall of *Staphylococcus aureus,* which interrupts peptidoglycan production and eventually leads to the death and lysis of the bacterium. These antibiotics, however, are unable to bind to PBP2a; therefore, *Staphylococcus* strains that have the mecA gene are not killed.[63] A number of different plasmids have also been found for an array of other widely used antibiotics against *S. aureus.*[63]

S. aureus are known to be found on healthy skin acting principally as a commensal and not as a pathogen. Some strains of *S. aureus* are known to produce a number of bacteriocins that include staphylococcin 462. These have been found to inhibit other *S. aureus*. Few studies have been done on its role as a "good" bacteria when compared to its role as a pathogen.

Propionibacterium spp.

These are Gram-positive rod-shaped bacteria, nonmotile, fermentative, and nonsporing. Often referred to as obligate anaerobes or more recently microaerophilic, they have a G+C content of 57 to 86 mol%. Four different species of the genus have been isolated from skin. These include *P. acnes, P. avidum,* and *P. granulosum* and *P. propionicum.* The most prevalent species found on skin is *P. acnes* and is found at high concentrations of 10^5 cfu/follicle. *P. acnes* is able to adhere to oleic acid, a component of sebum, as this fatty acid is found to aid coaggregation and this would help to keep the bacteria at the skin surface. *P. acnes* has also been shown to be able to bind to fibronectin. *P. acnes* is able to grow at pH ranges of 4.5 to 8 but is documented to grow optimally at pH 5.5 to 6. They are able to utilize fatty acids, glycerol, and many sugars as carbon and energy sources. As propionibacteria produce propionic acid and bacteriolytic enzymes and bacteriocins, they are able to inhibit the growth of other bacteria which is a selective advantage for these bacteria.

Group A Streptococcus (*Streptococcus pyogenes*)

Streptococci are Gram-positive, catalase-negative cocci that exist in short chains. They are transient skin microbiota. Their appearance on blood agar plates allows the pyogenic (or beta hemolytic streptococci) to be distinguished from the viridans group, and further characterization depends on analysis of cell wall antigens by Lancefield grouping. The pyogenic species include *Streptococcus pyogenes* (Lancefield group A), *Streptococcus agalactiae* (group B), *Streptococcus dysgalactiae* (group C) *Streptococcus equi, Streptococcus canis,* and *Streptococcus dysgalactiae* subsp *dysgalactiae* (group G). Many of these species have been implicated in wound infections, and their presence precludes successful grafting. The members of the viridans group are generally regarded as skin commensals and include small colony formers with variable hemolysis such as *Streptococcus anginosus, Streptococcus constellatus,* and *Streptococcus intermedius.* Occasionally these bacteria are isolated from abscesses.

Pseudomonas aeruginosa

Pseudomonas aeruginosa is a widely distributed Gram-negative, aerobic bacterium that is readily isolated from moist environments such as soil, vegetables, plants, river water, and drains. It can be recovered from mains water, too, and has been found in antiseptic solutions diluted with tap water. This hydrophilic bacterium is rarely found to colonize healthy individuals and does not normally inhabit skin. It prefers relatively low growth temperatures (between 4 and 36°C) but can tolerate body temperature and is increasingly being linked to large ulcers with delayed healing. It is an opportunist pathogen that is notoriously unresponsive to many antibiotics. Its ability to form biofilm has been suggested to cause failure to heal in chronic wounds.[64]

Acinetobacter species

Acinetobacter is a genus of Gram-negative bacteria generally considered to contain nonpathogenic members. Nineteen biotypes within the genus are known. *Acinetobacter* species have a ubiquitous distribution and have been isolated from soil, water, food, animals, and environmental surfaces. However, *Acinetobacter baumannii* has been recovered from hospital surfaces and has increasingly been implicated as a cause of healthcare-associated infections (HAIs) in debilitated patients resulting in septicemia, such as respiratory tract, urinary tract, and wound infections. Recently it was reported to give rise to severe wound infections in American military personnel injured in Iraq and Afghanistan, and some strains have been implicated in osteomyelitis. Its broad patterns of resistance to antibiotics make it a difficult organism to control.

Acinetobacter was first detected on human skin in 1963[65] and has since been accepted to be part of the indigenous microbiota of skin and mucous membranes. An epidemiological study to investigate the body sites colonized by *Acinetobacter* species was conducted with 40 in-patients on a cardiac ward and 40 healthy, non-hospitalized individuals. Nine samples from discrete body sites were collected from each person (forehead, ear, nose, throat, axilla, hand, groin, perineum, and toe web) and isolates were typed by phenotypic and genotypic methods. Thirty patients and 17 controls were colonized with *Acinetobacter* spp., and the sites most frequently colonized were hands (26%), groin (25%), toe webs (24%), forehead (23%), and ears (21%). *Acinetobacter lwoffii* (47%) and *A. johnsonii* (21%) were most frequently isolated. *A. baumannii* was rarely isolated from skin, and the natural reservoir was undefined.[66]

FUNGAL SPECIES

Both filamentous fungi and yeast may be isolated from skin, and it is not easy to determine whether each is part of either resident or transient flora. However, the lipophilic yeasts of the genus *Malassezia* (previously called *Pitryosporum*) are accepted to contribute to resident skin flora; their presence has been demonstrated by both conventional techniques and molecular analysis.[67] The distribution of these organisms has been shown to reflect the areas with highest sebum secretion, such as the back and chest.[68] They have also been linked to dandruff, seborrheic dermatitis, pityriasis versicolor, folliculitis, atopic dermatitis, and psoriasis.[69] Their role in psoriasis is not well established, but when affecting the scalp, antifungal agents have been shown to help resolve the condition

Nonlipophilic yeasts that have been isolated from the healthy skin belong to the following genera: *Candida, Rhodotorula, Debaryomyces, Torulopsis,* and *Cryptococcus. Candida albicans* is more likely to colonize mucous membranes than skin, but the risk of colonization is increased in immunocompromised and diabetic patients.

Dermatophytes are keratinophilic filamentous fungi capable of invading skin, hair, and nails. Three genera are associated with humans (anthropophilic): *Epidermophyton, Microsporum,* and *Trichophyton.* When isolated from clinical specimens, these organisms are usually implicated in superficial mycoses.

Bacterial Interactions on the Skin

Humans are constantly exposed to microorganisms and the populations that live on and in humans exceed the number of host cells by a factor of at least ten. The interactions between the human host and its natural flora are complex and variable. Most of the indigenous microorganisms are harmless, but not always. The microorganisms that comprise the indigenous microflora may be described as commensals. Yet in commensalism, one partner normally derives benefit, while the other remains unharmed. This seems a rather unsuitable definition here because advantages for both human hosts and their indigenous microbiota can be identified. Furthermore, the members of the normal flora of skin can occasionally become either opportunist or overt pathogens, and their interaction with their host is better described as parasitism because one member derives benefit at the expense of the other. So the relationship between the microflora of the skin and the host could be described as symbiotic, but the meaning of this term has changed as our perception of the nature of ecological interactions has changed. Traditionally, symbiosis was defined as an interaction where both dependent members of the partnership derived benefit from the association. Now the term has been replaced by mutualism or protocol operation. Although humans and their skin inhabitants do not display mutual dependency, subtle benefits for each arise. Whether the balance between the host and its indigenous skin flora is maintained and microbial colonization results, or whether it is displaced and infection develops depends upon the multiple determinants of host immunocompetency and microbial virulence.

As well as the resident microflora, skin is composed of a dead layer of keratinized cells, referred to as the stratum corneum, which aids in preventing bacterial attachment. As a food source, keratin can only be utilized by a small number of bacteria and as such does not constitute a good food source for colonizing bacteria. Found between these cells are fatty acids, waxes, sterols, and phospholipids, among others, which in combination with dead cells make the skin surface very dry and virtually uninhabitable by many bacteria. Combined also with a low pH, bacterial growth is inhibited. However, certain regions of the body have relatively high moisture content and a neutral pH aiding in bacterial adhesion. Other problems for bacteria reside on skin—namely, the ever shedding squames that are disseminated together with any adhering bacteria. There is also skin-associated lymphoid tissue involved with humoral and cell-mediated responses of the immune system and sweat production that is known to contain lysozyme, which is known to cleave beta 1-4 glycosidic bonds found in many Gram-positive (N-acetylglucosamine and N-acetylmuramic acid) and Gram-negative bacteria (peptidoglycan). Consequently, this armory of defensive mechanisms evident in the human body helps to substantially reduce microbial proliferation of the intact skin.[70,71]

Skin Flora and Infection

Culture techniques are simple and economical; however, there are intrinsic limitations related to this approach when applied to the study of microbial biodiversity. The use of selective media and environments and the difficulties associated with providing the specific nutrients and environmental requirements for all potential colonizing organisms can result in uncultivated microbial populations. Recent advances in PCR-based methodologies and their subsequent applications have allowed for more holistic microbial characterization of skin microbiota.

Dekio et al.[47] analyzed the forehead skin microbiota of five healthy volunteers by profiling 16s rRNA genes. This approach identified 13 potentially novel phylotypes and an overall increase in the number of organisms identified (based upon the nearest match in the DDBJ, EMBL, and GenBank databases) when compared to culture. These organisms included *Acinetobacter* spp., *Pseudomonas* spp., and *Stenotrophomonas maltophilia*. However, the study did not monitor the microbial population over a period of time; therefore, some of the organisms identified may represent transient rather than resident flora.

Fungal populations are notoriously difficult to culture because of slow growth and fastidiousness. Paulino et al.[72] utilized a broad range 18s rRNA PCR to analyze 25 skin samples from the flexor forearm of five healthy skin patients and three patients with psoriatic lesions. *Malassezia restricta, M. globosa,* and *M. sympodialis* were present in all eight subjects, whereas previous culture-based studies have isolated these in 38% to 55% of healthy persons and those with psoriasis.[72]

In general, the skin commensal flora is complex, diverse, and plays a symbiotic role in the protection of the host. When the skin becomes compromised via injury, surgery, or other underlying etiologies, the microbiota can rapidly colonize the wound. This flora may act as a reservoir for potentially pathogenic organisms or may play a role in the complex healing process. To understand these roles, we must first understand the wound healing process.

The balance of the skin barrier and innate immunity help to maintain a healthy skin. Disturbances in this balance may predispose the host to a number of infectious.[73–76] Skin infections have been shown to be more significant as humans age.[77] The "normal" skin flora is considered a significant source of serious infections. For example, micrococci, specifically the species *M. luteus*, has been associated with cases of pneumonia, septic arthritis, and meningitis. *Staphylococcus epidermidis* is a major inhabitant of the skin and generally comprises greater than 90% of the aerobic resident flora. They are often classified as contaminants of the skin when isolated during infections and are therefore thought of as mutual bacteria aiding the human's innate immune system. Antimicrobial peptides on the surface of the skin have recently been identified as originating from *S. epidermidis*.[78] However, in certain situations, *S. epidermidis* can be the cause of a number of life-threatening infections (i.e., biofilm formation on artificial heart valves, intravascular catheters). This is principally due to these bacteria being avid biofilm formers resulting in enhanced virulence and resistance to antimicrobial chemotherapy. A transient bacteria associated with skin infections is *S. aureus*. *S. aureus* is considered a normal component of the nasal microflora.[79–81]

86.9 million people (32.4% of the population) are considered to be colonized with *S. aureus*.[82] Twenty percent of the population are considered to be persistently colonized, 60% of the population intermittently carry the bacteria, and 20% are never colonized. *Staphylococcus aureus* found on healthy human skin generally acts as a commensal and rarely as a pathogen, but it is known to cause minor and self-limited skin infections. Skin infections due to *S. aureus* include impetigo, folliculitis, furuncles and subcutaneous abscesses, and scalded skin syndrome.[83,84]

Many diphtheroids are found on human skin. *Corynebacterium jeikeium* is the most frequently recovered and medically relevant member of the group, particularly in hospitalized patients.[85] In the last few years, *Corynebacterium* species have gained interest due to the increasing number of nosocomial infections with which they are being associated.[86,87]

Propionibacteria are prevalent on skin-colonizing bacteria. The most well-known ailment associated with *P. acnes* is acne vulgaris which affects up to 80% of adolescents in the United States.[88] In fact, *P. acnes* is able to initiate and also contribute to inflammation during acne episodes. Reports of *P. acnes* being associated with foreign device infections has also be highlighted. Gram-negative rods such as *Acinetobacter* are known to be a cause of skin infections, particularly in patients with wound infections and burns. *Acinetobacter* has been associated with, among others, a number of infections such as endocarditis and respiratory tract infections. *Pseudomonas aeruginosa* is another Gram-negative bacterium that lives innocuously on human skin. However, they are able to infect practically any tissue with which they come into contact. Infections due to *Pseudomonas aeruginosa* occur primarily in compromised patients.

Skin bacteria are generally very avid biofilm formers. Evidence of biofilms on skin has been documented and the actual architecture has been observed in dermatitis and eczema. The first reported incidence of biofilms on skin was by *Staphylococcus epidermidis*. It has been suggested that the severity of eczema is proportional to colonization resistance.[89]

SKIN MICROFLORA AND HUMAN IMMUNITY

The idea that "a little bit of dirt does you good" has long been debated. There is some evidence to suggest that the natural flora of the human gut and upper respiratory tract helps to elicit and perpetuate immunological responses that protect against microbial invasion. A review exploring whether the indigenous microflora of human skin contributes to this "hygiene hypothesis" has recently been published. The ability of *S. epidermidis* to produce several antibiotics and antimicrobial peptides that restrict the proliferation of potential pathogens such as *S. aureus* and *S. pyogenes* suggests a protective role for these indigenous bacteria that benefits the host. Furthermore, pheromones of *S. epidermidis* may interrupt cell-to-cell communication of *S. aureus* and so modulate their attachment and virulence mechanisms that are vital to pathogenicity. Another advantage of colonization by *S. epidermidis* was indicated by stimulating keratinocytes via Toll-like receptor signaling to facilitate rapid responses to the presence of pathogens.[90]

Although *S. aureus* is the organism most frequently isolated from wound infections, it is also found on healthy human skin. The ability to inhibit other strains of *S. aureus* by producing bacteriocins might be an important factor when it is non-pathogenic. Similarly, corynebacteria and propionibacteria produce inhibitory substances that might confer host protection against pathogenic colonization.[90] Despite involvement in a variety of skin infections, *S. pyogenes* and *P. aeruginosa* were also shown to have potential host benefits. In particular, streptolysin O at appropriate levels was able to enhance wound healing by stimulating keratinocyte function, and numerous antimicrobial products secreted by *Pseudomonas* were considered important in limiting the survival of bacteria and fungi. The protective value of these organisms can be inferred when their removal results in pathogenic invasion of previously suppressed species and justifies the term colonization resistance to describe their beneficial role.[90] The ability of some of the bacteria within the relatively stable communities that associate with human skin to form xenobiotic agents[91,92] and become involved in antagonistic interactions[93] confers an important defensive mechanism that helps to prevent skin infections.

CONCLUSION

Skin is considered quite uninhabitable. However, skin is colonized by a multitude of different microorganisms. Based on molecular studies, it is now clear that culturable techniques significantly underestimate the true microbial diversity of human skin. We require a deeper understanding of skin microbiology and a better understanding of the host factors that are known to affect the biofilm and its overall community and architecture on the skin. The National Institutes of Health Roadmap for Medical Research has launched the Human Microbiome Project (HMP) with a view to characterizing the human microbiota and defining its role in healthy and diseased states. It is probable that further investment in molecular studies will shed light into the enormously varied human skin microbiota and their interactions with different species and with the host.

REFERENCES

1. Wysocki, A.B. (2002) Evaluating and managing open skin wounds: Colonization versus infection. *AACN Clin Issues* 13(3):382–397.
2. Chiller, K., Selkin, B.A., and Murakawa, G.J. (2001) Skin microflora and bacterial infections of the skin. *J Invest Dermatol Symp Proc* 6:170–174.
3. Baumann, L., Weisberg, E., and Percival, S.L. (2009) Skin aging and microbiology. In: *Microbiology and Aging: Clinical Manifestations*. S.L. Pervical, editor. Springer, Humana Press, NY.
4. Savey, A., Fleurette, J., and Salle, B.L. (1992) An analysis of the microbial flora of premature neonates. *J Hosp Infect* 21(4):275–289.
5. Aly, R., and Maibach, H.I. (1977) Aerobic microbial flora of intertrigenous skin. *Appl Environ Microbiol* 33(1):97–100.
6. Sarkany, I., and Gaylarde, C.C. (1968) Bacterial colonisation of the skin of the newborn. *J Pathol Bacteriol* 95(1):115–122.
7. Schröder, J.M., Reich, K., Kabashima, K., Liu, F.T., Romani, N., Metz, M., Kerstan, A., Lee, P.H., Loser, K., Schön, M.P., Maurer, M., Stoitzner, P., Beissert, S., Tokura, Y., and Gallo, R.L. (2006) Who is really in control of skin immunity under physiological circumstances—Lymphocytes, dendritic cells or keratinocytes? *Exp Dermatol* 15(11):913–929.
8. Nizet, V., Ohtake, T., Lauth, X., Trowbridge, J., Rudisill, J., Dorschner, R.A., Pestonjamasp, V., Piraino, J., Hutter, K., and Gallo, R.L. (2001) Innate antimicrobial peptide protects the skin from invasive bacterial infection. *Nature* 414:454–457.
9. Schröder, J.M., and Harder, J. (2006) Innate antimicrobial peptides in the skin. *Med Sci (Paris)* 22(2):153–157.
10. Harder, J., and Schröder, J.M. (2005) Antimicrobial peptides in human skin. *Chem Immunol Allergy* 86:22–41.
11. Marples, R.R. (1982) Sex, constancy, and skin bacteria. *Arch Dermatol* 272:317–320.

12. Wilburg, J., Kasprowicz, A., and Heczko, P.B. (1984) Composition of normal bacterial flora of human skin in relation to the age and sex of examined persons. *Przegl Dermatol* 71(6):551–557.
13. Larson, E.L., Cronquist, A.B., Whittier, S., Lai, L., Lyle, C.T., and Latta, P.D. (2009) Differences in skin flora between inpatients and chronically ill outpatients. *Heart and Lung* 29:298–305.
14. Rebel, G., Pillsbury, D.M., Phalle, G., de Saint, M., and Ginsberg, D. (1950) Factors affecting the rapid disappearance of bacteria placed on the normal flora. *J Invest Dermatol* 14:247–263.
15. Price, P.B. (1938) The bacteriology of normal skin: A new quantitative test applied to a study of the bacterial flora and the disinfectant action of mechanical cleansing. *J Infect Dis* 63:301–318.
16. Somerville-Millar, D.A., and Noble, W.C. (1974) Resident and transient bacteria of the skin. *J Cutan Pathol* 1(6):260–264.
17. Williamson, P. (1965) Quantitative estimation of cutaneous flora. In: Maibach, H.I., and Hildick-Smith, G., editors. *Skin Bacteria and Their Role in Infection.* New York: McGraw-Hill, p. 3.
18. Marples, M.J. (1974) *The Ecology of Human Skin.* 1965. Thomas CC, Springfield, IL; Noble, W.C., and Somerville, C.A. *Microbiology of Human Skin,* pp. 50–76, 131, 212. 1974. WB Saunders, Philadelphia.
19. Noble, W.C. and Somerville, C.A. (1974) *Microbiology of Human Skin.* Philadelphia: WB Saunders, pp. 50–76, 131, 212.
20. Kligman, A.M. (1965) The bacteriology of normal skin. In: Maibach, H.I., and Hildick-Smith, G., editors. *Skin Bacteria and Their Role in Infection.* New York: McGraw-Hill, pp. 13–31.
21. Williamson, P., and Kligman, A.M. (1965) A new method for the quantitative investigation of cutaneous bacteria. *J Invest Dermatol* 45(6):498–503.
22. Kligman, A.M. (1965) The bacteriology of normal skin. In: Maibach, H.I., and Hildick-Smith, G., editors. *Skin Bacteria and Their Role in Infection.* New York: McGraw-Hill, pp. 13–31.
23. Paulino, L.C., Tseng, C.H., Strober, B.E., and Blaser, M.J. (2006) Molecular analysis of fungal microbiota in samples from healthy human skin and psoriatic lesions. *J Clin Microbiol* 44(8):2933–2941.
24. Bibel, D.J., and Lovell, D.J. (1976) Skin flora maps: A tool in the study of cutaneous ecology. *J Invest Dermatol* 67(2):265–269.
25. Noble, W.C. (1980) Carriage of micro-organisms on the skin. In: Newsom, S.W.B., and Caldwell, A.D.S., editors. *Problems in the Control of Hospital Infection.* London: Royal Society of Medicine, pp. 7–10.
26. Mackowiak, P.A. (1982) The normal microbial flora. *N Engl J Med* 307:83–93.
27. Gao, Z., Tseng, C.H., Pei, Z. *et al.* (2007) Molecular analysis of human forearm superficial skin bacterial biota. *Proc Natl Acad Sci USA* 104:2927–2932.
28. Somerville, D.A., and Noble, W.C. (1973) Microcolony size of microbes on human skin. *J Med Microbiol* 6(3):323–328.
29. Montes, L.F. (1970) Systemic abnormalities and the intracellular site of infections of the stratum corneum. *JAMA* 31:1469–1472.
30. Elsner, P. (2006) Antimicrobials and the skin physiological and pathological flora. In: Hipler, U.-C., and Elsner, P., editors. *Efficiency of Biofunctional Textiles and the Skin. Current Problems in Dermatology.* Basel: Karger, pp. 35–41.
31. Bojar, R.A., and Holland, K.T. (2002) The human cutaneous microbiota and factors controlling colonisation. *World J Microbiol Biotechnol* 18:889–903.

32. Seifert, H., Dijkshoorn, L., Gerner-Smidt, P., Pelzer, N., Tjernberg, I., and Vaneechoutte, M. (1997) Distribution of Acinetobacter species on human skin: Comparison of phenotypic and genotypic identification methods. *J Clin Microbiol* 53:2819–2825.

33. Leyden, J.J., McGinley, K.J., Nordstrom, K.M., and Webster, G.F. (1987) Skin microflora. *J Invest Dermatol* 88:65s–72s.

34. Kaźmierczak, A.K., Szarapińska-Kwaszewska, J.K., and Szewczyk, E.M. (2005) Opportunistic coryneform organisms—Residents of human skin. *Pol J Microbiol* 54(1):27–35.

35. Webster, G.F. (2007) Skin microecology: The old and the new. *Arch Dermatol* 143(1):105–106.

36. McGinley, K.J., Leyden, J.J., Marples, R.R., and Kligman, A.M. (1975) Quantitative microbiology of the scalp in non-dandruff, dandruff, and seborrheic dermatitis. *J Invest Dermatol* 64(6):401–405.

37. Chamberlain, A.N., Halablab, M.A., Gould, D.J., and Miles, R.J. (1997) Distribution of bacteria on hands and the effectiveness of brief and thorough decontamination procedures using non-medicated soap. *Zentralbl Bakteriol* 285(4):565–575.

38. Somerville, D.A. (1980) The normal flora of the skin in different age groups. *Br J Dermatol* 81:248–258.

39. Selwyn, S. (1980) Microbiology and ecology of human skin. *Practitioner* 224:1059–1062.

40. Marples, M.J. (1969) The normal flora of the human skin. *Br J Dermatol* 81(suppl. 1):2–13.

41. LeFrock, J.L., Ellis, C.A., and Weinstein, L. (1979) The impact of hospitalization on the aerobic fecal microflora. *Am J Med Sci* 277:269–274.

42. Larson, E.L. (1981) Persistent carriage of Gram-negative bacteria on hands. *Am J Infect Control* 9:112–119.

43. Larson, E.L., McGinley, K.J., Foglia, A.R., Talbot, G.H., and Leyden, J.J. (1986) Composition and antimicrobial resistance of skin flora in hospitalized and healthy adults. *J Clin Microbiol* 23:604–608.

44. Seifert, H., Dijkshoorn, L., Gerner-Smidt, P., Pelzer, N., Tjernberg, I., and Vaneechoutte, M. (1997) Distribution of Acinetobacter species on human skin: Comparison of phenotypic and genotypic identification methods. *J Clin Microbiol* 53:2819–2825.

45. Rennie, R.P., Jones, R.N., and Mutnick, A.H.; SENTRY Program Study Group (North America). (2003) Occurrence and antimicrobial susceptibility patterns of pathogens isolated from skin and soft tissue infections: Report from the SENTRY Antimicrobial Surveillance Program (United States and Canada, 2000). *Diagn Microbiol Infect Dis* 45(4):287–293.

46. Bek-Thomsen, M., Lomholt, H.B., and Kilian, M. (2008) Acne is not associated with yet-uncultured bacteria. *J Clin Microbiol* 46(10):3355–3360.

47. Dekio, I., Hayashi, H., Sakamoto, M., Kitahara, M., Nishikawa, T., Suematsu, M., and Benno, Y. (2005) Detection of potentially novel bacterial components of the human skin microbiota using culture-independent molecular profiling. *J Med Microbiol* 54:1231–1238.

48. Grice, E.A., Kong, H.H., Renaud, G., and Young, A.C. (2008) NISC Comparative Sequencing Program, Bouffard, G.G., Blakesley, R.W., Wolfsberg, T.G., Turner, M.L., Segre, J.A. A diversity profile of the human skin microbiota. *Genome Res* 18(7):1043–1050.

49. Barton, M.D., McBain, A.J., and Gilbert, P. (2007) Molecular and cultural analyses of human skin. *Proceedings of the 107th Meeting on the American Society for Microbiology.* May 21–25, Toronto, Canada

50. Dekio, I., Sakamoto, M., Hayashi, H., Amagai, M., Suematsu, M., and Benno, Y. (2007) Characterization of skin microbiota in patients with atopic dermatitis and in normal subjects using 16S rRNA gene-based comprehensive analysis. *J Med Microbiol* 56(Pt 12):1675–1683.

51. Gao, Z., Tseng, C.H., Pei, Z. *et al.* (2007) Molecular analysis of human forearm superficial skin bacterial biota. *Proc Natl Acad Sci USA* 104:2927–2932.
52. Sugita, T., Suzuki, M., Goto, S., Nishikawa, A., Hiruma, M., Yamazaki, T., and Makimura, K. (2009) Quantitative analysis of the cutaneous Malassezia microbiota in 770 healthy Japanese by age and gender using a real-time PCR assay. *Med Mycol* 21:1–5.
53. Suzuki, T., Kawamura, Y., Uno, T., Ohashi, Y., and Ezaki, T. (2005) Prevalence of Staphylococcus epidermidis strains with biofilm-forming ability in isolates from conjunctiva and facial skin. *Am J Ophthalmol* 140(5):844–850.
54. Funke, G., von Graevenitz, A., Clarridge, J.E., and Bernard, K.A. (1997) Clinical microbiology of coryneform bacteria. *Clin Micro Rev* 10(1):125–159.
55. Vuong, C., and Otto, M. (2002) *Staphylococcus epidermidis* infections. *Microbes and Infection* 4:481–489.
56. Harmory, B.H., and Parisi, J.T. (1987) *Staphylococcus epidermidis*: A significant nosocomial pathogen. *J Infect Control* 15:59–74.
57. Fekety, F.R. Jr. (1964) The epidemiology and prevention of staphylococcal infection. *Medicine* 43:593–613.
58. Nagase, N., Sasaki, A., Yamashita, K., Shimizu, A., Wakita, Y., Kitai, S., and Kawano, J. (2002) Isolation and species distribution of staphylococci from animal and human skin. *J Vet Med Sci* 64(3):245–250.
59. Barth, J.H. (1987) Nasal carriage of staphylococci and streptococci. *Int J Dermatol* 26:24–26.
60. Nobel, W.C. (1981) *Microbiology of Human Skin*. London: Lloyd-Luke.
61. Kloos, W.E., and Bannerman, T.L. (1994) Update on the clinical significance of coagulase-negative staphylococci. *Clin Rev Micro* 7(1):117–140.
62. O'Gara, J.P., and Humphreys, H. (2001) *Staphylococcus epidermidis* biofilms: Importance and implications. *J Med Micro* 50:582–587.
63. Foster, T.J. 2004. The *Staphylococcus aureus* "superbug." *J Clin Invest* 114:1693–1696.
64. Bjarnsholt, T., Kirketerp-Møller, K., Kristiansen, S., Phipps, R., Nielsen, A.K., Jensen, P.Ø., Høiby, N., and Givskov, M. (2007) Silver against *Pseudomonas aeruginosa* biofilms. *APMIS* 115(8):921–928.
65. Taplin, D., Rebell, A.M., and Zaiab, N. The human skin as a source of Mima-Herella infections. *JAMA* 186:166–168.
66. Seifert, H., Dijkshoorn, L., Gerner-Smidt, P., Pelzer, N., Tjernberg, I., and Vaneechoutte, M. (1997) Distribution of Acinetobacter species on human skin: Comparison of phenotypic and genotypic identification methods. *J Clin Microbiol* 53:2819–2825.
67. Paulino, L.C., Tseng, C.-H., Strober, B.E., and Blaser, M.J. (2006) Molecular analysis of fungal microbiota in samples from healthy human skin and psoriatic lesions. *J Clin Microbiol* 44:2933–2941.
68. Faergemann, J., Aly, R., and Maibach, H.I. (1983) Quantitative variations in distribution of *Pitryosporum orbiculare* on clinically normal skin. *Acta Derm Venereol* 63:346–348.
69. Gupta, A.K., and Kohli, Y. (2004) Prevalence of Malassezia species on various body sites in clinically healthy subjects representing different age groups. *Med Mycol* 10:125–159.
70. Barak, O., Treat, J.R., and James, W.D. (2005) Antimicrobial peptides: Effectors of innate immunity in the skin. *Adv Dermatol* 21:357–374.
71. Chiller, K., Selkin, B.A., and Murakawa, G.J. (2001) Skin microflora and bacterial infections of the skin. *J Invest Dermatol Symp Proc* 6:170–174.
72. Paulino, L.C., Tseng, C.H., Strober, B.E., and Blaser, M.J. (2006) Molecular analysis of fungal microbiota in samples from healthy human skin and psoriatic lesions. *J Clin Microbiol* 44(8):2933–2941.
73. Fredricks, D.N. (2001) Microbial ecology of human skin in health and disease. *J Invest Dermatol Symp Proc* 6:167–169.

74. Hadaway, L.C. (2003) Skin flora and infection. *J Infusional Nursing* 26:44–48.
75. Hadaway, L.C. (2005) Skin flora: Unwanted dead or alive. *Nursing* 35(7):20.
76. Roth, R.R., and James, W.D. (1989) Microbiology of the skin: Resident flora, ecology, infection. *J Am Acad Dermatol* 20:367–390.
77. Laube, S., and Farrell, A.M. (2002) Bacterial skin infections in the elderly: Diagnosis and treatment. *Drugs and Aging* 19:331–342.
78. Cogen, A.L., Nizet, V., and Gallo, R.L. (2007) *Staphylococcus epidermidis* functions as a component of the skin innate immune system by inhibiting the pathogen Group A Streptococcus. *J Invest Dermatol* 27:S131.
79. Lyon, G.J., and Novick, R.P. (2004) Peptide signaling in *Staphylococcus aureus* and other Gram-positive bacteria. *Peptides* 25:1389–1403.
80. von Eiff, C., Becker, K., Machka, K., Stammer, H., and Peters, G. (2001) Nasal carriage as a source of *Staphylococcus aureus* bacteremia. Study Group. *N Engl J Med* 344:11–16.
81. von Eiff, C., Peters, G., and Heilmann, C. (2002) Pathogenesis of infections due to coagulase negative staphylococci. *Lancet Infect Dis* 2:677–685.
82. Mainous, A.G. 3rd, Hueston, W.J., Everett, C.J., and Diaz, V.A. (2006) Nasal carriage of *Staphylococcus aureus* and methicillin-resistant *S. aureus* in the United States, 2001–2002. *Ann Fam Med* 4:132–137.
83. Iwatsuki, K., Yamasaki, O., Morizane, S., and Oono, T. (2006) Staphylococcal cutaneous infections: Invasion, evasion and aggression. *J Dermatol Sci* 42:203–214.
84. Bokarewa, M.I., Jin, T., and Tarkowski, A. (2006) *Staphylococcus aureus*: Staphylokinase. *Int J Biochem Cell Biol* 38(4):504–509.
85. Wichmann, S., Wirsing von Koenig, C.H., Becker-Boost, E., and Finger, H. (1985) Group JK corynebacteria in skin flora of healthy persons and patients. *Eur J Clin Microbiol* 4:502–504.
86. Kaźmierczak, A.K., and Szewczyk, E.M. (2005) Bacteria forming a resident flora of the skin as a potential source of opportunistic infections. *Pol J Microbiol* 54(1):27–35.
87. Kaźmierczak, A.K., Szarapińska-Kwaszewska, J.K., and Szewczyk, E.M. (2005) Opportunistic coryneform organisms: Residents of human skin. *Pol J Microbiol* 54(1):27–35.
88. Brüggemann, H., Henne, A., Hoster, F., Liesegang, H., Wiezer, A., Strittmatter, A., Hujer, S., Dürre, P., and Gottschalk, G. (2004) The complete genome sequence of *Propionibacterium acnes*, a commensal of human skin. *Science* 305:671–673.
89. Goodyear, H.M., Watson, P.J., Egan, S.A., Price, E.H., Kenny, P.A., and Harper, J.I. (1993) Skin microflora of atopic eczema in first time hospital attenders. *Clin Exp Dermatol* 18(4):300–304.
90. Cogen, A.L., Nizet, V., and Gallo, R.L. (2008) Skin microbiota: A source of disease or defence? *Br J Dermatol* 158:442–455.
91. Bickers, D.R., and Athar, M. (2006) Oxidative stress in the pathogenesis of skin disease. *J Invest Dermatol* 126(12):2565–2575.
92. Platzek, T., Lang, C., Grohmann, G., Gi, U.S., and Baltes, W. (1999) Formation of a carcinogenic aromatic amine from an azo dye by human skin bacteria *in vitro*. *Hum Exp Toxicol* 18(9):552–559.
93. Papacostas, G., and Gate, J. eds. (1928) *Les associations microbiennes, leurs applications therapeutiques*. Paris: Doin.

3 An Introduction to Wounds

Michel H.E. Hermans and Terry Treadwell

CONTENTS

Classification/Terminology 84
    Differences in Physiology 84
    Influence of Microorganisms 85
    Differences in Treatment 85
    Conversion 86
    Treatment Objectives and Outcomes 86
    Conclusion 86
General Guidelines of Surgical Wound Management 87
    Surgical Wounds 89
        Sutured Wounds 89
        Lesions with Pus 89
    Thermal Injuries 90
        Types of Burns 90
        Radiation Necrosis 91
        Depth of Burns 92
        Depth Diagnosis 93
        Physiology of the Burn Wound 93
        Size of the Burn, Inhalation Injury, and Burn Disease 94
        First Aid and Guidelines for Referral 95
        Long-Term Results 97
    Frostbite 98
    Chemical Injuries 99
    Conclusion 99
Skin Donor Sites 100
Necrotizing Fasciitis and Bacterial Myonecrosis 100
    Diagnosis 101
    Management 101
Purpura Fulminans 102
Toxic Epidermal Necrolysis and Stevens-Johnson Syndrome 102
Skin Tears 103
Pretibial Laceration 103
Zoonoses: Bite and Scratch Wounds 104
Road Rash, Abrasions, Mechanical Blisters 104

Chronic Wounds: Their Occurrence and Impact.. 105
    Venous Ulcer... 107
    Diabetic Ulcer 112
Pressure Ulcers.. 116
    Stage II Pressure Ulcers 118
    Stage III Pressure Ulcers.. 118
    Stage IV Pressure Ulcers.. 118
    Other Pressure Ulcers... 118
    Prevention of Pressure Ulcers 119
Arterial Ulcers... 120
References.. 124

CLASSIFICATION/TERMINOLOGY

In most wound care-related literature, a chronic wound is defined as one that has "failed to proceed through an orderly and timely process to produce anatomic and functional integrity or proceed through the repair process without establishing a sustained and functional result."[1] In contrast, an acute wound is defined as one that is acquired as a result of trauma or an operative procedure and that proceeds normally in a timely fashion along the healing pathway with at least external manifestations of healing apparent in the early postoperative period without complications.[1]

In the wound care community, virtually all skin lesions are now called wounds (diabetic, venous, pressure, surgical, fungating carcinoma, etc.). In this context, the often-used definition of a chronic wound is one that takes more than 3 weeks to heal.

General encyclopedias use a different definition. According to them, a *wound* is a break in the continuity of any bodily tissue due to violence, "where violence is understood to encompass any action of external agency, including, for example, surgery." These encyclopedias mention inflammation, a chronic nature and an internal factor in their definition of an *ulcer*.

Cynically enough, if these nonmedical definitions for wound and ulcer are used and refined, an ulcer can be defined as a gradual disturbance of tissues by an underlying (and thus, internal) etiology/pathology and a wound (trauma) as an acute disturbance of tissues by an external force. With the use of the these two definitions, the observed differences in behavior of the lesion and the required medical approach and treatment options, clinical appearance, demographics, anatomical locations, physiology, and pathology are more logical.

DIFFERENCES IN PHYSIOLOGY

Healing is a very complex process, but in principle, all wounds go through similar steps with similar cellular and humoral contributions. Reactive oxygen species (ROS), proteases, and many other soluble mediators and cells are crucial for dealing with necrosis, debris, and microbial invasion. These compounds need careful regulation because by nature many of them are aggressive and corrosive. "Normally," for every up-regulating mechanism, a down-regulating counteracting mechanism exists.

When regulation is not balanced, a situation of prolonged and persistent inflammation may occur. For example, in an ulcer, polymorphonuclear cells (PMNs) increase ROS production,[2] thus inducing a vicious circle.

An important category of "cleaning" compounds, normally not active in resting, nondamaged tissue, consists of the metalloproteases (MMPs). MMPs can be produced by many different types of cells. They break down components of the damaged extracellular matrix, which is a necessary step prior to the influx of "healing" cells and compounds. MMP activity is counterbalanced by tissue inhibitors of metalloproteases (TIMPs). As described later in this chapter, MMP and TIMP profiles in chronic lesions (and hypertrophic scars) differ from those in acute wounds.[3,4]

Cytokine profiles in ulcers differ from those in acute wounds (trauma),[5,6] and in fact, they may be inadequate in ulcers, which is assumed to contribute to the poor healing tendencies by disrupting optimal signaling pathways.[7] These phenomena (and many others) confirm the chronically inflamed status of an ulcer,[8] which, indeed, is caused by an underlying etiology/pathology.

INFLUENCE OF MICROORGANISMS

We live in a predominately microbial world with the human body containing an estimated 10^{14} microbial cells.[9] Although these microbiota have an important role to play in maintenance of health, they nonetheless have the potential to cause disease given the opportunity. The majority of cutaneous wounds are colonized (some heavily) with both aerobic and anaerobic indigenous microorganisms that are found colonizing the mucosal surfaces, such as the gut and oral cavity. The number of microbial species identified in cutaneous and soft tissue infections continues to increase, but the suggestion that a bacterial innoculum may provide a stimulus to healing[10,11] should not be ignored. The complexities in current thinking and management of wound infection are explored in upcoming chapters.

DIFFERENCES IN TREATMENT

There are many differences between "standard wound (trauma) care" and "standard ulcer care."

- Wounds usually are treated to heal by primary intention: They are closed with sutures (cuts, surgical incisions), grafts (burns), or flaps (deep defects), or, in the case of superficial partial thickness burns, heal with support of dressings before granulation tissue develops. Ulcers are most commonly treated with dressings, instead of surgical closure.
- Surgical wound bed preparation is usually more rigorous. Aggressive excision is more common than traditional debridement (with curette, scissors, and enzymes). The latter methods often are preferred in ulcer care.
- The use of modern dressings in burn care is not as common,[12] but in ulcer care, modern, nonsurgical therapies are more frequently used (personal observation, M. Hermans). Ulcers, even deep ones, usually are treated with a variety of dressings and heal by secondary intention.

- Ulcers also have a much poorer tendency to heal unless the underlying etiology is treated (i.e., offloading in diabetic foot ulcers,[13] compression or vein stripping in venous leg ulcers[14]).
- Skin grafting for ulcers is not very successful[15] when compared to wounds, and infection in grafted wounds is not as common as in grafted ulcers.[16]
- Pinch grafting generates even poorer results in ulcers and is not used in trauma care.[17]

CONVERSION

Ulcers can be changed into acute wounds by treating, and sometimes removing, the underlying etiology (i.e., compression/saphenous vein ablation in venous leg ulcers). This alone may lead to significant changes in wound bed properties and healing tendencies.[18,19] Alternatively, the ulcer may be extensively debrided or excised into healthy tissue and may be closed primarily with sutures (rare), a graft, or flap.[19]

Conversion of a trauma into an ulcer is not uncommon. A typical example would be a pretibial laceration in an older adult patient with diabetes or severe venous hypertension. The underlying etiologies turn a wound with good healing potential into an ulcer with poor healing tendencies. A Marjolin's ulcer[20] (carcinomatous degeneration) of a (burn) scar is an example of a late-term conversion.

TREATMENT OBJECTIVES AND OUTCOMES

The preferred objectives of trauma treatment are often different from those in ulcer treatment. Although, of course, healing is the overall outcome, large trauma may be immediately life threatening; thus the immediate and primary treatment objective in such a case is survival and prevention or treatment of shock, respiratory failure, and other life-threatening syndromes.

Ulcers are rarely acutely life threatening. Outcomes, in addition to healing, would be prevention of recurrence (i.e., through treatment of venous hypertension, surgical repair of a Charcot's foot). In burn care, after reepithelialization, prevention or treatment of hypertrophic scars, keloid and contracture formation is an important outcome as well. In contrast, because of the typical demography of patients with diabetic foot ulcers and venous leg ulcers, as well as the typical location of these ulcers, these sequelae are rare in these types of lesions.

CONCLUSION

Certain types of skin lesion do not fit any of the classifications mentioned earlier. For example, toxic epidermal necrolysis may be caused by an underlying etiology (i.e., drug anaphylaxis), but the lesions, provided they do not get infected, require the general approach of burn care. A hypertrophic scar or post-burn contracture is not the consequence of ulceration but still is insufficient with regard to anatomic and functional integrity.

For most types of skin lesions, though, classification based on etiology is more logical because an etiology/pathology-based definition is better related to appearance, demographics, therapy (topical versus systemic, surgical versus nonsurgical), outcomes, and so forth.

At the same time, there should be a bigger overlap in techniques used for ulcer care versus trauma care. Many wounds could benefit from the expertise of modern dressings, built up in ulcer care, and many ulcers could benefit from a more extensive surgical approach.

GENERAL GUIDELINES OF SURGICAL WOUND MANAGEMENT

The primary guiding principle of (surgical) wound management is the restoration of anatomical and functional integrity of a lesion. Depending on the type of wound or ulcer, this may be relatively easy (as is the case with a simple, small straight superficial cut, which can be closed with an adhesive closure strip, tissue glue, staples, or simple sutures), whereas a complicated surgical defect may pose challenges to the clinical team in both short- and long-term management. The latter is generally the case when major tissue loss or destruction has occurred as the consequence of, for example, trauma, certain infections, or malignancy.

Depending on the lesion, three different approaches to wound healing may (have to) be chosen.

- Healing by primary intention implies that restoration of tissue continuity occurs, without the development of granulation tissue: Uncomplicated healing of a sutured incision is an example.
- Healing by secondary intention includes wound healing through the development of granulation tissue and secondary reepithelialization: The healing of ulcers usually follows this route.
- Healing by tertiary intention happens when a wound initially is left open (i.e., because of contamination with foreign bodies) and is actively closed secondarily, usually after 4 or 5 days.

Dead space is defined as a hollow area within a lesion where tissues are not in contact with each other. Avoiding dead space is another guiding principle in any type of healing, because it may function as a nidus for the development of infection.

Burns have a central zone of necrosis, surrounded by a zone of stasis, which in turn is surrounded by a zone of hyperemia and inflammation.[21] A similar situation exists in all other wounds, albeit on a very small scale for small wounds: There is a central zone of dead tissue, which is surrounded by one with wounded, fragile tissue, and an exterior area of viable, normal tissue.

In a simple wound in a healthy patient, the likelihood of deterioration of the two outer zones is small. Approximation of different tissue layers is probably going to result in complete or nearly complete restoration of function and cosmesis (although late healing problems such as keloid formation or atrophy are always among the possibilities).

In wounds with a great deal of necrosis or lesions in which a large amount of tissue is missing or not viable, restoration requires a much more elaborate approach. Missing tissue may have to be replaced (i.e., by grafts, flaps, or transplants) or replacement may not be possible, in which case new anatomical boundaries have to be created (as in the closing of a stump after an amputation).

Necrosis will interfere with wound healing. In small trauma, though, the body will rid itself of the dead tissue, and any interference is, in fact, hardly noticeable. In lesions with a large amount of necrosis, not only will dead tissue make healing very slow or impossible, it may also serve as a nidus for infection, first local and sometimes culminating in systemic infection, sepsis. Thus the removal of dead tissue is central in surgical wound treatment.

Removal needs to be completed as quickly as possible, because necrosis does not serve any beneficial purpose. Large tissue defects may be the result of necrotectomy, and they may be difficult to bridge or close, but inappropriate debridement or excision only results in slower or ceased healing with an increased chance of infection. Certain tissue defects may be bridged by using specific techniques such as cultured skin, flaps, skin expanders, or free bone grafts.

The type of debridement or excision depends on the type of wound, its location, and the type of tissues involved. A full thickness flame burn "only" requires removal of dead skin, whereas an electrical burn or crush injury often involves deeper tissues; thus further deep exploration is necessary in these cases.

Sometimes immediate excisional surgery may not be possible (i.e., in mass casualties or because of the patient's general condition). However, particularly in circumferential lesions or crush injuries, compartment syndrome needs to be avoided by early and rapid fasciotomy. Rhabdomyolysis may accompany massive crush injuries and may result in renal damage. Thus major trauma virtually always requires not only dealing with the trauma but also with its systemic complications. A large trauma causes a systemic syndrome, SIRS (systemic inflammatory response syndrome).[22] SIRS has a continuum of severity, ranging from tachycardia, tachypnea, fever, and leukocytosis (all mild) to refractory hypotension, shock, multiple organ system failure, and death.[23]

Serious trauma often involves major vessels, fractures, injuries of internal organs, and extensive soft tissue injuries. The order of repair is dictated by the most acute, life- or limb-threatening injury. In these cases, temporary repair (i.e., of fractures by external fixation) may be necessary to allow for more urgently necessary permanent reconstruction (i.e., of vascularity). Permanent soft tissue repair often has to wait because the extent of the (ischemic) damage may not be immediately clear.

Specific types of injuries may require specific types of treatment and exploration, over and above "standard care." For example, high-velocity rifle bullets cause shock waves, and consequently damage tissue ahead of the bullet trajectory. In addition, cavitation may be responsible for damage: Even fractures may occur outside the direct pathway of such a bullet.[24] Automobile accidents often cause open, compound fractures with associated vascular and nerve injuries, in combination with deglovement. Often, a lot of debris (street dirt, clothing, paint flakes) may be found in the wound. In this type of lesion, extensive exploration is necessary.

SURGICAL WOUNDS

The term *surgical wounds* is used many times in articles and books about wound care, but it is often not clear what is meant. Is the term referring to sutured, uncomplicated wounds? Does it refer to wounds that had to be reopened, or to lesions that have to be incised (i.e., abscesses, according to the previously mentioned definition not really a wound in the first place)? The term *surgical wound* is often poorly defined.

Sutured Wounds

Whether sutures are used to close an incision, are made during an operation, or are used to close an accidental injury, the purpose of suturing (or stapling or gluing) a wound is to try to restore the anatomical integrity of the tissues as much as possible. Many suturing techniques and materials are available, and it is beyond the purpose of this book to describe them in detail.

Generally speaking, the different layers that may exist in an incision or cut (i.e., an abdominal incision) are put together by separate layers of sutures. In the example of an abdominal incision, the peritoneum, fasciae, and skin are approximated using different materials and techniques for each layer.

In uncomplicated healing of a sutured wound, depending on the location of the body, the skin sutures can usually be removed within 5 to 15 days. (The deeper sutures are usually made of resorbable material and do not have to be removed.)

However, several complications may occur. If hemostasis cannot be achieved or if some leakage is expected, different types of wound drains can be utilized. This is to avoid clot retention in the incision, which increases the chance of a postoperative wound infection. The same is true for the formation of a seroma.

A more serious complication is wound infection. Typically, after a few days the incision becomes more painful, erythema around the wound may flare up, the patient may develop fever, and pus may start leaking from the incision—but even without the last symptoms, sutures (sometimes of more layers) need to be removed to allow for drainage of the wound.

Lesions with Pus

A wound dehiscence with pus, an abscess, and other (skin or subcutaneous) lesions where pus is present are essentially all treated primarily according to one single principle, which is allowing the lesion to drain and get rid of the pus.

In a sutured wound, it is often enough to simply remove one or more layers of sutures. The wound may open (dehisce) and drainage starts. In larger, complex wounds, it is important to carefully explore for sinuses and pus pockets, which need to be opened as well.

An abscess is surgically opened or may burst spontaneously. If it is surgically opened, an elliptic slice of skin may have to be removed to prevent premature closure. Depending on the size, shape, and depth of the lesion (i.e., a small abscess versus a large dehisced abdominal lesion), drainage may further need to be assured by keeping the wound open using specific wound packaging techniques. In an abscess, for example, a piece of gauze or preferably another, more modern, nonadherent dressing is inserted as a drain. The drain is removed frequently to allow for inspection of the lesion/abscess cavity and to physically remove the pus collected in the drain. The lesion may have to be washed out prior to repacking.

Again, depending on the size and type of lesion, closure may be spontaneous, or secondary closure may be necessary, for example, with a spit skin graft. Large lesions, depending on their location, may have a higher chance of developing herniation; thus more extensive secondary wound repair may be necessary.

Chronic infections, such as hydradenitis suppurativa or pilonidal sinus, often have several small or large abscesses within an infected or inflamed field. Initial therapy is often via incision of the abscesses, sometimes accompanied by courses of antibiotics. If the infection persists, excision (with or without primary closure) of the entire infected area needs to be performed.[25] For hydradenitis suppurativa, recent reports indicate that the chronic inflammatory reaction, which may be one of the underlying etiologies of the disease, responds to antitumor necrosis factor monoclonal antibodies. However, this treatment has some serious side effects and further evaluation is necessary.[26] Pilonidal sinus formation is the result of a foreign body reaction (chronic inflammation) and particularly occurs in patients with hirsutism. Medically indicated hair removal using laser treatment has shown some promise.[27]

THERMAL INJURIES

Burns are among the most common types of trauma occurring in any society. Most burns are relatively small and consequently not life threatening, but large burns, even partial thickness ones, still pose a major threat when not treated properly. Even smaller burns may cause major morbidity because the injury is very painful and may lead to disfiguring scar formatting, primarily hypertrophic scarring.[28]

In the United States, an estimated 500,000 burn injuries require treatment every year,[29] and about 4000 deaths occur annually because of fire. Of those, an estimated 3500 deaths are caused by residential fires and 500 from motor vehicle and aircraft crashes, contact with electricity, chemicals, or hot liquids and substances, and other sources of injury.

Each year, 40,000 U.S. hospitalizations occur for burn injury, of which 25,000 (60%) are in hospitals with specialized burn centers. Burns rank as the fourth cause of unintentional child death in the United States.[30] Globally, 322,000 fire-related deaths occurred in 2002. Percentage-wise, serious burns occur more often in low- and middle-income countries.

Types of Burns

Burn (thermal) injuries can be categorized as follows:

- Scald: The injury is caused by contact with a hot fluid (i.e., hot tea, soup, coffee). In most cases, these injuries, when cooled quickly, are partial thickness.
- Flame: The injury is caused by exposure to flames (i.e., a house fire or a barbeque explosion with clothing catching fire). These burns are usually full thickness.

- Flash: The injury is caused by very short exposure to a burning gas or vapor (i.e.. a barbecue explosion without clothing catching fire). The injury is often partial thickness.
- Contact burn: The injury is caused by contact with a hot surface. Depending on the surface touched (i.e., an iron sole plate), in many cases these burns are not deep. However, the combination of pressure and prolonged exposure to the heat source may lead to major injuries, as is the case in patients who, after a seizure, remain in contact with a hot surface for a prolonged period.[31,32] In many areas of the world with a hot climate, asphalt can be hot enough during the day to cause a second-degree burn within 35 seconds.[33] Contact with molten metal, hot coals, or other high-temperature agents leads to very deep injuries.
- Electrical burns .are thermal injuries (though they may have many other effects as well). The burn is caused by contact with or strike-through of an electrical current: Electricity is converted to heat which causes coagulation and cell walls to explode. Extensive and deep tissue necrosis may be the result. The amount of heat is proportionate to the amperage and electrical resistance of the tissues through which the electricity passes.[34] Electric burns may lead to acidosis or myoglobinuria, which are life-threatening complications. Thus, early exploratory surgery is necessary. Sometimes the extent of the injury may not be immediately apparent, particularly when most of the damage done is on the subcutaneous or deeper levels.
- Radiation: The injury is caused by exposure to heat radiation. The typical example of this type of burn is sunburn. The injury is usually first degree.

Other types of injuries commonly treated in burn centers but having a different etiology include chemical injuries, frostbite, dermatological diseases such as toxic epidermal necrolysis, epidermolysis bullosa, as well as other skin diseases and conditions that are accompanied by, or lead to, major skin loss, such as necrotizing fasciitis,[35,36] and unusual infections such as phaeohyphomycosis.[37] Some of these conditions will be discussed elsewhere in this chapter.

Radiation Necrosis

Skin lesions caused by (ionizing) therapeutic radiation as part of treatment of malignancies are sometimes termed *radiation burns*. It is important to distinguish these lesions from the type of radiation injury caused by the sun (ultraviolet [UV] and infrared radiation) because for therapeutic purposes, different, much more powerful and, depending on the type of radiation, deeper penetrating radiation is used. Therefore *radiation necrosis* is the preferred term. This type of lesion behaves clinically like an ulcer and is a result of not only damage to the skin but also damage to the subcutaneous and deeper tissues. Often, direct damage is caused to the vascular structures,[38] leading to tissue atrophy and radiation necrosis.

Typically, the lesion starts with progressive erythema and continues to produce skin necrosis. For low-dose injuries, spontaneous resolution may occur over a 2-month period. However, the early skin symptoms give no indication of pathological changes in deeper tissues, and muscular radionecrosis may start early on.[39]

Studies indicate that ionizing radiation induces modulation of cytokine and chymokine expression by skin involving, among others, interleukin-1, -6, and -8; tumor necrosis factor alpha (TNFα), and transforming growth factor-beta (TGFβ).[40,41] The entire process is also called *cutaneous radiation syndrome,* and the expression of some of the cytokines is dose dependent.[42]

Because the underlying tissues are so much involved, it is clear that "simple excision" until a viable wound bed is reached is not necessarily the preferred option for treatment. Cutaneous radiation lesions should be classified as ulcers and not as examples of burn injury.

Depth of Burns

The depth of a burn is very important. It determines whether or not surgical intervention is necessary.

The depth classification is related to the anatomy of the skin. The upper layer, the epidermis, is separated from the dermis by the basal membrane. However, the dermis contains epidermal structures such as sebaceous glands, hair follicles, and sweat glands. If these structures are still intact the epidermis can, in principle, heal spontaneously from these epidermal remnants. Complete destruction of the epidermis and dermis, as is the case in a full-thickness lesion, makes reepithelialization possible only from the wound edges. This type of healing will take considerably longer and will not be successful in large burns.

Thicker skin can stand a given heat insult better than thin skin. Given a certain amount of heat exposure, a burn of the thin skin on the dorsum of the hand becomes deep more quickly than a burn of the lower back. Still, burns occur rapidly. In general, 25 seconds of exposure to water at 55°C results in a deep dermal or full thickness burn, whereas a 2-second exposure to water at 65°C leads to the same result.[43]

First Degree

The typical first-degree burn is sunburn. The skin is painful, but there is no breach of the epidermis. The skin looks red (inflammation). After a few days, desquamation may occur.

Superficial Partial Thickness (Superficial Second Degree)

In this type of burn, the superficial parts of the dermis are exposed because the epidermis is destroyed. Blisters may or may not occur. The skin (underneath the blisters) is moist, pink in color, and hypersensitive to the touch. Blanching with pressure is rapid and positive, and capillary refill is virtually immediate. When appropriate treatment is used, this type of burn will heal rapidly, within 10 to 14 days, with minimal scarring.

Deep Partial Thickness (Deep Second Degree)

Here, the deep dermis is exposed, because both the epidermis and the superficial dermis are destroyed. The exact depth of this type of burn may be very difficult to determine because the visual aspect may mimic a superficial partial thickness injury or a full thickness injury (see below). Capillary refill is slow or may not occur. The lesion may cause little pain but also can be painful. In contrast to a more superficial burn, most deep seating epidermal structures are destroyed; hence some healing from the depth may be possible, but this is not guaranteed. Healing is significantly longer and surgical intervention is often required.

Full Thickness (Third Degree)

In full-thickness burns, the entire skin (epidermis and dermis) is destroyed. Initially, these burns are not or are hardly painful because the nerve endings, residing in the dermis, have been destroyed as well. The visual aspect depends on the mode of injury and may be anywhere from white (a deep scald), to dark gray or black (a flame burn). The wound surface is usually dry and leather-like to the touch.

Fourth Degree

In fourth-degree burns, the entire skin is destroyed and substantial thermal damage also has been done to subcutaneous and deeper tissues (i.e., muscle). Extensive carbonization may be present if the injury was a flame burn.

Depth Diagnosis

The correct depth diagnosis can be established using the patient history as well as by judging the physical aspects of the burn. The patient history will provide a good indication of the burn depth. For example, short exposure to warm water with immediate cooling usually signals a partial thickness burn, whereas exposure to flames for only a short while will virtually always result in a full thickness burn.

Sometimes the patient history is not congruent with the injury. This may indicate abuse (i.e., a very deep burn on the palm of the hand, where the exposure was said to be very short and the contact surface was the sole plate of an iron).

The pinprick test is helpful to determine the pain level: A burned area is very gently touched with a sharp needle tip and the patient is asked about the intensity of the pain. Serious pain indicates a superficial burn, and minimal or no pain indicates a deep dermal or full thickness burn.

The level of blanching of the skin may also help establish a proper depth diagnosis, because it is the direct consequence of capillary refill. The slower is the refill, the deeper is the burn.

In an attempt to distinguish between dead and vital tissues, dyes such as fluorescein have been tested,[44] but these are not used in the clinical situation. Ultrasound has also been used but was shown not to be significantly different from clinical judgment.[45]

Laser Doppler flowmetry is promising. In experimental and clinical research, the technique was proven to be reliable.[46,47] Recently, devices have become available that make the technique practical in the day-to-day setting, allowing for an accurate and rapid diagnosis over large surfaces within a short time frame.[47–50]

Physiology of the Burn Wound

Burns are dynamic wounds. In the course of the first few days postinjury, they may change in depth. This phenomenon is known as conversion or secondary deepening.[51–53] Burns that were initially diagnosed as superficial partial thickness may actually turn out to be (or have become) deeper after a few days. Desiccation of the wound bed, as well as infection, may contribute or lead to wound conversion. On a cellular level, it has been shown that burn wound conversion is caused by the additive effects of inadequate tissue perfusion, free radical damage, and systemic alterations in the cytokine milieu of burn patients.[54] These phenomena lead to protein denaturation and necrosis and are the result of a number of events in and surrounding the wound and its environment, including infection, tissue desiccation, edema, circumferential eschar, impaired wound perfusion, metabolic derangements, advanced age, and poor general health.

Whatever the exact mechanisms, the dressing choice in partial thickness burns plays an important role in the prevention of conversion.[55] Still, in spite of the use of proper dressings and techniques, some burns may convert anyway. It also has been recognized that even experienced burn physicians and nurses sometimes misjudge the initial depth of a burn.

Size of the Burn, Inhalation Injury, and Burn Disease

Morbidity and mortality are largely determined by the size of the burn and whether or not an inhalation injury or other concomitant or preexisting diseases exist.[56] Even superficial but very large burns, particularly in older adults and young children, are still associated with a high level of morbidity and mortality.

Burn size is expressed as a percentage of the total body surface area (TBSA) and may be determined by the rule of nines:[57] The body is divided into areas of nine or multiples of 9%. In an adult, the head and arms each count for 9%, each side of the trunk and each leg count for 18%, and the remaining 1% is reserved for the genitalia and the perineum (Figure 3.1). For children these percentages are different. For

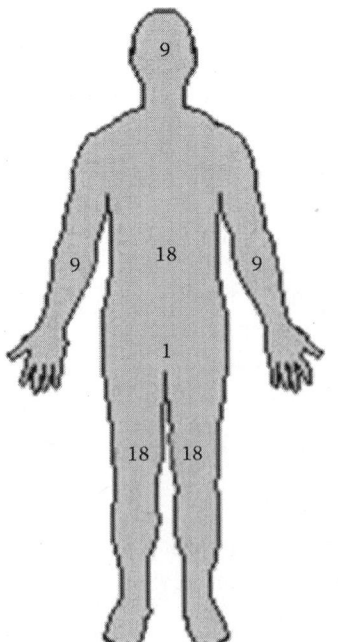

FIGURE 3.1   Rule of nines.

example, in a neonate the head counts for 18%. Burn centers use much more specific charts for determining the exact size of a burn. An easy guideline is that the patient's palm represents 0.5% TBSA.[58]

The amount of necrotic tissue, heat, and protein loss is directly related to the size of the burn injury and is responsible for systemic problems in large burns. Because of these secondary effects of the skin injury, a large burn is much more than just that: The systemic effects cause the "burn disease" associated with multiple complications.

The immediate threat of a larger burn is shock: A major change in capillary permeability is associated with massive fluid transport out of the circulation into the interstitium. Similar to other major trauma, larger burns cause SIRS:[22] A great deal of inflammatory mediators cause the systemic complications. Longer-term complications are the risk of sepsis and organ system responses to shock and to the "burn toxins"[59,60] that are released from the coagulation necrosis of the skin.

A specific, very serious complication that may accompany flame burns is inhalation injury, damage to the tracheal and pulmonary system caused by the inhalation of toxic or hot gasses and fumes.[61] This condition, which is still associated with a high level of morbidity and mortality,[62,63] often requires artificial ventilation.

Mortality in burn care has dropped significantly during recent decades because of a much better understanding of the physiology of burn disease and the systemic responses. Better prevention and management of complications, in combination with better topical therapies and more aggressive surgical approaches, have improved outcomes. Survival of patients with full thickness burns of more than 95% TBSA is not uncommon.[64,65]

First Aid and Guidelines for Referral

Guidelines for referral are fairly straightforward (Figure 3.2). With respect to simple measures (i.e., cooling and cleaning of the wound, IV administration of fluids in larger burns), initial care essentially is identical and independent of whether or not a patient is referred to a burn center.

- Dissipating the heat, by cooling with running tap water for a minimum of 10 minutes, is essential because this removes as much heat as possible, helps reduce the initial pain,[66–68] and decreases edema in the wound [69] Rapid cooling is important because tissue temperatures above 45°C will continue to cause local injury.[70] Particularly in young children, the risk of overcooling, with an associated dangerous drop in core temperature exists; thus burn patients should not be immersed in a bath with ice cold water.
- Rings on fingers and toes have to be removed because they will act as a tourniquet when edema occurs.
- Wounds may be gently cleaned with a neutral soap. Some prefer chlorhexidine gluconate soap because it has antimicrobial activity against regular skin flora.[71]
- Tar and asphalt burns should be cooled first. The causing agents will stick to the skin. Using a solvent[72] is preferred over physically peeling the materials off because peeling may do further (mechanical) harm to the skin and the wound.

BURN CENTER REFERRAL CRITERIA

- Partial-thickness burns of greater than 10% of the total body surface area
- Burns that involve the face, hands, feet, genitalia, perineum, or major joints
- Third-degree burns in any age group
- Electrical burns, including lightning injury
- Chemical burns
- Inhalation injury
- Burn injury in patients with preexisting medical disorders that could complicate management, prolong recovery, or affect mortality
- Any patients with burns and concomitant trauma (such as fractures) in which the burn injury poses the greatest risk of morbidity or mortality. In such cases, if the trauma poses the greater immediate risk, the patient's condition may be stabilized initially in a trauma center before transfer to a burn center. Physician judgment will be necessary in such situations and should be in concert with the regional medical control plan and triage protocols.
- Burned children in hospitals without qualified personnel or equipment for the care of children
- Burn injury in patients who will require special social, emotional, or rehabilitative intervention

Source: Guidelines for the Operation of Burn Centers (pp. 79–86), *Resources for Optimal Care of the Injured Patient 2006*, Committee on Trauma, American College of Surgeons.

FIGURE 3.2   Burn center referral criteria.

- Many chemical lesions may benefit from rinsing with water, because for most, water just serves to dilute the agent. However, first aid may be agent specific.[73–79] Therefore it is always important to identify the chemical agent that caused the injury.
- Prior to transportation, clothing may be removed, but this has to be done carefully because it may be stuck to the wound. A nonadherent dressing should be used to cover the burned areas. Silver sulfadiazine should not be used if the patient is referred, because painful removal of the cream upon arrival in the burn center will have to take place to assess wound aspect and size.
- In larger burns, administration of intravenous (IV) fluids is indicated prior to transporting the patient to a burn center, if transportation is expected to take longer than 60 minutes. Ringer's lactate should be infused at 2 to 4 mL/kg/percentage TBSA per 24 hours[80] to provide circulatory volume support. However, burn centers may prefer a different regimen, and consulting them should be considered. Intravenous access should be achieved

using larger veins (central lines are preferable) and, if possible, should not penetrate through burned skin.

- Narcotics may be used as pain medication but may only be given IV. Other methods of administration should be avoided because the pattern of uptake is unpredictable.
- If an inhalation injury is suspected, 100% humidified oxygen should be provided during transportation. However, given the possible acute onset of edema, it is wise to consult with the burn center to which the patient will be transported about possible intubation prior to putting the patient in the ambulance or helicopter.
- Similarly, guidelines from a burn center are advisable when one is considering escharotomies.[81,82] These are release excisions that may need to be made in patients with circumferential deep burns that may restrict respiratory excursion of the chest, circulation into the limbs, or cause postburn intra-abdominal hypertension.[83]

Before transporting a patient to a burn center, it is always wise to call the center about general and specific measures they would like to be taken before the patient is sent off.

Long-Term Results

After reepithelialization is complete, the wound healing process continues into the remodeling phase. During this phase, the extracellular matrix is reorganized. However, in many burns, the remodeling phase goes awry. Macroscopically, this results in hypertrophic scarring and the formation of contractures or keloid. Biochemically, a deregulation of a number of inflammatory mediators, such as TNFα, PDGF, and TGFβ, seems to play a major role in the development of hypertrophy.[84]

Hypertrophic scars are raised above the skin level and are very inflamed in the beginning. They can be extremely debilitating and will interfere negatively with quality of life, because they may limit movement, can be painful, and virtually always are very pruritic. The psychological aspects of "being ugly" are extremely important in this context as well.

Hypertrophic scarring is virtually certain to occur in burns that have taken a long time to heal spontaneously.[85] However, rapidly healing burns may also result in serious scar formation, because scarring is largely genetically determined. Dark-skinned patients have a significantly higher risk of serious scarring.[28,86] Scarring also depends on other factors, such as the location of the wound (a sternotomy incision, for example, virtually always results in a hypertrophic scar). During the reepithelialization process, not much can be done to prevent hypertrophic scarring. However, because the chance of developing a hypertrophic scar is, to a certain degree, linked to the time to reepithelialization, using dressings and techniques that are proven to reduce time to healing may contribute indirectly to reducing the incidence of hypertrophic scarring.[85]

In patients who are prone to scarring (based on the results of previous injury and wound healing time), preventative measures must be taken after reepithelialization is complete. These measures include the use of customized pressure garments,[87–90] with or without silicon sheeting as a contact layer on the wound.[91–93] Corticosteroid injections are also used,[94,95] and other therapies are being developed as well, among them the use of different types of laser,[96,97] and possibly the use of pharmacological agents.[98]

The results of hypertrophy prevention are often not truly satisfactory and a visible scar may remain, although in the long term, hypertrophic scars will become flatter and less inflamed. However, surgical scar revision is often necessary, particularly when scar formation leads to contractures.[99–101]

Keloid formation is different from hypertrophic scarring, both physiologically as well as macroscopically. A typical keloid extends beyond the borders of the original wound and has a cauliflower-type appearance.[102–104] Prevention and treatment of keloids is even more difficult than that of hypertrophic scarring[105–109] and lies beyond the scope of this chapter.

Other, non-skin-related post-burn complications are not uncommon (i.e., heterotopic ossifications in periarticular tissue[110]) but also are beyond the scope of this chapter.

FROSTBITE

Frostbite is the result of low temperatures leading to direct injury to the cells as well as to ischemic injuries to the tissues. Injury to the cell is initially caused by the formation of ice crystals in the extracellular space. The crystals cause injury to the cellular membranes, which results in cellular dehydration because of changes in the osmotic gradient.[111] Through different mechanisms (i.e., pH and protein changes), cellular homeostasis is lost and cell death is the result. If cooling occurs more rapidly, ice crystals are also quickly formed intracellularly, which leads to more rapid cell death.[112] Microvascular changes happen at the same time as well: Vasoconstriction and the formation of microemboli can be observed. Thawing may restore circulation, but only temporarily.

From a pathobiochemical point of view, many similarities between the inflammatory response to frostbite and burn wounds have been observed.[113] Depth classification is similar to that used for burns, although the appearance of frostbite injuries may be different. Full thickness lesions typically show dark blisters that turn into black eschar in the course of 1 or 2 weeks postinjury.

The exact depth and extension of serious frostbite lesions are often very difficult to determine. Many wait until spontaneous, complete demarcation has taken place prior to reconstruction. The use of technetium scanning is now being tested with respect to the involvement of bone in the frozen tissues.[114]

With respect to first aid, the old adage of rubbing a frozen skin area with snow is now considered outdated. This often leads to a thawing and refreezing cycle with an overall worsening outcome.[115] Therefore, partial or slow rewarming during transport should be avoided. Instead, elaborate rewarming protocols, including rapid rewarming at a temperature of 40°C, are recommended. Good results have also been obtained with protocols aimed at rapid rewarming while using medication to reduce local thromboxane and systemic prostaglandin.[116]

CHEMICAL INJURIES

Chemical injuries are often treated by burn centers but only have limited similarities to thermal lesions. The type and depth of injuries depend on the nature of the chemical agent and its concentration, as well as on the length of exposure. In addition, the type of tissue exposed to the agent plays a major role. For example, strong alkali agents cause very rapid scarring, opacification and perforation of the cornea,[117] and liquefaction necrosis of the skin. Acids, on the other hand, cause a hard, dry eschar that often prevents the inflicting agent from deep penetration and systemic absorption.

Generally speaking, rinsing with copious amounts of water is a good first aid measure for many chemical injuries. Patients should not be put in a tub, though, because this might expose previously normal skin to the agent and is also known to lead to hypothermia if the water is too cold. Neutralizing an alkaline agent with an acid is not recommendable because proper titration is virtually impossible and the reaction is exothermic.

Some common household materials can cause chemical injuries, and often the general public is not aware of this. For example, regular cement may lead to fairly serious lesions, though the exact mechanism is not clear. Cement has a high pH but also contains a number of compounds (i.e., calcium ions, chromium salts) that may be harmful when absorbed through abraded skin, and skin abrasions occur frequently in people who work with cement or concrete.[118]

Some chemical injuries require very specific first aid and treatment. Phosphorus, for example, used extensively in civilian factories as well as in the military, ignites in air. The agent continues to burn until it is completely oxidized or until the oxygen source is removed. The latter can be done by immersing the patient in water and by keeping the phosphorus wet until it is physically removed in the operating room (OR). Ultraviolet light can be used to identify imbedded pieces.[119]

Many chemicals not only cause serious skin injuries but also have serious systemic effects when absorbed. Hydrogen fluoride (HF) is one of those compounds. It is an acid that requires specific topical treatment (i.e., with calcium gluconate cream or intra-arterial injections). However, HF is also a metabolic toxin. When absorbed in high quantities, the fluoride may lead to hypocalcemia, ventricular arrhythmias, and respiratory failure. To remove it, hemodialysis or peritoneal dialysis may be indicated.[120]

Oral ingestion of a toxic or poisonous compound often leads to systemic effects but also may have very serious local effects, causing constrictures and sometimes perforation of the esophagus. This requires specific treatment.

Generally, the treatment of chemical injuries should be guided by the insulting agent; thus it is important to find out as quickly as possible what the causative compound is.

CONCLUSION

The treatment of serious burns and related injuries, whether large, deep, or located in functional areas, should be done in a burn center. In these centers, an entire team (physicians, nurses, occupational therapists [OTs], physical therapists [PTs] dieticians, psychologists, etc.) is dedicated to burn care, and their treatment options often lead to impressive results.

However, the large majority of burn victims suffer from lesions that do not need this high level of care and that are small enough to be treated outside a burn center, in a general hospital or an outpatient clinic, provided that wound management is done in line with burn care guidelines.

SKIN DONOR SITES

Skin donor sites, from which skin grafts are taken, can be virtually anywhere in the body and they can be full thickness or partial thickness. Full thickness donor sites are not commonly used in burn management, because the donor site, unless very small (when it could be closed by suturing), would have to be covered with epithelium. Partial thickness donor sites still contain deep-seated epithelium remnants, from which they can regrow. In extensive burns, a reepithelialized donor site may be reused (reharvested) a number of times, although the quality of the skin diminishes over time.

In large burns, the scalp is often used because the site reepithelializes rapidly (and thus can be reharvested quickly and often).[121] Alopecia is usually not a problem because there are many deep-seated hair follicles. In principle, though, the quality of the skin of a donor site should resemble the skin of the recipient site as much as possible.

Donor sites can cause considerable morbidity, primarily because they are very painful.[122–126] In large burns, and particularly in older adult patients with frail skin, donor sites may also be difficult to heal.[127] A donor site may also be surrounded by burned areas (in large burns, when no other locations are available), and thus become more prone to infection and difficulties with regard to dressing regimes.

Donor sites are made with hand driven, air driven, or electric dermatomes, are very painful, and bleed profusely. Thus different types of hemostats (i.e., epinephrine, thrombin, hemostatic dressings) are often used in combination with, or prior to, application of a dressing. Superficial donor sites usually heal without serious scarring, but some patients develop serious hypertrophic scars on their donor site.

Many different types of dressings are used for the treatment of donor sites.[12] One that provides a pain-reducing environment reduces the morbidity of the patient.

NECROTIZING FASCIITIS AND BACTERIAL MYONECROSIS

Necrotizing fasciitis used to have many different names, such as necrotizing cellulitis, hemolytic streptococcal gangrene, and necrotizing erysipelas. All these terms have been replaced by necrotizing fasciitis, although the term *Fournier's gangrene* is still used for necrotizing fasciitis of the perineal area and genitalia.

The disease is characterized by a rapidly spreading soft-tissue infection, typically causing necrosis of subcutaneous tissues and fascia. The necrosis may extend into muscles and skin. Among others, subjects with diabetes mellitus, intravenous drug users, and subjects older than 50 years have a significantly higher risk, as do people with hypertension, malnutrition, and the obese.[128]

Often, the primary cause for the infection is a deep contaminated or infected wound, and the subsequent necrotizing fasciitis is often polymicrobial in nature. *Staphylococcus, Streptococcus, Enterococcus,* and *Bacteriodes* can commonly be cultured.

Infections with *Clostridium* species cause gas gangrene and have an even higher mortality rate than "regular" necrotizing fasciitis. The infections result in severe systemic toxicity. The mortality of Streptococcal myositis is described as between 80% and 100%, and streptococcal toxic shock syndrome (strep TSS) with associated necrotizing fasciitis is a rapidly progressive process that kills 30% to 60% of patients in 72 to 96 h.[129]

Streptococcal and staphylococcal infections are not predictors of mortality but infections with *Aeromonas* or *Vibrio*, the presence of cancer, hypotension, and band form white blood cell count greater than 10% were found to be independent positive predictors of mortality in patients with necrotizing fasciitis. The presence of hemorrhagic bullae is an independent negative predictor of mortality.[130]

DIAGNOSIS

Early diagnosis is of extreme importance and consequence. The initial signs often are not very impressive and may be limited to localized pain and edema. Later, induration and erythema may be evident. Paresthesia of overlying skin and, later on, discoloration and blistering frequently occur as the next stage of the infection. Generally, though, the local signs seem not to be in proportion to the severe toxemia.

Gas inclusion may be evident in subcutaneous tissues on x-ray. Computerized tomography (CT) and magnetic resonance imaging (MRI) may help in the diagnosis and provide information on the nature and extent of the infection.[131] Obviously, frozen section biopsies may provide early histological evidence of infection. Gram stains and microbiological testing are important diagnostic tools and guide antibiotic treatment. However, a definite distinction between necrotizing fasciitis, myonecrosis, and other soft-tissue infections often only can be established during surgery.

MANAGEMENT

Critical to successful management of necrotizing infections is early diagnosis and radical surgical intervention.[132] Surgical exploration involves complete excision of all necrotic tissue; simple drainage is not enough.[132] If more than one operation for debridement of infected necrotic tissue is needed, mortality increases from 43% to 71%. This indicates the importance of adequate and complete necrotectomy as early as possible. Wounds are left open, and regular and frequent inspection for recurrence or extension of infection is necessary. In patients with many risk factors, early amputation of the extremity may be the best choice, especially in cases of myonecrosis. Comprehensive infection control is required before wound closure is attempted.

Broad-spectrum antibiotics are started preoperatively, and shock treatment, other resuscitative measures, and nutritional support virtually always are necessary.

Once the lesions are free of infection, secondary closure can be obtained using the techniques described elsewhere in this book (i.e., the application of split skin grafts, flaps, etc.).

Some authors advocate the use of hyperbaric oxygen and claim that it results in decreased mortality and reduced need for debridement.[133] However, any adjunct therapy is only secondary to good surgical excision.

PURPURA FULMINANS

Purpura fulminans is a rare, acute syndrome of rapidly progressive hemorrhagic necrosis of the skin. The disease occurs primarily in children but also has been diagnosed in adults. The necrosis is caused by dermal vascular thrombosis, vascular collapse, and disseminated intravascular coagulation (DIC). Neonatal purpura fulminans is associated with a hereditary deficiency of the natural anticoagulants antithrombin III and proteins S and C. Idiopathic purpura fulminans, probably linked to a protein S deficiency, usually follows an initiating febrile illness that manifests with rapidly progressive purpura. The most common type of purpura fulminans is the acute infectious form,[134] which has been associated with systemic infection by *Meningococcus*, Gram-negative bacilli, *Staphylococcus*, *Streptococcus*, and *Rickettsia*. Skin necrosis begins with a region of discomfort, which is rapidly followed by a short period of flush, and subsequently, petechiae. The next stage is the formation of hemorrhagic bullae, which progress to skin necrosis. The process generally involves the skin and subcutaneous tissues, without involvement of muscle. Because skin involvement is frequently an early manifestation, a skin biopsy increases the chance of early diagnosis.

In the infectious form, management is directed at stopping the progression of the underlying infection, while secondary infections have to be prevented as well. Necrotic and nonviable tissues have to be removed. Early heparin administration and replacement of clotting factors and protein C have proven useful to stop intravascular clotting.[135] Shock and sepsis frequently occur and require urgent treatment. Early escharotomy or fasciotomy may be necessary to prevent compartmental syndrome.

In a recent study, full-thickness skin and soft-tissue necrosis was extensive, leading to the need for skin grafting and amputations in 90% of the patients, with 25% of the patients requiring amputations of all extremities.[136]

TOXIC EPIDERMAL NECROLYSIS AND STEVENS-JOHNSON SYNDROME

Toxic epidermal necrolysis (TEN, Lyell's disease) and Stevens-Johnson syndrome (SJS) are rare severe blistering skin diseases, which are mainly caused by drugs (although idiopathic forms also exist). About 2 cases per million per year are estimated to occur. The mortality rate is high and ranges from 20% to 30%.[137] About one-half of survivors will have sequelae, especially on the eyes.[138]

The two conditions are distinguished on the basis of the degree and extent of blistering and other symptoms. A genetic predisposition might exist.[139] The diseases are, in fact, a form of immune complex-mediated hypersensitivity (allergic reaction) in an extreme form, and are characterized by widespread erythema, necrosis, and bullous detachment of the epidermis and mucous membranes. On a cellular level, massive keratinocyte apoptosis is the hallmark of TEN. Cytotoxic T lymphocytes appear to be the main effector cells.

The disease is characterized by extensive exfoliation, and if not treated properly and promptly, secondary sepsis and death may occur. Mucous membrane involvement can result in respiratory failure, gastrointestinal hemorrhage, and ocular (corneal ulcers, which may lead to blindness) and genitourinary complications.

A dozen "high-risk" medications account for 50% of cases. Anticonvulsants, antibiotics (penicillins), NSAIDs (nonsteroidal anti-inflammatory drugs), sulfonamides, and allopurinol are often quoted as the drugs most commonly responsible for TEN and SJS.[140,141]

TEN needs to be distinguished from staphylococcal scalded skin syndrome. The appearance of both diseases is similar, but the etiology is different. The latter disease is caused by staphylococcal toxins and needs to be treated as such. A biopsy shows histological differences.[142]

Typical symptoms include painful, rapidly expanding rashes and exfoliation of skin with considerable pain. As mentioned, mucosal areas are often involved.

Because the treatment of the disease is similar to that of burn victims, hospitalization in a burn unit is typically required. Treatment includes discontinuation of the offending drug, isolation of the affected areas to prevent infection, protective bandages, antibiotics, and general prevention of shock. Prompt referral to a burn unit is recommended.

More specific treatment (i.e., administration of systemic corticosteroids, the use of immunoglobuline intravenously) is still somewhat controversial.[140]

SKIN TEARS

Skin tears are small avulsion injuries. They occur most commonly in older adults and are the result of friction and tear. The pretibial area is most commonly involved, but the lesions also occur on other anatomical areas with thin or fragile skin, such as the dorsum of the hand or the elbows. Skin tears may be caused by an accident, but in the frail skin of older adults they may even be caused by rapid removal of an adhesive dressing. The exact incidence of skin tears is not known, but data gathered in a long-term care institution indicate that more than 95% of all injuries not related to falls were actually skin tears and bruises.[143]

Skin tears may involve not only the epidermis but can also be full thickness. Usually, the lesion retracts somewhat and part of the underlying tissues is visible. The diagnosis is straightforward.

Although a small percentage of patients will develop wound healing problems, most lesions will heal fairly quickly with proper wound care, including moisture-retentive dressings.

If a lesion is retracted, approximation may be attempted. However, too much tension has to be avoided because this may result in flap necrosis.[144]

PRETIBIAL LACERATION

A pretibial laceration may resemble (and actually be) a skin tear. However, the classical pretibial laceration is the result of knocking the leg or falling against a hard object. The typical lesion occurs in older adults and has a V shape. If the vascularization of the flap is minimal, flap necrosis may occur. Specific algorithms for treatment have been developed.[145] Treatment depends on the level of bacterial contamination, the viability of the skin flap, and general (overall health) and local (wound area) conditions and stretches from the use of simple dressing to reconstructive surgery.

ZOONOSES: BITE AND SCRATCH WOUNDS

Animal (and human) bites may cause minimal to very serious diseases. Aside from the numerous types of neurotoxins and other specific systemic venoms that poisonous animals may inject through their bite, being bitten may also result in serious local symptoms, including infection,[146] crush injuries, and serious tissue defects, and being scratched may lead to infections such as cat-scratch disease.[147]

Generally speaking, superficial bites and scratches, without deep puncturing through the skin, cause lesions that will rapidly heal. However, deep puncturing wounds need exploration and opening, because saliva of animals as well as humans contains a great deal of potentially harmful bacteria. Large animals may also cause crush injuries, tissue loss, and other lesions such as open fractures, tendon and nerve laceration, persistent deep infection including osteomyelitis, and so forth, which all require exploration.[148] Primary closure of a deep bite is never to be recommended, because with the bacterial contamination, serious infectious complications may readily occur. For deep bites, prophylactic antimicrobial therapy and tetanus prophylaxis are indicated.

The consequences of bites from animals such as arachnoids (i.e., spiders and scorpions) range from few symptoms to gangrenous skin necrosis or even death.[149,150] The severity of the symptoms depends on the species of the animal and the type of venom. Many cytotoxic venoms cause a dermonecrotic reaction, leading to ulceration that sometimes requires extensive treatment.

ROAD RASH, ABRASIONS, MECHANICAL BLISTERS

A sliding scale exists on the size and seriousness of abrasions and road rash. The morbidity depends on size, depth, and contamination. However, most of these injuries are small.

Depending on their depth, some punctate bleeding may be observed, but lesions may also be deeper with partial exposure of subcutaneous tissues.

Treatment primarily consists of cleansing and removing debris. Although this is certainly not always possible, an attempt is necessary because nonremoval may lead to infection or, later, a traumatic tattoo. Disinfection is often necessary, particularly in deeper lesions, and tetanus prophylaxis may be necessary as well. Prophylactic antibiotics are usually not indicated.

A simple moisture-retentive dressing may be sufficient, but in deeper lesions grafting may be necessary.

Mechanical blisters usually have only superficial lesions underneath the blister roof, and they will heal quickly. In healthy patients lesions rarely infect. If the blister roof is intact, a simple support dressing (i.e., a polyurethane film or a thin hydrocolloid) may be used to protect and support it. The blister roof will function as a biologic dressing. The blister may also be punctured (after disinfection) to allow for drainage of the wound fluid. Again, a support dressing may be needed after drainage.

CHRONIC WOUNDS: THEIR OCCURRENCE AND IMPACT

Ever since man first injured himself, the question of wound healing and how to treat wounds has been discussed. Most wounds heal promptly and without a problem, but some are very slow to heal and some just do not heal. It has been suggested that there are more than 6,000,000 persons in North America alone with wounds that have delayed healing. What is meant by a wound that does not heal promptly? The Wound Healing Society defines a "chronic wound" as one "that has failed to proceed through an orderly and timely repair process to produce anatomic and functional integrity."[1] In simpler terms, a chronic wound is one in which the healing is delayed longer than 3 weeks.

The most prevalent chronic wounds are diabetic foot ulcers, venous leg ulcers, ischemic ulcers, and pressure (decubitus) ulcers (Table 3.1), although numerous other types may be found related to medical conditions, such as sickle cell disease, rheumatoid arthritis, and inflammatory bowel disease. Certain infections and tropical diseases can result in hard-to-heal ulcers. Wounds caused by trauma or operative procedures can become chronic wounds. Even malignancies can present as wounds of the skin that are not healing. Unfortunately, the cost of treating these wounds in the United States and worldwide is estimated to be in the billions of dollars, and consumes large portions of the annual health care budget of the United States and many underdeveloped countries.[151-156] These are just the direct costs of treating patients with these chronic wounds and ulcers. The indirect costs are much higher when considering the lost workdays, inconvenience to the families, and social and mental toll on the patient and family. It has been reported that chronic leg ulcers result in 2 million lost workdays in the United States per year.[157] In the treatment of Buruli ulcer patients in Ghana, Africa, the indirect costs can be up to 70% of the total treatment cost for the patient.[158] Chronic wounds are the new global epidemic in medicine.

There are a number of known factors that influence wound healing (Table 3.2). If these factors are not recognized or addressed, many wounds will not heal appropriately or in a timely manner. Studies have shown that if a wound heals within 2 weeks, there is a good chance of a favorable cosmetic outcome,[159] but if the wound takes longer than 2 weeks to heal, there is a significant tendency for unsightly scarring[160] to occur. Unfortunately, these are not the only factors involved in recalcitrant wounds.

Chronic wounds are generally considered acute wounds that have failed to progress through the normal phases of wound healing (Table 3.3). When wounds fail to progress, they become "stuck" in the inflammatory phase of wound healing. This

TABLE 3.1
Chronic Wounds

Diabetic foot ulcers	Vasculitic ulcers
Venous leg ulcers	Traumatic wounds
Ischemic ulcers	Operative wounds
Pressure (decubitus) ulcers	Radiation wounds
Malignant ulcers	Hematologic wound
Inflammatory ulcers	Infectious wounds

TABLE 3.2
Factors Influencing Wound Healing

Systemic Factors	Wound Factors
1. Inadequate perfusion (ischemia)	1. Mechanical injury/pressure
2. Inflammatory conditions	2. Ischemic/necrotic tissue
3. Nutritional factors	3. Edema/lymphedema
4. Metabolic diseases	4. Infection/bioburden
5. Immunosuppression	5. Wound moisture balance
6. Connective tissue disorders	6. Topical agents
7. Social issues (smoking, alcohol abuse, drug abuse)	7. Ionizing radiation
8. Age	8. Low oxygen tension
9. Mental status	9. Foreign bodies
10. Hormones	10. Molecular factors
11. Ethnicity	11. Cellular factors
12. Peripheral neuropathy	12. Microcirculatory factors
13. Genetic factors	
14. Medications	

TABLE 3.3
Four Phases of Normal Wound Healing

1. Hemostatic phase—Occurs after injury, stopping bleeding and resulting in a fibrin matrix being established. Platelets are brought into the wound releasing growth factors to continue the healing process.
2. Inflammatory phase—Occurs when white blood cells and macrophages arrive in the wound. Allows for removal of foreign material and damaged matrix components. Other growth factors are delivered into the wound bed.
3. Proliferative phase—Occurs when new cells migrate into the wound and repair of the defect begins. New capillaries and matrix develop in the wound bed. Keratinocytes proliferate and close the wound by migrating over the newly formed granulation tissue bed.
4. Remodeling phase—Occurs as the cellular density of the wound returns to normal levels. The matrix is remodeled to a more normal structure with tensile strength approximating normal skin.

prolonged inflammatory state results in a wound microenvironment that is not conducive to healing. The levels of proinflammatory cytokines are significantly elevated in chronic wounds compared to healing wounds, and once the inflammatory cytokine level returns to a more normal level, the wounds begin to heal.[161,162] A second finding in the microenvironment of wounds that are not healing is that of increased levels of proteases. Numerous proteases are necessary for normal wound healing, but in the poorly healing wound, the levels of proteases remain elevated and the level of protease inhibitors is low.[5,163] Ladwig and associates showed that the ratio of proteases to the protease inhibitors is critical in the healing of pressure ulcers.[164] If this imbalance persists and the protease levels remain elevated, damage to the forming tissue matrix, reduced growth factor and growth factor receptor levels, and delayed wound healing will result.[165]

Cells in the abnormal microenvironment of the chronic wound show a decrease in mitogenic activity, fail to respond normally to growth factor stimulation, and show all the signs of senescence.[162,166] A significant number of fibroblasts from the margin of chronic venous ulcers show signs of senescence compared to fibroblasts in the skin of the thigh of the same patient.[167] Interestingly, when grown in tissue culture, neonatal foreskin fibroblasts did not show signs of senescence until exposed to wound fluid collected from the ulcer bed.[168] This implies that the inflammatory environment and cellular damaging component of the chronic wound may reside in the chronic wound fluid.

Another difference in the healing wound and the chronic wound is the difference in growth factor levels. In the chronic wound, the level of a number of growth factors has been found to be reduced.[169] Decreased production of growth factors and increased destruction of them by the increased proteases result in a wound that is not prepared to heal.

A general profile of the chronic wound can be derived from the following observations. The wound that is slow to heal has high levels of inflammatory cytokines and high levels of proteases which results in a destroyed, disrupted matrix, low mitogenic activity, and cells that are not responsive to the environment (senescent). Strategies to treat these basic problems are needed to correct these cellular imbalances and return the wound to a healing phenotype.[170]

VENOUS ULCER

Each of the particular types of chronic wounds presents its own particular problem. Venous leg ulcers can be seen in up to 3% of the population in the United States.[171,172] These ulcers have a major economic impact on the healthcare budget. Recent studies have shown that the cost to heal a single venous ulcer ranged from $1873 to $15,052.[173]

In addition to the financial cost to the patient and society, there are many uncounted costs in the patient's impaired quality of life. A large, open, draining, foul-smelling hole in the leg may well lead to social isolation, depression, and loss of positive self-image. It is unlikely that the patient with such an ulcer will want to attend or be invited to a social activity. Venous leg ulcers account for over 2 million days lost from work annually.[157] Decreased mobility and productivity are sources of additional disability, requiring family or home health care provider involvement.

The incidence of chronic venous insufficiency or postphlebitic syndrome is twice as common in women as in men. Women may develop the problem at an earlier age than men (55 years compared to 61 years of age). In the patients who have a history of thrombophlebitis, the clinical findings and symptoms of chronic venous insufficiency or the postphlebitic syndrome will develop in 20% to 50% of them within 5 years of having the disease. This unfortunately is true even if the thrombophlebitis is appropriately treated with anticoagulation or thrombolytic therapy. If a venous ulcer develops, it will do so an average of 7 years following the episode of thrombophlebitis. This is not absolute and may range from as short a period as 2 years to as long as "several" decades.[174]

The cause of chronic venous insufficiency is damage to the valves in the veins of the lower extremity. The damaged venous valves allow the blood to pool in the legs, resulting in ambulatory venous hypertension. This increased venous pressure causes the signs and symptoms of chronic venous insufficiency or the postphlebitic syndrome. It has been suggested that up to 76% of venous ulcers are the result of ambulatory venous hypertension.[175]

Thrombophlebitis is a common cause of valvular incompetence, often occurring postpartum or postoperatively. The inflammatory reaction in the vein results in scarring and fibrosis of the valves, causing incompetence. Because there is no history of thrombophlebitis in 20% to 40% of patients, other causes of damaged valves and venous insufficiency need to be investigated. Valve damage can occur after saphenous vein stripping procedures. The increase in access to the central venous system by femoral venous catheterization has resulted in valve damage. Contrast phlebography can cause valve problems. Trauma to the legs and fractures can lead to valve incompetency. Recently, an increased incidence of chronic venous insufficiency and valve damage following total knee replacement surgery has been noted (personal observation, T. Treadwell). Investigations are ongoing to define the nature of this relationship. Unfortunately, a number of patients with chronic venous insufficiency and valve incompetence have no cause that can be identified.

Although venous insufficiency is implicated in the development of venous ulceration, what actually happens at a cellular level to cause skin breakdown remains unclear. Theories postulating a causal link in respect of venous ulcer development have been proposed. Three of these are briefly described to provide an overview of current thinking.

The fibrin cuff theory involves the leakage of fibrinogen through endothelial pores that have enlarged due to the venous hypertension. The fibrinogen in the perivascular space becomes activated and converted to fibrin, forming a thick cuff around the capillary. This fibrin cuff can prevent the diffusion of oxygen and nutrients to the surrounding tissues, resulting in death of the cells and an ulcer.[176] It is interesting that dermal pericapillary fibrin cuffs have been noted in nonulcerated but lipodermatosclerotic skin.[177] Recent work has shown that these cuffs of fibrin may not be a continuous barrier around the dermal vessels. They may, however, act as a physiological barrier affecting oxygen and nutrient perfusion in the dermis.[178] Controversially, Balslev et al. stated that fibrin deposits do not play a major causal role in chronic leg ulcers following a study of 19 patients with venous insufficiency and 14 patients with ischemic ulcers.[179]

The "trap" theory involves macromolecules present in the tissues. With venous hypertension, the macromolecules become active and "trap" (inactivate) growth factors and other cytokines essential for skin repair and maintenance of skin integrity. If the skin becomes damaged in any way, the factors necessary for repair are unavailable. The result is that the skin breaks down, forming an ulcer.[180]

The leukocyte trapping theory proposes that venous hypertension causes leukocytes to accumulate in the vessels of the lower limbs, aggregate, and occlude capillaries. The leukocytes (inflammatory cells) become activated and release toxic metabolites—free radicals and proteolytic enzymes that damage the capillary endothelium and spill out into the surrounding tissue space, damaging these cells as well. The continuing destruction results in breakdown of the skin and the formation of an ulcer.[181]

Genetic factors have been noted to increase the susceptibility of some patients to develop venous ulcers. Polymorphisms of certain genes have been associated with venous ulcers.[182,183] Abnormalities of gene expression may also account for the development of venous ulcers in other patients.[184–186] Work is continuing on the clinical importance of these findings in the majority of patients with chronic venous ulcers.

Venous ulcers can be associated with other conditions. A recent study has shown the association of venous ulcers with HIV/AIDS, especially if the patient had a history of intravenous drug abuse.[187] This would suggest that the evaluation for a venous ulcer that is not responding to therapy might need to include a test for HIV, especially if there is a history of intravenous drug abuse. None of the HIV-positive patients with venous ulcers who have been treated at the Institute for Advanced Wound Care, Montgomery, Alabama, have had a history of intravenous drug abuse, and in our experience these patients must have their primary disease treated and under control before the venous ulcer will respond to any therapy.

Interesting associations between venous ulcers and other conditions have been noted. In a study by Margolis and his associates of 44,195 postmenopausal women, the incidence of venous leg ulcers and pressure ulcers was 30% to 40% less if the patients were taking estrogen replacement therapy.[188] It has been shown that estrogen therapy will decrease neutrophil chemotaxis and localization in the wound bed, decrease wound elastase activity, increase fibronectin and collagen in the wound, and increase the healing rate.[189,190] It has been suggested that treatment with intermittent topical estrogen therapy may be as effective as systemic therapy.[190]

All of these theories and associations may play a role in the ultimate formation of the ulcer, but regardless of the mechanism, the clinical picture is the same. Patients with chronic venous insufficiency or the postphlebitic syndrome all have edema of the legs (Figure 3.3), pain in the legs (usually a tightness or bursting type of pain), dermatitis of the skin of the lower legs (Figure 3.4), and subcutaneous fibrosis (lipodermatosclerosis) (Figure 3.5). Many patients will have varicose veins (Figure 3.6). A venous ulcer can follow (Figure 3.7).

In the evaluation of the lower extremity ulcer, it is most important to be sure of the etiology. Not all ulcers of the lower leg are venous ulcers. Ulcers due to vasculitis or other causes can masquerade as venous ulcers and thus not respond to therapy. Malignancy can present as a lower leg ulcer or even develop in a long-standing ulcer (Figure 3.8), and has been reported to be as frequent as 2.2 per 100 patients.[191] The clinician is encouraged to biopsy any wound that has not responded to therapy within 3 months, or one that just does not "look right" (Table 3.4).[191–193] When undertaking a chronic wound biopsy for diagnostic purposes, it is recommended that a wedge biopsy of the edge of the lesion be taken to include a portion of the normal

TABLE 3.4
Indications for Biopsy

Ulcer present more than 3 months
Ulcer unresponsive to therapy
Ulcer that "just doesn't look right"

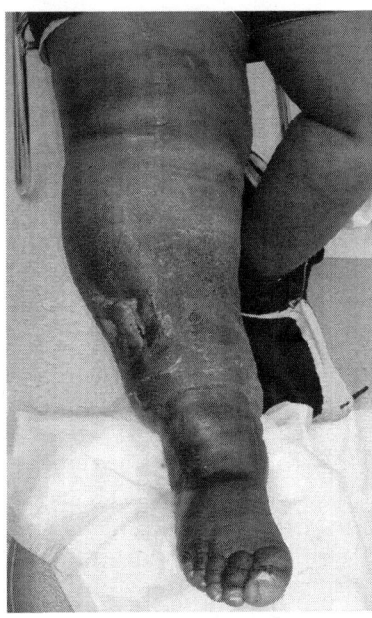

FIGURE 3.3   (Please see color insert following page 114.) Edema of the lower limb.

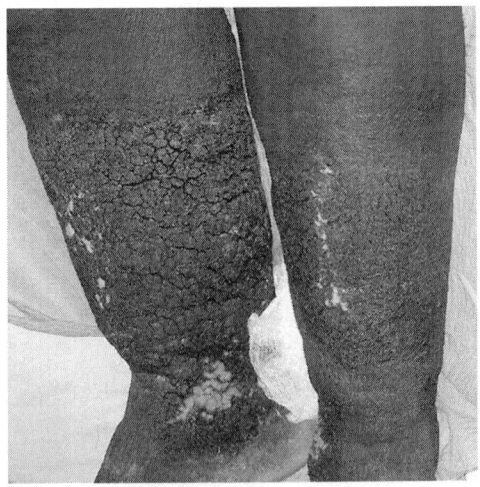

FIGURE 3.4   (Please see color insert following page 114.) Varicose eczema.

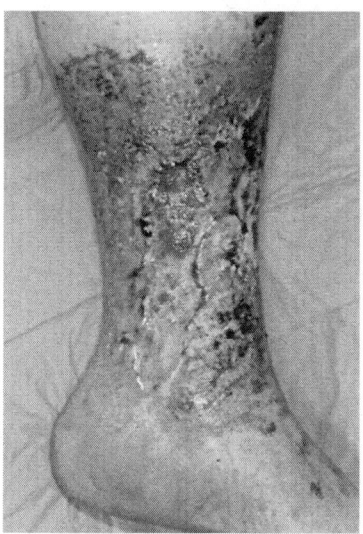

FIGURE 3.5   (Please see color insert following page 114.) Lipodermatosclerosis.

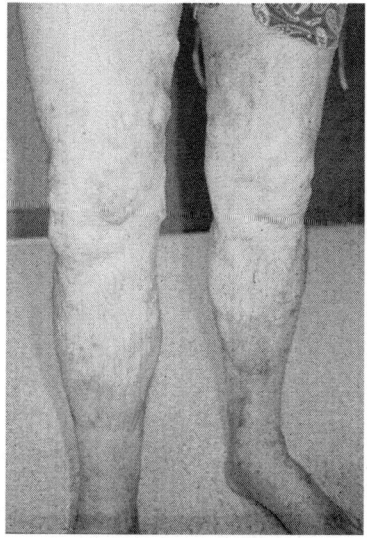

FIGURE 3.6   (Please see color insert following page 114.) Extensive varicose veins.

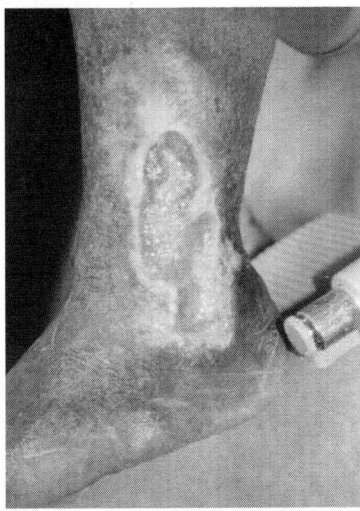

FIGURE 3.7   (Please see color insert following page 114.) Venous ulceration.

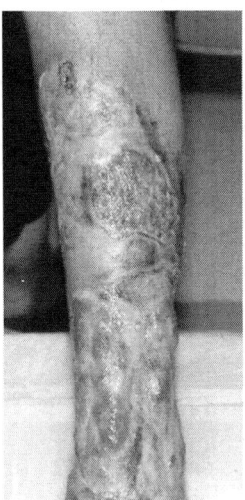 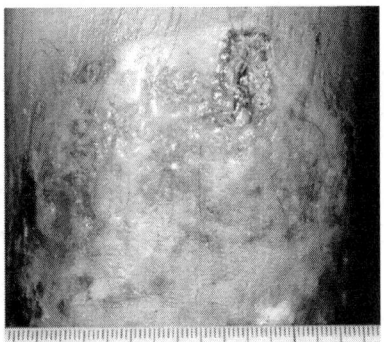

FIGURE 3.8   (Please see color insert following page 114.) Malignant ulcer masquerading as a venous ulcer.

tissue around the wound margin in addition to a portion of the wound bed. This will allow the pathologist a chance to compare the normal tissue to that in the wound bed so that an accurate diagnosis will be possible.

DIABETIC ULCER

The diabetic foot is a common and serious problem. There are approximately 16,000,000 diabetics in the United States, and approximately 15% will develop a foot problem during the course of their disease.[194,195] Diabetic foot problems account for two-thirds of the major amputations in the United States each year, or about 82,000 amputations annually.[196] This is even more important from a worldwide standpoint because the incidence of diabetes varies among ethnic groups. In Mexico 4 to 6 million people have diabetes—roughly 8% to 12% of the population. Over 75,000 limbs were lost to diabetic foot ulcers in Mexico in 2000.[197] There are 7 to 8 million people in Sub-Saharan Africa with diabetes mellitus.[198] Diabetes is the leading cause of morbidity and mortality in this area of Africa.[199] Up to 9.5% of these patients will develop a foot ulcer.[200] In Cameroon the incidence of diabetic ulcers ranges from 11% of outpatients to 25% of hospital inpatients.[201] Unfortunately, the foot ulcer is the presenting symptom of diabetes in 26% of patients.[202] Up to 33% of patients admitted to the hospital in Tanzania with a diabetic foot ulcer will have a major amputation during that visit.[203] The mortality rate in this group of people can be as high as 54%.[201] From a financial standpoint, diabetic foot problems cost the United States healthcare system $1.45 billion between 1995 and 1996.[204] In some underdeveloped countries, diabetes care and foot ulcer management can consume a significant percent of the health resource budget per year. The figures mentioned only include the direct cost to the healthcare system. As previously mentioned, the indirect cost in lost days from work, disability, and other items can add up to 70% of the cost of treating lower extremity ulcers.[158]

The etiology of diabetic foot ulcers can vary to some degree. Diabetic peripheral neuropathy has been cited as being the most important factor leading to the development of a diabetic foot ulcer. Diabetic angiopathy accounts for diabetic foot ulcers that are mostly ischemic. There is a group of mixed neuropathic and ischemic ulcers as well. In the United States, 45% to 60% of diabetic foot ulcers are neuropathic with 40% to 50% being of ischemic or mixed etiology.[205] In Africa, 87% of diabetic foot ulcers are due to diabetic neuropathy, with only 13% of foot ulcers being of ischemic or mixed etiology.[202] This is due to the very low incidence of peripheral vascular disease in Africa. Only 36% of patients in one African study had peripheral vascular disease,[202] but this number is increasing.[199]

Diabetic foot problems can present in a variety of forms, including heavy callous, infected toe, ingrown toenail, a foreign body in the foot, a seemingly insignificant injury that has not healed, an infected foot with an abscess, or gangrenous toes or foot. All of these can be seen when managing diabetic foot problems. No matter how insignificant the problem may seem when the patient is first assessed, each problem could potentially result in the loss of the patient's limb and should be taken very seriously.

In the approach to the patient with a diabetic foot problem, the whole patient must be evaluated, not just the foot, if good results are to be realized. The status of the patient's diabetes must be evaluated, because good blood sugar control is essential for healing of a diabetic foot problem. We have known for years that patients with diabetes mellitus experience wound recalcitrance. Studies now have shown the effects hyperglycemia has on the wound environment and identified the cells causing deficient wound healing. Hyperglycemia results in a decreased proliferation and differentiation of keratinocytes and fibroblasts in the wound.[206,207] Keratinocytes and fibroblasts exposed to an environment of high glucose levels do not migrate into the wound normally.[206,208] Cells in the diabetic patients do not respond normally to growth factors, thus impeding wound healing.[206,209] Cells exposed to abnormal levels of glucose fail to produce normal levels of DNA and protein, and eventually stop replicating.[210] Dr. William Marston and his group showed that lowering the hemoglobin A1c in patients with diabetic foot ulcers will result in a 16% increase in healing when treating the wound with control of infection and a moist healing environment.[211] This emphasizes the importance of control of patients with diabetes, especially those who are being treated for a diabetic foot ulcer.

Because peripheral neuropathy is a common problem in the diabetic, it is important to test for this in the patient with diabetes or a diabetic foot ulcer. Unfortunately, the peripheral neuropathy involves the sensory, autonomic, and motor nerves. The most common test to determine the presence of diabetic neuropathy is the Semmes-Weinstein nylon monofilament test. The basis for the test is to press a monofilament nylon that requires 10 grams of pressure to flex against the sole of the foot. If the patient cannot feel the filament at the time it flexes at two or three locations on the bottom of the foot, the diagnosis of peripheral neuropathy can be made. It is important to avoid scars and calluses when doing the test to avoid false-positive results. This is a simple and reproducible test that should be done for all diabetic patients at every visit until the diagnosis is made. If a monofilament is not available, as is the case in many underdeveloped countries, our experience has shown that an extended point of a ballpoint pen will substitute. Gently press a ballpoint pen on the skin of the foot until it indents the skin just slightly. If the patient is unable to feel the point of the pen, a diagnosis of peripheral neuropathy can reasonably be made.

Another method of detecting peripheral neuropathy is by using a 128 Hz tuning fork. Touch the vibrating tuning fork to the toes or to the metatarsal head area of the foot. If the patient does not feel the vibration, the diagnosis of peripheral neuropathy can be made.[212]

Because the neuropathy involves all of the nerves, symptoms may be widespread and varied. Symptoms from sensory nerve involvement include paresthesias in a stocking-like distribution, superficial and deep shooting pains, loss of vibration sense and proprioception, and eventually a totally insensate foot. This is a major problem because the patient cannot feel injuries or other problems with the foot that would be very painful to anyone without neuropathy. This leads to delays in recognition of potential problems, in underestimating the severity of the injury or problem, and in delays seeking medical help. Education for patients with diabetic neuropathy is critical if the morbidity of diabetic foot ulcers and injuries is to be reduced.[213]

The changes in the autonomic nerves are just as severe but not as dramatic. Autonomic neuropathy results in dryness and fissuring of the skin of the foot and toes which can predispose the patient to an infection. It also results in loss of autonomic nerve control of the small vessels of the extremities and other microcirculatory disruptions. This can result in the patient's feet taking on a deep red hue when they are in a dependent position. This condition is called dependent rubor and is seen in patients with lower extremity ischemia or autonomic neuropathy. To the uneducated eye, this may have the look of cellulitis of the foot. To distinguish between the two, elevate the patient's foot. In patients with ischemia or autonomic neuropathy, the foot will lose its red color on elevation, whereas in patients with cellulitis, the limb will remain discolored.

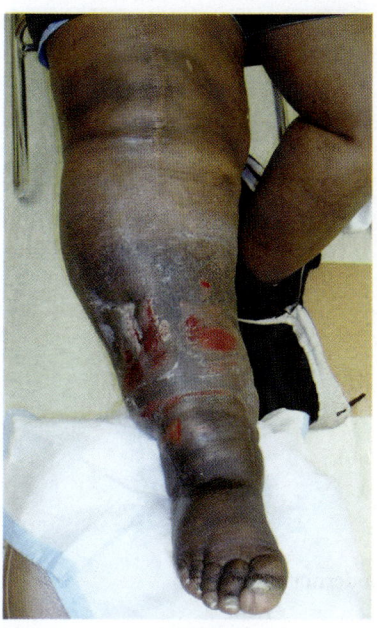

COLOR FIGURE 3.3   Edema of the lower limb.

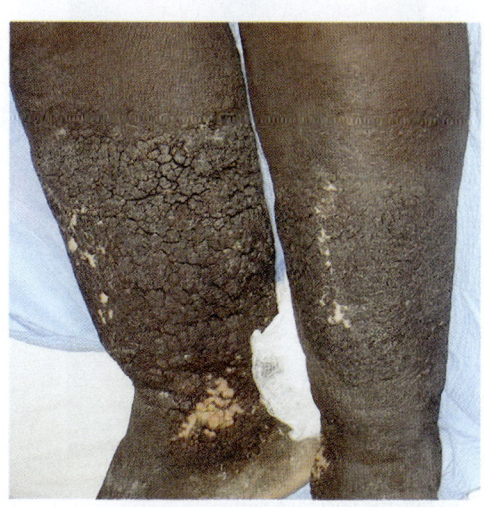

COLOR FIGURE 3.4   Varicose eczema.

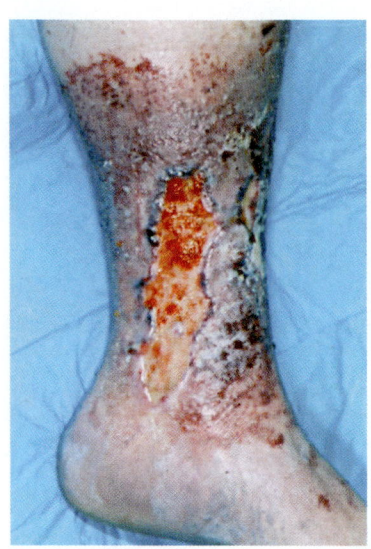

COLOR FIGURE 3.5 Lipodermatosclerosis.

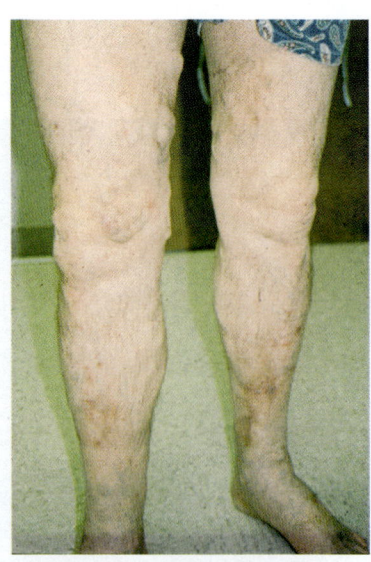

COLOR FIGURE 3.6 Extensive varicose veins.

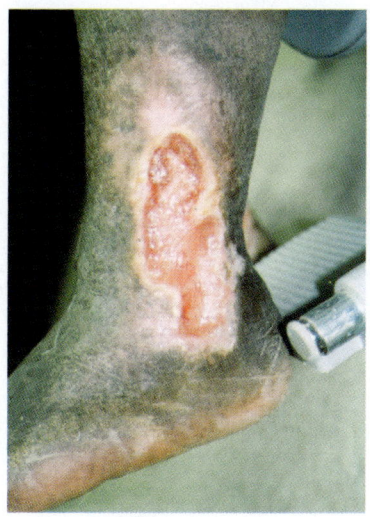

COLOR FIGURE 3.7   Venous ulceration.

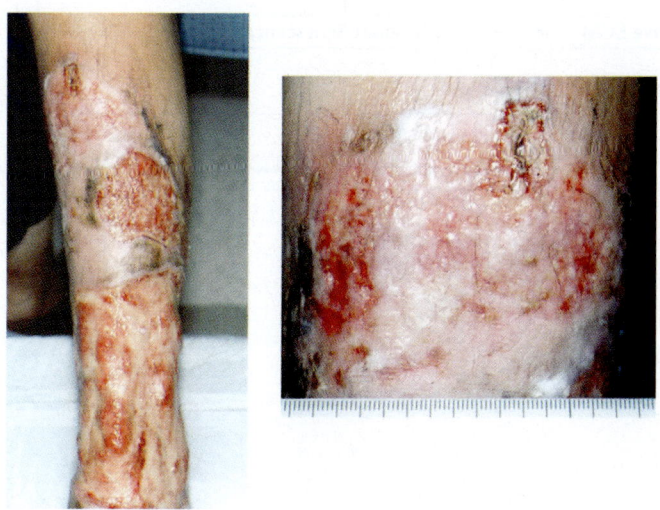

COLOR FIGURE 3.8   Malignant ulcer masquerading as a venous ulcer.

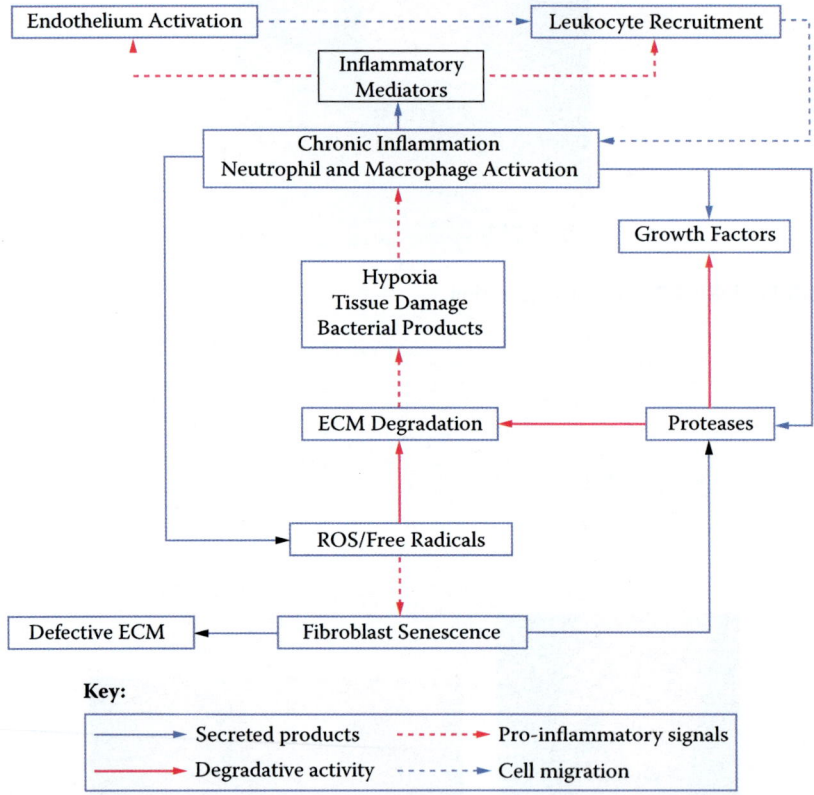

FIGURE 5.2 Generation of wound chronicity.

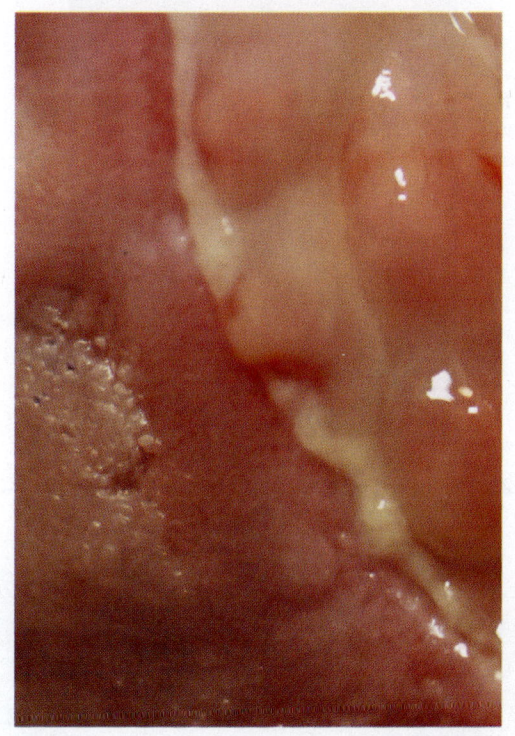

COLOR FIGURE 8.1 Slough overlapping the wound edge.

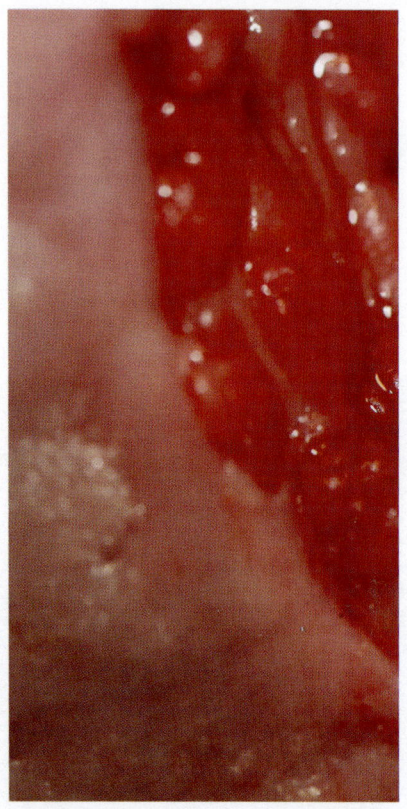

COLOR FIGURE 8.2   Bright red wound border is a positive sign of host control of wound biofilm.

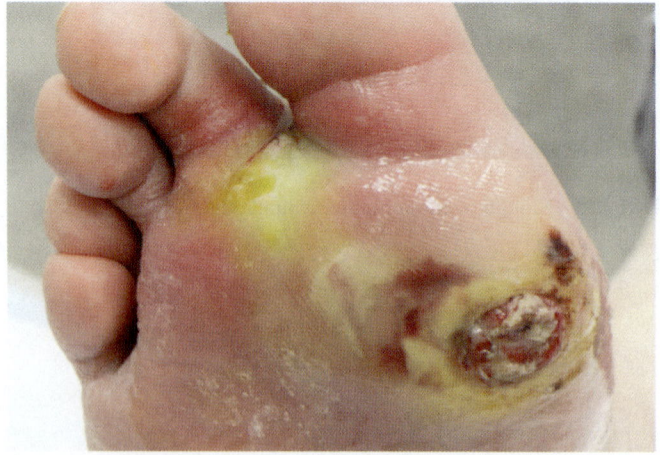

COLOR FIGURE 8.3   Plantar hyperkeratosis in an immobile patient.

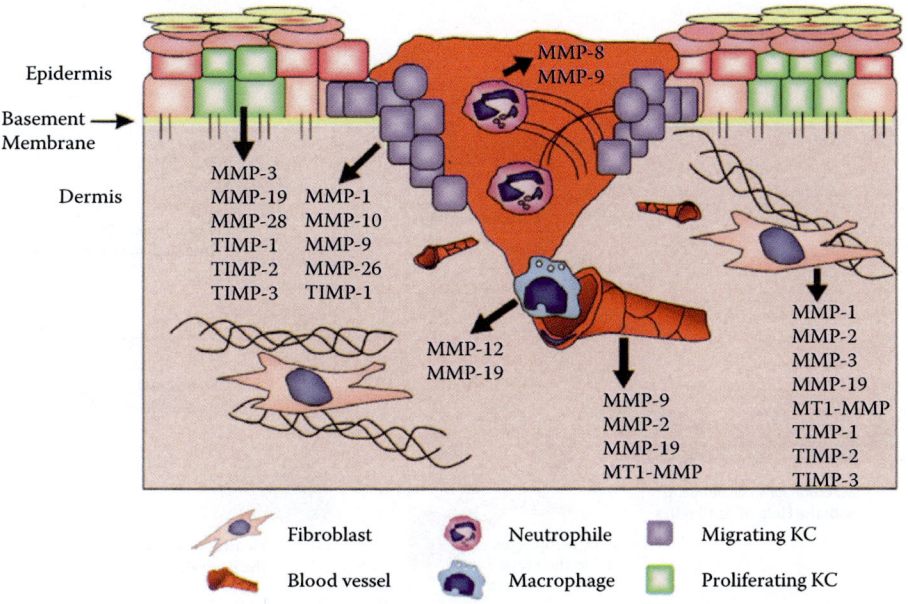

COLOR FIGURE 9.2 The expression and cellular source of matrix metalloproteinases (MMP) and tissue inhibitors of MMPs (TIMP) in acute wounds. (Toriseva, M., and Kahari, V.M. (2009) Proteinases in cutaneous wound healing. *Cell Mol Life Sci* 66:203–224. With permission)

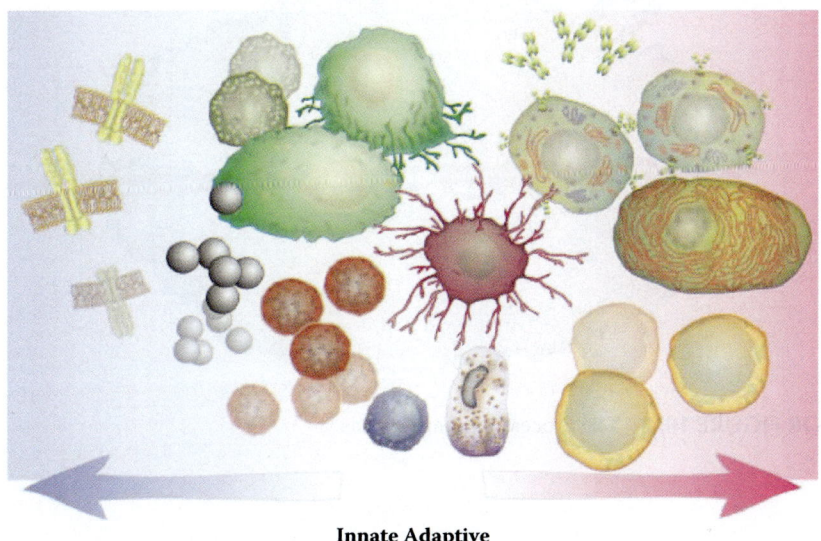

Innate Adaptive

COLOR FIGURE 10.1 Innate and adaptive compartments of the immune system.

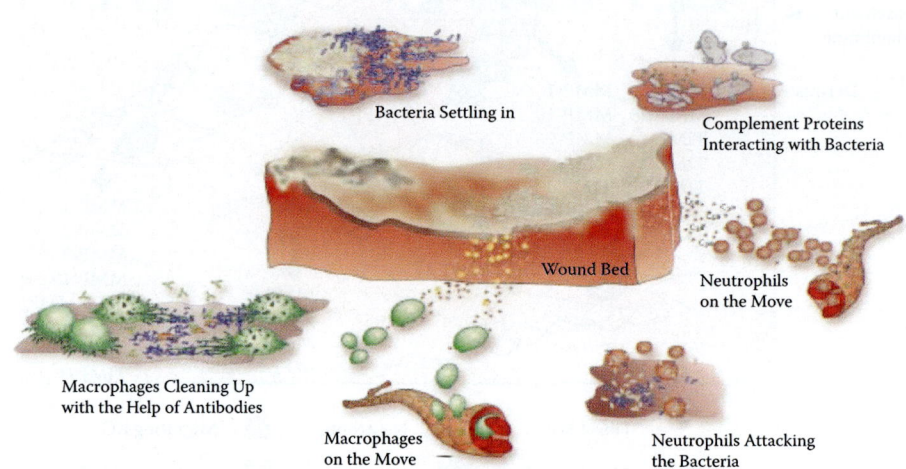

COLOR FIGURE 10.3   Battle plan: the immune defense of the wound.

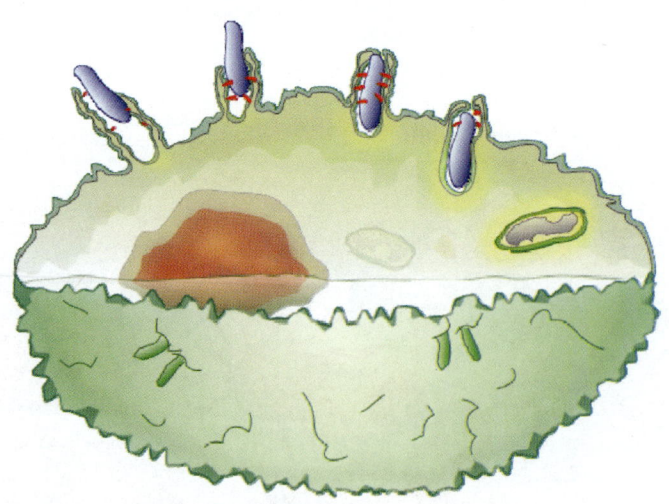

COLOR FIGURE 10.4   The process of phagocytosis.

A most devastating and severe outcome of autonomic and sensory neuropathy is the neuropathic or Charcot foot. This occurs when the effects of the neuropathy result in weakening of the bones of the mid-foot, with fracture and collapse. Many deformed feet are the result of simple fractures that have been ignored because of the lack of pain.[214] Many physicians mistakenly believe that once a Charcot deformity develops, treatment is not possible. Treatment is not only possible but can be successful. With complete immobilization of the limb, most will heal. Patience is a virtue because neuropathic fractures require two to three times longer to heal than bone with normal sensation.[214] After the bones become stable, reconstruction by a surgeon skilled in the reconstructive techniques can result in limb salvage and mobility.[215]

The involvement of the motor nerves results in wasting of the intrinsic muscles of the foot with a dorsal subluxation of the digits due to a flexor–extensor imbalance. This causes an abnormal distribution of the weight on the plantar aspect of the foot causing pressure points under the metatarsal heads. This can lead to development of ulceration (mal perforans ulceration).

Once an ulcer develops, the most serious complication is an infection. Unfortunately, the classical signs of infection are masked by the effect of the diabetes mellitus resulting in delay in diagnosis and delay in seeking medical attention. If the infection progresses, osteomyelitis can result. The incidence of osteomyelitis can vary widely, but the probability of a patient developing osteomyelitis is close to 15%.[216]

How is the diagnosis of osteomyelitis made? The "gold standard" for the diagnosis of osteomyelitis is a bone biopsy and culture.[217] Swab cultures have no diagnostic significance in the osteomyelitis.[218] Magnetic resonance imaging is the radiologic procedure of choice, being accurate in 89% of the cases.[219] For a comprehensive review of osteomyelitis of the diabetic foot, the reader is referred to the article by Sonia Butalia and coauthors.[217]

Circulation in the diabetic foot must be evaluated because up to 50% of patients with a diabetic foot ulcer will have peripheral vascular occlusive disease.[205] The vascular disease seen in the diabetic involves both the large vessels (macroangiopathy) and the small vessels (microangiopathy) in the extremity. The macroangiopathy in the diabetic patient is more diffuse, involving multiple arterial segments compared to the patient without diabetes. It frequently involves collateral vessels, making the ischemia more profound. The occlusive disease involves the infrapopliteal vessels more frequently than in the nondiabetic. Unfortunately, the involvement is frequently bilateral, enough so that it has been named the "second limb syndrome."[220–222] The microvascular disease noted in the small vessels of the diabetic involves thickening of the basement membrane of the vessel. The thickening extends to the adventitial side (outside) of the blood vessel. This results in a thicker vessel wall, which may impede the diffusion of oxygen and nutrients to the tissue but does not occlude the lumen of the vessel. Even despite the presence of the microvascular disease common in the lower extremity of the diabetic patient, limb revascularization is just as successful in the diabetic as in the nondiabetic.[223] The current use of endovascular techniques has made the risk of revascularization in these seriously ill patients safer.

The role of prevention is rarely mentioned in the management of the diabetic foot; however, it should be one of the most important topics discussed. Because the treatment of diabetic foot ulcers and their complications is such a large economic burden, perhaps more time and effort should be spent preventing them. To accomplish that, many factors have to be considered. A surveillance system must be put in place to collect data about the complications of the diabetic foot. Identification of the most important factors for each population is critical. Characterization of individual populations at higher risk for the development of diabetes mellitus and its complications must be done. Once these programs are in place, only then can appropriate, population-specific preventative measures be outlined.[224] For example, an association between smoking and the onset of Type II diabetes mellitus has been noted. Patients who currently smoke or have smoked in the past and quit still make up 12% of the newly diagnosed cases of Type II diabetes mellitus.[225] This information should lead to more stringent screening of smokers for the disease. Another example is that the diabetic with predominately peripheral vascular disease needs a different intervention and prevention strategy than does the diabetic with predominately peripheral neuropathy. Until the importance of this information is appreciated, the reduction in the morbidity, mortality, and cost of treatment of diabetic foot problems is unlikely to significantly decrease.

PRESSURE ULCERS

Pressure ulcers have been a problem through the ages. In early times these ulcers were most commonly seen in young people with wasting diseases, mostly because people with chronic diseases did not live long enough to develop one. Originally these ulcers were called decubitus ulcers from the Latin *decumbere*, which means "to lie down," echoing the most frequent etiological link. The majority of the pressure ulcers today are found in immobile patients and spinal cord injury patients. The problem is of major magnitude today. Two and one half million patients in acute care facilities are treated for pressure ulcers each year.[226] Sixty thousand patients die each year of complications of pressure ulcers.[227] Amazingly, the incidence of pressure ulcers in hospitalized patients increased 63% between 1993 and 2003.[228] A seemingly forgotten group of patients suffering from pressure ulcers are children. Pressure ulcers can be a problem in 3% to 4% of pediatric hospital admissions.[229] The economic burden of pressure ulcers is tremendous, with estimated cost of $11 billion per year to treat them.[227]

The development of a pressure ulcer in the hospital setting is a significant problem for the patient in that it can delay discharge from the hospital, predispose the patient to potential serious infections, and cause pain related to the ulcer and its treatment.[227] There is a significant mortality associated with developing a pressure ulcer in an acute care setting.[227]

The factors predisposing one to develop a pressure ulcer are limited mobility, reduced sensation, moisture, friction, and shear forces (Table 3.5). Friction and shear forces are the factors most commonly responsible for superficial (Stage I and Stage II) pressure ulcers. This is compounded by the presence of moisture on the skin in the form of perspiration or incontinence. Friction occurs when the skin slides or is pulled

TABLE 3.5
Predisposing Factors for Pressure Ulcer Development

Limited mobility	Friction
Reduced sensitivity	Shear forces
Moisture	Malnutrition

along a surface, oftentimes a bed sheet. The presence of moisture makes the skin more likely to be damaged during this event. Pressure, limited mobility, and shear forces are the factors most responsible for the deep tissue injury causing Stage III and Stage IV ulcers.

Identifying patients at risk for developing a pressure ulcer is important so that preventative measures can be directed at the population most likely to benefit. The various assessment tools include the Braden Scale and the Norton Scale.[226] Each of these assesses different parameters in an attempt to identify and grade the patients at risk of developing pressure ulcers. The parameters include sensation, activity, mental state, mobility, moisture and continence, friction, and nutrition in various combinations. This will be part of the overall patient assessment for formulation of a "Prevention Pathway" to address each risk area and use individualized interventions to minimize the risk of developing a pressure ulcer.[230]

One of the difficult problems encountered in the discussion of pressure ulcers is the staging. Staging is defined as "an assessment system that classifies pressure ulcers based on anatomic depth of soft tissue damage."[231] It is important to note that the main measurement is the depth of the wound and not necessarily the size. It is important to stage ulcers so that treatment outcomes can be fairly determined and so that communication can be accurate. Unfortunately, significant confusion and limitations arise when attempts at staging are made. There are currently four stages recognized by the National Pressure Ulcer Advisory Panel (Table 3.6). Stage I is a nonblanching erythema of the skin. This is supposed to represent a minor skin problem that should respond to local care; however, many times the nonblanching sign means the skin has lost its blood supply and will continue to necrosis. At other times it may imply a lesion in which the subcutaneous tissue and possibly the muscle have already been damaged (a deep tissue injury). It is known that the first tissues damaged by pressure are the deep tissues.[232] New evidence supports the fact that pressure ulcers resulting from deep tissue injury develop from the inside out as opposed to the outside in.[233] As a result, patient documentation records that the patient has a

---

TABLE 3.6
Stages of Pressure Ulcers

Stage I	Redness or discoloration of the intact skin that does not blanch on pressure
Stage II	Partial thickness loss of the epidermis, dermis, of both
Stage III	Full thickness skin loss including subcutaneous tissue down to but not through the fascia
Stage IV	Full thickness skin loss with damage to muscle, bone, and associated structures
Deep tissue injury	Purple or maroon localized area of discolored intact skin or blood-filled blister caused by damage to the underlying soft tissue
	May evolve and become covered with eschar
	Even with optimal treatment, the wound may evolve rapidly exposing additional layers wissue
Unstageable	Ulcer with intact skin or covered with necrotic tissue
	Unless the skin has been violated, no depth can be determined thus no staging can be done

---

Stage I pressure ulcer and that within 48 hours the skin has turned black, the edge of the eschar loosens, and a large amount of foul-smelling drainage pours out. Much to everyone's dismay, the patient now has a Stage IV pressure ulcer. This also implies that the Stage I ulcer progressed to a Stage IV ulcer. Stage I or Stage II ulcers *do not* progress to become Stage III or Stage IV ulcers.[233] The pathophysiology of each type of ulcer is different because friction and moisture, the etiology of Stage I and Stage II ulcers, do not contribute to the development of Stage III or Stage IV ulcers. Because of the deficiencies in staging, the patient had a Stage IV or deep tissue injury ulcer from the beginning. The National Pressure Ulcer Advisory Panel has recently published a revised staging scheme that should help identify those patients with deep tissue injury.[234]

STAGE II PRESSURE ULCERS

These ulcers are partial-thickness ulcers that involve damage to the epidermis and superficial dermis. This is seen as an abrasion, blister, or shallow crater in the skin. As noted, these lesions do not progress to deeper wounds because they are caused by friction and moisture and not pressure.

STAGE III PRESSURE ULCERS

This type of ulcer is full thickness, involving loss of all layers of the skin and subcutaneous tissue. The damage can go down to but not through the fascia. Even patients with Stage III ulcers can have impressively large defects, especially if there is a large amount of subcutaneous tissue which has been lost.

STAGE IV PRESSURE ULCERS

These ulcers are full thickness, involving all layers of the skin, the underlying subcutaneous tissue, the muscle, and the bone. The the ligaments and joint capsule can be involved, resulting in an open joint.

OTHER PRESSURE ULCERS

There are pressure ulcers that should be declared "unstageable." Ulcers with intact skin or eschar cannot be and are not to be staged. Staging is based on the depth of the wound. The appropriate time for staging of this type of pressure ulcer is after debridement when the true extent of the tissue damage can be assessed.

Reverse staging is a concept noted to occur as the wound heals. The ulcer is a Stage IV ulcer but with treatment begins to heal. Soon the bone is covered and the healing is above the fascial level. According to the staging definitions, the ulcer is now a Stage III. Unfortunately, the current staging systems do not allow for healing or for reverse staging.[231] The healing wound must be identified as a "healing Stage IV pressure ulcer" until it completely closes.

PREVENTION OF PRESSURE ULCERS

Prevention of pressure ulcers is the goal for healthcare providers and institutions. Unfortunately, the optimal methods of pressure ulcer prevention have yet to be identified. One of the more obvious methods is to decrease the pressure on the bony prominences when the patient is lying or sitting. Pressure-reducing surfaces may be static in which the mattress, mattress overlay, or wheelchair pad are filled with a deformable material to shift the weight off the bony prominences. They may be dynamic support surfaces in which the surface automatically shifts the pressure for the patient.[227] In a large randomized control trial of 447 patients, no difference was found in the incidence of pressure ulcers between the patients on static surfaces or those on dynamic surfaces,[235] although others found the dynamic surfaces to be better.[236]

Repositioning has been the mainstay of pressure ulcer prevention. Turning the patient at regular intervals has been thought to reduce the pressure and maintain the circulation to the tissues under susceptible bony prominences. Unfortunately, there is no science to confirm the utility of repositioning the patient every 2 hours even though that is the most frequent interval selected. The only randomized control trial evaluating different positions in pressure ulcer reduction compared patients placed in the 30° tilt position with use of pillows compared to patients placed on their sides. There was no decrease in the incidence of pressure ulcer development between the two groups.[237]

The moisture environment of the skin is thought to play a major role in the development of pressure ulcers. Moisture associated with friction is one of the risk factors for developing Stage I or Stage II pressure ulcers. It is interesting that when taken alone, urinary incontinence, a chief cause of moisture in patients at risk for pressure ulcer, is not an independent predictor for the development of an ulcer.[238] Moisture does not contribute to the formation of Stage III and Stage IV ulcers.[233] On the other hand, it is suggested that dry sacral skin can lead to the development of a presacral ulcer.[239] A study involving skin treatment with a hyperoxygenated fatty acid preparation reduced the incidence of pressure 10% compared to the placebo group.[240] The effect of the fatty acid preparation on the skin is thought to be protection against friction and pressure.[240] Other special formula moisturizers did not show any statistical improvement in pressure ulcer development compared to standard, commonly used lotions.

Nutritional support for patients at risk for pressure ulcer development seems reasonable but is often ignored. Unfortunately, the evidence for nutritional status and the prevention and healing of pressure ulcers is marginal at best. Only one trial has shown a decreased risk of developing a pressure ulcer with nutritional supplementation.[241] Even if nutritional supplements are found to be of benefit to undernourished patients, the exact nutrients that would be of the most benefit are not known.[241]

There are other factors that play a role in the prevention of pressure ulcers. Margolis et al. showed a 30% to 40% decrease in pressure ulcer and venous ulcer development in postmenopausal patients taking estrogen replacement.[188] Are there other factors yet unknown that will help decrease the development of pressure ulcers in the high-risk patient? One can see that a significant amount of research needs to be undertaken before we can determine which individuals require intensive preventative measures to ensure a reduction in pressure ulcer development. There is also a need for more information as to which preventative measures will be the most beneficial and cost-effective for patients at high risk for developing pressure ulcers. It has been said, "In the case of pressure ulcers, coexistent impairments in mobility, nutritional status, and skin health often conspire together to produce ulcers."[227]

Our objective should be education of healthcare providers about the state-of-the-art evaluation and prevention measures while we forge ahead until the best preventative measures are identified.

ARTERIAL ULCERS

Ulcers due to peripheral vascular occlusive disease result when the narrowed (stenotic) or occluded arteries cannot supply the amount of blood needed by the tissues of an extremity. Once the tissue becomes damaged, usually from a minor injury, the skin may undergo necrosis to a greater extent than would be expected from the magnitude of the injury. Because of a deficiency of blood flow, the wound fails to heal. Arterial or ischemic ulcers usually form at the most distal point on an extremity, generally the fingers or toes. They may be extremely painful as the tissue is dying from lack of circulation. If the circulation worsens or infection occurs, gangrene of the extremity can occur.

The incidence of peripheral vascular disease is difficult to determine, because a number of patients have asymptomatic vascular disease.[242] Peripheral vascular disease rarely exists as a separate entity and coexists or is a comorbidity of many diseases including diabetes mellitus, hypertension, cardiac disease, and renal disease, to name a few. For this reason, evaluating the patient with a suspected arterial ulcer and peripheral vascular disease must include a complete history and physical evaluation (Table 3.7). This must not only include evaluation of the current problem, but also all risk factors associated with vascular disease and its many manifestations currently and in the past.

Symptoms from stenosis or occlusion of the arteries only occur when the demand for blood by the tissue is greater than the amount of blood that can be supplied. The first symptom of which patients with peripheral vascular disease complain is intermittent claudication. This comes from the Latin, *claudicare*, which means to limp.[243] The pain occurs when the arteries cannot supply the blood needed by the tissue, usually during exercise. The symptoms of claudication, which include an aching pain, tightness, cramping, or tiredness, can occur in the calf, thigh, or buttock depending on the site of the arterial stenosis or occlusion. With rest the pain will be relieved. One of the more interesting facts is that the symptoms will occur when the patient walks the same distance each time. After stopping to rest, the patient may again walk the same distance before the pain returns. The distance

TABLE 3.7
Evaluation of Patient with Peripheral Vascular Disease

History of present illness
Past medical history
Medications and other drugs
Family history
Social history
Occupational history
Complete review of systems
Complete physical examination
Assessments of circulation
- Noninvasive
- Invasive

the person can walk is partly determined by the severity of the arterial occlusion, the adequacy of the collateral circulation, and the intensity of the activity. The more intense the activity, the shorter the time or distance until symptoms occur.

If the disease progresses and the arterial insufficiency worsens, the patient may develop a constant pain in the extremity called rest pain. This pain is generally felt to be in the toes or foot. The pain is made worse by cold or limb elevation. It is not unusual for the patient to describe the onset of severe, unrelenting pain in the extremity within an hour or so of going to bed. After the pain develops, the only relief will be if the patient gets up and walks around for a period of time or hangs the leg over the side of the bed. This allows gravity to assist the lower limb circulation and relieve the pain. If the examiner closely questions the patient about sleeping habits, he may find that the patient sleeps in a chair with his foot dependent so that he can get more than a few minutes of sleep. Rest pain is an indication of a limb that is in danger of being lost. Prompt evaluation and revascularization are indicated.

Physical examination of the patient is that part of the evaluation when the skill of the healthcare provider is demonstrated in detecting signs of a disease. Unfortunately, it is becoming a lost art because of the tendency to skip the physical examination and start ordering tests. This shows a lack of understanding of the patient's disease and a lack of skill on the part of the examiner (a deficiency called hyposkillia).[244]

The physical examination of the patient suspected of having peripheral vascular disease must include astute observation. The condition and color of the skin of the extremities should be noted. One sign that is noted in many patients with significant peripheral vascular disease is rubor. The limb has a dark red color, especially when it is in the dependent position. The color is due to the increased oxygen extraction from the blood as its flow through the extremity is slowed due to the arterial disease and the dependent position. Unfortunately, many times the untrained eye will mistake this for cellulitis. The way to tell the difference is to elevate the extremity. If the color is due to vascular disease, it will fade quickly and the limb will become pale. If the color is due to cellulitis, the limb will remain red on elevation.

The simple technique of feeling the legs to determine the temperature can be instructive. If one limb is cooler that the other, a decrease in the blood flow in that leg may be found, especially if that is the symptomatic leg. If both legs are cool, the finding may not have any clinical significance, but additional evaluation is needed to be sure bilateral vascular disease does not exist. If the extremity is warm to the touch compared to the opposite extremity, this may indicate the presence of inflammation. When checking the temperature of the extremity, it is most accurately done using the back of the examiner's fingers or hand rather than the palmar surface. One should be able to detect differences in temperature between extremities of 2 degrees with this technique.[245]

The visual examination of the extremities should include noting the presence or absence of ulcers or gangrene, the exact location of the ulceration or gangrenous tissue, obvious foot deformities, calluses, previous operations, or infections. If ulceration is present, the exact location and characteristic of the ulcer should be noted and recorded. The presence and extent of edema in the extremity should be noted.

The next task should be to evaluate the circulation in the extremity. The goal in wound care is to determine if the patient has circulation adequate to support wound healing response to therapy. This can be done by palpating the arterial pulses. One should not confine the evaluation to the ankle pulses. Include the upper extremity pulses, the carotids, the abdominal aortic pulse, and all the lower extremity pulses (femoral, popliteal, dorsalis pedis, posterior tibial, and peroneal). The reader should refer to a suitable textbook of physical diagnostic techniques for the exact methods of palpating arterial pulses. The dorsalis pedis artery on the dorsum of the ankle and the posterior tibial artery behind the medial malleolus are the vessels commonly examined. One should not forget to evaluate the distal branches of the peroneal artery, which can be found just anterior to the lateral malleolus at the ankle. This vessel is generally overlooked in evaluations of the vascular tree, but it is a main source of collateral blood flow to the dorsalis pedis and posterior tibial arteries. This lower extremity vessel has a different embryologic origin than the anterior and posterior tibial arteries and is one of the last to be affected by atherosclerosis, especially in the diabetic.[246] If the posterior tibial artery is occluded, flow in the peroneal artery is increased significantly, and its pulse often can be palpated anterior to the lateral malleolus. A strong Doppler signal can generally be found in this area even in the presence of a patent posterior tibial artery.

Because palpation of pulses is not an exact science, the next step in the evaluation of the vascular tree usually involves a noninvasive Doppler study. This evaluation can be done at the bedside with a small, handheld Doppler or in a vascular laboratory with a more complex Doppler device. The basic information collected with either device will provide useful information about the status of the circulation. The Doppler can be placed over the sites of arterial pulses and the signal noted. An experienced evaluator can identify the normal triphasic waveform heard with each heartbeat. As the circulation in the vessels decreases, the waveform may become biphasic or even monophasic, indicating only minimal flow in the vessel. The reader should again refer to a textbook of vascular evaluation to learn the details of the noninvasive Doppler examination.[243]

After the Doppler signals are located and identified, the next step is to use the Doppler to take a systolic blood pressure at the ankle with a standard blood pressure cuff. This pressure measurement, along with the brachial artery blood pressure in the arm, will allow the calculation of the ankle/brachial index (ABI). This index number is used to determine the relative severity of the ischemia in the extremity. An index above 0.8 is generally considered to be acceptable, an index below 0.5 is considered critical ischemia, and an index between 0.5 and 0.8 is considered to be reflective of symptomatic ischemia. Unfortunately, there are factors that can make the ABI less than accurate. Any factor that would invalidate the taking of the blood pressure in either the arm or leg will make the numbers suspect. The most frequent problem is a lack of skill in the examiner. Inappropriate placement of the blood pressure cuff on the arm or leg can lead to false readings. When determining the brachial systolic blood pressure, the examiner should use the Doppler over the brachial artery instead of a stethoscope. The Doppler is more sensitive in detecting blood flow and will prevent an inaccurate reading of the brachial pressure.

The most frequently discussed problem causing a potential error in the ABI is the rigidity of the blood vessels in the extremity. For an accurate pressure to be measured, the blood pressure cuff must compress the vessels, arresting the flow of blood. If the vessels are more rigid than normal (usually due to calcification of the vessel walls), it requires more pressure to compress the vessels and arrest the flow. This can result in an erroneously high blood pressure reading. At times the vessels are so calcified that the blood flow cannot be stopped even with the blood pressure cuff fully inflated.

If there are questions about the accuracy of the ankle blood pressure readings, pressures can be determined in the toes, allowing for a calculation of a toe/brachial index. The vessels in the toes are generally compressible, making this measurement a reasonable alternative. The pressure in the toes is generally less than that in the arm or ankle. For that reason a toe/brachial index of <0.7 is considered abnormal.[247]

Other tests for evaluation of the circulation are available and can provide useful information if used appropriately. Transcutaneous oxygen measurements are helpful in evaluating the microcirculation and its ability to deliver oxygen to the tissues in question. It is imperative to utilize the transcutaneous oxygen measurement determination before and after administering oxygen to the patient to see if hyperbaric oxygen therapy might be useful. Plethysmographic techniques are useful in detecting blood flow when the Doppler is not adequate. Arterial duplex scanning can provide information about the levels of arterial obstruction. Magnetic resonance angiography (MR angiography) and certain computed tomography (CT) techniques can show areas of arterial stenosis or occlusion. Even though they do not provide the detail of contrast angiography (considered the "gold standard" for evaluation of the arterial tree), MR angiography and CT angiography do have a place because arterial catheterization is avoided. It must be remembered that contrast material is still involved in these techniques just as it is in contrast angiography.

Once the patient with an extremity ulcer is found to have significant peripheral vascular disease, revascularization is indicated. Numerous endovascular and operative options for the treatment of the ischemic extremity are currently available. The patient should be referred to the vascular surgeon for treatment. I would suggest one word of caution in the patient with peripheral vascular disease and an extremity ulcer. Once the patient has a revascularization procedure, do not forget to have him return to the wound center to have his or her ulcer treated. Too many times it is assumed that just restoring the circulation to the extremity will result in the wound healing. This could not be further from the truth. Appropriate therapy of the ulcer is necessary if the ulcer is to heal.[248] What has happened with the revascularization procedure is that you have converted a patient with a chronic wound and poor circulation to a patient with a chronic wound and good circulation. There is nothing more frustrating than to have a patient with a successful revascularization procedure eventually undergo a major amputation just because the wound was not appropriately managed.

As we continue to search for evidence-based care for patients with chronic wounds, it is imperative that we remember to treat the patient and not just the wound. A quotation from Dr. James Peck emphasizes this, "It is the individual patient who we treat, not the disease. It is the patient who recovers or dies, not the illness."[249] Treating the patient and not just the wound will mean that the patient is involved in deciding what the best care is for him or her. We, as the caregiver, may not feel the patient has chosen wisely, but we must have the flexibility to manage the patient's wound within the best interests of the patient. We should remember the quotation from the title of a recent article, "It is me who endures, but my family that suffers."[250] We must never forget the influence a chronic wound or other chronic illness has on the entire family. Every therapeutic approach to the treatment of a chronic wound must be designed with the effect it will have on the entire family in mind. When we can approach a patient with a chronic wound with those thoughts in mind and combine the best care for the patient with the best science available, then we shall be truly ready to face the "new epidemic in medicine"—the chronic wound.

REFERENCES

1. Lazarus GS, Cooper DM, Knighton DR, et al. Definitions and guidelines for assessment of wounds and evaluation of healing. *Wound Repair Regen* 1994; 2(3):165–170.
2. Wlaschek M, Scharffetter-Kochanek K. Oxidative stress in chronic venous leg ulcers. *Wound Repair Regen* 2005; 13(5):452–461.
3. Saarialho-Kere UK, Kovacs SO, Pentland AP, et al. Cell-matrix interactions modulate interstitial collagenase expression by human keratinocytes actively involved in wound healing. *J Clin Invest* 1993; 92(6):2858–2866.
4. Ulrich D, Lichtenegger F, Unglaub F, et al. Effect of chronic wound exudates and MMP-2/-9 inhibitor on angiogenesis in vitro. *Plast Reconstr Surg* 2005; 116(2):539–545.
5. Tarnuzzer RW, Schultz GS. Biochemical analysis of acute and chronic wound environments. *Wound Repair Regen* 1996; 4(3):321–325.
6. Yager DR, Zhang LY, Liang HX, et al. Wound fluids from human pressure ulcers contain elevated matrix metalloproteinase levels and activity compared to surgical wound fluids. *J Invest Dermatol* 1996; 107(5):743–748.
7. Baker EA, Kumar S, Melling AC, et al. Temporal and quantitative profiles of growth factors and metalloproteinases in acute wound fluid after mastectomy. *Wound Repair Regen* 2008; 16(1):95–101.
8. Loots MA, Lamme EN, Zeegelaar J, et al. Differences in cellular infiltrate and extracellular matrix of chronic diabetic and venous ulcers versus acute wounds. *J Invest Dermatol* 1998; 111(5):850–857.
9. Teitelbaum JE, Walker WA. Nutritional impact of pre- and probiotics as protective gastro-intestinal organisms. *Annu Rev Nutr* 2002; 22:107–138.
10. Gruber DK, Gruber C, Seifter E. Acceleration of wound healing by a strain of *Staphylococcus aureus*. *Surg Forum* 1980; 31:232–234.
11. Raju DR, Jindrak K, Weiner M, et al. A study of the critical bacterial inoculum to cause a stimulus to wound healing. *Surg, Gynecol Obstet* 1977; 144(March):347–350.
12. Hermans MH. Results of an internet survey on the treatment of partial thickness burns, full thickness burns, and donor sites. *J Burn Care Res* 2007; 28(6):835–847.
13. Brem H, Jacobs T, Vileikyte L, et al. Wound-healing protocols for diabetic foot and pressure ulcers. *Surg Technol Int* 2003; 11:85–92.
14. Kunimoto B, Cooling M, Gulliver W, et al. Best practices for the prevention and treatment of venous leg ulcers. *Ostomy Wound Manage* 2001; 47(2):34–46, 48–50.
15. Turczynski R, Tarpila E. Treatment of leg ulcers with split skin grafts: Early and late results. *Scand J Plast Reconstr Surg Hand Surg* 1999; 33(3):301–305.
16. Jewell L, Guerrero R, Quesada AR, et al. Rate of healing in skin-grafted burn wounds. *Plast Reconstr Surg* 2007; 120(2):451–456.
17. Ahnlide I, Bjellerup M. Efficacy of pinch grafting in leg ulcers of different aetiologies. *Acta Derm Venereol* 1997; 77(2):144–145.

18. Borges EL, Caliri MH, Haas VJ. Systematic review of topic treatment for venous ulcers. *Rev Lat Am Enfermagem* 2007; 15(6):1163–1170.
19. Steed DL, Donohoe D, Webster MW, et al. Effect of extensive debridement and treatment on the healing of diabetic foot ulcers. Diabetic Ulcer Study Group. *J Am Coll Surg* 1996; 183(1):61–64.
20. Phillips TJ, Salman SM, Bhawan J, et al. Burn scar carcinoma. Diagnosis and management. *Dermatol Surg* 1998; 24(5):561–565.
21. Jackson DM. [The diagnosis of the depth of burning.]. *Br J Surg* 1953; 40(164):588–596.
22. Jeschke MG, Mlcak RP, Finnerty CC, et al. Burn size determines the inflammatory and hypermetabolic response. *Crit Care* 2007; 11(4):R90.
23. Sherwood N, Traber D. Total burn care. In Herndon DN, ed., Vol. *The systemic inflammatory response syndrome.* New York: Saunders, 2002. pp. 257–270.
24. Ordog GJ, Wasserberger J, Balasubramanium S. Wound ballistics: Theory and practice. *Ann Emerg Med* 1984; 13(12):1113–1122.
25. Buimer MG, Ankersmit MF, Wobbes T, et al. Surgical treatment of hidradenitis suppurativa with gentamicin sulfate: A prospective randomized study. *Dermatol Surg* 2008; 34(2):224–227.
26. Fardet L, Dupuy A, Kerob D, et al. Infliximab for severe hidradenitis suppurativa: Transient clinical efficacy in 7 consecutive patients. *J Am Acad Dermatol* 2007; 56(4):624–628.
27. Sadick NS, Yee-Levin J. Laser and light treatments for pilonidal cysts. *Cutis* 2006; 78(2):125–128.
28. Rockwell WB, Cohen IK, Ehrlich HP. Keloids and hypertrophic scars: A comprehensive review [see comments]. *Plast Reconstr Surg* 1989; 84(5):827–837.
29. American Burn Association. Burn incident fact sheet. Available at http://www.ameriburn.org/resources_factsheet.php accessed July 10, 2009.
30. World Health Organization. Facts about injuries—Burns. Available at http://www.who.int/violence_injury_prevention/publications/other_injury/en/burns_factsheet.pdf accessed July 10, 2009.
31. DeToledo JC, Lowe MR. Microwave oven injuries in patients with complex partial seizures. *Epilepsy Behav* 2004; 5(5):772–774.
32. Josty IC, Mason WT, Dickson WA. Burn wound management in patients with epilepsy: Adopting a multidisciplinary approach. *J Wound Care* 2002; 11(1):31–34.
33. Harrington WZ, Strohschein BL, Reedy D, et al. Pavement temperature and burns: Streets of fire. *Ann Emerg Med* 1995; 26(5):563–568.
34. Sances A, Jr., Myklebust JB, Larson SJ, et al. Experimental electrical injury studies. *J Trauma* 1981; 21(8):589–597.
35. Redman DP, Friedman B, Law E, et al. Experience with necrotizing fasciitis at a burn care center. *South Med J* 2003; 96(9):868–870.
36. Barillo DJ, McManus AT, Cancio LC, et al. Burn center management of necrotizing fasciitis. *J Burn Care Rehabil* 2003; 24(3):127–132.
37. Arnoldo BD, Purdue GF, Tchorz K, et al. A case report of phaeohyphomycosis caused by cladophialophora bantiana treated in a burn unit. *J Burn Care Rehabil* 2005; 26(3):285–287.
38. Berger ME, Hurtado R, Dunlap J, et al. Accidental radiation injury to the hand: Anatomical and physiological considerations. *Health Phys* 1997; 72(3):343–348.
39. Lefaix JL, Daburon F. Diagnosis of acute localized irradiation lesions: Review of the French experimental experience. *Health Phys* 1998; 75(4):375–384.
40. Muller K, Meineke V. Radiation-induced alterations in cytokine production by skin cells. *Exp Hematol* 2007; 35(4 Suppl 1):96–104.

41. Vozenin-Brotons MC, Gault N, Sivan V, et al. Histopathological and cellular studies of a case of cutaneous radiation syndrome after accidental chronic exposure to a cesium source. *Radiat Res* 1999; 152(3):332–337.

42. Beetz A, Messer G, Oppel T, et al. Induction of interleukin 6 by ionizing radiation in a human epithelial cell line: Control by corticosteroids. *Int J Radiat Biol* 1997; 72(1):33–43.

43. Artz C, Moncrief J, Pruitt BA, Jr. *The body's response to heat. Burns, a team approach.* New York: Saunders, 1979. pp. 23–44.

44. Black KS, Hewitt CW, Miller DM, et al. Burn depth evaluation with fluorometry: Is it really definitive? *J Burn Care Rehabil* 1986; 7(4):313–317.

45. Wachtel TL, Leopold GR, Frank HA, et al. B-mode ultrasonic echo determination of depth of thermal injury. *Burns Incl Therm Inj* 1986; 12(6):432–437.

46. Park DH, Hwang JW, Jang KS, et al. Use of laser Doppler flowmetry for estimation of the depth of burns. *Plast Reconstr Surg* 1998; 101(6):1516–1523.

47. Holland AJ, Martin HC, Cass DT. Laser Doppler imaging prediction of burn wound outcome in children. *Burns* 2002; 28(1):11–17.

48. Hemington-Gorse SJ. A comparison of laser Doppler imaging with other measurement techniques to assess burn depth. *J Wound Care* 2005; 14(4):151–153.

49. Mileski WJ, Atiles L, Purdue G, et al. Serial measurements increase the accuracy of laser Doppler assessment of burn wounds. *J Burn Care Rehabil* 2003; 24(4):187–191.

50. Riordan CL, McDonough M, Davidson JM, et al. Noncontact laser Doppler imaging in burn depth analysis of the extremities. *J Burn Care Rehabil* 2003; 24(4):177–186.

51. Saranto JR, Rubayi S, Zawacki BE. Blisters, cooling, antithromboxanes, and healing in experimental zone- of-stasis burns. *J Trauma* 1983; 23(10):927–933.

52. Zawacki BE. The natural history of reversible burn injury. *Surg Gynecol Obstet* 1974; 139(6):867–872.

53. Zawacki BE. Reversal of capillary stasis and prevention of necrosis in burns. *Ann Surg* 1974; 180(1):98–102.

54. Singh V, Devgan L, Bhat S, et al. The pathogenesis of burn wound conversion. *Ann Plast Surg* 2007; 59(1):109–115.

55. Hermans MHE. Treatment of burns with occlusive dressings: Some pathophysiological and quality of life aspects. *Burns* 1992; 18(Suppl 2):S15–18.

56. Herndon DN. *Total burn care.* 2nd ed. New York: Saunders, 2002.

57. Lund CC, Browder, NC. The estimate of area of burns. *Surg Gynecol Obstet* 1944; 79:352–358.

58. Sheridan RL, Petras L, Basha G, et al. Planimetry study of the percent of body surface represented by the hand and palm: Sizing irregular burns is more accurately done with the palm. *J Burn Care Rehabil* 1995; 16(6):605–606.

59. Rosenthal SR. Burn toxin and its competitin. *Burns Incl Therm Inj* 1982; 8(3):215–219.

60. Allgower M, Stadtler K, Schoenenberger GA. Burn sepsis and burn toxin. *Ann R Coll Surg Engl* 1974; 55(5):226–235.

61. Enkhbaatar P, Traber DL. Pathophysiology of acute lung injury in combined burn and smoke inhalation injury. *Clin Sci (Lond)* 2004; 107(2):137–143.

62. Smith DL, Cairns BA, Ramadan F, et al. Effect of inhalation injury, burn size, and age on mortality: A study of 1447 consecutive burn patients. *J Trauma* 1994; 37(4):655–659.

63. Barrow RE, Spies M, Barrow LN, et al. Influence of demographics and inhalation injury on burn mortality in children. *Burns* 2004; 30(1):72–77.

64. Herndon DN, Gore D, Cole M, et al. Determinants of mortality in pediatric patients with greater than 70% full-thickness total body surface area thermal injury treated by early total excision and grafting. *J Trauma* 1987; 27(2):208–212.

65. Herndon DN, LeMaster J, Beard S, et al. The quality of life after major thermal injury in children: An analysis of 12 survivors with greater than or equal to 80% total body, 70% third-degree burns. *J Trauma* 1986; 26(7):609–619.

66. King TC, Zimmerman JM. First-Aid Cooling of the Fresh Burn. *Surg Gynecol Obstet* 1965; 120:1271–1273.

67. King TC, Zimmerman JM. Optimum temperatures for postburn cooling. *Arch Surg* 1965; 91(4):656–657.

68. King TC, Zimmerman JM, Price PB. Effect of immediate short-term cooling on extensive burns. *Surg Forum* 1962; 13:487–488.

69. Demling RH, Mazess RB, Wolberg W. The effect of immediate and delayed cold immersion on burn edema formation and resorption. *J Trauma* 1979; 19(1):56–60.

70. Moritz ARH, F.C. Studies of thermal injury. The relative importance of time and surface area in the causation of cutaneous burns. *Am J Pathol* 1947; 23:695–720.

71. Demling RH. Burns. *N Engl J Med* 1985; 313(22):1389–1398.

72. Stratta RJ, Saffle JR, Kravitz M, et al. Management of tar and asphalt injuries. *Am J Surg* 1983; 146(6):766–769.

73. Leonard LG, Scheulen JJ, Munster AM. Chemical burns: Effect of prompt first aid. *J Trauma* 1982; 22(5):420–423.

74. Hermans MHE, Vloemans AFPM. [A patient with a subungual burn caused by hydrofluoric acid]. *Ned Tijdschr Geneeskd* 1985; 129(52):2510–2511.

75. Eldad A, Chaouat M, Weinberg A, et al. Phosphorous pentachloride chemical burn—A slowly healing injury. *Burns* 1992; 18(4):340–341.

76. Eldad A, Simon GA. The phosphorous burn—A preliminary comparative experimental study of various forms of treatment. *Burns* 1991; 17(3):198–200.

77. Iverson RE, Laub DR. Hydrofluoric acid burn therapy. *Surg Forum* 1970; 21:517–519.

78. Murao M. Studies on the treatment of hydrofluoric acid burn. *Bull Osaka Med Coll* 1989; 35(1–2):39–48.

79. Mangion SM, Beulke SH, Braitberg G. Hydrofluoric acid burn from a household rust remover. *Med J Aust* 2001; 175(5):270–271.

80. ABA. *Advanced life support providers manual.* Chicago, IL: American Burn Association, 1994.

81. Pegg SP. Escharotomy in burns. *Ann Acad Med Singapore* 1992; 21(5):682–684.

82. Wong L, Spence RJ. Escharotomy and fasciotomy of the burned upper extremity. *Hand Clin* 2000; 16(2):165–174, vii.

83. Tsoutsos D, Rodopoulou S, Keramidas E, et al. Early escharotomy as a measure to reduce intraabdominal hypertension in full-thickness burns of the thoracic and abdominal area. *World J Surg* 2003; 27(12):1323–1328.

84. Castagnoli C, Stella M, Berthod C, et al. TNF production and hypertrophic scarring. *Cell Immunol* 1993; 147(1):51–63.

85. Deitch EA, Wheelahan TM, Rose MP, et al. Hypertrophic burn scars: Analysis of variables. *J Trauma* 1983; 23(10):895–898.

86. Worley CA. The wound healing process: Part III—The finale. *Dermatol Nurs* 2004; 16(3):274, 295.

87. Roques C. Pressure therapy to treat burn scars. *Wound Repair Regen* 2002; 10(2):122–125.

88. Rochet JM, Zaoui A. [Burn scars: Rehabilitation and skin care]. *Rev Prat* 2002; 52(20):2258–2263.

89. Puzey G. The use of pressure garments on hypertrophic scars. *J Tissue Viability* 2002; 12(1):11–15.

90. Jordan RB, Daher J, Wasil K. Splints and scar management for acute and reconstructive burn care. *Clin Plast Surg* 2000; 27(1):71–85.

91. Ayhan M, Gorgu M, Silistreli KO, et al. Silastic sheet integrated polymethylmetacrylate splint in addition to surgery for commissure contractures complicated with hypertrophic scar. *Acta Chir Plast* 2004; 46(4):132–135.

92. Van den Kerchhove E, Boeckx W, Kochuyt A. Silicone patches as a supplement for pressure therapy to control hypertrophic scarring. *J Burn Care Rehabil* 1991; 12(4):361–369.

93. Kavanagh GM, Page P, Hanna MM. Silicone gel treatment of extensive hypertrophic scarring following toxic epidermal necrolysis. *Br J Dermatol* 1994; 130(4):540–541.

94. Haedersdal M, Poulsen T, Wulf HC. Laser induced wounds and scarring modified by antiinflammatory drugs: A murine model. *Lasers Surg Med* 1993; 13(1):55–61.

95. Beldon P. Management of scarring. *J Wound Care* 1999; 8(10):509–512.

96. Lupton JR, Alster TS. Laser scar revision. *Dermatol Clin* 2002; 20(1):55–65.

97. Liew SH, Murison M, Dickson WA. Prophylactic treatment of deep dermal burn scar to prevent hypertrophic scarring using the pulsed dye laser: A preliminary study. *Ann Plast Surg* 2002; 49(5):472–475.

98. Xiang J, Wang XQ, Qing C, et al. The influence of dermal template on the expressions of signal transduction protein Smad 3 and transforming growth factorbeta1 and its receptor during wound healing process in patients with deep burns. *Zhonghua Shao Shang Za Zhi* 2005; 21(1):52–54.

99. Suliman MT. Experience with the seven flap-plasty for the release of burns contractures. *Burns* 2004; 30(4):374–379.

100. Deb R, Giessler GA, Przybilski M, et al. [Secondary plastic surgical reconstruction in severely burned patients]. *Chirurg* 2004; 75(6):588–598.

101. Lu KH, Guo SZ, Ai YF, et al. [Management of severe postburn scar contracture in the lower extremities]. *Zhonghua Shao Shang Za Zhi* 2004; 20(2):69–71.

102. Selezneva LG. Keloid scars after burns. *Acta Chir Plast* 1976; 18(2):106–111.

103. Iudenich VV, Pal'tsyn AA, Zalugovskii OG. Electron microscopic and autoradiographic study of keloid scars. *Arkh Patol* 1982; 44(1):44–49.

104. Bang RL, Dashti H. Keloid and hypertrophic scars: Trace element alteration. *Nutrition* 1995; 11(5 Suppl):527–531.

105. Morison WL. Oral treatment of keloid. *Med J Aust* 1968; 1(10):412–413.

106. Donati L, Taidelli Palmizi GA. [Treatment of hypertrophic and keloid cicatrices with thiomucase]. *Minerva Chir* 1975; 30(6):326–333.

107. Tammelleo AD. Suit for post-operative scar: Keloid or burn? *Regan Rep Nurs Law* 1996; 37(4):4.

108. Ahlering PA. Topical silastic gel sheeting for treating and controlling hypertrophic and keloid scars: Case study. *Dermatol Nurs* 1995; 7(5):295–297, 322.

109. Sizov VM. [The diagnosis and treatment of hypertrophic and keloid scars]. *Klin Khir* 1994(9):41–43.

110. Siemers F, Lohmeyer JA, Machens HG, et al. [Heterotopic ossifications: A severe complication following extensive burn injury]. *Handchir Mikrochir Plast Chir* 2007; 39(5):360–363.

111. Merryman H. Mechanics of freezing in living cells and tissue. *Science* 1956; 124:515–521.

112. Merryman H. The exceeding of a minimum tolerable cell colume in hypertonic suspension as a cause of freezing injury. In: Wasterholme G, O'Conner M, eds. *The frozen cell.* London: Churchill; 1970.

113. Robson MC, Heggers JP. Evaluation of hand frostbite blister fluid as a clue to pathogenesis. *J Hand Surg [Am]* 1981; 6(1):43–47.

114. Lisbona R, Rosenthall L. Assessment of bone viability by scintiscanning in frostbite injuries. *J Trauma* 1976; 16(12):989–992.

115. Grace TG. Cold exposure injuries and the winter athlete. *Clin Orthop Relat Res* 1987; Mar(216):55–62.

116. McCauley RL, Hing DN, Robson MC, et al. Frostbite injuries: A rational approach based on the pathophysiology. *J Trauma* 1983; 23(2):143–147.

117. Pfister RR. Chemical corneal burns. *Int Ophthalmol Clin* 1984; 24(2):157–168.

118. Pike J, Patterson A, Jr., Arons MS. Chemistry of cement burns: Pathogenesis and treatment. *J Burn Care Rehabil* 1988; 9(3):258–260.

119. Barillo DJ, Cancio LC, Goodwin CW. Treatment of white phosphorus and other chemical burn injuries at one burn center over a 51-year period. *Burns* 2004; 30(5):448–452.

120. Yolken R, Konecny P, McCarthy P. Acute fluoride poisoning. *Pediatrics* 1976; 58(1):90–93.

121. Barret JP, Dziewulski P, Wolf SE, et al. Outcome of scalp donor sites in 450 consecutive pediatric burn patients. *Plast Reconstr Surg* 1999; 103(4):1139–1142.

122. Madden MR, Nolan E, Finkelstein JL, et al. Comparison of an occlusive and a semi-occlusive dressing and the effect of the wound exudate upon keratinocyte proliferation. *J Trauma* 1989; 29(7):924–930; discussion 930–931.

123. Zapata-Sirvent R, Hansbrough JF, Carroll W, et al. Comparison of Biobrane and Scarlet Red dressings for treatment of donor site wounds. *Arch Surg* 1985; 120(6):743–745.

124. Innes ME, Umraw N, Fish JS, et al. The use of silver coated dressings on donor site wounds: A prospective, controlled matched pair study. *Burns* 2001; 27(6):621–627.

125. Hansbrough W. Nursing care of donor site wounds. *J Burn Care Rehabil* 1995; 16(3 Pt 1):337–339; discussion 339–340.

126. Hyland WT. A painless donor-site dressing. *Plast Reconstr Surg* 1982; 69(4):703–704.

127. Smith DJ, Jr., Thomson PD, Garner WL, et al. Donor site repair. *Am J Surg* 1994; 167(1A):49S–51S.

128. Hefny AF, Eid HO, Al-Hussona M, et al. Necrotizing fasciitis: A challenging diagnosis. *Eur J Emerg Med* 2007; 14(1):50–52.

129. Stevens DL. Streptococcal toxic shock syndrome associated with necrotizing fasciitis. *Annu Rev Med* 2000; 51:271–288.

130. Hsiao CT, Weng HH, Yuan YD, et al. Predictors of mortality in patients with necrotizing fasciitis. *Am J Emerg Med* 2008; 26(2):170–175.

131. Falasca GF, Reginato AJ. The spectrum of myositis and rhabdomyolysis associated with bacterial infection. *J Rheumatol* 1994; 21(10):1932–1937.

132. Bilton BD, Zibari GB, McMillan RW, et al. Aggressive surgical management of necrotizing fasciitis serves to decrease mortality: A retrospective study. *Am Surg* 1998; 64(5):397–400; discussion 400–401.

133. Kaide CG, Khandelwal S. Hyperbaric oxygen: Applications in infectious disease. *Emerg Med Clin North Am* 2008; 26(2):571–595, xi.

134. Edlich RF, Cross CL, Dahlstrom JJ, et al. Modern concepts of the diagnosis and treatment of purpura fulminans. *J Environ Pathol Toxicol Oncol* 2008; 27(3):191–196.

135. Schellongowski P, Bauer E, Holzinger U, et al. Treatment of adult patients with sepsis-induced coagulopathy and purpura fulminans using a plasma-derived protein C concentrate (Ceprotin). *Vox Sang* 2006; 90(4):294–301.

136. Warner PM, Kagan RJ, Yakuboff KP, et al. Current management of purpura fulminans: A multicenter study. *J Burn Care Rehabil* 2003; 24(3):119–126.

137. Pereira FA, Mudgil AV, Rosmarin DM. Toxic epidermal necrolysis. *J Am Acad Dermatol* 2007; 56(2):181–200.

138. Roujeau JC. [Toxic epidermal necrolysis and Stevens-Johnson syndrome]. *Rev Prat* 2007; 57(11):1165–1170.

139. Pirmohamed M, Arbuckle JB, Bowman CE, et al. Investigation into the multidimensional genetic basis of drug-induced Stevens-Johnson syndrome and toxic epidermal necrolysis. *Pharmacogenomics* 2007; 8(12):1661–1691.

140. Sharma VK, Sethuraman G, Minz A. Stevens Johnson syndrome, toxic epidermal necrolysis and SJS-TEN overlap: A retrospective study of causative drugs and clinical outcome. *Indian J Dermatol Venereol Leprol* 2008; 74(3):238–240.

141. Cac NN, Messingham MJ, Sniezek PJ, et al. Stevens-Johnson syndrome induced by doxycycline. *Cutis* 2007; 79(2):119–122.

142. Schwartz RA. Toxic epidermal necrolysis. *Cutis* 1997; 59(3):123–128.

143. Gurwitz JH, Sanchez-Cross MT, Eckler MA, et al. The epidemiology of adverse and unexpected events in the long-term care setting. *J Am Geriatr Soc* 1994; 42(1):33–38.

144. Sutton R, Pritty P. Use of sutures or adhesive tapes for primary closure of pretibial lacerations. *Br Med J (Clin Res Ed)* 1985; 290(6482):1627.

145. Dunkin CS, Elfleet D, Ling C, et al. A step-by-step guide to classifying and managing pretibial injuries. *J Wound Care* 2003; 12(3):109–111.

146. Wouters EG, Ho HT, Lipman LJ, et al. Dogs as vectors of *Streptobacillus moniliformis* infection? *Vet Microbiol* 2008; 128(3–4):419–422.

147. Chretien JH, Garagusi VF. Infections associated with pets. *Am Fam Physician* 1990; 41(3):831–845.

148. Benson LS, Edwards SL, Schiff AP, et al. Dog and cat bites to the hand: Treatment and cost assessment. *J Hand Surg [Am]* 2006; 31(3):468–473.

149. Farace F, Lissia M, Mele A, et al. Local cutaneous arachnidism: A report of three cases and their management. *J Plast Reconstr Aesthet Surg* 2006; 59(2):197–201.

150. Radmanesh M. Clinical study of Hemiscorpion lepturus in Iran. *J Trop Med Hyg* 1990; 93(5):327–332.

151. Bergstrom N, Bennett N, Carlson C. Clinical practice guideline number 15: Treatment of pressure ulcers. Rockville, MD: US Department of Health and Humans Services, 1994.

152. Boulton AJ, Vileikyte L, Ragnarson-Tennvall G, et al. The global burden of diabetic foot disease. *Lancet* 2005; 366(9498):1719–1724.

153. Msambichakga L. *Economic Adjustment Policies and Health Care in Tanzania.* Dar Es Salaam: Economic Research Bureau, University of Dar Es Salaam, 1997.

154. Olin JW, Beusterien KM, Childs MB, et al. Medical costs of treating venous stasis ulcers: Evidence from a retrospective cohort study. *Vasc Med* 1999; 4(1):1–7.

155. Rolfe M, Tang CM, Walker RW, et al. Diabetes mellitus in The Gambia, west Africa. *Diabet Med* 1992; 9(5):484–488.

156. Smith A. *Etiology of the Problem Wound.* Vol. 1. Flagstaff, AZ: Best, 2007.

157. McGuckin M, Kerstein MD. Venous leg ulcers and the family physician. *Adv Wound Care* 1998; 11(7):344–346.

158. Asiedu K, Etuaful S. Socioeconomic implications of Buruli ulcer in Ghana: A three-year review. *Am J Trop Med Hyg* 1998; 59(6):1015–1022.

159. Liu CJ, Tahara S, Gao S. Phosphorylation of extracellular signal-regulated protein kinase in cultured keloid fibroblasts when stimulated by platelet-derived growth factor BB. *Scand J Plast Reconstr Surg Hand Surg* 2003; 37(6):321–324.

160. Greenhalgh DG. Consequences of excessive scar formation: Dealing with the problem and aiming for the future. *Wound Repair Regen* 2007; 15 Suppl 1:S2–5.

161. Harris IR, Yee KC, Walters CE, et al. Cytokine and protease levels in healing and non-healing chronic venous leg ulcers. *Exp Dermatol* 1995; 4(6):342–349.

162. Trengove NJ, Bielefeldt-Ohmann H, Stacey MC. Mitogenic activity and cytokine levels in non-healing and healing chronic leg ulcers. *Wound Repair Regen* 2000; 8(1):13–25.

163. Yager DR, Nwomeh BC. The proteolytic environment of chronic wounds. *Wound Repair Regen* 1999; 7(6):433–441.

164. Ladwig GP, Robson MC, Liu R, et al. Ratios of activated matrix metalloproteinase-9 to tissue inhibitor of matrix metalloproteinase-1 in wound fluids are inversely correlated with healing of pressure ulcers. *Wound Repair Regen* 2002; 10(1):26–37.

165. Chin G, GF. S, Diegelmann RF. *Biochemistry of wound healing in wound care practice.* Flagstaff AZ: Best, 2007.

166. Bucalo B, Eaglstein WH, Falanga V. Inhibition of cell proliferation by chronic wound fluid. *Wound Repair Regen* 1993; 1(3):181–186.

167. Stanley A, Osler T. Senescence and the healing rates of venous ulcers. *J Vasc Surg* 2001; 33(6):1206–1211.

168. Mendez MV, Raffetto JD, Phillips T, et al. The proliferative capacity of neonatal skin fibroblasts is reduced after exposure to venous ulcer wound fluid: A potential mechanism for senescence in venous ulcers. *J Vasc Surg* 1999; 30(4):734–743.

169. Robson MC, Smith P. Topical use of growth factors to enhance healing. In: Falanga V, ed. *Cutaneous wound healing.* London, UK; 2001:pp. 479–493.

170. Schultz GS, Sibbald RG, Falanga V, et al. Wound bed preparation: A systematic approach to wound management. *Wound Repair Regen* 2003; 11 Suppl 1:S1–28.

171. Heit JA, Rooke TW, Silverstein MD, et al. Trends in the incidence of venous stasis syndrome and venous ulcer: A 25-year population-based study. *J Vasc Surg* 2001; 33(5):1022–1027.

172. Jorgensen B, Price P, Andersen KE, et al. The silver-releasing foam dressing, Contreet Foam, promotes faster healing of critically colonised venous leg ulcers: A randomised, controlled trial. *Int Wound J* 2005; 2(1):64–73.

173. Kerstein MD, Gemmen E, Van Rijswijk L. Cost and cost effectiveness of venous and pressure ulcer protocols of care. *Dis Manage Health Outcomes* 2001; 9:651–653.

174. Treadwell T. *Chronic venous insufficiency, postphlebitic syndrome, and venous leg ulcers.* Montgomery, AL, 203.

175. Callam MJ, Harper DR, Dale JJ, et al. Chronic ulcer of the leg: Clinical history. *Br Med J (Clin Res Ed)* 1987; 294(6584):1389–1391.

176. Browse NL, Burnand KG. The cause of venous ulceration. *Lancet* 1982; 2(8292):243–245.

177. Falanga V, Moosa HH, Nemeth AJ, et al. Dermal pericapillary fibrin in venous disease and venous ulceration. *Arch Dermatol* 1987; 123(5):620–623.

178. Kobrin KL, Thompson PJ, van de Scheur M, et al. Evaluation of dermal pericapillary fibrin cuffs in venous ulceration using confocal microscopy. *Wound Repair Regen* 2008; 16(4):503–506.

179. Balslev E, Thomsen HK, Danielsen L, et al. The occurrence of pericapillary fibrin in venous hypertension and ischaemic leg ulcers: A histopathological study. *Br J Dermatol* 1992; 126(6):582–585.

180. Falanga V, Eaglstein WH. The "trap" hypothesis of venous ulceration. *Lancet* 1993; 341(8851):1006–1008.

181. Coleridge Smith PD, Thomas P, Scurr JH, et al. Causes of venous ulceration: A new hypothesis. *Br Med J (Clin Res Ed)* 1988; 296(6638):1726–1727.

182. Nagy N, Szolnoky G, Szabad G, et al. Tumor necrosis factor-alpha-308 polymorphism and leg ulceration—Possible association with obesity. *J Invest Dermatol* 2007; 127(7):1768–1769; author reply 1770–1771.

183. Wallace HJ, Vandongen YK, Stacey MC. Tumor necrosis factor-alpha gene polymorphism associated with increased susceptibility to venous leg ulceration. *J Invest Dermatol* 2006; 126(4):921–925.

184. Kim BC, Kim HT, Park SH, et al. Fibroblasts from chronic wounds show altered TGF-beta-signaling and decreased TGF-beta Type II receptor expression. *J Cell Physiol* 2003; 195(3):331–336.

185. Lauer G, Sollberg S, Cole M, et al. Expression and proteolysis of vascular endothelial growth factor is increased in chronic wounds. *J Invest Dermatol* 2000; 115(1):12–18.

186. Nagy N, Nemeth IB, Szabad G, et al. The altered expression of syndecan 4 in the uninvolved skin of venous leg ulcer patients may predispose to venous leg ulcer. *Wound Repair Regen* 2008; 16(4):495–502.

187. Pieper B, Templin T, Ebright JR. Chronic venous insufficiency in HIV-positive persons with and without a history of injection drug use. *Adv Skin Wound Care* 2006; 19(1):37–42.

188. Margolis DJ, Knauss J, Bilker W. Hormone replacement therapy and prevention of pressure ulcers and venous leg ulcers. *Lancet* 2002; 359(9307):675–677.

189. Ashcroft GS, Greenwell-Wild T, Horan MA, et al. Topical estrogen accelerates cutaneous wound healing in aged humans associated with an altered inflammatory response. *Am J Pathol* 1999; 155(4):1137–1146.

190. Hardman M, Ashcroft G. Hormonal influence in wound healing: A review a current experimental data. *Wounds* 2005; 17(11):313–320.

191. Yang D, Morrison BD, Vandongen YK, et al. Malignancy in chronic leg ulcers. *Med J Aust* 1996; 164(12):718–720.

192. Ackroyd JS, Young AE. Leg ulcers that do not heal. *Br Med J (Clin Res Ed)* 1983; 286(6360):207–208.

193. Hansson C, Andersson E. Malignant skin lesions on the legs and feet at a dermatological leg ulcer clinic during five years. *Acta Derm Venereol* 1998; 78(2):147–148.

194. Frykberg R. *Epidemiology of the diabetic foot: Ulcerations and amputations.* Totowa, NJ: Humana Press, 1998.

195. Palumbo P, Melton LJ, 3rd. Peripheral vascular disease and diabetes. *NIH Publication* 2003; 85–1468:1.

196. Reiber G. Lower extremity foot ulcers and amputations in diabetes. In: Harris M, Cowie C, Stern M, eds. *Diabetes in America*, 2nd edition, Vol. NIH: 95–1468. Washington DC: US Goverment Printing Office; 1995.

197. Contreras-Ruiz J. Mexico's Wound Care Statistics. *Adv Skin Wound Care* 2007; 20(2):96–98.

198. Wild S, Roglic G, Green A, et al. Global prevalence of diabetes: Estimates for the year 2000 and projections for 2030. *Diabetes Care* 2004; 27(5):1047–1053.

199. Abbas ZG, Archibald LK. Epidemiology of the diabetic foot in Africa. *Med Sci Monit* 2005; 11(8):RA262–270.

200. Ogbera AO, Fasanmade O, Ohwovoriole AE, et al. An assessment of the disease burden of foot ulcers in patients with diabetes mellitus attending a teaching hospital in Lagos, Nigeria. *Int J Low Extrem Wounds* 2006; 5(4):244–249.

201. Tchakonte B, Ndip A, Aubry P, et al. [The diabetic foot in Cameroon]. *Bull Soc Pathol Exot* 2005; 98(2):94–98.

202. Ogbera OA, Osa E, Edo A, et al. Common clinical features of diabetic foot ulcers: Perspectives from a developing nation. *Int J Low Extrem Wounds* 2008; 7(2):93–98.

203. Gulam-Abbas Z, Lutale JK, Morbach S, et al. Clinical outcome of diabetes patients hospitalized with foot ulcers, Dar es Salaam, Tanzania. *Diabet Med* 2002; 19(7):575–579.

204. Harrington C, Zagari MJ, Corea J, et al. A cost analysis of diabetic lower-extremity ulcers. *Diabetes Care* 2000; 23(9):1333–1338.

205. Kerdel FA. Inflammatory ulcers. *J Dermatol Surg Oncol* 1993; 19(8):772–778.

206. Hehenberger K, Hansson A. High glucose-induced growth factor resistance in human fibroblasts can be reversed by antioxidants and protein kinase C-inhibitors. *Cell Biochem Funct* 1997; 15(3):197–201.

207. Spravchikov N, Sizyakov G, Gartsbein M, et al. Glucose effects on skin keratinocytes: Implications for diabetes skin complications. *Diabetes* 2001; 50(7):1627–1635.

208. Lerman OZ, Galiano RD, Armour M, et al. Cellular dysfunction in the diabetic fibroblast: Impairment in migration, vascular endothelial growth factor production, and response to hypoxia. *Am J Pathol* 2003; 162(1):303–312.

209. Fard A, Tuck CH, Donis JA, et al. Acute elevations of plasma asymmetric dimethylargi-nine and impaired endothelial function in response to a high-fat meal in patients with type 2 diabetes. *Arterioscler Thromb Vasc Biol* 2000; 20(9):2039–2044.
210. Terashi H, Izumi K, Deveci M, et al. High glucose inhibits human epidermal keratinocyte proliferation for cellular studies on diabetes mellitus. *Int Wound J* 2005; 2(4):298–304.
211. Marston WA. Risk factors associated with healing chronic diabetic foot ulcers: The importance of hyperglycemia. *Ostomy Wound Manage* 2006; 52(3):26–8, 30, 32 passim.
212. Meijer JW, Smit AJ, Lefrandt JD, et al. Back to basics in diagnosing diabetic polyneu-ropathy with the tuning fork! *Diabetes Care* 2005; 28(9):2201–2205.
213. Benskin L. *Handbook for health care.* Searcy AR: International Health Care Foundation, 2002.
214. Warren G, Nade S. *The care of the neuropathic limbs.* New York: Parthenon, 1999.
215. Thomas J, Huffman L. Charcot foot deformity: Surgical treatment options. *Wounds* 2008; 20(3):67–73.
216. Ramsey SD, Newton K, Blough D, et al. Incidence, outcomes, and cost of foot ulcers in patients with diabetes. *Diabetes Care* 1999; 22(3):382–387.
217. Butalia S, Palda VA, Sargeant RJ, et al. Does this patient with diabetes have osteomyeli-tis of the lower extremity? *JAMA* 2008; 299(7):806–813.
218. Sackett DL. The rational clinical examination. A primer on the precision and accuracy of the clinical examination. *JAMA* 1992; 267(19):2638–2644.
219. Kapoor A, Page S, Lavalley M, et al. Magnetic resonance imaging for diagnosing foot osteomyelitis: A meta-analysis. *Arch Intern Med* 2007; 167(2):125–132.
220. Beach KW, Strandness DE, Jr. Arteriosclerosis obliterans and associated risk factors in insulin-dependent and non-insulin-dependent diabetes. *Diabetes* 1980; 29(11):882–888.
221. Jude EB, Oyibo SO, Chalmers N, et al. Peripheral arterial disease in diabetic and nondiabetic patients: A comparison of severity and outcome. *Diabetes Care* 2001; 24(8):1433–1437.
222. King TA, DePalma RG, Rhodes RS. Diabetes mellitus and atherosclerotic involvement of the profunda femoris artery. *Surg Gynecol Obstet* 1984; 159(6):553–556.
223. Taylor LM, Jr., Porter JM. The clinical course of diabetics who require emergent foot surgery because of infection or ischemia. *J Vasc Surg* 1987; 6(5):454–459.
224. Abbas ZG, Archibald LK. Challenges for management of the diabetic foot in Africa: Doing more with less. *Int Wound J* 2007; 4(4):305–313.
225. Willi C, Bodenmann P, Ghali WA, et al. Active smoking and the risk of type 2 diabetes: A systematic review and meta-analysis. *JAMA* 2007; 298(22):2654–2664.
226. Lyder CH. Pressure ulcer prevention and management. *JAMA* 2003; 289(2):223–226.
227. Reddy M, Gill SS, Rochon PA. Preventing pressure ulcers: A systematic review. *JAMA* 2006; 296(8):974–984.
228. Russo CA Elixhauser A. Hospitalizations Related to Pressure Sores, 2003. Available at www.hcup-us.ahrq.gov/reports/statbriefs/sb3.pdf. Accessed 8/08.
229. McLane KM, Bookout K, McCord S, et al. The 2003 national pediatric pressure ulcer and skin breakdown prevalence survey: A multisite study. *J Wound Ostomy Continence Nurs* 2004; 31(4):168–178.
230. Armstrong DG, Ayello EA, Capitulo KL, et al. New opportunities to improve pres-sure ulcer prevention and treatment: Implications of the CMS inpatient hospital care present on admission indicators/hospital-acquired conditions policy: A consensus paper from the International Expert Wound Care Advisory Panel. *Adv Skin Wound Care* 2008; 21(10):469–478.
231. NPUAP Position Statement on Reverse Staging. Available at www.npuap.org
232. Stekelenburg A, Oomens CW, Strijkers GJ, et al. Compression-induced deep tissue injury examined with magnetic resonance imaging and histology. *J Appl Physiol* 2006; 100(6):1946–1954.

233. Fife C. *Towards a new understanding of an old problem.* Vol. 1. Flagstaff, AZ: Best, 2007.

234. National Pressure Ulcer Advisory Panel. Pressure Ulcer Stages Revised by NPUAP. Available at http://www.npuap.org/pr2.htm. Accessed 4/08.

235. Vanderwee K, Grypdonck MH, Defloor T. Effectiveness of an alternating pressure air mattress for the prevention of pressure ulcers. *Age Ageing* 2005; 34(3):261–267.

236. Dunlop V. Preliminary results of a randomized, controlled study of a pressure ulcer prevention system. *Adv Wound Care* 1998; 11(3 Suppl):14.

237. Young T. The 30 degree tilt position vs the 90 degree lateral and supine positions in reducing the incidence of non-blanching erythema in a hospital inpatient population: A randomised controlled trial. *J Tissue Viability* 2004; 14(3):88, 90, 92–96.

238. Horn SD, Bender SA, Ferguson ML, et al. The National Pressure Ulcer Long-Term Care Study: Pressure ulcer development in long-term care residents. *J Am Geriatr Soc* 2004; 52(3):359–367.

239. Allman RM, Goode PS, Patrick MM, et al. Pressure ulcer risk factors among hospitalized patients with activity limitation. *JAMA* 1995; 273(11):865–870.

240. Torra i Bou JE, Segovia Gomez T, Verdu Soriano J, et al. The effectiveness of a hyper-oxygenated fatty acid compound in preventing pressure ulcers. *J Wound Care* 2005; 14(3):117–121.

241. Bourdel-Marchasson I, Barateau M, Rondeau V, et al. A multi-center trial of the effects of oral nutritional supplementation in critically ill older inpatients. GAGE Group. Groupe Aquitain Geriatrique d'Evaluation. *Nutrition* 2000; 16(1):1–5.

242. Criqui MH. Peripheral arterial disease—Epidemiological aspects. *Vasc Med* 2001; 6(3 Suppl):3–7.

243. Lee B, Trainor F, Thoden W, et al. *Handbook of noninvasive diagnostic techniques in vascular surgery.* New York: Appleton-Century-Crofts; 1981:pp. 33.

244. Fred HL. Hyposkillia: Deficiency of clinical skills. *Tex Heart Inst J* 2005; 32(3):255–257.

245. Lee B, Trainor F, Thoden W, et al. *Handbook of noninvasive diagnostic techniques in vascular surgery.* New York: Appleton-Century-Crofts; 1981:pp. 41.

246. Giordano J. Development of the vascular system. In: Giordano J, Trout H, DePalma R, eds. *The basic science of vascular surgery.* Mount Kisco, NY: Futura; 1988:pp. 3–30.

247. Lee B, Trainor F, Thoden W, et al. *Handbook of noninvasive diagnostic techniques in vascular surgery.* New York: Appleton-Century-Crofts; 1981:p. 188.

248. Chang DW, Sanchez LA, Veith FJ, et al. Can a tissue-engineered skin graft improve healing of lower extremity foot wounds after revascularization? *Ann Vasc Surg* 2000; 14(1):44–49.

249. Peck J. The art of surgery. *Am J Surg* 2004; 187(5):569–574.

250. Peeters Grietens K, Um Boock A, Peeters H, et. Al. "It is me who endures but my family that suffers:" Social isolation as a consequence of the household cost burden of Buruli ulcer free of charge hospital treatment. *PLos Negl Trop Dis.* 2008; 2(10):e321.doi.1371/journal.pntd.0000321. www.plosntds.org. Accessed 10/16/08.

4 Burn Wound Management

Michel H.E. Hermans

CONTENTS

Principles of Wound Management in Burn Care 135
Burn Wound Infection... 135
First-Degree Burns.. 136
Superficial Partial Thickness Burns 137
Deep Partial Thickness Burns 139
Mixed Partial Thickness Burns 139
Full Thickness Burns 139
Donor Sites.. 141
Chemical Lesions.. 142
Long-Term Results.. 142
References.. 144

PRINCIPLES OF WOUND MANAGEMENT IN BURN CARE

In addition to wound management, in patients with larger burns or complications (whether burn injury related or not), burn care involves the management of the burn disease. However, this chapter concentrates only on the skin injury per se. At the same time, wound closure in burn care is the primary treatment of the disease process.

Treating a burn wound is based on the general surgical principles of wound management: removal of necrosis and providing optimal circumstances for reepithelialization, either by creating the optimal wound bed for spontaneous reepithelialization or by closing the wound with a graft. General objectives to be met are reduction of pain, prevention of infection, desiccation and conversion, rapid healing, and, for the long term, minimization of the chance of scarring, particularly hypertrophic scarring and contractures.

BURN WOUND INFECTION

Burn wound infection is always a serious complication. On a local level it leads to conversion: The wound becomes deeper and an initially superficial partial thickness burn may become deep partial or even full thickness, requiring excision and grafting. An invading infection may rapidly lead to sepsis, which is still a leading cause of mortality in burn patients.

Many different types of organisms cause infection in patients with burns. *Pseudomonas aeruginosa*, *Staphylococcus aureus*, and *Enterobacter* species are pathogens found in many types of wounds, and they are among the most common microorganisms to cause infection in burns.[1] Less common in general wound care, but cultured frequently from infected burns are microorganisms such as *Acinetobacter baumanii.*[2,3] In patients with extensive burns, even "normally" nonvirulent microorganisms, such as *Streptococcus epidermidis,* may cause serious infections, and virtually every type of microorganism may become invasive and lead to sepsis, including anaerobes, yeasts, and fungi.[4]

Many topical agents are used, either for prophylactic purposes or for the treatment of infection. The most commonly used material in burn care, both for partial thickness burns and as a temporary protective cream prior to excision in full thickness burns, is silversulfadiazine.[5] Although many different types of this cream exist, the differences are primarily in the cream base: The active ingredient is always silversulfadiazine 1% (with or without chlorhexidine [Australia] or cerium nitrate [Europe]). Silversulfadiazine is fairly broad spectrum, although it is more active against Gram-negative microorganisms. In spite of its popularity, it has some acknowledged side effects, including allergies to the sulfa-compound,[6] a relatively slow reepithelialization time,[7] and the development of a pseudo eschar.[8] This thin layer of "precipitate" makes it very difficult to see and judge the underlying wound bed. Resistance to the drug occurs as well.[9,10] According to the manufacturer's guidelines, the cream has to be applied twice per day, which makes its use labor intensive. Some contribute initial leukopoenia to the material, but others consider transient leukopoenia a normal consequence of the burn injury itself.[11]

Other compounds more or less exclusively used in burn care are silver nitrate solutions ($AgNO_3$) and mafenide cream or solution (Sulfamylon®, UDL Laboratories, Inc., Rockford, Illinois). Although both agents have good antimicrobial properties, they also have some serious side effects. For example, mafenide may cause serious metabolic acidosis, and a high percentage of patients becomes allergic.[12] Silver nitrate may cause methemoglobinemia and, being hypotonic, may lead to hyponatremia and hypochloremia.[13,14] When the solutions are used, (re)application needs to occur frequently to avoid drying out of the dressings.

Dressings that incorporate a topical antimicrobial agent are frequently used as well. Silver dressings seem to be the material of choice.[5] Pure silver (as opposed to silver salts) has few side effects and is broad spectrum. Many silver dressings are now available, and it is important to realize that they have different chemical and physical properties.[15,16]

Although not extensively used in Western medicine, a number of materials of botanical origin are used for the treatment of partial thickness burns. MEBO (moist exposed burn ointment) is probably the best-known example and is used primarily in the Middle and Far East. Although not all articles on this material report the same results, several of them indicated clearly positive effects.[17,18]

FIRST-DEGREE BURNS

Most first-degree burns require virtually no real medical treatment and certainly no dressings (in spite of what many commercial package inserts state). A nonmedicated, soothing, moisturizing cream, in combination with a mild (anti-inflammatory) analgesic usually provide sufficient patient comfort.

SUPERFICIAL PARTIAL THICKNESS BURNS

Superficial partial thickness burns are characterized by a very thin layer of necrosis that does not require extensive, specific debridement or excision. There are enough viable epidermal elements left in the deeper dermis for them to heal within approximately 2 weeks. Skin grafting is not necessary, and significant scarring is rare. However, even superficial partial thickness may convert and become deeper, most commonly through desiccation or infection.[19,20] Thus, even when in the initial evaluation a superficial burn was diagnosed, healing times longer than 2 weeks are a reason to check for conversion. Virtually always, secondary tangential excision (see below) will be necessary.

Superficial partial thickness burns often have blisters. The literature is not consistent with regard to whether or not a blister should be removed. Some state that the blister roof acts as a biological occlusive dressing and that blister fluid is beneficial to wound healing and has antimicrobial properties.[21] Other research indicates the opposite, particularly with respect to the fluid being detrimental to fibroblasts.[22]

Most blisters will break within a few days postburn and the advantage of the blister roof as a biological dressing would cease to exist. Therefore, many advocate carefully evacuating the fluid with a syringe after the blister roof has been painted with povidone iodine or chlorhexidine, and after evacuation, supporting the blister with a thin hydrocolloid dressing or polyurethane film. For large blisters, the entire blister and its fluid may be removed and replaced with a synthetic occlusive or moisture-retentive dressing.

To prevent infection (which in smaller superficial partial thickness burns is not common), to reduce pain, and to prevent desiccation, the burn should be covered with an appropriate dressing. However, facial burns are often treated just with an ointment, medicated or not, without a cover dressing.

Many different types of dressings have been advocated, and different burn care specialists prefer different materials for different reasons. The perfect dressing/skin substitute has specific properties (Table 4.1). None of the currently available synthetic or biologic materials fills all these requirements, but a number of required properties are combined in each of them.

Allograft (from a different individual but the same species) is either human cadaver skin[23,24] or amnion membrane.[25,26] Both types of tissues are available as fresh grafts and in freeze dried/cryopreserved form, and cadaver skin is also available in glycerol preserved form.[27,28] Fresh allograft or amnion is generally considered to provide clinically superior performance because the cells, assumingly, deliver the appropriate growth factors to the wound; however, a direct comparative clinical trial has never been performed, and when published clinical results, obtained in noncomparative trials, are compared,[29–32] real and clinically relevant differences do not seem to exist.

TABLE 4.1

The Perfect Dressing/Skin Substitute

Prevents water loss

Acts as a barrier to bacteria

Helps prevent against infection

Does not transmit diseases

Does not incite an inflammatory response

Does not become hypertrophic itself or prevents the development of a hypertrophic scar

Is durable

Is flexible and thus conforms to irregular surfaces

Can be used off the shelf

Does not require refrigeration or other complex means of storage

Does not require complex preparations (i.e., thawing in the case of liquid nitrogen stored allografts)

Can be applied in one session

Is easy to secure

Is inexpensive

Has a long shelf life

Source: Amended from Sheridan and Thomkins. In: Herndon, D.N. *Total Burn Care*. 2nd ed. New
            York: Saunders, 2002.

Xenografts (from a different species) are used in burn care as well, primarily in the form of porcine skin,[29,33] although other animals (i.e., Brazilian bullfrogs), also have been used as donors.[34]

Both allografts and xenografts are rejected after a number of days, and thus are used only as a temporary coverage, to enhance reepithelialization, to protect the wound, and to provide pain relief.

Other biological dressings used with good results for the treatment of partial thickness burns include potato peels[34] and banana leaves.[35] All these materials are (semi)occlusive and thus prevent the wound from desiccation.

(Semi)biological dressings used for treating partial thickness burns include a number of materials based on components of the normal extracellular skin matrix, such as collagen and hyaluronic acid.[36-39] The durability of these materials depends on their exact nature and, for example, to what extent they are cross-linked. Similar to allografts and xenografts, most of them become temporarily adherent to the wound bed, thus providing a biological barrier and, in some cases, a specific matrix for cells to migrate into and over the wound bed.

Other dressings used for the treatment of partial thickness burns are those used for the management of chronic lesions as well. This large group includes materials such as hydrocolloids, foam dressings, honey dressings, hydrofibers, and impregnated gauze. In principle, they provide the same advantages and disadvantages as they do for chronic wounds,[16,40,41] and many have been shown to work well in burns.

Impregnated gauze is also used quite extensively for superficial partial thickness burns, and again, this type of dressing has been shown to have the same positive and negative aspects for burn wounds as for chronic lesions.[42]

Some burn surgeons use collagenase to remove the thin layer of necrosis that exists in partial thickness burns, although this is not standard practice in most burn centers.[43]

DEEP PARTIAL THICKNESS BURNS

These wounds may heal spontaneously, similar to superficial partial thickness burns, but because fewer epithelial remnants are left in the wound, they take longer. In fact, many deep partial thickness burns will take more than 2 to 3 weeks and consequently will result in significant scarring.

Many burn centers, therefore, are more aggressive with this type of burn and perform tangential excision.[44,45] Using a specially designed dermatome, necrosis is excised in thin layers until a viable wound bed (punctate bleeding) is reached. The depth of the excision indicates the depth of the burn. Deeper lesions are grafted, but the more superficial burns may be treated with a dressing. Different centers take different approaches here. A specific alternative for the dermatome is VersaJet® (Smith & Nephew Wound Management, Hull, UK), a device that uses high-pressure water to excise layers of necrosis.[46]

When the depth of a burn is difficult to determine, tangential excision also is used as a diagnostic tool; again, the burn depth is judged from the appearance of the wound bed.

MIXED PARTIAL THICKNESS BURNS

Mixed partial thickness burns often pose a problem in that superficial and deeper areas cannot be easily distinguished and often are confluent. Tangential excision is used as a diagnostic and therapeutic tool, and deeper excised areas may be grafted while the superficial ones are left to heal with the help of a dressing. Timing of the excision depends on the burn center—many will perform an early tangential excision but, for example, on the lower back, often treatment with dressings is the primary choice and excision (and grafting) is delayed for about 14 days.

Tangential excision is a serious procedure, primarily because a great deal of blood is lost. Therefore, on the limbs, a tourniquet is often used. The procedure also requires experience because it is easy not to excise enough but also to excise too much.

FULL THICKNESS BURNS

Unless the lesions are really small, the preferred treatment for full thickness burns is excision and grafting. Specific enzymes that may replace surgical excision are being evaluated[47] but at the time of this writing are not standard treatment.

Spontaneous desloughing will occur, and this was the preferred method some decades ago. However, this takes considerable time, and the wounds often infect in the meantime. Moreover, if spontaneous debridement does occur, the resulting wound bed will certainly be contaminated, and it is of poor quality. Thus, excision and grafting is the preferred treatment. In addition to the dermatome and the VersaJet, excision to fascia or fat (avulsion) is used in large, undoubtedly full thickness burns: The burn perimeter is incised and the dead tissue is pulled off. Avulsion had specific advantages. It causes less blood loss than tangential excision of the same size wound surface because of retraction of the arterioles, and the procedure may save time. In addition, the separation plane more or less assures a wound bed that will accept grafts very well.[48] However, avulsion usually results in less satisfying cosmesis, and there is a "dip" into the depth where the perimeter of the excised skin is situated. From a cosmetic point of view, excising to fat seems to have better results than excision to fascia (personal observation). To reduce blood loss, some use the tumescent technique:[49,50] Preexcision or prior to harvesting a split skin graft, epinephrine is injected subdermally into the operation site.

Reepithelialization over large surfaces will simply not occur. Instead, contraction will become one of the ways the body tries to decrease the wound surface, and this results in seriously debilitating contractures. Contraction is caused by the wound edges drawn toward each other as a consequence of cellular changes in the wound bed and the extracellular matrix. If it happens over a joint, flexing of the joint will occur, and the rigidity of the wound bed and its skin will prevent extension of the joint.

The lack of spontaneous healing and the contraction and scarring problems that result are the primary reason why excision is the only real option for the treatment of full thickness burns. Early excision also has been shown to reduce morbidity and mortality significantly,[51,52] because it removes the breeding grounds for microorganisms and the source of the burn toxins that lead to the burn disease.

Excision will result in an open wound bed that needs to be covered as soon as possible, preferably with autografts. Full sheet autografting offers the best cosmetic results but is only possible in small burns. Large burns simply lack the donor sites. In major burns, meshing is used, although for cosmetically and functionally important areas such as the face, the neck, and the hands, full sheet grafting is still preferred. Larger, cosmetically less important areas are covered with meshed grafts: The autograft is incised with a dermatome, and the incisions in the graft allow it to be expanded.[53] The interstices will be covered through lateral reepithelialization from the meshed autograft. Different mesh sizes are used and depending on the type of dermatome can go up to 1:9. Techniques other than meshing also are available. All aim at expansion of the obtained split skin autograft.[53–60] Ratios of expansion depend on equipment used, personal preference of the surgeon, and again the total amount of grafts available in comparison to the size of the burn.

Cultured epithelium may also be used to cover excised areas.[61] Biopsies are taken from unburned skin. The different cell layers of the biopsy are separated and keratinocytes put in culture. After 10 to 14 days (depending on the culturing technique), cultured confluent sheets of the patient's own epidermis are ready for application. Some use dispersed, nonconfluent cells that allow for earlier application.[62] Although these techniques are not new,[61,63,64] several disadvantages (biochemical and physical fragility, odd aspects of the grafted areas, the lack of dermis, economical factors[65]) still have to be overcome to make them widely used, although their life saving properties have been described as well.[66]

Some use temporary coverage materials after the initial excision: This depends on the size of the burn (i.e., the available donor sites) and the quality of the wound bed (i.e., older burns, secondary excision), but in some centers it is part of standard treatment, even in early excised, small burns.

A number of different temporary cover materials are available: The most common ones are allografts[26,28] (cadaver skin from a skin bank), amnion membrane,[30–32,67–74] xenografts (primarily porcine skin,[29,75] less commonly used nowadays) or (semi)synthetic materials.

Allografts are also used as an overlay over widely meshed autografts. This technique allows for undisturbed outgrowth of the autografts, and the allograft serves as a biological, protective dressing. Two different techniques are used. The sandwich technique uses wide-mesh autograft under small-mesh allograft,[68] and the intermingled technique[76,77] uses small, nonmeshed pieces of autograft embedded in large sheets of nonmeshed allograft.

One of the semisynthetic materials, Integra® (Integra Life Sciences, Plainsboro, NJ), is bilayered and designed in such a way that the wound bed grows into the wound side layer of the material. It thus becomes a "neodermis."[78,79] The outer layer of the dressing is a thin, protective, silicone sheet that is peeled off and replaced with autografts once the donor sites can be reharvested and the neodermis is vascularized. This allows for coverage of large excised areas even when not enough donor sites are available.

A new material with a different biochemical structure has just become available for the same purpose.[80,81] With this material, immediate grafting on the dermal replacement material seems to lead to a good take rate. Thus, waiting for ingrowth of tissues into the dressing does not seem to be necessary.

Proper fixation of the graft is important, and different techniques, including stapling, suturing, synthetic glue,[82] fibrin glue,[83] and specially designed fixation materials are used. The use of a negative pressure wound therapy is also advocated by some for fixation purposes,[84] and the negative pressure may also have a positive effect on wound healing.

DONOR SITES

Donor sites from which the skin grafts are taken can be nearly anywhere on the body. For the coverage of excised burns, they are virtually always split skin thickness: The recipient site grows into the split skin graft and enough deeper-dermis epidermal remnants remain in the donor site for it to heal quickly, and sometimes to be reharvested.

In extensive burns, the scalp is a good donor site. It reepithelializes rapidly and can be reharvested quickly and often.[85] Alopecia is usually not a problem.

To prevent excessive blood loss from the donor sites, topical agents such as epinephrine are applied after the donor site has been made. Other agents include thrombin and fibrin, and hemostatic dressings such as alginates are used as well.[86–90] To provide a better and flatter surface for the donor site excision, curved surfaces are sometimes injected subcutaneously with saline.

Donor sites can cause considerable morbidity,[91] primarily because they are very painful. They also may be difficult to heal.[92] In addition to the hemostatic requirements, dressings for donor sites generally need to have the same properties as those for the burns.

CHEMICAL LESIONS

As mentioned previously in this book, chemical lesions are not real burns. The injury is not thermal in nature, and the type of tissue damage is therefore different and dependent on the type of offending agent.

In general, though, partial thickness and full thickness chemical lesions are treated in the same way as partial and full thickness burns, although specific agents require additional measures. Depending on the nature of the chemical, systemic treatment may be necessary as well.

LONG-TERM RESULTS

Wound healing progresses into the remodeling phase after reepithelialization is complete. The amount of collagen and other extracellular matrix compounds is always a function of continuing lysis and production, and this is also true during the remodeling phase. However, the collagen produced during earlier stages of wound healing is disorganized, and during the remodeling phase it is reorganized into its fibrillar structure.[93] This highly complex process is regulated by a large number of cytokines.[94] Transforming growth factor β (TGF β) seems to play an early and central role.[94–96] The early formation of types I, III, and V collagen fibrils provides initial tensile strength to the wound while many other extracellular matrix (ECM) compounds, such as hyaluronic acid and fibronectin, also play a role in this process.

However, in many burns, the remodeling phase goes awry both with respect to the type of collagen as well as its orientation.[97] Macroscopically this results in hypertrophic scarring. A hypertrophic scar is raised above the skin level and very inflamed at the beginning. The scars often are debilitating and will interfere with the quality of life, because they may limit movement, can be painful, and virtually always are very pruritic. In addition, being disfigured is known to have a major impact on a patient's psychological well-being.

Hypertrophic scarring is virtually certain to occur in burns that have taken a long time to heal,[98,99] but rapidly healing burns may also result in serious scar formation, because scarring is largely genetically determined.[99–102] Dark-skinned patients have a significantly higher risk of scar formation. However, other factors contribute as well, including the location of the lesion (a sternotomy incision, e.g., virtually always results in a hypertrophic scar) and the age of the patient.

During the reepithelialization process, not much can be done to prevent hypertrophic scarring. However, because the chance of hypertrophic scar formation correlates with the time to reepithelialization, the use of dressings and techniques that are proven to reduce time to healing may contribute indirectly to reducing the incidence of hypertrophic scarring.[99]

In addition, in patients who are prone to scarring (based on the results of previous injury and wound healing time), preventative measures should be taken after reepithelialization is complete. Customized pressure garments are used,[103–105] sometimes with silicon sheeting as the primary contact layer,[106,107] as are steroid injections into the lesion.[108,109,110] Other therapies, such as the use of pharmacological agents[111] and different types of laser,[112] are currently being evaluated. Surgical scar revision is sometimes necessary, particularly when scar formation leads to contractures.

The results of hypertrophy prevention are often not truly satisfactory, and a visible scar may remain. In the long term, though, a hypertrophic scar will always become flatter and less inflamed.

Keloid formation is different from hypertrophic scarring, both physiologically as well as macroscopically. A typical keloid extends beyond the borders of the original wound and has a cauliflower-type aspect.[113] Prevention and treatment of keloid is even more difficult than that of hypertrophic scars,[114–117] and its discussion lies beyond the scope of this chapter.

Because of the scarring and the subsequent contractures and other healing problems, many patients have to undergo numerous reconstructive procedures. Facial and neck reconstruction, eyelid surgery, reconstruction of the hand's mobility and functions, and reconstructive surgery to increase mobility of joints in general are among the common procedures a burn victim has to undergo after reepithelialization is complete. The saying "once a burn patient, always a burn patient" is therefore not far beyond the truth.

An uncommon but important long-term complication in a burn wound is the development of a malignant tumor in an old scar, known as Marjolin's ulcer. Usually the tumor is a squamous cell carcinoma,[118] but basal cell carcinomas and melanomas have also been reported.[119] One has to be suspicious of the development of a malignancy when (part of) a burn lesion does not reepithelialize or when previously reepithelialized areas start to ulcerate. A biopsy or a series of biopsies is usually necessary to confirm the diagnosis. Therapy, of course, depends on the findings.

Heterotopic ossification is another uncommon but serious long-term complication, with a reported incidence of 1% to 3%,[120] particularly in patients with large burns.[121,122] The most frequent location of heterotopic bone is in joints with overlying deep burns.[123] The exact pathogenesis of heterotopic ossification is not known, but the fact that this complication occurs also in uninjured areas[124] supports the hypothesis that it is the consequence of the hypermetabolic[125] state that causes systemic changes in the connective tissues. Prevention thus may be linked to getting the patient back to a normometabolic state by early removal of all dead tissue and rapid wound closure. In addition, early mobilization seems to contribute to prevention of heterotopic bone formation.

If heterotopic ossification significantly interferes with joint motion, surgical exploration is indicated. This should preferably be done when the burn is healed, the scars are beyond their inflammatory phase, and the bone is roentgenongraphically well defined and mature.[125]

REFERENCES

1. Agnihotri N, Gupta V, Joshi RM. Aerobic bacterial isolates from burn wound infections and their antibiograms—A five-year study. *Burns* 2004;30(3):241–243.
2. Babik J, Bodnarova L, Sopko K. Acinetobacter—Serious danger for burn patients. *Acta Chir Plast* 2008;50(1):27–32.
3. Herruzo R, de la Cruz J, Fernandez-Acenero MJ, Garcia-Caballero J. Two consecutive outbreaks of *Acinetobacter baumanii* 1-a in a burn intensive care unit for adults. *Burns* 2004;30(5):419–423.
4. Girao E, Levin AS, Basso M, Gobara S, Gomes LB, Medeiros EA, et al. Seven-year trend analysis of nosocomial candidemia and antifungal (fluconazole and caspofungin) use in intensive care units at a Brazilian university hospital. *Med Mycol* 2008;46(6):581–588.
5. Hermans MH. Results of an internet survey on the treatment of partial thickness burns, full thickness burns, and donor sites. *J Burn Care Res* 2007;28(6):835–847.
6. McKenna SR, Latenser BA, Jones LM, Barrette RR, Sherman HF, Varcelotti JR. Serious silver sulphadiazine and mafenide acetate dermatitis. *Burns* 1995;21(4):310–312.
7. Hermans MH. A general overview of burn care. *Int Wound J* 2005;2(3):206–220.
8. Gear AJ, Hellewell TB, Wright HR, Mazzarese PM, Arnold PB, Rodeheaver GT, et al. A new silver sulfadiazine water soluble gel. *Burns* 1997;23(5):387–391.
9. Hendry AT, Stewart IO. Silver-resistant Enterobacteriaceae from hospital patients. *Can J Microbiol* 1979;25(8):915–921.
10. Modak SM, Fox CL, Jr. Sulfadiazine silver-resistant pseudomonas in burns. New topical agents. *Arch Surg* 1981;116(7):854–857.
11. Jarrett F, Ellerbe S, Demling R. Acute leukopenia during topical burn therapy with silver sulfadiazine. *Am J Surg* 1978;135(6):818–819.
12. Liebman PR, Kennelly MM, Hirsch EF. Hypercarbia and acidosis associated with carbonic anhydrase inhibition: A hazard of topical mafenide acetate use in renal failure. *Burns Incl Therm Inj* 1982;8(6):395–398.
13. Bondoc CC, Morris PJ, Wee T, Burke JF. Metabolic effects of 0.5 per cent silver nitrate therapy for extensive burns in children. *Surg Forum* 1966;17:475–477.
14. Monafo WW, Jr. Use of silver nitrate in the treatment of burns. *J Iowa Med Soc* 1966;56(9):927–932.
15. Innes ME, Umraw N, Fish JS, Gomez M, Cartotto RC. The use of silver coated dressings on donor site wounds: A prospective, controlled matched pair study. *Burns* 2001;27(6):621–627.
16. Caruso DM, Foster KN, Hermans MH, Rick C. Aquacel Ag in the management of partial-thickness burns: Results of a clinical trial. *J Burn Care Rehabil* 2004;25(1):89–97.
17. Atiyeh BS, El-Musa KA, Dham R. Scar quality and physiologic barrier function restoration after moist and moist-exposed dressings of partial-thickness wounds. *Dermatol Surg* 2003;29(1):14–20.
18. Zhang HQ, Yip TP, Hui I, Lai V, Wong A. Efficacy of moist exposed burn ointment on burns. *J Burn Care Rehabil* 2005;26(3):247–251.
19. Zawacki BE. The natural history of reversible burn injury. *Surg Gynecol Obstet* 1974;139(6):867–872.
20. Zawacki BE. Reversal of capillary stasis and prevention of necrosis in burns. *Ann Surg* 1974;1180(1):98–102.
21. Uchinuma E, Koganei Y, Shioya N, Yoshizato K. Biological evaluation of burn blister fluid. *Ann Plast Surg* 1988;20(3):225–230.
22. Wilson AM, McGrouther DA, Eastwood M, Brown RA. The effect of burn blister fluid on fibroblast contraction. *Burns* 1997;23(4):306–312.
23. Herndon DN. Perspectives in the use of allograft. *J Burn Care Rehabil* 1997;18(1 Pt 2):S6.

24. Horch RE, Jeschke MG, Spilker G, Herndon DN, Kopp J. Treatment of second degree facial burns with allografts—Preliminary results. *Burns* 2005;31(5):597–602.

25. Kesting MR, Wolff KD, Hohlweg-Majert B, Steinstraesser L. The role of allogenic amniotic membrane in burn treatment. *J Burn Care Res* 2008;29(6):907–916.

26. Lineen E, Namias N. Biologic dressing in burns. *J Craniofac Surg* 2008;19(4):923–928.

27. Hussmann J, Russell RC, Kucan JO, Hebebrand D, Bradley T, Steinau HU. Use of glycerolized human allografts as temporary (and permanent) cover in adults and children. *Burns* 1994;120(1):S61–65; discussion S65–66.

28. de Backere AC. Euro Skin Bank: Large scale skin-banking in Europe based on glycerol-preservation of donor skin. *Burns* 1994;20 Suppl 1:S4–9.

29. Hassan Z, Shah M. Porcine xenograft dressing for facial burns: Meshed versus non-meshed. *Burns* 2004;30(7):753.

30. Dahinterova J, Dobrkovsky M. Treatment of the burned surface by amnion and chorion grafts. *Sb Ved Pr Lek Fak Karlovy Univerzity Hradci Kralove* 1969;Suppl:513–515.

31. Chang CJ, Yang JY. Frozen preservation of human amnion and its use as a burn wound dressing. *Changgeng Yi Xue Za Zhi* 1994;17(4):316–324.

32. Rose JK, Desai MH, Mlakar JM, Herndon DN. Allograft is superior to topical anti-microbial therapy in the treatment of partial-thickness scald burns in children. *J Burn Care Rehabil* 1997;18(4):338–341.

33. Choukairi F, Hussain A, Rashid A, Moiemen N. Re: xenoderm dressing in the treatment of second degree burns. *Burns* 2008;34(6):896; author reply 97.

34. Keswani MH, Patil AR. The boiled potato peel as a burn wound dressing: A preliminary report. *Burns Incl Therm Inj* 1985;11(3):220–224.

35. Gore MA, Akolekar D. Evaluation of banana leaf dressing for partial thickness burn wounds. *Burns* 2003;29(5):487–492.

36. Delatte SJ, Evans J, Hebra A, Adamson W, Othersen HB, Tagge EP. Effectiveness of beta-glucan collagen for treatment of partial-thickness burns in children. *J Pediatr Surg* 2001;36(1):113–118.

37. Purna SK, Babu M. Collagen based dressings—A review. *Burns* 2000;26(1):54–62.

38. Smith DJ, Jr. Use of Biobrane in wound management. *J Burn Care Rehabil* 1995;16(3 Pt 1):317–320.

39. Voinchet V, Vasseur P, Kern J. Efficacy and safety of hyaluronic acid in the management of acute wounds. *Am J Clin Dermatol* 2006;7(6):353–357.

40. Sezer AD, Hatipoglu F, Cevher E, Ogurtan Z, Bas AL, Akbuga J. Chitosan film containing fucoidan as a wound dressing for dermal burn healing: Preparation and in vitro/in vivo evaluation. *AAPS PharmSciTech* 2007;8(2).Article 39.

41. Eisenbud D, Hunter H, Kessler L, Zulkowski K. Hydrogel wound dressings: Where do we stand in 2003? *Ostomy Wound Manage* 2003;49(10):52–57.

42. Hoekstra MJ, Hermans MH, Richters CD, Dutrieux RP. A histological comparison of acute inflammatory responses with a hydrofibre or tulle gauze dressing. *J Wound Care* 2002;11(3):113–117.

43. Ozcan C, Ergun O, Celik A, Corduk N, Ozok G. Enzymatic debridement of burn wound with collagenase in children with partial-thickness burns. *Burns* 2002;28(8):791–794.

44. Jancekovic Z. The burn wound from a surgical point of view. *J. Trauma* 1975;15(1):42–62.

45. Jackson D. Primary excision and grafting in deep burns of the hands. *Ned Tijdschr Geneeskd* 1960;104:1861–1862.

46. Gravante G, Delogu D, Esposito G, Montone A. Versajet hydrosurgery versus classic escharectomy for burn debridment: A prospective randomized trial. *J Burn Care Res* 2007;28(5):720–724.

47. Koller J, Bukovcan P, Orsag M, Kvalteni R, Graffinger I. Enzymatic necrolysis of acute deep burns—Report of preliminary results with 22 patients. *Acta Chir Plast* 2008;50(4):109–114.

48. Levine BA, Sirinek KR, Pruitt BA, Jr. Wound excision to fascia in burn patients. *Arch Surg* 1978;113(4):403–407.

49. Husain S, Ofodile FA. Tumescent technique for burn wound debridement: A cost effective method of reducing transfusion requirements. *J Coll Physicians Surg Pak* 2006;16(3):227–228.

50. Robertson RD, Bond P, Wallace B, Shewmake K, Cone J. The tumescent technique to significantly reduce blood loss during burn surgery. *Burns* 2001;27(8):835–838.

51. Herndon DN, Barrow RE, Rutan RL, Rutan TC, Desai MH, Abston S. A comparison of conservative versus early excision. Therapies in severely burned patients. *Ann Surg* 1989;209(5):547–552; discussion 52–53.

52. Heimbach DM. Early burn excision and grafting. *Surg Clin North Am* 1987;67(1):93–107.

53. Tanner JC, Jr., Vandeput J, Olley JF. The Mesh Skin Graft. *Plast Reconstr Surg* 1964;34:287–292.

54. Kreis RW, Mackie DP, Hermans RR, Vloemans AR. Expansion techniques for skin grafts: Comparison between mesh and Meek island (sandwich-) grafts. *Burns* 1994;20 Suppl 1:S39–42.

55. Raff T, Hartmann B, Wagner H, Germann G. Experience with the modified Meek technique. *Acta Chir Plast* 1996;38(4):142–146.

56. Vandeput J, Nelissen M, Tanner JC, Boswick J. A review of skin meshers. *Burns* 1995;21(5):364–370.

57. Vandeput JJ, Tanner JC, Boswick J. Implementation of parameters in the expansion ratio of mesh skin grafts. *Plast Reconstr Surg* 1997;100(3):653–656.

58. Meek CP. Microdermagrafting: The Meek technic. *Hosp Top* 1965;43:114–116.

59. Kreis RW, Mackie DP, Vloemans AW, Hermans RP, Hoekstra MJ. Widely expanded postage stamp skin grafts using a modified Meek technique in combination with an allograft overlay. *Burns* 1993;19(2):142–145.

60. Lari AR, Gang RK. Expansion technique for skin grafts (Meek technique) in the treatment of severely burned patients. *Burns* 2001;27(1):61–66.

61. Gallico GG, 3rd, O'Connor NE, Compton CC, Kehinde O, Green H. Permanent coverage of large burn wounds with autologous cultured human epithelium. *N Engl J Med* 1984;311(7):448–451.

62. Magnusson M, Papini RP, Rea SM, Reed CC, Wood FM. Cultured autologous keratinocytes in suspension accelerate epithelial maturation in an in vivo wound model as measured by surface electrical capacitance. *Plast Reconstr Surg* 2007;119(2):495–499.

63. Gallico GG, 3rd, O'Connor NE. Cultured epithelium as a skin substitute. *Clin Plast Surg* 1985;12(2):149–157.

64. Teepe RG, Kreis RW, Koebrugge EJ, Kempenaar JA, Vloemans AF, Hermans RP, et al. The use of cultured autologous epidermis in the treatment of extensive burn wounds. *J Trauma* 1990;30(3):269–275.

65. Boyce ST, Warden GD, Holder IA. Cytotoxicity testing of topical antimicrobial agents on human keratinocytes and fibroblasts for cultured skin grafts. *J Burn Care Rehabil* 1995;16(2 Pt 1):97–103.

66. Sheridan RL, Tompkins RG. Recent clinical experience with cultured autologous epithelium. *Br J Plast Surg* 1996;49(1):72–74.

67. Kreis RW, Hoekstra MJ, Mackie DP, Vloemans AF, Hermans RP. Historical appraisal of the use of skin allografts in the treatment of extensive full skin thickness burns at the Red Cross Hospital Burns Centre, Beverwijk, The Netherlands. *Burns* 1992;18(Suppl 2):S19–22.

68. Kreis RW, Vloemans AF, Hoekstra MJ, Mackie DP, Hermans RP. The use of non-viable glycerol-preserved cadaver skin combined with widely expanded autografts in the treatment of extensive third-degree burns. *J Trauma* 1989;29(1):51–54.

69. Alsbjorn BF. Clinical results of grafting burns with epidermal Langerhans' cell depleted allograft overlay. *Scand J Plast Reconstr Surg Hand Surg* 1991;25(1):35–39.

70. Leicht P, Muchardt O, Jensen M, Alsbjorn BA, Sorensen B. Allograft vs. exposure in the treatment of scalds—A prospective randomized controlled clinical study. *Burns Incl Therm Inj* 1989;15(1):1–3.

71. Eldad A, Din A, Weinberg A, Neuman A, Lipton H, Ben-Bassat H, et al. Cryopreserved cadaveric allografts for treatment of unexcised partial thickness flame burns: Clinical experience with 12 patients. *Burns* 1997;23(7–8):608–614.

72. Gajiwala K, Gajiwala AL. Evaluation of lyophilised, gamma-irradiated amnion as a biological dressing. *Cell Tissue Bank* 2004;5(2):73–80.

73. Ravishanker R, Bath AS, Roy R. "Amnion Bank"—The use of long term glycerol preserved amniotic membranes in the management of superficial and superficial partial thickness burns. *Burns* 2003;29(4):369–374.

74. Maral T, Borman H, Arslan H, Demirhan B, Akinbingol G, Haberal M. Effectiveness of human amnion preserved long-term in glycerol as a temporary biological dressing. *Burns* 1999;25(7):625–635.

75. Chiu T, Pang P, Ying SY, Burd A. Porcine skin: Friend or foe? *Burns* 2004;30(7):739–741.

76. Yeh FL, Yu GS, Fang CH, Carey M, Alexander JW, Robb EC. Comparison of scar contracture with the use of microskin and Chinese-type intermingled skin grafts on rats. *J Burn Care Rehabil* 1990;11(3):221–223.

77. Qu MM. [Dual grafting of autologous epidermal cells and skin allografts in the full-thickness burn wounds]. *Zhonghua Zheng Xing Shao Shang Wai Ke Za Zhi* 1988;4(1):25–26.

78. Orgill DP, Straus FH, 2nd, Lee RC. The use of collagen-GAG membranes in reconstructive surgery. *Ann N Y Acad Sci* 1999;888:233–248.

79. Winfrey ME, Cochran M, Hegarty MT. A new technology in burn therapy: INTEGRA artificial skin. *Dimens Crit Care Nurs* 1999;18(1):14–20.

80. Ryssel H, Gazyakan E, Germann G, Ohlbauer M. The use of MatriDerm in early excision and simultaneous autologous skin grafting in burns—A pilot study. *Burns* 2008;34(1):93–97.

81. Schneider J, Biedermann T, Widmer D, Montano I, Meuli M, Reichmann E, et al. Matriderm versus Integra: A comparative experimental study. *Burns* 2009;35(1):51–57.

82. Kilic A, Ozdengil E. Skin graft fixation by applying cyanoacrylate without any complication. *Plast Reconstr Surg* 2002;110(1):370–371.

83. Parry JR, Minton TJ, Suryadevara AC, Halliday D. The use of fibrin glue for fixation of acellular human dermal allograft in septal perforation repair. *Am J Otolaryngol* 2008;29(6):417–422.

84. Isago T, Nozaki M, Kikuchi Y, Honda T, Nakazawa H. Skin graft fixation with negative-pressure dressings. *J Dermatol* 2003;30(9):673–678.

85. Mimoun M, Chaouat M, Picovski D, Serroussi D, Smarrito S. The scalp is an advantageous donor site for thin-skin grafts: A report on 945 harvested samples. *Plast Reconstr Surg* 2006;118(2):369–373.

86. Basse P, Siim E, Lohmann M. Treatment of donor sites—Calcium alginate versus paraffin gauze. *Acta Chir Plast* 1992;34(2):92–98.

87. Brezel BS, McGeever KE, Stein JM. Epinephrine v thrombin for split-thickness donor site hemostasis. *J Burn Care Rehabil* 1987;8(2):132–134.

88. Cartotto R, Musgrave MA, Beveridge M, Fish J, Gomez M. Minimizing blood loss in burn surgery. *J Trauma* 2000;49(6):1034–1039.

89. Drake DB, Wong LG. Hemostatic effect of Vivostat patient-derived fibrin sealant on split-thickness skin graft donor sites. *Ann Plast Surg* 2003;50(4):367–372.

90. Nervi C, Gamelli RL, Greenhalgh DG, Luterman A, Hansbrough JF, Achauer BM, et al. A multicenter clinical trial to evaluate the topical hemostatic efficacy of fibrin sealant in burn patients. *J Burn Care Rehabil* 2001;22(2):99–103.

91. Zapata-Sirvent R, Hansbrough JF, Carroll W, Johnson R, Wakimoto A. Comparison of Biobrane and Scarlet Red dressings for treatment of donor site wounds. *Arch Surg* 1985;120(6):743–745.

92. Smith DJ, Jr, Thomson PD, Garner WL, Rodriguez JL. Donor site repair. *Am J Surg* 1994;167(1A):49S–51S.

93. Garcia-Filipe S, Barbier-Chassefiere V, Alexakis C, Huet E, Ledoux D, Kerros ME, et al. RGTA OTR4120, a heparan sulfate mimetic, is a possible long-term active agent to heal burned skin. *J Biomed Mater Res A* 2007;80(1):75–84.

94. Tang S, Pang S, Cao Y. [Changes in TGF-beta 1 and type I, III procollagen gene expression in keloid and hypertrophic scar]. *Zhonghua Zheng Xing Shao Shang Wai Ke Za Zhi* 1999;15(4):283–285.

95. Polo M, Smith PD, Kim YJ, Wang X, Ko F, Robson MC. Effect of TGF-beta2 on proliferative scar fibroblast cell kinetics. *Ann Plast Surg* 1999;43(2):185–190.

96. Smith P, Mosiello G, Deluca L, Ko F, Maggi S, Robson MC. TGF-beta2 activates proliferative scar fibroblasts. *J Surg Res* 1999;82(2):319–323.

97. Shakespeare PG, van Renterghem L. Some observations on the surface structure of collagen in hypertrophic scars. *Burns Incl Therm Inj* 1985;11(3):175–180.

98. McDonald WS, Deitch EA. Hypertrophic skin grafts in burned patients: A prospective analysis of variables. *J Trauma* 1987;27(2):147–150.

99. Deitch EA, Wheelahan TM, Rose MP, Clothier J, Cotter J. Hypertrophic burn scars: Analysis of variables. *J Trauma* 1983;23(10):895–898.

100. Bayat A, Bock O, Mrowietz U, Ollier WE, Ferguson MW. Genetic susceptibility to keloid disease and transforming growth factor beta 2 polymorphisms. *Br J Plast Surg* 2002;55(4):283–286.

101. Brissett AE, Sherris DA. Scar contractures, hypertrophic scars, and keloids. *Facial Plast Surg* 2001;17(4):263–272.

102. Lewis WH, Wan KC, Luk SC. Hypertrophic scar and prolonged reaction to BCG vaccination—Evidence for common genetic control. *Br J Biomed Sci* 1997;54(3):224–225.

103. Roques C, Teot L. The use of corticosteroids to treat keloids: A review. *Int J Low Extrem Wounds* 2008;7(3):137–145.

104. Roques C. Massage applied to scars. *Wound Repair Regen* 2002;10(2):126–128.

105. Rochet JM, Hareb F. Burns and rehabilitation. *Pathol Biol (Paris)* 2002;50(2):137–149.

106. Van den Kerckhove E, Stappaerts K, Boeckx W, Van den Hof B, Monstrey S, Van der Kelen A, et al. Silicones in the rehabilitation of burns: A review and overview. *Burns* 2001;27(3):205–214.

107. Van den Kerchhove E, Boeckx W, Kochuyt A. Silicone patches as a supplement for pressure therapy to control hypertrophic scarring. *J Burn Care Rehabil* 1991;12(4):361–369.

108. Grisolia GA, Danti DA, Santoro S, Panozzo G, Bonini G, Pampaloni A. Injection therapy with triamcinolone hexacetonide in the treatment of burn scars in infancy: Results of 44 cases. *Burns Incl Therm Inj* 1983;10(2):131–134.

109. Haedersdal M, Poulsen T, Wulf HC. Laser induced wounds and scarring modified by antiinflammatory drugs: A murine model. *Lasers Surg Med* 1993;13(1):55–61.

110. Beldon P. Management of scarring. *J Wound Care* 1999;8(10):509–512.

111. Xiang J, Wang XQ, Qing C, Liao ZJ, Lu SL. The influence of dermal template on the expressions of signal transduction protein Smad 3 and transforming growth factorbeta1 and its receptor during wound healing process in patients with deep burns]. *Zhonghua Shao Shang Za Zhi* 2005;21(1):52–54.

112. Asilian A, Darougheh A, Shariati F. New combination of triamcinolone, 5-Fluorouracil, and pulsed-dye laser for treatment of keloid and hypertrophic scars. *Dermatol Surg* 2006;32(7):907–915.
113. Selezneva LG. Keloid scars after burns. *Acta Chir Plast* 1976;18(2):106–111.
114. Demling RH, DeSanti L. Scar management strategies in wound care. *Rehab Manag* 2001;14(6):26–30.
115. Eisenbeiss W, Peter FW, Bakhtiari C, Frenz C. Hypertrophic scars and keloids. *J Wound Care* 1998;7(5):255–257.
116. Horswell BB. Scar modification. Techniques for revision and camouflage. *Atlas Oral Maxillofac Surg Clin North Am* 1998;6(2):55–72.
117. Wagner W, Alfrink M, Micke O, Schafer U, Schuller P, Willich N. Results of prophylactic irradiation in patients with resected keloids—A retrospective analysis. *Acta Oncol* 2000;39(2):217–220.
118. Dupree MT, Boyer JD, Cobb MW. Marjolin's ulcer arising in a burn scar. *Cutis* 1998;62(1):49–51.
119. Christopher JWL, Galzalez BD, Valuilis, JP. Non healing ulcer in a diabetic foot *Wounds* 2004;16(6):212–217.
120. Elledge ES, Smith AA, McManus WF, Pruitt BA, Jr. Heterotopic bone formation in burned patients. *J Trauma* 1988;28(5):684–687.
121. Chen HC, Yang JY, Chuang SS, Huang CY, Yang SY. Heterotopic ossification in burns: Our experience and literature reviews. *Burns* 2009; in press.
122. Siemers F, Lohmeyer JA, Machens HG, Eisenbeiss W, Mailander P. Heterotopic ossifications: A severe complication following extensive burn injury. *Handchir Mikrochir Plast Chir* 2007;39(5):360–363.
123. Jay MS, Saphyakhajon P, Scott R, Linder CW, Grossman BJ. Bone and joint changes following burn injury. *Clin Pediatr (Phila)* 1981;20(11):734–736.
124. Vanden Bossche L, Vanderstraeten G. Heterotopic ossification: A review. *J Rehabil Med* 2005;37(3):129–136.
125. Evans EB. Heterotopic bone formation in thermal burns. *Clin Orthop Relat Res* 1991;263:94–101.
126. Sheridan RLT. *Total Burn Care.* 2nd ed. New York: Saunders, 2002.

5 Cell Biology of Normal and Impaired Healing

Keith Moore

CONTENTS

Introduction 152
Principal Cells Involved in Healing 152
    Platelets 152
    Neutrophils 153
    Macrophages 154
    Endothelial Cells 154
    Fibroblasts 155
    Keratinocytes 155
Cell Interactions during Normal Healing 156
    Hemostasis 156
    Inflammation 157
    Neutrophils 157
    Macrophages 158
    Resolution of Inflammation 159
    Lymphocytes 161
Granulation Tissue Formation 162
Reepithelialization 164
Remodeling 165
Cytokines and Growth Factors in Regulation of Healing 165
Properties of Key Growth Factors and Cytokines 166
    Platelet-Derived Growth Factor 166
    Transforming Growth Factor-β 167
    Fibroblast Growth Factor 167
    Epidermal Growth Factor 168
    Vascular Endothelial Growth Factor 168
    Insulin-Like Growth Factor 169
    Cytokines and Chemokines 169
    Interleukin-1 169
    Tumor Necrosis Factor-α 170
Chronic and Impaired Wounds 171
    General Factors Impacting on Healing 171
        Aging 171
        Menopausal Effects 172

Nutrition 172
Other Factors 173
Chronic Wound Etiology... 173
Venous Leg Ulcers 173
Diabetic Ulcers.. 173
Pressure Ulcers... 174
Cellular Defects within the Chronic Wound 174
Inflammation 174
Granulation Tissue Formation.. 176
Epithelialization 178
Modulation of Wound Cell Biology by Clinical Interventions............ 178
References.. 180

INTRODUCTION

Healing of dermal wounds by secondary intention (i.e., without suturing), may be considered the natural process that has evolved to restore dermal integrity. The mechanisms involved predate medical interventions such as dressing the wound site to prevent infection and represent a process functioning as rapidly as possible to minimize blood loss and limit bacterial ingress to underlying tissues. Cosmetic consequences are of little consideration. Evolution has provided a complex system of interacting cells whose interrelated functions are controlled by multiple signaling systems. To aid in description and understanding of its apparent complexity, healing has been rationalized into a framework of sequential phases.[1] These lead in a temporal sequence from injury to wound closure and finally organized scar tissue. The phases are defined as hemostasis, inflammation, proliferation, epithelialization, and finally scar formation. They are not distinct, and events within one initiate and regulate events within other phases so that a number may be occurring simultaneously. Each phase of the sequence can be considered to be dominated by the functions of particular cell types (Figure 5.1).

PRINCIPAL CELLS INVOLVED IN HEALING

PLATELETS

Platelets, also called thrombocytes, are nonnucleated cellular fragments derived from megakaryocytes in the bone marrow. They circulate in large numbers in the peripheral blood and enter a wound site by release from damaged vasculature. Platelets possess two types of granules. Alpha granules that contain hemostatic proteins such as von Willebrand Factor, Factor V, Factor XIII, and fibrinogen plus growth factors. The latter are particularly important for initiation of healing. They include platelet derived growth factor (PDGF), vascular endothelial growth factor (VEGF), and platelet factor 4, a chemokine chemotactic for neutrophils, monocytes, and fibroblasts. They also contain proteins such as fibronectin that are involved in healing by promoting cell adhesion, migration, and differentiation. The second type of granule, called a dense granule, contains adenosine di- and triphosphate, calcium, and serotonin.

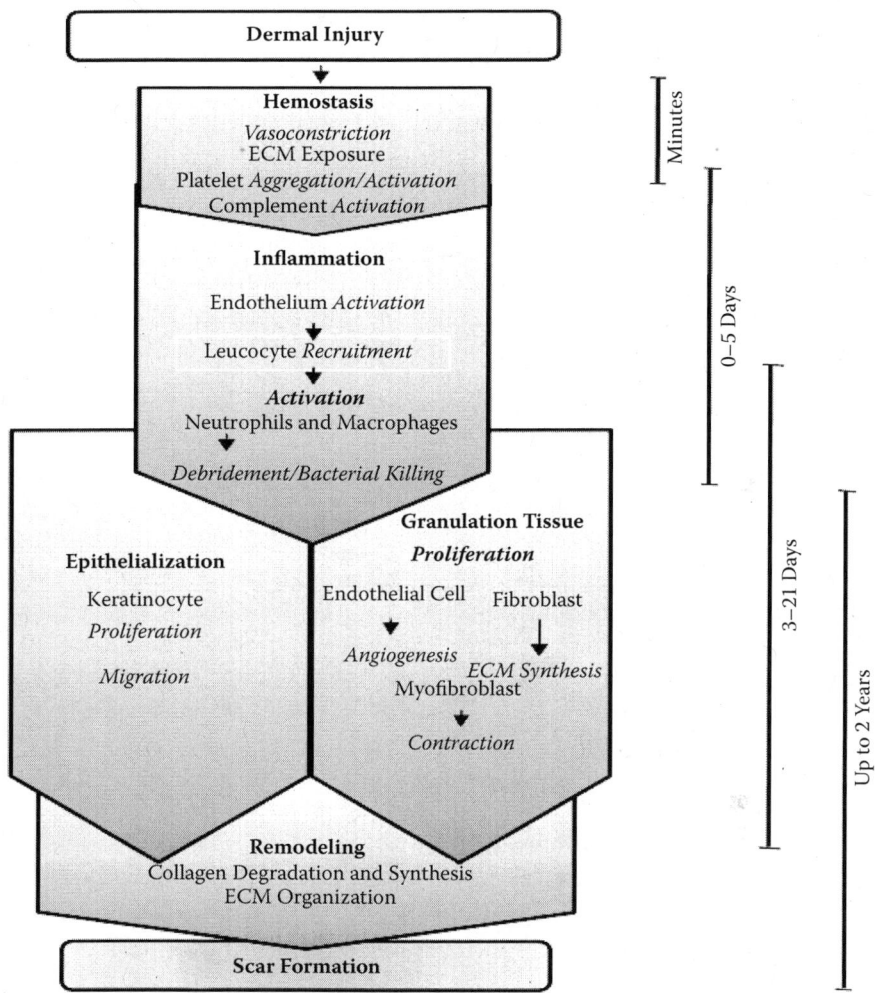

FIGURE 5.1 Normal dermal healing.

NEUTROPHILS

Along with the basophil and eosinophil, the neutrophil is part of the bone marrow-derived polymorphonuclear cell family found in the blood. It is the most numerous leukocyte, and in the nonactivated state has a half life of approximately 12 hours. Neutrophils are phagocytic cells that are a component of the innate immune system playing a primary role in defense against bacterial infection. Following binding to vascular endothelium at sites of inflammation, they extravasate and migrate down concentration gradients of chemotactic cytokines/chemokines such as interleukin-8 (IL-8), interleukin-1 (IL-1), and the complement component C5a. This chemotactic response is triggered by G-protein coupled membrane receptors that control reorganization of the neutrophil actin cytoskeleton to generate directed cell motility. Neutrophils are activated as a consequence of interaction with chemotactic agents, complement components, and phagocytosis of opsonized bacteria. Activation results in a lengthened life span to approximately 2 days and generation of bactericidal oxygen radical intermediates at the phagosome membrane following phagocytosis of bacteria.[2] Neutrophils are also able to generate extracellular fibers composed of DNA that form a meshwork called an NET (neutrophil extracellular trap) because of its ability to trap bacteria. NETs contain a number of enzymes (neutrophil elastase, myeloperoxidase, cathepsin G, and gelatinase) found in PMN cytoplasmic granules and the histones H1, H2A, H2B, H3, and H4. They are found at inflammatory sites and may serve as a physical antibacterial barrier and act to prevent tissue damage.[3]

MACROPHAGES

Macrophages found at the wound site lie at the end of a differentiation pathway originating in the bone marrow, which generates circulating monocytes that transmigrate into the peripheral tissues to form tissue macrophages. Under conditions of inflammation such as that resulting from tissue damage, they extravasate in a similar manner to neutrophils by initially binding via selectin ligands expressed by endothelial cells activated by inflammatory mediators. Membrane integrin-type receptors such as LFA-1, Mac-1, and VLA-4 then bind endothelial intracellular adhesion molecule-1 and -2 (ICAM-1 and -2) and vascular cell adhesion molecule-1 (VCAM-1), respectively. Resulting signal transduction then generates the monocyte chemotactic response to cytokines/chemokines such as IL-8 and monocyte chemotactic protein.

As their name suggests, macrophages are phagocytic scavenger cells capable of internalizing and digesting necrotic tissue and opsonized bacteria. Phagocytosed bacteria are contained within the phagosome where they can be killed by reactive oxygen species (ROS) to complement the antibacterial activity of neutrophils at the inflammatory site. Cells of the monocyte lineage are exquisitely responsive to their environment, which allows multiple functional, or activation, states to be achieved.[4] Phagocytosis, or interaction with the cytokine environment within wound tissue, influences the differentiation pathway that a monocyte will follow after leaving the vasculature. This confers an ability to secrete a diverse range of bioactive molecules including cytokines, growth factors, complement components, coagulation factors, enzymes and their inhibitors, ROS, ECM proteins, and many other molecules.[5] This diversity of functional states and the activity of the cytokines and growth factors secreted supports the concept proposed in 1975[6] that macrophages play a central role in regulating healing in addition to their role in innate immunity.

ENDOTHELIAL CELLS

Endothelial cells lining the lumen of dermal capillaries perform two major functions in the context of wound healing.

1. They act as gatekeepers regulating the rate of plasma exudation and the transmigration of leukocytes from circulating blood into wound tissue. By virtue of their ability to respond to proinflammatory cytokines such as IL-1β, they modulate surface adhesion molecules that bind leukocytes rolling along the capillary lumen and stimulate their migration in response to chemotactic cytokines generated in wound tissue.

2. In response to growth factors such as fibroblast growth factor (FGF) and VEGF they are stimulated to initiate angiogenesis. The growth factors induce endothelial activation and production of proteolytic enzymes that degrade the surrounding basement membrane. Endothelial cells then migrate away from the capillary and initiate proliferation. As cell numbers increase, they aggregate and differentiate into capillary sprouts to form new blood vessels within the surrounding tissue matrix.

Fibroblasts

Fibroblasts are mesenchymal cells found within the connective tissue of the dermis. They represent the activated state of fibrocytes and are found in large numbers in wound granulation tissue. They are a major source of ECM components such as collagens, glycosaminoglycans, elastin, and glycoproteins. Interaction of fibroblasts with factors present early in healing, such as PDGF, and proinflammatory cytokines, such as IL-1 and tumor necrosis factor-α). (TNFα induces secretion of keratinocyte growth factor, also known as FGF-7, which allows them to communicate with keratinocytes in a reciprocal paracrine fashion during healing.[7])

There is a large phenotypic variability of fibroblasts within the body, and in wound tissue they are considered to differentiate into a subpopulation of myofibroblasts under the influence of the local cytokine environment and mechanical signals received from the extracellular matrix (ECM). *In vitro* experiments indicate that transforming growth factor-β (TGF-β) drives toward myofibroblast formation, and IL-1 and TNFα modulate fibroblast responsiveness so that differentiation is a balance between a TGF-β-dominated or a proinflammatory environment.[7] Myofibroblasts are identified on the basis of ultrastructural features such as the presence of contractile microfilaments, many cell-to-matrix attachment sites and intercellular adherins and gap junctions.[8] They are able to exert tension using cytoplasmic actin stress fibers that generate contraction forces applied to their local environment.

Keratinocytes

Skin keratinocytes form a constantly renewing cell population in the stratum corneum. Stem cells are found at a frequency of 1 in 35,000 in the basal layer of the epidermis. Under normal conditions of homeostasis, daughter cells undergo terminal differentiation as they migrate upwards forming the stratum spinosum, stratum granulosum, and finally the stratum corneum. The migration is associated with increasing cornification of the keratinocyte followed by nonapoptotic death to form corneocytes.

Following dermal wounding, keratinocytes proliferating at the wound margin follow an alternative differentiation pathway by assuming a migratory phenotype characterized by changes in keratin subtype expression[9] to allow them to cover and reepithelialize the wound bed. They express the CXCR2 receptor that interacts with the chemokines IL-8 and growth-related oncogene-α (GROα). IL-8 is expressed at the wound margin and stimulates keratinocyte migration and proliferation[10] to suggest it has a role in promotion of keratinocyte migration from the wound margin to the wound bed that is additional to the chemotactic response induced by factors such as epidermal growth factor (EGF). The migratory phenotype is also associated with production of proteases that allow keratinocytes to dissect a migratory pathway through tissue and debris that would otherwise obstruct their movement. Additional to their role in providing a barrier function, keratinocytes play a role in resistance to infection by secreting antimicrobial peptides such as defensins and proinflammatory cytokines after interaction of toll-like receptors (TLRs) with bacteria.[11]

CELL INTERACTIONS DURING NORMAL HEALING

Hemostasis

Immediately following dermal injury, thromboxanes and prostaglandins are released from damaged cells to induce vasoconstriction and limit blood loss. As blood flows from damaged vasculature, platelets respond to the change in their environment. Platelets may be considered to act as monitors of the vascular system. Under normal conditions of homeostasis, they circulate freely. They do not interact with intact vasculature but will become adherent as soon as functional modification to endothelial cells is detected or they are exposed to ECM by tissue damage. Adhesion requires the synergistic interaction of platelet ligands that can bind to endothelial cell receptors such as P- and E-selectins and platelet receptors that can bind to ECM components including collagen types I, III, and VI, von Willebrand factor, fibronectin, laminin, fibulin, and thrombospondin. Fibrinogen, fibrin, and vitronectin bound to ECM at sites of injury may also induce platelet adherence which is followed by activation[12] and further aggregation.

Hemostasis is achieved by formation of a platelet-fibrin plug and the wound space fills with a fibrin clot composed of cross-linked fibrin fibers, plasma fibronectin, vitronectin, and thrombospondin. The clot provides temporary cover for the wounded dermis and acts as a provisional matrix that supports cell migration during healing. The local environment is modified by release of a diverse range of bioactive molecules following platelet degranulation and lysosomal and cytosol leakage. In addition to components required for coagulation, a number of factors are released that may be considered to initiate healing by regulating endothelial cell phenotype to allow leukocyte transmigration and chemotaxis to the wound site. Platelets coaggregate with leukocytes[13] to promote adherence to the endothelial cell wall and stimulate their transmigration into the wound site. Simultaneously, growth factors are released that induce chemotaxis of fibroblasts and keratinocytes to migrate from the wound periphery and stimulate their proliferation (Table 5.1). In addition to recruiting cells via chemotaxis, platelet-derived molecules influence diverse biological functions

TABLE 5.1
Platelet-Derived Bioactive Molecules
Involved in Initiation of Healing
Growth Factors

Platelet-derived growth factor

FGF-2

Transforming growth factor-β

Epidermal growth factor

Chemokines

Interleukin-8

CCL-5 (RANTES)

CXCL-1 (Cytokine-induced neutrophil chemoattractant-1)

CXCL-4 (Platelet factor-4)

CXCL-5 (Epithelial neutrophil-activating protein 78, ENA-78

Cytokines

Interleukin-1β

such as cell proliferation, survival, differentiation, and proteolysis, all of which are important in healing. This functional diversity has led to the development of platelet releasate–derived products for the stimulation of healing in chronic wounds.[14]

INFLAMMATION

Following adherence, platelet surface expressed IL-1β activates endothelial cells to produce interleukin-6 (IL-6), IL-8, and monocyte chemoattractant protein-1.[15] Simultaneously, IL-1β-dependent surface expression of the endothelial cell adhesion molecules ICAM-1 and αvβ₃ integrin increase via activation of the transcription factor NF-κB. Cytokine production and up-regulation of adhesion molecule expression synergize to induce neutrophil and monocyte adhesion to endothelial cells, forming the lumen of endothelium at the wound site and thus initiating the inflammatory phase of healing shortly after tissue injury.

NEUTROPHILS

Endothelial cell activation at the wound site effectively generates an inflammatory response in the local microenvironment. Neutrophils roll along the endothelium of postcapillary venules where they receive cytokine delivered inflammatory signals, bind to endothelium, and transmigrate into the wound space now filled with a fibrin clot. Neutrophil rolling along walls of inflammatory venules is mediated by β2-integrins, and conversion from rolling to binding is dependent on the time spent in close contact with endothelium.[16] Following binding neutrophil surface ligands such as P-selectin glycoprotein ligand-1 bind selectins and transmit transmembrane signals to activate the neutrophil for transmigration through vascular endothelium and generate a respiratory burst and degranulation. These events occur shortly after wounding and give rise to the early inflammatory phase that is neutrophil dominated. Activated neutrophils are highly phagocytic, a source of antimicrobial peptides and ROS, and may be considered a first-line defense against infection developing from bacterial contamination during wounding. Other antibacterial activity is mediated by proteases such as elastase, cathepsins, and urokinase-type plasminogen activator. Once phagocytosed, bacteria are sequestered in lysozomes where they are killed by ROS. Additionally, devitalized tissue is phagocytosed and proteolytically degraded to initiate the process of debridement.

Neutrophil degranulation and cell death result in loss of proteolytic enzymes and ROS to the wound tissue, and if neutrophil infiltration is prolonged, can contribute to delayed healing (*vide infra*). However for a non-infected wound, early inflammation resolves with 1–2 days of injury and the number of neutrophils decreases. The absence of further tissue damage and eradication of contaminating bacteria leads to cessation of synthesis of pro-inflammatory mediators, their catabolism and down-regulation of pro-inflammatory signaling pathways. This leads to resolution of the acute type inflammatory response, and endothelial cells revert to a non-inflammatory phenotype. Following their apoptosis, remaining neutrophils are then cleared by macrophage phagocytosis. This scenario implies that in the absence of a prolonged stimulus, early inflammation simply "winds down." Recent evidence suggests that this is not the case and that resolution of inflammation is a highly coordinated and active process.[17]

Initiation of inflammation during infection or tissue injury involves TLRs and nod-like receptors (NLRs). Both recognize bacterial products and products of inflamed tissue such as low molecular weight (MW) hyaluronic acid. Numerous inhibitors of TLR and NLR activity are induced by TLRs and trigger negative feedback inhibition. Thus, once triggered, the activity of these receptors is essentially self-limiting. Dysregulation of this system can lead to chronic inflammation. For example, fragments of hyaluronic acid generated during tissue injury can stimulate TLR2 and TLR4 receptors that are not activated by high MW hyaluronic acid and stimulate a pro-inflammatory feedback loop that may lead to chronic inflammation. T lymphocytes may also act to down-regulate inflammation via their TLR2. Although ligands for TLR2 are available for binding during an active inflammatory response, T lymphocyte activity is inhibited. Once TLR2 ligands have been eliminated, T lymphocytes produce the anti-inflammatory cytokine interleukin-10 (IL-10). Interestingly IL-10 levels are elevated in wound fluid taken from VLUs healing in response to compression therapy.[18]

MACROPHAGES

Monocyte transmigration from the blood to wound space is initiated at the same time as neutrophil transmigration and driven by the same source of proinflammatory mediators. However, whereas neutrophil numbers peak after 1–2 days and in the absence of infection rapidly decline, wound macrophages derived from blood monocytes continue to increase until day 5 after wounding.[19] Macrophage numbers then decrease slowly as healing proceeds and a significant population is retained during granulation tissue formation.

Induction of monocytopenia to abrogate extravasation of inflammatory mono-cytes and formation of wound macrophages has been demonstrated to impair heal-ing of experimental wounds.[6] Debridement and neutrophil clearance were delayed and fibroblast proliferation was diminished as granulation tissue formation slowed. These early observations led to the concept that the macrophage "plays a pivotal role in the transition between wound inflammation and wound repair."[20] Thus in addition to its role as a phagocytic scavenger removing cell debris and apoptotic neutrophils, the wound macrophage plays a role in regulation of healing. Macrophages are a major source of cytokines and growth factors, many of which play a role in regula-tion of both inflammation and healing (Table 5.2).

A number of these factors are counterregulatory (e.g., IL-10 and interleu-kin-12 [IL-12]), requiring that temporal and probably spatial separation be applied to their secretion by macrophages. Thus, not all factors are secreted all of the time and the cytokine/growth factor profile at any time point is deter-mined by macrophage activation status. Macrophage functional programming is achieved by their ability to respond to microenvironmental stimuli via mem-brane receptors. Differing combinations of stimuli generate different activation pathways and functional states.[21] Classically, activated macrophages are gener-ated by priming with interferon-γ (IFNγ) followed by stimulation with lipopoly-saccharide (LPS). This is analogous to an antibacterial response *in vivo* and is described as classical because it has been used for many years as an experimental model of macrophage activation. Type-II activated macrophages are generated by IFNγ priming and activation with LPS plus immune complexes and alterna-tively primed macrophages are generated by priming with interleukin-4 (IL-4). The resultant macrophage subpopulations have differing functional activities. Classically activated macrophages produce IL-12 mRNA, whereas Type II mac-rophages express IL-10 mRNA to respectively generate pro-inflammatory and anti-inflammatory activities.[22]

RESOLUTION OF INFLAMMATION

In the context of wound healing, classically activated macrophages would be required early in healing or when infection occurs, whereas type II macrophages would be required later as the inflammatory phase subsides and granulation pro-ceeds. Additional down-regulation of the inflammatory response is achieved when a state of hyporesponsiveness to stimuli is generated as an end stage of the classic activation pathway. Cellular reprogramming to achieve this refractory state, known as LPS tolerance, is dependent upon the cytokine context in which macrophages encounter stimuli such as LPS.[23] The consequence of LPS-tolerance is suppression of production of pro-inflammatory cytokines such as TNF-α, IL-6, and IL12 with no effect on IL-10, IL-1 receptor antagonist, or nitric oxide production.[24] The pri-mary role of nitric oxide in mediating vascular permeability changes and antibac-terial activity is during the inflammatory phase of healing. However its continued production as inflammation diminishes may be important in regulating granulation tissue formation as a consequence of its ability to stimulate angiogenesis and col-lagen deposition.[25]

TABLE 5.2

Macrophage-Derived Products Relevant to Healing

Product	Principal Activities
	Growth Factors
EGF	Mitogen: fibroblasts, keratinocytes
FGF-1, FGF-2	Mitogen: endothelial cells, fibroblasts, keratinocytes
	Stimulates angiogenesis
EGF	Mitogen: fibroblasts, keratinocytes
GM-CSF	Chemotaxis, proinflammatory, enhances neutrophil microbicidal activity
HGF	Mitogen: keratinocytes
	Activates neutrophils
	Antagonist to TGFβ
IGF-1	Mitogen: keratinocytes, fibroblasts, endothelial cell activation, angiogenesis
PDGF	Mitogen: fibroblasts
	Chemotactic: fibroblasts, macrophages, neutrophils
	Stimulates production of ECM
TGFβ	Regulates fibroblast proliferation and production of ECM proteins
VEGF	Mitogen: endothelial cells
	Synergy with FGF-2 promotes angiogenesis
	Chemokines
eg MIP-1 (CCL-1), MIP-1a (CCL3), CINC-1(CXL3), ENA78(CXL5)	Chemotactic for neutrophils and macrophages, proinflammatory
IL-8	Chemotactic for leukocytes; specific neutrophil activator; pro-angiogenic
	Cytokines
IL-1β	Chemotactic for leukocytes, proinflammatory, stimulates T lymphocytes
IL-2	Stimulates proliferation of T lymphocytes
IL-4	Down-regulates macrophage cytokine production
IL-6	Stimulates T-lymphocyte differentiation, synergy with IL-2
IL-10	Anti-inflammatory
	Inhibits TNFα production, T-lymphocyte proliferation
IL-12	Proinflammatory counterregulatory to IL-10
INFγ	Induces cytokine production by T lymphocytes
	Enzymes
MMP-1, -2, -3, -7, -9, -11, -12, -13, -18,-19	Proteolytic enzymes with wide substrate specificity
	Enzyme Inhibitors
PAI-1, PAI-2, TIMP-1, -2, -3, -4	Down-regulation of proteolytic enzyme activity

TABLE 5.2 (continued)
Macrophage-Derived Products Relevant to Healing

Product	Principal Activities
Antimicrobial	
Defensins	Bactericidal by insertion into cell membrane
Calcium binding proteins (S100A4, S100A8, S100A9)	Proinflammatory May be involved in keratinocytes differentiation
Other	
Fibronectin	ECM component
Nitric Oxide	Proinflammatory, antibacterial Stimulates endothelial cell and keratinocyte proliferation, collagen deposition

Note: FGF-1, -2, fibroblast growth factors-1,-2; EGF, epidermal growth factor; G-CSF, granulocyte colony stimulating factor; GM-CSF, granulocyte monocyte colony stimulating factor; HGF, hepatocyte growth factor; IGF-1, insulin-like growth factor-1; PDGF, platelet-derived growth factor; TGFβ, transforming growth factor–β; VEGF, vascular endothelial growth factor; MIP-1, monocyte chemoattractant protein-1 (CCL1); MIP-1α, macrophage inflammatory protein-1-alpha (CCL3); CINC-1, cytokine-induced neutrophil chemoattractant-1 (CXL3); ENA78, epithelial neutrophil-activating protein 78 (CXL5); IL-, interleukin-; IFγ, interferon-γ; MMP-, matrix metalloprotease-; PAI-1, -2, plasminogen activator inhibitor-1, -2; TIMP-1, -2, -3, -4, tissue inhibitor of metalloprotease-1, -2, -3, -4; ECM, extracellular matrix.

Macrophage deactivation can also be achieved by membrane interaction with apoptotic cells, CD36, $\alpha_v\beta_3$ integrin, and the phosphatidyl serine receptor. The latter is exposed on the outer surface of apoptotic cells. If neutrophils undergo necrotic death, they release proteases and ROS that may lead to tissue damage and delayed healing. Additionally, phagocytosis of necrotic cells will lead to macrophage activation and generation of more pro-inflammatory mediators and proteolytic enzymes. Neutrophil apoptosis therefore serves the dual purpose in the early phase of inflammation to prevent release of excess proteases and ROS and contribute to resolution of inflammation by reprogramming macrophages to a low-inflammatory phenotype.[26]

LYMPHOCYTES

Lymphocytes comprise a further component of the wound inflammatory infiltrate. They are predominantly T lymphocytes, but B lymphocytes have been identified in human acute wound tissue. Experiments in rodents demonstrate that T-lymphocyte depletion by systemic administration of anti-CD3 monoclonal antibody[27] or inactivation of Tγδ lymphocytes with Rapamycin[28] impairs healing of surgical wounds.

In acute wounds, lymphocytes accumulate at the wound margin adjacent to the migrating epidermal tip. Using immunohistochemical phenotyping, T lymphocytes can be divided phenotypically into CD4+ and CD8+ subpopulations. Within 1 week of surgery, human wound tissue contains raised levels of CD4+ cells which decrease with time so that immediately prior to wound closure CD8+ T lymphocytes predominate.[29] An increase in B lymphocytes was also observed as the wound closed in this study. A low CD4:CD8 ratio has also been documented for nonhealing chronic wounds.[30] Support for a role of CD8+ T lymphocytes in down-regulating healing is provided by the observation that depletion of these cells in a murine acute wound model enhances healing.[31] No effect on healing is observed when CD4+ cells are depleted.

T-lymphocyte interaction with the healing process is likely to be mediated by secreted cytokines. Both CD8[+32] and CD4[+33] subsets are heterogeneous with respect to their secreted cytokine profiles. For example, CD4+ TH1 cells produce IL-2, IFNγ, and TNFα, whereas CD4+ TH2 cells produce IL-4, IL-5, IL-6, and IL-10. The majority of studies defining T lymphocyte subset functionality are intended to define their role in generating immune responses. Until the cytokine profiles of wound tissue lymphocytes are characterized either by *in situ* hybridization or their isolation and functional characterization, it remains difficult to define their precise role in regulating healing.

GRANULATION TISSUE FORMATION

This phase of cellular proliferation initiates as the inflammatory phase peaks and numbers of neutrophils start to decline. Cytokines and growth factors in the inflammatory environment stimulate capillary endothelial cells to undergo angiogenesis and generate chemotactic gradients that attract fibroblasts from the surrounding dermis to migrate into the wound provisional matrix. They migrate by initially binding to matrix components such as collagen, vitronectin, fibronectin, or fibrin and then elongating until another attachment site is found down the chemotactic gradient. The original attachment site is then disengaged by proteolysis using matrix metalloproteases (MMPs) and the fibroblast moves forward by cytoskeletal contraction.

Although the majority of endothelial cells, fibroblasts, and keratinocytes involved in the postinflammatory stages of healing are derived from the local skin, a number of cells found within the wound are derived from circulating pluripotent mesenchymal stem cells originating from the bone marrow.[34] They are capable of differentiating into mesenchymal cells required for wound healing and can accelerate healing of normal and diabetes-impaired healing[35] suggesting their potential as a wound healing therapy.

Angiogenesis is regulated by tissue oxygen levels[36] and is crucial to wound healing because the high metabolic activity of cells within granulation tissue demands an adequate supply of oxygen and nutrients to be supplied from the peripheral circulation. Under hypoxic conditions, macrophages and fibroblasts that have been activated by TGF-β produce VEGF that activates capillary endothelial cells to detach and, using proteases to dissect a pathway, migrate into the surrounding clot and provisional matrix. Under the mitogenic influence of FGF and VEGF, they proliferate and form new vessels, or capillary sprouts as they are often referred to, that can grow at the rate of a few millimeters per day. When examined with low power magnification, the proliferating capillaries give the wound bed a characteristic red granular appearance.

In addition to proliferating endothelial cells granulation tissue is a mixture of inflammatory leukocytes and proliferating fibroblasts within a matrix of collagen, fibronectin, proteoglycans, and glycosaminoglycans. As fibroblasts increase in number, they synthesize matrix components to replace the fibrin clot with a provisional ECM that is continually degraded by proteolysis and resynthesized as the wound continues to heal. Controlled degradation of ECM components fulfils a number of functions. It is important for digestion of devitalized tissue and degraded proteins. It contributes to regulation of healing by release of bioactive molecules[37] from ECM components. For example, peptides derived from collagen, elastin, and fibronectin can mediate chemotaxis and modulate cell proliferation; fibrin-derived peptides can stimulate angiogenesis. Proteolysis is also important in the process of remodeling where fibers of collagen are degraded and resynthesized and cross-linked as the healed wound achieves greater strength during the later maturation phase.

Large amounts of collagen are synthesized by fibroblasts during the granulation phase. Collagen is secreted as procollagen which is cleaved at the n-terminal end to form tropocollagen. This forms the basic collagen molecule and comprises three polypeptides possessing a covalently bonded left-handed helical structure. Collagen is rich in hydroxyproline and hydroxylysine involved in forming cross-linkages. Tropocollagen self-aggregates into overlapping units to form collagen fibrils that then further aggregate to form collagen fibers. The whole process takes place within an extracellular space filled with a high water content gel formed from hyaluronic acid, glycosaminoglycans, chondroitin sulfate, dermatan sulfate, and heparan sulfate, all of which are produced by fibroblasts. Seven different collagens are found in significant quantities within the skin. Collagens I, III, V, XII, and XIV form structural fibrils, collagen VI forms microfilaments, collagen V forms reticular fibers, and collagen VII forms fibrils anchoring the epidermis to the dermis.

As the granulation phase progresses, the wound undergoes the process of contraction. Wound contraction is mediated by the myofibroblast that exerts contractile forces via anchor points at the epidermal margin to draw the wound margins inward and decrease wound area.

The myofibroblast develops by differentiation from the fibroblast.[38] Quiescent dermal fibroblasts acquire a migratory or activated phenotype in response to cytokines and signals received from wound ECM which differs in composition and mechanical properties from normal dermal ECM. This phenotype is characterized by the presence of contractile bundles composed of actin. Activated fibroblasts remodel their microenvironment and generate increasing stress in the ECM. They have also been described as protomyofibroblasts as they develop into myofibroblasts by expressing α-smooth muscle actin under the influence of TGFβ1, ECM proteins, and the mechanical environment. Interleukin-1 counteracts the stimulatory effect of TGFβ1, and its role has been suggested to prevent premature development of myofibroblasts during healing.[39] α-Smooth muscle actin, which acts as a marker for myofibroblasts, is incorporated into stress fibers and increases the contractile potential of the myofibroblast. Following wound re-epithelialization and possibly in response to repaired

ECM regaining its original mechanical properties, myofibroblasts are removed from the wound environment by apoptosis.

REEPITHELIALIZATION

Following injury, normal keratinocyte differentiation is perturbed and redirected toward covering the exposed wound tissue. Keratinocytes at the wound margin created by mechanical injury respond to factors such as TNFα, TGFβ, EGF, and IFNγ and convert to an activated phenotype. Keratinocyte proliferative activity increases in cells distal to the wound margin. The normal differentiation pathway is modified and daughter cells migrate from the margin to the wound bed in response to signals received via surface integrin receptors that recognize fibrin, fibronectin, and collagen present in the provisional wound ECM. Keratinocyte hemidesmosomes that attach to ECM and desmosomes that form cell–cell junctions are dissolved and surface membrane molecules such as vitronectin, fibronectin receptors, and the integrin α5β1 are expressed.[40] Intracellular actin filaments are synthesized and migration from the wound margin can commence before the proliferative program that generates large numbers of cells for migration over neogranulation tissue is initiated. This generates a migrating sheet of keratinocytes that are characterized by intercellular gaps, retraction of keratin filaments, and cytoplasmic vacuoles resulting from phagocytic activity. Migration is an organized process, and two theories have been suggested to describe its mechanics.[41] "Leap-frogging" proposes that daughter cells generated away from the wound margin migrate up to the suprabasal compartment and are pushed over the basal layer to the wound margin where they revert to the basal phenotype prior to migration. The "tractor-tread" theory explores the possibility that basal keratinocytes migrate over the wound bed and maintain desmosomal junctions to pull the epidermis inward from the margin.

As with other cell types that migrate in wound tissue, proteolytic action is required to dissect a pathway for keratinocyte migration. Migration is dependent on the keratinocyte cell membrane interacting with the ECM of developing granulation tissue. To maintain ECM contact keratinocytes have to migrate underneath any scab and phagocytose and digest any debris they may encounter. This is achieved by secretion of plasminogen activator, a serine protease that converts plasminogen to plasmin, to promote lysis of the scab. ECM components are also degraded by the actions of keratinocyte secreted MMPs. The whole process is executed more efficiently in a moist environment, and the accelerated healing found when wounds are prevented from air drying[42] led to development of the concept of moist wound healing and wound dressings that manage wound moisture content.

Re-epithelialization is terminated when keratinocytes migrating from the wound margins achieve a density over the wound bed that allows cell–cell interaction and contact inhibition induces a cessation of migration. Intact normal skin is characterized by a basement membrane at the dermal–epidermal junction and to restore skin integrity the arrested keratinocytes have to reestablish attachment to the granulation tissue surface. This process is partially controlled by TGFβ that can up-regulate synthesis of attachment molecules such as $\alpha_6\beta_4$ integrin and basement membrane components such as collagen type IV. The basement membrane also contains anchoring structures such as collagen type VII fibrils which are produced locally by keratinocytes or macrophages. The now stationary keratinocytes down-regulate MMP synthesis.

For an incisional, sutured wound or a partial thickness wound where keratinocytes can migrate from surviving hair follicles re-epithelialization is rapid; for a larger full thickness wound healing by secondary intention re-epithelialization is slower because of the requirement for proliferation at the margin to provide sufficient cells to cover the wound surface.

REMODELING

Collagen is constantly being degraded and resynthesized even in normal skin, and during healing, the rate of both increases. The process of remodeling where blood vessels are resorbed and fibroblasts decrease in number initiates approximately 21 days after injury when the overall collagen content of the wound has stabilized. Resynthesis allows collagen cross-linking and reorientation of the fibers to increase wound tensile strength. Remodeling is controlled by achieving a balance between synthetic and degradative activity. In part, this is achieved by regulating MMP activity by specific inhibitors known as tissue inhibitors of matrix metalloproteases (TIMPs). MMPs are produced by macrophages, fibroblasts, and keratinocytes. Their production is inducible and regulated by proinflammatory cytokines, growth factors, and contact with ECM. Transforming growth factor-β down-regulates MMP synthesis and up-regulates production of TIMP.

As ECM remodeling progresses, collagen fibers become more organized, fibronectin decreases, and hyaluronic acid and glycosaminoglycans are replaced by proteoglycans. Tensile strength of the healed wound increases dramatically from 1 to 8 weeks after injury and slowly thereafter. Remodeling continues for many months or years after wound closure but tensile strength never usually achieves more than 80% of that of nonwounded skin. The resultant scar tissue is brittle and less elastic than normal skin with an absence of hair follicles and sweat glands. Scar tissue is often aesthetically unsatisfactory, but it fulfils the major desired result, in evolutionary terms, of restoring skin barrier function and preventing the ingress of bacteria and allowing maintenance of homeostasis.

CYTOKINES AND GROWTH FACTORS IN REGULATION OF HEALING

Historically the term *cytokine* has been used to refer to molecules such as interleukins, interferons, and molecules such as TNFα which are involved in cell–cell signaling during the generation of immune responses. Growth factor generally refers to polypeptides capable of stimulating cell proliferation and differentiation. However, in the context of wound healing, there is much overlap between the two and they can be considered as intercellular messengers that regulate cell function mediated by interaction with cell membrane receptors. In this section, the generic term *cytokine* will be used to describe both.

From the foregoing section on normal healing, it is self-evident that cytokines play a major role in regulating the interactions between the various cells involved in healing. They may act in a paracrine manner on cells in the local environment of the secreting cell, as autocrine factors on the secreting cell, and also as endocrine factors when bound to carrier proteins.

However, consideration of their bioactivity in relation to particular cell functions in isolation can be misleading and obscure the true complexity of the interactions involved in healing. By attempting to understand this complexity, it may be possible to eventually identify the key factors leading to the pathological disruption to cytokine networks that occur within chronic wounds without falling into the trap of trying to gain insight by the criterion of "guilt by localisation, timing and properties."[43] In any consideration of the role of cytokines, it is also necessary to recognize that they are only one of many cell modulatory signals that can be found within the wound environment, including peptides, lipids, hypoxia, nitric oxide, pH change, or ECM stress.

The importance of the inflammatory phase in initiating and regulating granulation tissue formation and re-epithelialization has been described earlier. These observations were derived from experimental studies designed to investigate the cell biology of healing wounds. In passing, many of these studies highlight the importance of cytokine involvement in regulating the process. The importance of immune cell–mesenchymal cell interaction has been highlighted by using informatics analysis (neural networks) to identify functional cytokine networks.[44] This approach demonstrates that cells do not function alone and that cytokine connectivity of immune cells and other cells involved in wound healing is compatible with the concept that the cells are collectively integrated into a maintenance system that restores dermal function after injury.

PROPERTIES OF KEY GROWTH FACTORS AND CYTOKINES

PLATELET-DERIVED GROWTH FACTOR

A number of homodimeric PDGF molecules are formed from the polypeptides PDGF-A, -B, -C, -D, and the heterodimeric PDGF-AB. They are released early in healing from platelets and are bound by serum and ECM proteins. Having a short (~2 minutes) half life in the circulation, PDGF functions as a local autocrine and paracrine factor mediating its effect via transmembrane tyrosine kinase receptors whose synthesis is autoregulated by PDGF.

PDGF is chemotactic for neutrophils, macrophages, and fibroblasts and thus can be considered important for initiating healing after platelet aggregation. This source of PDGF is no longer available after hemostasis is achieved and continuing production is performed by macrophages, endothelial cells, and fibroblasts. This allows PDGF to play a role in augmenting fibroblast proliferation, ECM production, myofibroblast differentiation, and wound contraction. Such diversity of function supports the established concept that PDGF plays an important role in regulation of healing[45] and has led to the use of recombinant PDGF-BB for the treatment of nonhealing wounds.[46]

TRANSFORMING GROWTH FACTOR-β

TGF-β is considered to play a significant role in regulating granulation tissue formation, ECM production, and angiogenesis.[47] It exists as at least five isoforms although TGF-β1 is the most prevalent. As with PDGF it is released during hemostasis and is synthesized by macrophages, lymphocytes, and endothelial cells. Production can be autoregulated by FGF and stimulated by EGF and IL-1. It is secreted as an inactive form complexed with the n-terminal portion of the TGF-β precursor molecule that has to be cleaved by proteolysis at the cell surface to release the bioactive form of TGF.

Following hemostasis, TGF-β duplicates the chemoattraction exerted by PDGF to attract neutrophils, macrophages, and fibroblasts to the wound site. Initial elevated levels of TGF-β are found after the first day following injury. Synthesis of the -1 and -2 isoforms are induced early in healing with TGF-β3 being found at the later stages of healing to contribute to a second peak of TGF-β activity 5 days after wounding. This may also be contributed to by release from ECM where TGF-β is bound via latent-TGFβ-binding-protein from which it is released and activated by proteases such as thrombin and plasmin.

In addition to its role in the earlier phases of healing, TGF-β has been shown to play a role in control of scar formation.[48] Inhibition of the activity of the -1 and -2 isoforms in rodent incisional wounds decreased ECM deposition and reduced scarring, suggesting that endogenous TGFs contribute to scarring. In contrast, treatment with recombinant TGF-β3 reduced scarring to demonstrate that this isoform acts as an antagonist for TGF-β1 and -β2.

FIBROBLAST GROWTH FACTOR

There are more than 20 members of the FGF family of 16–18 kDa polypeptides. FGF-2, previously known as b-FGF, is considered a prototype for the family. They are bound by heparin sulfate–containing proteoglycans which allows them to be stored in an inactive form in the ECM. They are synthesized by endothelial cells and fibroblasts that also store FGF to allow rapid release following injury.

Bioactivities of FGFs suggest a primary role in granulation tissue formation and epithelialization and a possible role during inflammation.[49] Involvement in the inflammatory phase may be via FGF-1 activating T lymphocytes to produce the proinflammatory cytokine interleukin-2.

Both FGF-1 and FGF-2 stimulate proliferation of fibroblasts and endothelial cells to act as positive regulators of granulation tissue synthesis. FGF-2 can also facilitate migration of these cells by inducing production of urokinase-type plasminogen activator, an enzyme that converts plasminogen to plasmin. This proteolytic enzyme can then cleave a pathway through the fibrin clot for migrating cells. Cell surface αVβ3 integrin expression is up-regulated by FGF-2 to mediate binding to ECM components such as vitronectin and fibrinogen. FGF-7 is also produced by fibroblasts and acts a potent mitogen specific for keratinocytes.

EPIDERMAL GROWTH FACTOR

EGF is a 64 kDa molecular mass member of the EGF family whose members exhibit sequence homology and act as strong mitogens. The family includes TGF-α, which binds to the same receptor as EGF and shares similar biological activities. EGF is widely distributed in body fluids, being found in saliva, milk, plasma, and urine as well as wound fluid. At the wound site potential cellular sources for both EGF and TGF-α include macrophages during the inflammatory phase of healing and wound margin keratinocytes.

EGF receptors are transmembrane glycoproteins found on the majority of mammalian cells. They are present in highest numbers on epithelial cells and also present in high numbers on fibroblasts and endothelial cells. The receptor exists in one of two conformational states that bind EGF with either a high or low affinity. Its activity can be modulated by other growth factors such as PDGF and FGF. In addition to its mitogenic effect on epithelial cells, fibroblasts, and endothelial cells, EGF stimulates synthesis of the ECM components fibronectin, collagen, and glycosaminoglycans.

VASCULAR ENDOTHELIAL GROWTH FACTOR

The VEGF family is composed of seven members: VEGF-A, -B, -C, -D, -E, -F, and PLGF (placental growth factor). VEGF-A appears to be the dominant isoform in wound healing.[50] It is a 34-to-42- kDa dimeric, disulfide bound glycoprotein that exists in at least seven homodimeric isoforms. Three isoforms express a heparin-binding domain that binds strongly to heparin-containing proteoglycans in the ECM. The remaining VEGF-A isoforms that lack the domain are diffusible within the wound matrix. Proteolysis of the ECM liberates bound VEGF-A to generate highly diffusible fragments.

VEGF-A only binds to either type 1 and 2 receptors tyrosine kinase membrane receptors. VEGF receptor-1 (VEGFR-1) binds VEGF-A with greater affinity than VEGFR-2 and each receptor mediates different bioactivities. VEGFR-2 binding is more important for endothelial cell mitogenesis and stimulates production of platelet-activating factor. Ligand binding by VEGFR-1 is unable to initiate mitogenesis but modulates cell division immediately before formation of primitive blood vessels. The receptor is expressed by leukocytes and induces transmigration and activation of monocytes and chemotaxis of neutrophils.

VEGF-A which is secreted by macrophages, endothelial cells, and keratinocytes, acts in a paracrine and autocrine manner to stimulate angiogenesis. Its expression is induced when cells are subjected to hypoxia. This response is dependent upon production of hypoxia-induced protein complex, HIF-1, which binds to the enhancer sequence of the VEGF-A gene. In the proinflammatory environment of the wound, TNF-α triggers release of VEGF-A and increases transcription of the VEGFR-2 gene in endothelial cells. VEGF-A synthesis is also augmented by TGF-β, EGF, and PDGF. IL-1β and TNF-α are also able to induce VEGF-A gene expression in wound margin keratinocytes.

INSULIN-LIKE GROWTH FACTOR

Two insulin-like growth factors (IGFs) have been described.[51] They share 47% sequence homology with insulin and, like it, can cause hypoglycemia. They are widely distributed throughout the body and were originally identified as endocrine factors although they can act in an autocrine and paracrine manner. They are potent mitogens, anti-apoptotic, and stimulate keratinocyte chemotaxis and fibroblast ECM production. The insulin-like effects of IGF-1 from which it derives its name include glucose uptake, glycolysis, and glycogen synthesis. These biological effects are transduced via the IGF-1 transmembrane tyrosine kinase receptor.

Within the wound, IGF may be produced by macrophages, fibroblasts, and keratinocytes as well as being derived from plasma. Maximum levels are observed in wound fluid or tissue early during healing and correlate with increased cell migration and proliferation. Its potentially important role in healing is demonstrated by the observation that IGF-1 is absent or reduced in recalcitrant diabetic wounds and that fibroblasts isolated from these wounds are resistant to its mitogenic effect.[52]

CYTOKINES AND CHEMOKINES

The inflammatory process generates a diverse range of cytokines[53] and chemokines[54] that can be found within wound tissue. These are involved in a complex regulatory network that controls wound inflammation and the outcome, healed or chronic wound, of the healing process. The precise identity of the component cytokines and their interaction remains to be elucidated. Although we have to be cognizant of the earlier mentioned caveat regarding guilt by association, the data available suggest that IL-1 and TNFα may play significant roles in regulation of healing and, in particular, pathogenesis of the chronic wound. However, many other cytokines and chemokines have been identified in wound fluid taken from healing and nonhealing wounds. Although their potential roles are less well defined, they will no doubt contribute in some part to the overall regulation of healing.

INTERLEUKIN-1

Secreted IL-1 exists as the –alpha or –beta 17 kDa forms that share 27% c-terminal homology and have approximate functional equivalence. IL-1β is predominantly secreted by monocytes and activated macrophages and keratinocytes are a major source of Il-1α. Other sources include neutrophils, endothelial cells, and fibroblasts. Both forms of IL-1 are synthesized as precursors that require enzymatic cleavage by the proteases elastase, cathepsin G, and collagenase that are found in the inflammatory environment.

Synthesis of IL-1 can be induced by many stimuli including bacterial endotoxins, TNF-α, IFN-α, -β, -γ. Macrophages are particularly responsive to stimulation with bacterial endotoxin and fibroblasts produce IL-1 in response to IL-1α and TNF-α. IL-1 production is subject to positive or negative feedback regulation depending upon the functional status of the secreting cell. Activity of IL-1 is controlled by IL-1 receptor antagonist (IL-1RA). Multiple isoforms of this molecule are produced by macrophages and fibroblasts that bind to the IL-1 receptor to competitively inhibit the local pro-inflammatory effects of IL-1. Overall tissue pro-inflammatory activity with respect to IL-1 is thus a balance between concentration determined by level of synthetic stimuli and receptor availability regulated by receptor expression and binding of IL-1RA.

The major biological functions of IL-1 are promotion of immune and inflammatory responses—the former by stimulation of T-helper cells to secrete IL-2 and express IL-2 receptors, and promote proliferation of B lymphocytes and immunoglobulin production; and the latter by enhancing arachidonic acid metabolism in inflammatory cells, fibroblasts, and endothelial cells. It modulates endothelial cell function promoting thrombotic processes, enhancing adhesion molecule expression, and leukocyte adhesion, chemotaxis, and activation. It enhances production of proinflammatory proteins including collagenase and elastase that are considered to play a role in the pathogenesis of chronic wounds (*vide infra*).

Tumor Necrosis Factor-α

TNF-α is a nonglycosylated 17 kDA protein that has many features in common with IL-1β. It is primarily produced by monocytes, macrophages, neutrophils, and T-cells but may also be secreted by fibroblasts and keratinocytes. Production is up-regulated by many stimuli including bacterial endotoxins, IFNs, IL-2, and granulocyte monocyte colony stimulating factor (GM-CSF). Production is inhibited by IL-6 and TGF-β.

Two receptors of 55–60 kDa and 75–80 kDa have been described and identified on all cells found within wound tissue except erythrocytes. The 55-kDa receptor has a c-terminal region that has been described as the "death domain" as it is involved in induction of apoptosis. Interaction of TNF-α with its receptor is modulated in a number of ways. Receptor density is increased by IFN-α, -β, -γ and decreased by IL-1. Soluble TNF binding factors derived from the TNF-binding domain of membrane receptors also inhibit bioactivity by binding TNF before it can interact with the membrane expressed receptor.

TNF-α exhibits a wide spectrum of biological activities, notably induction of cytolysis and cytostasis of many tumor cell lines *in vitro* and hemorrhagic necrosis of experimental tumors *in vivo*. Its pro-inflammatory properties synergize with IL-1 *in vivo* to enhance inflammation. It is prothrombotic, a chemoattractant for neutrophils, and induces synthesis of chemotactic chemokines. TNF-α also induces synthesis of IL-1 and prostaglandin E_2 (PGE_2) by nonactivated macrophages and augments immune responses by enhancing the response of T-cells to IL-2 and promotion of B-cell differentiation.

Although the promotion of potent inflammatory responses may be considered to potentiate wound chronicity, TNF-α possibly contributes to healing as it can promote angiogenesis and act as a growth factor for fibroblasts. It can also induce fibroblast secretion of collagenase and PGE_2. The relative roles of TNF-α in healing/nonhealing may be determined by concentration and timing of production during the healing process.

CHRONIC AND IMPAIRED WOUNDS

As described earlier, normal wound healing for the healthy individual in the absence of complicating factors such as wound infection is a predictable sequence of events leading to wound closure and restoration of dermal integrity.[55] However, comorbidities and associated medication may exacerbate delayed healing and development of chronic nonhealing wounds.

Wounds exhibiting defective healing are found in surprisingly high numbers, particularly in an aged population.[56] Venous leg ulcers (VLUs) occur with a frequency >1%, 15% of diabetics may develop foot ulcers, and the prevalence of pressure ulcers (PUs) can reach 15% in an acute setting. Following treatment of underlying pathologies, compression therapy for VLU, pressure relief for PU, and diabetic foot ulcers (DFUs) plus metabolic control for diabetics, many of these wounds will respond positively to standard good wound care practice. However, a substantial minority will remain refractory to treatment and require adjunctive therapies to achieve wound closure, or in some cases may never heal.

GENERAL FACTORS IMPACTING ON HEALING

Aging

Aging is associated with many well-documented changes to tissue structures and biological functions essential to maintaining skin integrity and tissue repair. However, the process of normal aging is not necessarily considered detrimental to epidermal healing and may even confer some benefits. Thus Eaglestein[57] concluded that "The ability of the aged to heal so well illustrates not that their healing processes are equal to those of the young but rather that our healing capacity is far in excess of what is needed." (p. 183)

Beneficial changes to some aspects of the healing process are indicated by the demonstration that increased levels of fibrillin and elastin found in aged human experimental dermal wounds confer improved scar quality.[58] This may be due to age-related changes in wound inflammatory processes as decreased inflammatory responses are also associated with scarless fetal healing. The lack of a detrimental effect of aging on normal wound healing in the healthy is not restricted to the skin, as experimental wounds of stomach and duodenum in aged rats heal as well as those in the young.

The aged individual may be more susceptible to wound healing problems resulting from interactions of body systems with environmental stresses, progressive diseases, concomitant medication, and the general aging process. Although these changes may not directly induce suboptimal healing in older adults, they may exacerbate pathological changes leading to a poorer wound prognosis than the same pathology may induce in the young. This concept is illustrated by comparison of models of impaired healing in young and aged rats.[59] The healing rate of incisional wounds in 3-month-old and 24-month-old rats is identical. An ischemic wound can be created by using an H-shaped double skin flap, where the test wound is below the horizontal line in the H. Blood flow in the ischemic wound the first day after wounding is only 7% of that found in normal skin. Although healing is delayed in both the young and old rats by ischemia, a further delay of 40% to 65% is found in aged rats.

Menopausal Effects

The decrease in estrogen levels post-menopause may have a deleterious effect on normal skin function and either topical estrogen or hormone replacement therapy (HRT) has been indicated as a means of at least partially restoring normal function. It prevents a decrease in skin collagen content and administration either systemically or applied topically to healthy skin may increase collagen content.[60] This may explain to some extent the observed lower incidence of PUs (relative risk 0.68) and VLUs (relative risk 0.65) in women treated with HRT); however, another contributory factor may be the effect of estrogen directly on the healing process. Aging in healthy females is accompanied by a reduced rate of cutaneous healing with improved scar quality.[61] These effects are reversed by HRT treatment or topical estrogen treatment of both aged females and males[62] and may be related to an anti-inflammatory effect of estrogen decreasing the numbers of neutrophils and amount of the proteolytic enzyme elastase in the wound tissue.

Nutrition

The most important dietary components considered essential for healing have been defined as carbohydrates, proteins and amino acids (particularly arginine and glutamine), fats, polyunsaturated fats, zinc, vitamin A, and vitamin C.[63] All these are supplied by a well-balanced diet, but malnutrition (undernutrition, overnutrition, or dietary imbalance) is prevalent in the aging population prone to chronic wounds. For a metabolically active lesion such as a wound, it is almost axiomatic that overt malnutrition would be associated with impaired healing. Poor nutritional status is considered a predisposing factor in impaired ulcer healing, but a causal relationship to wound chronicity has not been demonstrated.[64]

Patients with chronic wounds have been demonstrated to have a number of specific nutritional defects. In a study of 25 patients[65] with large VLUs, 18 (72%) had vitamin C deficiency, 10 (40%) decreased albumin, 3 (12%) zinc deficiency, 2 (8%) folate deficiency, and 2 (8%) iron deficiency. These deficiencies were identified by comparison to normal biochemical values for the whole population. Comparison of age- and sex-matched controls[66] also demonstrated significant deficiencies of zinc and also vitamin A and vitamin E. Zinc has received much attention in this context and up to 50% of patients with leg ulcers have decreased plasma zinc levels. Healing of gastric ulcers is delayed in zinc-deficient animals with decreased cell proliferation rates.[67] However, restoration of zinc levels immediately after ulcer formation did not restore healing to normal. Attempts to enhance healing with dietary zinc supplementation in humans has met with limited success, with some studies demonstrating a positive effect and others demonstrating no benefit.

In the context of wound healing, nutritional data of the type quoted in this section have to be interpreted with caution for two reasons:

1. For example, vitamin C is required for collagen formation and a deficiency leads to lower levels of skin hydroxyproline, raising the theoretical possibility of impaired healing. However, vitamin C deficiency is common in the aged population and is not necessarily related to defective healing.
2. Nutritional status is usually assessed by measuring systemic factors such as blood parameters or anthropometrics. Localized lesions such as chronic wounds create unique microenvironments and plasma levels may not reflect bioavailability at a tissue level.

Other Factors

A number of other factors such as vascular disease, diabetes, obesity, hematologic diseases, or cytotoxic chemotherapy contribute to defective healing.

CHRONIC WOUND ETIOLOGY

The most prevalent chronic wounds are VLUs, DFUs, and PUs. Once formed they exhibit many common features, yet they arise from differing etiologies.

VENOUS LEG ULCERS

Venous leg ulcers arise as a severe consequence of chronic venous insufficiency. The precise mechanism for development of the VLU remains to be defined, as there are a number of proposed hypothetical mechanisms available, each of which may contribute in part.[68] VLUs are characterized by pericapillary deposition of fibrin ("fibrin cuffs") associated with extravasation of neutrophils and monocytes from the capillary circulation. Whether the fibrin cuff is deposited first leading to leukocyte trapping or whether increased venous pressure leads to a capillary pressure decrease causing leukocyte trapping and activation that in turn leads to increased capillary permeability and fibrin deposition is the subject of debate. However, once formed, the fibrin cuffs decrease oxygen diffusion and the trapped monocytes and neutrophils activate and release mediators that further damage capillaries and dermal tissue. Leakage of plasma macromolecules such as fibrinogen and α_2-macroglobulin into the dermis may also dysregulate growth factor function so that the tissue damage cannot be repaired.

DIABETIC ULCERS

Diabetic foot ulcers may be divided into the two groups of those in neuropathic feet (neuropathic ulcers) and those found in ischemic feet associated with neuropathy (neuroischemic ulcers).[69] Most commonly, neuropathic ulcers are caused by repetitive mechanical forces of walking leading to hyperkeratosis and formation of a callus on the plantar aspect of the foot. The callus presses on underlying soft tissue, damaging capillaries with consequent inflammation and hematoma formation leading to tissue necrosis and ulceration. Neuroischemic ulcers are frequently found on the margins of the foot, the tips of the toes, and beneath toenails. They are often caused by friction arising from poorly fitting shoes and first arise as a superficial blister that develops into a shallow ulcer.

Patients with diabetes have multiple physiological defects that can impair healing[70] including growth factor dysregulation, impaired angiogenic response, impaired keratinocyte, fibroblast, macrophage, and neutrophil function. Hyperglycemia results in a poor inflammatory response to bacterial invasion, and tight metabolic control combined with good wound care practice is essential in preventing wound infection that may lead to lower limb amputation.

PRESSURE ULCERS

Pressure ulcers develop as a consequence of the application of pressure, shear forces, and friction to the skin.[71] Pressure is the most important factor and produces local ischemia when the exerted pressure exceeds that within dermal capillaries. Under normal circumstances, pressure is relieved by movement and blood flows back into the area of skin to reoxygenate the tissue. However, if the pressure persists due to continued immobility, then tissue necrosis and ulcer formation ensue. The effects of pressure can be exacerbated by reduced skin elasticity in aged patients, requiring less applied pressure to cause tissue damage. Shear is a mechanical stress applied parallel to a plane of interest, and its application to the skin can deform skin structures and cause capillary damage. Shear and pressure often occur simultaneously. Friction applied to the skin, for example when a patient is dragged across a sheet, also contributes to skin damage.

CELLULAR DEFECTS WITHIN THE CHRONIC WOUND

Chronic wounds appear to exist in a state that is characterized by non-resolving inflammation driven by elements in the chronic wound environment such as tissue degradation products, hypoxia, and bacterial products that establish a positive feedback loop (Figure 5.2). This results in a disordered cytokine regulatory network where fibroblasts and endothelial cells within the wound bed fail to produce normally functioning granulation tissue and keratinocytes proliferate at the wound margin yet fail to migrate over a defective ECM.

INFLAMMATION

Deposition of fibrin around capillaries is considered to be one of the contributory factors in VLU formation. The cuffs are a predominant histologic feature of VLUs and, in addition to fibrin, are composed of actin, collagen IV, extravasated factor XIIIa, and α2-macroglobulin.[72] Immunohistochemical localization of TGF-β is also observed, possibly as a consequence of α2-macroglobulin acting as a scavenger for TGF-β.[73] Typically, neutrophils and pro-collagen-I positive fibroblasts are also associated with the cuffs. The complement system is activated in chronic wounds as demonstrated by the presence of terminal complement complex in association with C3D in the capillary walls of VLUs.[74] This, combined with the absence of the membrane

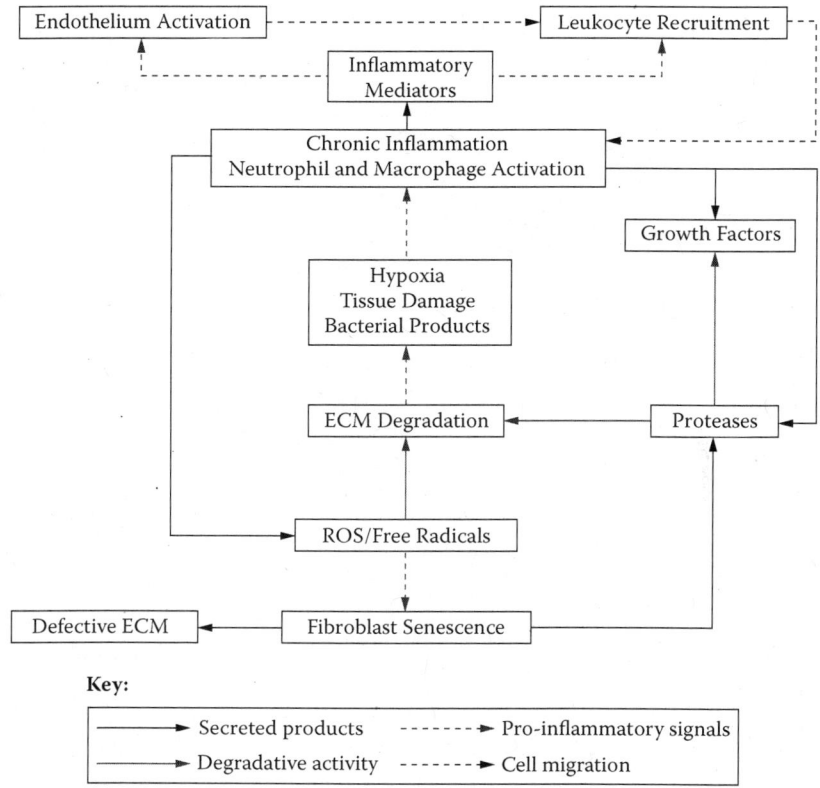

FIGURE 5.2 (Please see color insert following page 114.) Generation of wound chronicity.

glycoprotein protectin, may contribute to persisting inflammation. Protectin (also known as CD59) is a complement regulatory protein widely expressed by cells in all tissues. It acts to protect cells from the membrane attack complex generated by inappropriate complement activation. Generation of the complement components C3a, C4a, and C5a following complement activation acts as a powerful stimulus to inflammation as they increase vascular permeability and act as neutrophil chemoattractants. In combination with the up-regulation of another key mediator of inflammation, prostaglandin,[75] in chronic wounds they act to induce a persisting acute-type inflammatory response.

In contrast to the rapid decrease in the inflammatory response seen in a noninfected healing wound, the non-resolving inflammation found in chronic wounds is characterized histologically by a leukocyte infiltrate in which macrophages, neutrophils, and T lymphocytes are the predominant cell types.[30] The dense infiltration of T lymphocytes and macrophages found at the margin of VLUs is associated with a strong up-regulation of ICAM-1[76] and VCAM-1[77] by co-located blood vessels giving a picture of a typical chronic inflammatory response. Expression of leukocyte-associated adhesion molecules LFA-1 and VLA-4 is also enhanced on cells within the microvasculature.

Wound exudate is the fluid that can be harvested from the surface of open granulating wounds. It is assumed to represent the interstitial fluid of the underlying wound fluid and therefore to contain factors secreted by cells that represent their functional state. Data generated by wound exudate analysis have to be interpreted with caution when attempting to deduce cellular functional defects that may contribute to wound chronicity. The intracellular message delivered by cytokines is short ranged, and location of a secreting cell may be more important than the overall level of cytokine detected at the cell surface in exudate. Measurement of a few cytokines in isolation may be misleading, as considerable redundancy exists between the biological effects of many cytokines and growth factors (e.g., compare IL-1β and TNFα). The presence of cytokine inhibitors also has to be considered as they will contribute to the actual bioactivity as opposed to the absolute concentration of the cytokine. For example, total levels of immunoreactive TNFα are significantly higher in exudate from nonhealing VLUs than from healing ulcers, yet levels of bioactive TNF-α are not elevated.[78] Levels of soluble TNFα_p75 receptors are significantly higher in nonhealing exudate compared to healing exudates; however these levels are theoretically inadequate to substantially neutralize the bioactivity of the accompanying TNF-α.

Notwithstanding these caveats, wound exudate contains a diverse array of cytokines and growth factors that can be quantitated by immunoassay or bioassay. Many of the measurable components are present as a consequence of the chronic inflammatory process. Recognition of the complexity of interacting cytokine networks has led investigators to analyze multiple factors in the context of healing versus nonhealing wounds. One example is a study[18] where multiple chemokines and cytokines were quantified by immunoassay in wound exudates and tissue extracts from VLUs undergoing compression therapy. As the ulcers healed, conversion of an overall non-resolving chronic inflammatory cytokine profile to a resolving acute inflammatory one was observed. In contrast, other studies found that measurement of PDGF, GM-CSF, IL-1α, IL-1β, IL-6, and bFGF from healing and nonhealing ulcers demonstrated no statistically significant differences between the two healing states.[79]

Increased oxidative stress[80] resulting from a perturbation of the balance of oxidants and antioxidants at the chronic wound site may also be a manifestation of continuing inflammation. Excess iron derived from hemoglobin present at the wound site as a consequence of capillary leakage of red blood cells can generate hydroxyl free radicals via the Fenton reaction. Ferritin is known to be elevated in VLU exudate and is found with a concomitant increase in 8-isoprostane, a measure of lipid peroxidation.[81] The oxidative stress marker allantoin:uric acid ratio is also elevated in VLUs.[82]

GRANULATION TISSUE FORMATION

Histological examination of chronic wound granulation tissue suggests that there is a disordered regulation of the healing process rather than a diminution in cellular activity involved in defective healing. Aberrant distribution and amounts of growth factors, cytokines, and enzymes such as inducible nitric oxide synthase are found. For example, TGFβ-1, -2, and -3 can be identified in peri-ulcer skin, but they are absent from nonhealing wounds. The receptor for TGFβ-1 is strongly expressed in VLU, but no TGFβ-2 receptor is present at the protein level.[83] By contrast, in healing VLUs positive immunostaining for all the TGFs and the type I and type II receptors are detected.

Disordered proteolytic enzyme activity and consequential matrix degradation is also present. Levels of neutrophil-derived elastase, membrane-type MMP, extracellular MMP inducer, soluble MMPs, and urokinase-type plasminogen activator (uPA), which is an activator of MMP-2, are all elevated. Tissue inhibitors of matrix metalloproteases, in contrast, are decreased, leading to an overall net increase in proteolytic activity. This is associated with defects in the ECM, most notably a deficiency of fibronectin that is required as a provisional matrix for keratinocyte migration.[84] The importance of this defect in maintaining the nonhealing state is shown by the observation that within two weeks of initiating compression bandaging for VLUs, an acute inflammatory response is initiated followed by deposition of fibronectin and epithelial migration.[85]

Chronic wound exudate degrades fibronectin *in vitro,* and this activity is prevented by inhibitors of neutrophil elastase. In contrast to acute wound exudate that contains intact fibronectin, chronic wound exudates contain low molecular mass fragments found with elevated levels of elastase and decreased proteinase inhibitors.[86] MMPs are required for normal healing, but their levels rapidly decrease after the acute inflammation resolves and healing is initiated. Proteolytic activity persists in the chronic wound and concentrations of inhibitors are decreased. However, when chronic wounds respond to treatment, MMP levels fall as healing proceeds.[87]

Overall, protease activity is determined by a balance between enzyme activity and enzyme inhibitors. Urokinase-type plasminogen activator occupies a potentially significant role in regulation of chronic wound protease activity because it activates plasminogen to form plasmin that in turn activates collagenases. Its activity is controlled by a specific inhibitor that is found in higher levels at the ulcer edge and base in healing compared to nonhealing wounds.[88] When uPA expression in wound exudate switches from an active to inhibitor bound form, there follows a decrease in MMP-9 expression. This infers the existence of a proteolytic cascade initiated by the plasminogen activator/plasmin system during wound healing that is directly linked to activation of MMP-9.

Although the increased levels of proteolytic enzymes in the chronic wound derive mainly from inflammatory macrophages and neutrophils, another source is that from fibroblasts. Under normal conditions, for example during scar formation, fibroblasts will eventually undergo apoptosis. However, factors such as oxidative stress and pro-inflammatory cytokines such as TNFα and IL-1 can induce a senescent phenotype. This is characterized by telomere shortening, resistance to apoptotic death, and elevated production of collagenase, elastase, stromelysin, and decreased levels of the metalloproteinase inhibitors TIMP-1 and TIMP-3.[89] The senescent fibroblast exhibits an ECM degrading phenotype and the higher levels of these cells found in chronic wounds[90] may in some part contribute to defective ECM that does not support keratinocyte migration.

EPITHELIALIZATION

Epithelialization of the wound bed is the crucial step in the healing process that reestablishes barrier function of the epidermis to prevent ingress of bacteria and loss of interstitial fluid. In the normal wound, epithelialization progresses at maximal speed, yet despite keratinocyte proliferation at the wound margin, it is inhibited in chronic wounds.

Defective epithelialization is likely to be the consequence of a number of features of the chronic wound that act to inhibit migration over the chronic wound bed. Key among these is the absence of a suitable substrate to support migration as a consequence of proteolytic damage to the ECM so that it can no longer provide the required cues to migration via appropriate integrin ligand expression for keratinocyte receptors.[91] There is also evidence that keratinocyte functional defects may be induced in the chronic wound environment. Hypoxia generated as a consequence of underlying pathologies such as venous hypertension or diabetes may reduce the ability of keratinocytes to migrate and produce MMPs.[92]

The majority of chronic wounds are colonized to some extent by bacteria[93] that are considered to contribute to nonhealing by delivering a stimulus to development of a non-resolving inflammatory response. Additionally, Gram-negative bacteria may inhibit epithelialization directly by lipopolysaccharide interaction with keratinocyte TLRs and subsequent inhibition of migration.[94]

MODULATION OF WOUND CELL BIOLOGY BY CLINICAL INTERVENTIONS

Even though the chronic wound does not achieve healing, it is an active lesion where inflammation, proliferation, and ECM synthesis and degradation occur in a disordered manner that prevents an appropriate coordination of the sequence of events required for healing. It is a convenient working hypothesis that for chronic wounds, the healing process exists in some form of steady-state[95] or self-reinforcing loop that has to be interrupted for the healing to be initiated. This is supported by observations that as VLUs initiate a healing response during compression therapy, components of a dysregulated cytokine regulatory system are modulated. TGFβ and its receptors are up-regulated,[83] pro-inflammatory cytokines such as IL-1 are down-regulated with up-regulation of their inhibitors, and pro-inflammatory IL-12 is down-regulated with simultaneous up-regulation of anti-inflammatory IL-10.[18] Activity of proteolytic enzymes that degrade ECM and growth factors is also down-regulated by decreased synthesis and increased production of their inhibitors.[96] Following from these events, chronic inflammation is replaced with an acute type inflammation and normalization of ECM characterized by deposition of fibronectin that allows epithelialization to proceed.[85]

Knowledge of the underlying cellular events that characterize the nonhealing state and how they change in response during healing has led to development of treatment paradigms such as wound bed preparation (WBP). This concept recognizes that chronic wounds such as DFUs and VLUs are arrested at the inflammatory and proliferative phase[97] and that treatment strategies should be developed that focus on restoration of a balance between growth factors, cytokines, proteases, and their inhibitors as found in healing wounds. Wound bed preparation infers preparation for healing by removing barriers to healing and initiation of the endogenous healing process or promotion of the efficacy of other interventions. These objectives are achieved by debridement to minimize necrotic burden, management of exudate with advanced wound dressings to maintain a moisture level optimal for healing, and minimization of wound bacteria to eliminate them as a source of inflammatory stimuli. None of these interventions alone are novel, but the concept of WBP integrates them into a knowledge-based scheme that relies on reevaluation and modification of treatment strategy based on an understanding of the cell biology of healing.

Characterization of the differences between healing and nonhealing wounds has identified potential therapeutic targets and led to the development of a number of treatments that are designed to modulate healing by interacting with key aspects of biological events regulating the healing process. Examples include recombinant growth factors such as PDGF,[98] KGF,[99] and GM-CSF,[100] tissue-engineered dermal replacements,[101] synthetic protease inhibitors,[102] protease modulating dressings,[103] and pH-modulating ointments.[104]

No single "magic bullet" has been identified to treat the chronic wound, and early enthusiasm for the use of growth factors has been tempered with a realization that "there is no convincing evidence that growth factors may substitute for good wound care."[105] Many clinical trials have used single growth factors, and in retrospect, it is not surprising that this strategy is not of universal value for a lesion with multifactorial defects. Also, application of growth factors to a wound with high levels of proteases may simply result in degradation of the growth factor,[106] possibly requiring co-administration of a protease inhibitor. A further limitation is that growth factors may be required over a period of time not achieved with a single application because of a relatively short half life *in vivo* that requires multiple or continuous delivery systems. Additionally, the target cell has to express the membrane receptor for the delivered factor and be in the correct physiological state to respond appropriately.

A potentially more successful approach may lie in modification of cell functions within the wound tissue as individual cells produce a multiplicity of often counterregulatory factors. In part, this is achieved with tissue-engineered dermal equivalents that when applied to the wound surface after WBP will provide temporary barrier function and the cellular component will deliver growth factors to the wound bed.[107]

The future holds the possibility that gene therapy may overcome current drawbacks and poor efficacy of current growth factor treatment strategies.[108] *Ex vivo* and *in vivo* gene transfer has been achieved for cells present within the wound bed and gene overexpression has demonstrated that cell differentiation can be achieved and wound healing enhanced in animal models. However, the key to successful application of this technology will be identification of appropriate target cell functions that prevent healing in the chronic wound even after treatment of underlying pathologies combined with good wound management practice.

REFERENCES

1. Clark, R.A. (1985) Cutaneous tissue repair: Basic biologic considerations. *I J Am Acad Dermatol* 13:701–725.
2. Zarbock, A. and Ley, K. (2008) Mechanisms and consequences of neutrophil interaction with the endothelium. *Am J Pathol* 172:1–7.
3. Brinkmann, V. and Zychlinsky, A. (2007) Beneficial suicide: Why neutrophils die to make NETs. *Nat Rev Microbiol* 5:577–582.
4. Gordon, S. (2007) The macrophage: Past, present and future. *Eur J Immunol* 37 Suppl 1:S9–17.
5. Nathan, C.F. (1987) Secretory products of macrophages. *J Clin Invest* 79:319–326.
6. Leibovich, S.J. and Ross, R. (1975) The role of the macrophage in wound repair. A study with hydrocortisone and antimacropahge serum. *American J Pathol* 78:71–91.
7. Werner, S., Krieg, T., and Smola, H. (2007) Keratinocyte-fibroblast interactions in wound healing. *J Invest Dermatol* 127:998–1008.
8. Hinz, B. (2007) Formation and function of the myofibroblast during tissue repair. *J Invest Dermatol* 127:526–537.
9. Patel, G.K., Wilson, C.H., Harding, K.G., Finlay, A.Y., and Bowden, P.E. (2006) Numerous keratinocyte subtypes involved in wound re-epithelialization. *J Invest Dermatol* 126:497–502.
10. Michel, G., Kemeny, L., Peter, R.U., Beetz, A., Ried, C., Arenberger, P., and Ruzicka, T. (1992) Interleukin-8 receptor-mediated chemotaxis of normal human epidermal cells. *FEBS Lett* 305:241–243.
11. Miller, L.S. and Modlin, R.L. (2007) Human keratinocyte Toll-like receptors promote distinct immune responses. *J Invest Dermatol* 127:262–263.
12. Gawaz, M., Langer, H., and May, A.E. (2005) Platelets in inflammation and atherogenesis. *J Clin Invest* 115:3378–3384.
13. McEver, R.P. (2001) Adhesive interactions of leukocytes, platelets, and the vessel wall during hemostasis and inflammation. *Thromb Haemost* 86:746–756.
14. Anitua, E., Aguirre, J.J., Algorta, J., Ayerdi, E., Cabezas, A.I., Orive, G., and Andia, I. (2008) Effectiveness of autologous preparation rich in growth factors for the treatment of chronic cutaneous ulcers. *J Biomed Mater Res B Appl Biomater* 84:415–421.
15. Hawrylowicz, C.M., Howells, G.L., and Feldmann, M. (1991) Platelet-derived interleukin 1 induces human endothelial adhesion molecule expression and cytokine production. *J Exp Med* 174:785–790.
16. Zarbock, A. and Ley, K. (2008) Mechanisms and consequences of neutrophil interaction with the endothelium. *Am J Pathol* 172:1–7.
17. Serhan, C.N., Brain, S.D., Buckley, C.D., Gilroy, D.W., Haslett, C., O'Neill, L.A.J., Perretti, M., Rossi, A.G., and Wallace, J.L. (2007) Resolution of inflammation: State of the art, definitions and terms. *FASEB J* 21:325–332.
18. Fivenson, D.P., Faria, D.T., Nickoloff, B.J., Poverini, P.J., Kunkel, S., Burdick, M., and Strieter, R.M. (1997) Chemokine and inflammatory cytokine changes during chronic wound healing. *Wound Repair Regen* 5:310–322.
19. Dyson, M., Young, S., Pendle, C.L., Webster, D.F., and Lang, S.M. (1988) Comparison of the effects of moist and dry conditions on dermal repair. *J Invest Dermatol* 91:434–439.
20. Clark, R.A.F. (1988) Overview and general considerations of wound repair. In *The Molecular and Cellular Biology of Wound Repair.* R.A.F. Clark and P.M. Henson, editors. New York: Plenum Press, pp. 3–33. Clark, R.A.F. 1993. Regulation of fibroplasia in cutaneous wound repair.

21. Edwards, J.P., Zhang, X., Frauwirth, K.A., and Mosser, D.M. (2006) Biochemical and functional characterization of three activated macrophage populations. *J Leukoc Biol* 80:1298–1307.
22. Mosser, D.M. (2003) The many faces of macrophage activation. *J Leukoc Biol* 73:209–212.
23. Shnyra, A., Brewington, R., Alipio, A., Amura, C., and Morrison, D.C. (1998) Reprogramming of lipopolysaccharide-primed macrophages is controlled by a counterbalanced production of IL-10 and IL-12. *J Immunol* 160:3729–3736.
24. Schade, F., Flash, R., Flohe, S., Majetshack, M., Kreuzfelder, E., Dominguez, E., et al. (1999) Endotoxin tolerance. In *Endotoxin in Health and Disease.* H. Brade, S. Opal, S. Vogel, and D. Morrison, editors. New York: Marcel Dekker, pp. 50, 751–768.
25. Curran, J.N., Winter, D.C., and Bouchier-Hayes, D. (2006) Biological fate and clinical implications of arginine metabolism in tissue healing. *Wound Repair Regen* 14:376–386.
26. Duffield, J.S. (2003) The inflammatory macrophage: A story of Jekyll and Hyde. *Clin Sci (Lond)* 104:27–38.
27. Barbul, A. (1990) Immune aspects of wound repair. *Clin in Plastic Surg* 17:433–442.
28. Mills, R.E., Taylor, K.R., Podshivalova, K., McKay, D.B., and Jameson, J.M. (2008) Defects in skin gamma delta T cell function contribute to delayed wound repair in rapamycin-treated mice. *J Immunol* 181:3974–3983.
29. Boyce, D.E., Jones, W.D., Ruge, F., Harding, K.G., and Moore, K. (2000) The role of lymphocytes in human dermal wound healing. *Br J Dermatol* 143:59–65.
30. Loots, M.A., Lamme, E.N., Zeegelaar, J., Mekkes, J.R., Bos, J.D., and Middelkoop, E. (1998) Differences in cellular infiltrate and extracellular matrix of chronic diabetic and venous ulcers versus acute wounds. *J Invest Dermatol* 111:850–857.
31. Barbul, A., Breslin, R.J., Woodyard, J.P., Wasserkrug, H.L., and Efron, G. (1989) The effect of in vivo T helper and T suppressor lymphocyte depletion on wound healing. *Ann Surg* 209:479–483.
32. De Panfilis, G. (1998) CD8+ cytolytic T lymphocytes and the skin. *Exp Dermatol* 7:121–131.
33. Romagnani, S., Parronchi, P., D'Elios, M.M., Romagnani, P., Annunziato, F., Piccinni, M.P., Manetti, R., Sampognaro, S., Mavilia, C., De Carli, M., Maggi, E., and Del Prete, G.F. (1997) An update on human Th1 and Th2 cells. *Int Arch Allergy Immunol* 113:153–156.
34. Sun, X., Fu, X., and Sheng, Z. (2007) Cutaneous stem cells: Something new and something borrowed. *Wound Repair Regen* 15:775–785.
35. Wu, Y., Chen, L., Scott, P.G., and Tredget, E.E. (2007) Mesenchymal stem cells enhance wound healing through differentiation and angiogenesis. *Stem Cells* 25:2648–2659.
36. Jeon, S.H., Chae, B.C., Kim, H.A., Seo, G.Y., Seo, D.W., Chun, G.T., Kim, N.S., Yie, S.W., Byeon, W.H., Eom, S.H., Ha, K.S., Kim, Y.M., and Kim, P.H. (2007) Mechanisms underlying TGF-beta1-induced expression of VEGF and Flk-1 in mouse macrophages and their implications for angiogenesis. *J Leukoc Biol* 81:557–566.
37. Mott, J.D. and Werb, Z. (2004) Regulation of matrix biology by matrix metalloproteinases. *Curr Opin Cell Biol* 16:558–564.
38. Desmouliere, A., Chaponnier, C., and Gabbiani, G. (2005) Tissue repair, contraction, and the myofibroblast. *Wound Repair Regen* 13:7–12.
39. Hinz, B.J. (2007) Formation and function of the myofibroblast during tissue repair. *Invest Dermatol* 127:526–537.
40. Morasso, M.I. and Tomic-Canic, M. (2005) Epidermal stem cells: The cradle of epidermal determination, differentiation and wound healing. *Biol Cell* 97:173–183.

41. Patel, G.K., Wilson, C.H., Harding, K.G., Finlay, A.Y., and Bowden, P.E. (2006) Numerous keratinocyte subtypes involved in wound re-epithelialization. *J Invest Dermatol* 126:497–502.

42. Winter, G.D. (1962) Formation of the scab and the rate of epithelization of superficial wounds in the skin of the young domestic pig. *Nature* 193:293–294.

43. Coulombe, P.A. (2003) Wound epithelialization: Accelerating the pace of discovery. *J Invest Dermatol* 121:219–230.

44. Frankenstein, Z., Alon, U., and Cohen, I.R. (2006) The immune-body cytokine network defines a social architecture of cell interactions. *Biol Direct* 1:32.

45. Pierce, G.F., Mustoe, T.A., Altrock, B.W., Deuel, T.F., and Thomason, A. (1991) Role of platelet-derived growth factor in wound healing. *J Cell Biochem* 45:319–326.

46. Nagai, M.K. and Embil, J.M. (2002) Becaplermin: Recombinant platelet derived growth factor, a new treatment for healing diabetic foot ulcers. *Expert Opin Biol Ther* 2:211–218.

47. O'Kane, S. and Ferguson, M.W.J. (1997) Transforming growth factors and wound healing. *Int J Biochem Cell Biol* 29:63–78.

48. Shah, M., Foreman, D.M., and Ferguson, M.W.J. (1995) Neutralisation of TGF-ß$_1$ and TGF-ß$_2$ or exogenous addition of TGF-ß$_3$ to cutaneous rat wounds reduces scarring. *J Cell Sci* 108:985–1002.

49. Powers, C.J., McLeskey, S.W., and Wellstein, A. (2000) Fibroblast growth factors, their receptors and signaling. *Endocr Relat Cancer* 7:165–197.

50. Hoeben, A., Landuyt, B., Highley, M.S., Wildiers, H., Van Oosterom, A.T., and De Bruijn, E.A. (2004) Vascular endothelial growth factor and angiogenesis. *Pharmacol Rev* 56:549–580.

51. Edmondson, S.R., Thumiger, S.P., Werther, G.A., and Wraight, C.J. (2003) Epidermal homeostasis: The role of the growth hormone and insulin-like growth factor systems. *Endocr Rev* 24:737–764.

52. Loot, M.A., Kenter, S.B., Au, F.L., van Galen, W.J., Middelkoop, E., Bos, J.D., and Mekkes, J.R. (2002) Fibroblasts derived from chronic diabetic ulcers differ in their response to stimulation with EGF, IGF-I, bFGF and PDGF-AB compared to controls. *Eur J Cell Biol* 81:153–160.

53. Hübner, G., Brauchle, M., Smola, H., Madlener, M., Fässler, R., aand Werner, S. (1996) Differential regulation of pro-inflammatory cytokines during wound healing in normal and glucocorticoid-treated mice. *Cytokine.* 8:548–556.

54. Zaja-Milatovic, S. and Richmond, A. (2008) CXC chemokines and their receptors: A case for a significant biological role in cutaneous wound healing. *Histol Histopathol* 23:1399–1407.

55. Marks, J., Hughes, L.E., Harding, K.G., Campbell, H., and Ribeiro, C.D. (1983) Prediction of healing time as an aid to the management of open granulating wounds. *World J Surg* 7:641–645.

56. Moore, K., McCallion, R., Searle, R.J., Stacey, M.C., and Harding, K.G. (2006) Prediction and monitoring the therapeutic response of chronic dermal wounds. *Int Wound J* 3:89–96.

57. Eaglstein, W.H. (1989) Wound healing and aging. *Clin Geriatr Med* 5:183–188.

58. Ashcroft, G.S., Kielty, C.M., Horan, M.A., and Ferguson, M.W. (1997) Age-related changes in the temporal and spatial distributions of fibrillin and elastin mRNAs and proteins in acute cutaneous wounds of healthy humans. *J Pathol* 183:80–89.

59. Quirinia, A. and Viidik, A. (1991) The influence of age on the healing of normal and ischemic incisional skin wounds. *Mech Ageing Dev* 58:221–232.

60. Shah, M.G. and Maibach, H.I. (2001) Estrogen and skin. An overview. *Am J Clin Dermatol* 2:143–150.

61. Ashcroft, G.S., Dodsworth, J., van Boxtel, E., Tarnuzzer, R.W., Horan, M.A., Schultz, G.S., and Ferguson, M.W. (1997) Estrogen accelerates cutaneous wound healing associated with an increase in TGF-beta1 levels. *Nat Med* 3:1209–1215.

62. Ashcroft, G.S., Greenwell-Wild, T., Horan, M.A., Wahl, S.M., and Ferguson, M.W. (1999) Topical estrogen accelerates cutaneous wound healing in aged humans associated with an altered inflammatory response. *Am J Pathol* 155:1137–1146.

63. Williams, L. (2002) Assessing patients' nutritional needs in the wound-healing process. *J Wound Care* 11:225–228.

64. Thomas, D.R. (2001). Improving outcome of pressure ulcers with nutritional interventions: A review of the evidence. *Nutrition* 17: 121–125.

65. Balaj, P. and Mosley, J.G. (1995) Evaluation of vascular and metabolic deficiency in patients with large leg ulcers. *Ann R Coll Surg Engl* 77:270–272.

66. Rojas, A.I. and Phillips, T.J. (1999) Patients with chronic leg ulcers show diminished levels of vitamins A and E, carotenes, and zinc. *Dermatol Surg* 25:601–604.

67. Watanabe, T., Arakawa, T., Fukuda, T., Higuchi, K., and Kobayashi, K. (1995) Zinc deficiency delays gastric ulcer healing in rats. *Dig Dis Sci* 40:1340–1344.

68. Abbade, L.P. and Lastoria, S. (2005) Venous ulcer: Epidemiology, physiopathology, diagnosis and treatment. *Int J Dermatol* 44:449–456.

69. Edmonds, M.E. and Foster, A.V. (2006) Diabetic foot ulcers. *BMJ* 332(7538):407–410.

70. Brem, H. and Tomic-Canic, M. (2007) Cellular and molecular basis of wound healing in diabetes. *J Clin Invest* 117:1219–1222.

71. Collier, M. and Moore, Z. (2005) Etiology and risk factors. In *Science and Practice of Pressure Ulcer Management.* Romanelli, M., Clark, M., Cherry, G., Colin, D., and Defloore, T., editors. New York: Springer, pp. 27–36.

72. Higley, H.R., Ksander, G.A., Gerhardt, C.O., and Falanga, V. (1995) Extravasation of macromolecules and possible trapping of transforming growth factor-beta in venous ulceration. *Br J Dermatol* 132:79–85.

73. O'Connor-McCourt, M.D. and Wakefield, L.M. (1987) Latent transforming growth factor-beta in serum. A specific complex with alpha 2-macroglobulin. *J Biol Chem* 262:14090–14099.

74. Balslev, E., Thomsen, H.K., Danielsen, L., Sheller, J., and Garred, P. (1999) The terminal complement complex is generated in chronic leg ulcers in the absence of protectin (CD59). *Apmis* 107:997–1004.

75. Abd-El-Aleem, S.A., Ferguson, M.W., Appleton, I., Bhowmick, A., McCollum, C.N., and Ireland, G.W. (2001) Expression of cyclooxygenase isoforms in normal human skin and chronic venous ulcers. *J Pathol* 195:616–623.

76. Hahn, J., Junger, M., Friedrich, B., Zuder, D., Steins, A., Hahn, M., and Klysez, T. (1997) Cutaneous inflammation limited to the region of the ulcer in chronic venous insufficiency. *Vasa* 26:277–281.

77. Weyl, A., Vanscheidt, W., Weiss, J.M., Peschen, M., Schopf, E., and Simon, J. (1996) Expression of the adhesion molecules ICAM-1, VCAM-1, and E-selectin and their ligands VLA-4 and LFA-1 in chronic venous leg ulcers. *J Am Acad Dermatol* 34:418–423.

78. Wallace, H.J. and Stacey, M.C. (1998) Levels of tumor necrosis factor-alpha (TNF-alpha) and soluble TNF receptors in chronic venous leg ulcers—Correlations to healing status. *J Invest Dermatol* 110:292–296.

79. Harris, I.R., Yee, K.C., Walters, C.E., Cunliffe, W.J., Kearney, J.N., Wood, E.J., and Ingham, E. (1995) Cytokine and protease levels in healing and nonhealing chronic venous leg ulcers. *Exp Dermatol* 4:342–349.

80. Wlaschek, M. and Scharffetter-Kochanek, K. (2005) Oxidative stress in chronic venous leg ulcers. *Wound Repair Regen* 13:452–461.

81. Yeoh-Ellerton, S. and Stacey, M.C. (2003) Iron and 8-isoprostane levels in acute and chronic wounds. *J Invest Dermatol* 121:918–925.

82. James, T.J., Hughes, M.A., Cherry, G.W., and Taylor, R.P. (2003) Evidence of oxidative stress in chronic venous ulcers. *Wound Repair Regen* 11:172–176.

83. Cowin, A.J., Hatzirodos, N., Holding, C.A., Dunaiski, V., Harries, R.H., Rayner, T.E., Fitridge, R., Cooter, R.D., Schultz, G.S., and Belford, D.A. (2001) Effect of healing on the expression of transforming growth factor beta(s) and their receptors in chronic venous leg ulcers. *J Invest Dermatol* 117:1282–1289.

84. Clark, R.A., Folkvord, J.M., and Wertz, R.L. (1985) Fibronectin, as well as other extracellular matrix proteins, mediate human keratinocyte adherence. *J Invest Dermatol* 84:378–383.

85. Herrick, S.E., Sloan, P., McGurk, M., Freak, L., McCollum, C.N., and Ferguson, M.W. (1992) Sequential changes in histologic pattern and extracellular matrix deposition during the healing of chronic venous ulcers. *Am J Pathol* 141:1085–1095.

86. Grinnell, F., Ho, C.H., and Wysocki, A. (1992) Degradation of fibronectin and vitronectin in chronic wound fluid: Analysis by cell blotting, immunoblotting, and cell adhesion assays. *J Invest Dermatol* 98:410–416.

87. Trengove, N.J., Stacey, M.C., MacAuley, S., Bennett, N., Gibson, J., Burslem, F., Murphy, G., and Schultz, G. (1999) Analysis of the acute and chronic wound environments: The role of proteases and their inhibitors. *Wound Repair Regen* 7:442–452.

88. Stacey, M.C. and Mata, S.D. (2000) Lower levels of PAI-2 may contribute to impaired healing in venous ulcers—A preliminary study. *Cardiovasc Surg* 8:381–385.

89. Campisi, J. (1998) The role of cellular senescence in skin aging. *J Invest Derm Symp Proc* 3:1–5.

90. Telgenhoff, D. and Shroot, B. (2005) Cellular senescence mechanisms in chronic wound healing. *Cell Death Differ* 12:695–698.

91. Agren, M.S. and Werthén, M. (2007) The extracellular matrix in wound healing: A closer look at therapeutics for chronic wounds. *Int J Low Extrem Wounds* 6:82–97.

92. Xia, Y.P., Zhao, Y., Tyrone, J.W., Chen, A., and Mustoe, T.A. (2001) Differential activation of migration by hypoxia in keratinocytes isolated from donors of increasing age: Implication for chronic wounds in the elderly. *J Invest Dermatol* 116:50–56.

93. Zmudzinska, M., Czarnecka-Operacz, M., and Silny, W. (2005) Bacterial flora of leg ulcers in patients admitted to Department of Dermatology, Poznan University of Medical Sciences, during the 1998–2002 period. *Acta Dermatovenerol Croat* 13:168–172.

94. Loryman, C. and Mansbridge, J. (2008) Inhibition of keratinocyte migration by lipopolysaccharide. *Wound Repair Regen* 16:45–51.

95. Moore, K. (1999) Cell biology of chronic wounds: The role of inflammation. *J Wound Care* 8:345–348.

96. Tarlton, J.F., Bailey, A.J., Crawford, E., Jones, D., Moore, K., and Harding, K.D. (1999) Prognostic value of markers of collagen remodeling in venous ulcers. *Wound Repair Regen* 7:347–355.

97. Schultz, G.S., Sibbald, R.G., Falanga, V., Ayello, E.A., Dowsett, C., Harding, K., Romanelli, M., Stacey, M.C., Teot, L., and Vanscheidt, W. (2003) Wound bed preparation: A systematic approach to wound management. *Wound Repair Regen* 11(1):S1–28.

98. Guzman-Gardearzabal, E., Leyva-Bohorquez, G., Salas-Colin, S., Paz-Janeiro, J.L., Alvarado-Ruiz, R., and Garcia-Salazar, R. (2000) Treatment of chronic ulcers in the lower extremities with topical becaplermin gel .01%: A multicenter open-label study. *Adv Ther* 17:184–189.

99. Robson, M.C., Phillips, T.J., Falanga, V., Odenheimer, D.J., Parish, L.C., Jensen, J.L., and Steed, D.L. (1999) Randomized trial of topically applied repifermin (recombinant human keratinocyte growth factor-2) to accelerate wound healing in venous ulcers. *Wound Repair Regen* 9:347–352.

100. Da Costa, R.M., Ribeiro Jesus, F.M., Aniceto, C., and Mendes, M. (1999) Randomized, double-blind, placebo-controlled, dose- ranging study of granulocyte-macrophage colony stimulating factor in patients with chronic venous leg ulcers. *Wound Repair Regen* 7:17–25.

101. Marston, W.A. (2004) Dermagraft, a bioengineered human dermal equivalent for the treatment of chronic nonhealing diabetic foot ulcer. *Expert Rev Med Devices* 1:21–31.

102. Fray, M.J., Dickinson, R.P., Huggins, J.P., and Occleston, N.L. (2003) A potent, selective inhibitor of matrix metalloproteinase-3 for the topical treatment of chronic dermal ulcers. *J Med Chem* 46:3514–3525.

103. Cullen, B., Smith, R., McCulloch, E., Silcock, D., and Morrison, L. (2002) Mechanism of action of PROMOGRAN, a protease modulating matrix, for the treatment of diabetic foot ulcers. *Wound Repair Regen* 10:16–25.

104. Greener, B., Hughes, A.A., Bannister, N.P., and Douglass, J. (2005) Proteases and pH in chronic wounds. *J Wound Care* 14:59–61.

105. Fu, X., Li, X., Cheng, B., Chen, W., and Sheng, Z. (2005) Engineered growth factors and cutaneous wound healing: Success and possible questions in the past 10 years. *Wound Repair Regen* 13:122–130.

106. Trengove, N.J., Stacey, M.C., MacAuley, S., Bennett, N., Gibson, J., Burslem, F., Murphy, G., and Schultz, G. (1999) Analysis of the acute and chronic wound environments: The role of proteases and their inhibitors. *Wound Repair Regen* 7:442–452.

107. Sabolinski, M.L., Alvarez, O., Auletta, M., Mulder, G., and Parenteau, N.L. (1996) Cultured skin as a "smart material" for healing wounds: Experience in venous. *Biomaterials* 17:311–320.

108. Eming, S.A., Krieg, T., and Davidson, J.M. (2007) Gene therapy and wound healing. *Clin Dermatol* 25:79–92.

6 The Microbiology of Wounds

Steven L. Percival and Scot E. Dowd

CONTENTS

Introduction 187
Stages Involved in Microbiology Progression of a Chronic Wound Infection 189
   Contamination Stage 189
   Colonization: Reversible Adhesion 191
   Colonization: Irreversible Adhesion 191
   Critical Colonization 192
   Infection—Local and Systemic 192
Role of Microorganisms in Wound Healing 193
   Numbers 194
Species Diversity in Acute and Chronic Wounds 196
   Staphylococcus aureus 197
   Beta-Hemolytic Streptococcus 197
   Pseudomonas aeruginosa 198
Anaerobic Bacteria 198
Community Hypothesis and Wound Infections/Healing 198
   The Role of a Community 199
   Unculturable Bacteria and Wounds 201
Sampling of a Wound 203
   Superficial Wound Swabs 205
   Aspirate/Tissue 205
   Wound Sampling Overview 207
   Microbiological Specimen Handling for Laboratory Culture-Based
   Diagnostics 207
Microbiological Specimen Handling for Laboratory Molecular-Based
Diagnostics 208
Conclusion 209
References 210

INTRODUCTION

Acute wounds are formed during a trauma, such as a surgical procedure, a cut/incision, or a bite (e.g., insect, snake, human) and heal through specific defined phases (see Figure 6.1). Healing rates in acute wounds are affected by a number of factors including

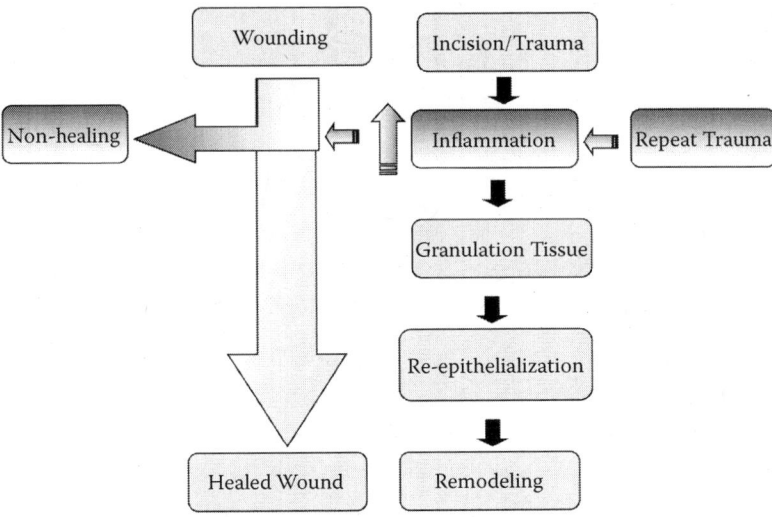

FIGURE 6.1 Phases of wound healing.

TABLE 6.1
Factors That Delay Wound Healing

Microbial numbers/pathogenicity/virulence/synergy

Granulation tissue hemorrhagic/fragile

Inflammatory mediators

Inactivated state of neutrophils

Bacterial and human toxins

Tissue hypoxia

Metabolic wastes

Reduce fibroblast number/collagen production

---

wound type, site of formation, area, and depth, repeated traumas, and underlying pathologies (see Table 6.1). Complicated acute wounds (e.g., gunshot wounds, in particular) have
an increased susceptibility of developing an infection[1] and a prolonged healing process.

Chronic wounds occur when the "normal" wound healing process is hindered,
often resulting in the formation of an ulcer.[2] Wounds that are classified as "chronic"
do not heal in a timely manner and do not respond effectively to traditional wound
management practices and standard protocols of care. Patients who have poor blood
or lymph circulation, due to an impaired arterial supply or venous drainage or
patients with underlying metabolic diseases (e.g., diabetes mellitus) have been found
to be highly susceptible to the formation of chronic wounds.[3–5] The most commonly
encountered chronic wounds include venous leg ulcers, diabetic extremity ulcers,
and decubitous ulcers.[6] Together these types of chronic wounds are known to affect
a large percentage of individuals, especially in Western communities.[7]

All open wounds are considered to be contaminated with microorganisms. Consequently, as wounds become necrotic, the wound bed environment favors microbial proliferation.[8] As the wound microbial bioburden increases, the demand on the host immune system increases. As a result, if the wound bioburden is not effectively suppressed by the host's immune defenses, it will continue to increase. This increasing wound bioburden will enhance the risk of clinical infection occurring[9–11] unless appropriate interventions such as debridement of devitalized tissue and topical/systemic antimicrobial agents are administered.

STAGES INVOLVED IN MICROBIOLOGY PROGRESSION OF A CHRONIC WOUND INFECTION

The six stages involved in the development of a wound "microbiota" are considered individual but often overlapping. Visual classification of these stages, apart from stages 5 and 6, cannot be clearly determined. Attempting to determine the changing microbiological status macroscopically without the aid of appropriate microbiological techniques is not possible.

The stages involved in the development of a wound microbial bioburden include the following:

Stage 1: Contamination or Transient Stage
Stage 2: Colonization Stage 1—Reversible Adhesion
Stage 3: Colonization Stage 2—Irreversible Adhesion
Stage 4: Critical Colonization Stage—Climax Community or "Biofilm"
Stage 5: Local Infection Stage
Stage 6: Systemic Infection Stage

These stages above constitute the "wound-microbiology life cycle" (see Figures 6.2 and 6.3).

CONTAMINATION STAGE

The contamination stage of a wound refers to the presence of microorganisms in the wound bed, without multiplication. No wound is sterile, therefore all wounds are considered to harbor transient microorganisms—contamination. In the "transient stage" of the wound-microbiology life cycle, microorganisms enter a lag phase or adaptive phase in their growth cycle where an initial biochemical assessment of their new surroundings takes place.

The bacteria contaminating the wound environment are a collection of bacteria derived from different sources. These often transient bacteria are not expected to have a major role to play in delaying wound healing and may originate from either the mucous membranes found around the body (endogenous microorganisms), surrounding skin, or from external environmental sources (exogenous microorganisms).

At the contamination stage, the majority of bacteria are suppressed or regulated at an appropriate level by the host defenses (provided the individual is not

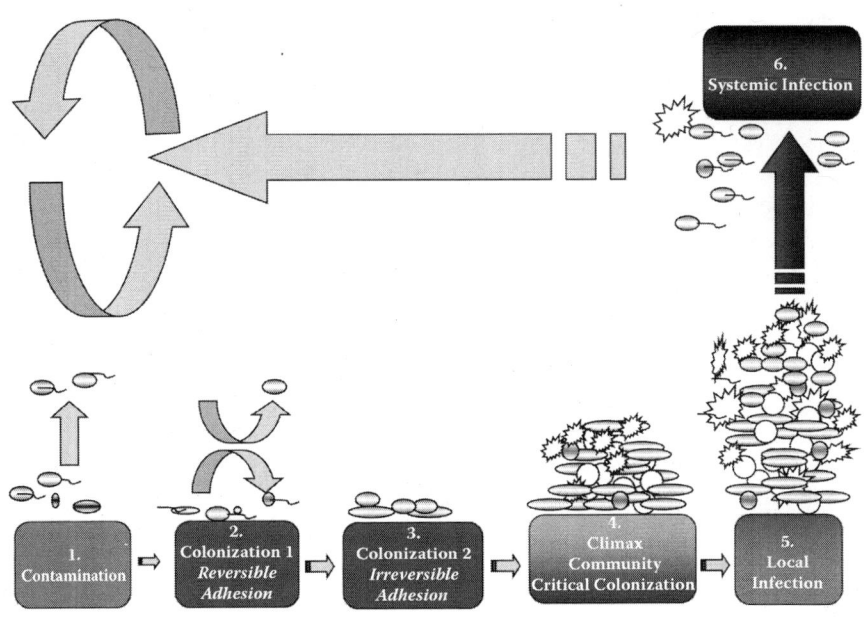

FIGURE 6.2   The wound–microbiology life cycle.

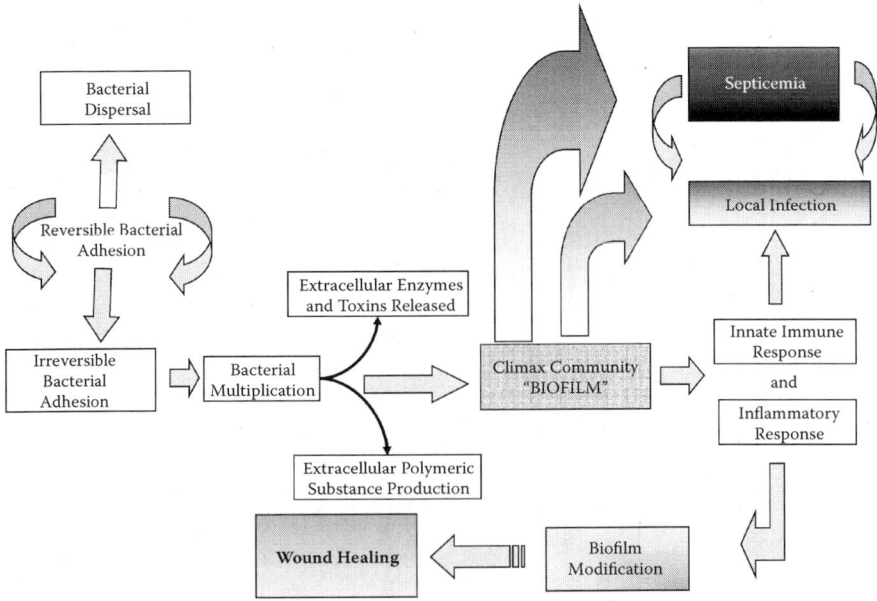

FIGURE 6.3   The role of bacteria in wound healing.

physiologically or immunologically compromised). Because contaminating bacteria are considered to be in a planktonic (nonattached) phenotype, they are typically more susceptible and vulnerable to eradication.

COLONIZATION: REVERSIBLE ADHESION

Once a bacterium is present on a wound surface, it will assess the local environment before attachment occurs. This assessment is made possible by the use of chemical receptors found on the cell wall of many bacteria. Utilizing the environmental signals, the bacteria will either attach or remain planktonic (free floating). If the environmental signals promote attachment, the bacteria then interact with a surface using adhesins (e.g., fimbriae or surface proteins). At this stage in the wound-microbiology life cycle, the bacterium is considered reversibly attached and can be easily washed away from a surface if low levels of shear force (e.g., saline rinses) are applied. The bacteria are also more susceptible to antimicrobial agents and host defenses. Once a surface interaction occurs, the subcutaneous tissue of the wound would provide an adequate nutrient source, shelter, and an environment conducive to microbial irreversible adhesion, and colonization. As reversible attachment progresses, the bacteria begin to develop stronger surface interaction forces leading to irreversible adhesion and move into a biofilm phenotype. Furthermore, bacteria begin to become adapted to their new surroundings and produce intracellular and extracellular substances for polymeric encasement. Secretion of EPS acts not only to strengthen the adhesion of the bacteria to a surface but aids to initiate maturation of the biofilm.

Early colonizing bacteria in a wound, particularly chronic wounds, have included Staphylococcus and beta hemolytic Streptococci.

COLONIZATION: IRREVERSIBLE ADHESION

Once a favorable microenvironment for microbial growth is established, the bacteria undergo a transition to an irreversible adhesion state. At this stage, bacteria become firmly attached to the wound surface and eventually become encased within EPS. In the irreversible state, bacteria become more resistant to removal and require reasonable amounts of shear stress to be removed. Over time the EPS encapsulated bacteria begin to multiply. As the biofilm phenotype of these colonies progresses, inherent resistance to therapeutics such as antibiotics develops. In addition to this, the biofilm bacteria begin to modify their genetic expression profiles and proliferate.

As bacteria multiply, they are able to influence their local microenvironment to their own benefit. This encourages further multiplication of the bacteria and supports the adhesion of different species of secondary and tertiary colonizing microorganisms as nutritional and environmental conditions change. Consequently, as time progresses, increasing numbers of bacteria begin to colonize the wound surface. The community of microorganisms then develops into discrete aggregates or "micro-colonies."

Within a relatively short period of time (within days), the microbiology of the wound will become more complex and oxygen availability will significantly be reduced throughout the microbial community. Consequently, specific areas in the biofilm will become hypoxic. This will locally enhance the proliferation of many anaerobic bacteria which will increase the microbial diversity within the wound bed.

CRITICAL COLONIZATION

Critical colonization refers to a stage where bacteria have colonized, multiplied, and induced a state of delayed healing with no visible host reaction or obvious clinical signs of infection.

The term "critical colonization" is used frequently in wound care, but its overall usefulness to wound management is misleading, as the majority of (if not all) chronic wounds should be considered to be *critically* colonized. Recently, the concept of critical colonization has been akin to a wound biofilm, as the biofilm model of infection helps to explain many of the complex challenges that are clinically observed. From a clinical perspective, critical colonization of the wound is the stage where the wound has become compromised but as yet is demonstrating no clinical signs of infection other than nonhealing. From a microbiological perspective, critical colonization occurs when levels of microorganisms are at a "critical" level in the wound. However, this critical level is specific to wound type, and patient, and should not be considered universal to all wounds. Consequently, the value of 10^5 cfu/g, or mL, of tissue or tissue fluid is arbitrary and overgeneralized.

As the microorganisms at the wound surface continue to multiply, their numbers increase and the complexity of the microbial community develops. This is aided by the many agonistic, antagonistic, commensal, and mutalistic relationships that occur between the sessile microorganisms. Continuation of microbial interactions in the wound bed leads to the development of a dynamic ecosystem culminating in the formation of an array of discrete, functional, and versatile microbial niches that promote survival and persistence.

Overall, within a short period of time at the critical colonization stage, the whole wound microbial ecosystem will climax and stabilize (i.e., achieve a "host–biofilm balance"). The wound microbial ecosystem that has climaxed and stabilized is best described as a "mature" biofilm. Such a stabilized system will induce an effect on the wound that will delay healing and induce a local infection. Critical colonization and infection are two similar concepts, and distinguishing between the two is clinically and microbiologically unachievable.[12]

INFECTION—LOCAL AND SYSTEMIC

A wound is deemed to be locally infected when multiplication of bacteria within the fully mature biofilm continues exponentially. Once the level of bacteria exceeds a critical mass, it is more likely to induce notable host immune reactions. A local (acute) wound infection will present clinically with redness (erythema), excessive pain, swelling, heat generation, wound breakdown, increased temperature, and sometimes evidence of cellulitis. In this situation, the bacteria may be considered predatory in nature. Other more subtle clinical signs of infection have been documented and have included, among others, alteration in exudate, friable granulation tissue that bleeds easily, malodor, and discolored granulation tissue. Granulation tissue has been shown to become discolored by certain bacteria. Colors have included yellow, green, or blue when bacteria such as *Pseudomonas aeruginosa*, Streptococci, and *Bacteroides fragilis* have been cultured.

It is important to appreciate that the numbers and types of bacteria found in a wound are not predictors of a local wound infection. If the local infection is not correctly managed and the microbial bioburden increases, a systemic infection may develop. A systemic infection occurs when the bacterial wound bioburden exceeds the capacity of its local environment and through biochemical signals seeks to invade new tissue which can lead to bacterial ingress into the bloodstream (bacteremia) leading to sepsis or septicemia (multiplication of bacteria in the blood and toxin production), potential organ failure, and in extreme cases, death.

At stages 5 and 6 of the wound-microbiology continuum, intervention is vital to reestablish the bacterial balance or homeostasis in the wound. This can hopefully be achieved by utilizing correct management strategies and protocols of care. Such strategies typically include aggressive debridement to remove devitalized tissue and fully mature biofilm, the use of appropriate wound dressings and topical antimicrobials, for local infections, and topical antiseptic/systemic antibiotics for systemic infections.

Diagnosing a wound infection is a judgment made clinically, but microbiological assay results may also have a role to play in diagnosis. Diagnosing a wound infection using microbiology data alone without reference to the clinical condition of the patient is considered to be poor practice.[13]

To help reduce the cost and risk of infection in chronic wounds, it is appropriate for clinicians to be able to "promptly and accurately recognise wounds that may progress to active infection and to treat them appropriately."[14] Older and immunocompromised individuals are more prone to infection.[15,16]

In general, based upon the principles of biofilm-based wound care, the wound must be kept from reaching critical colonization. In addition, the biofilm must certainly be kept from reaching maturity as it will become highly resistant to therapeutics.

ROLE OF MICROORGANISMS IN WOUND HEALING

Fundamental to healing is the chronic wound's "established microbiota" or biofilm.[17] In particular, the more diverse the microbiology of the chronic wound, the more recalcitrant the microbial populations are to therapeutics and the greater the risk of infection developing.[18] However, the numerous scientific and clinical studies that have been reported on the microbial diversity of chronic wounds, have been based solely on the results of agar culture-based technologies. The conclusions drawn from these studies have suggested that the majority of chronic wounds are colonized with bacteria such as staphylococci, streptococci, enterococci, and facultative Gram-negative bacteria.[18,19] However, such studies have rarely taken into account the recovery of the more fastidious, slow-growing bacteria, or unculturable bacteria, in particular anaerobes. The incidence and prevalence of anaerobic bacteria in most chronic wounds are not reported despite being acknowledged as significant to nonhealing and infection. Anaerobic bacteria that have been isolated from many chronic wounds have included *Peptostreptococcus* and *Bacteroides* spp.[19–21] Recent research has shown that wounds that are critically colonized have anaerobes as their predominant populations.[22,23]

Historically, significant amounts of research has been undertaken to investigate the association of particular bacterial groups and the relationship between microbial bioburden and healing in chronic wounds,[24] although in many of these studies, the patient groups have been poorly characterized. In addition to this, little reference has been made to areas such as the presence of infection, duration of ulceration/wound, or previous antimicrobial therapy that had been administered to the patients. As a consequence, the area of wound microbiology is one of intensive research that reflects the growing technologies of polymerase chain reaction, denaturing gradient electrophoresis, fluorescent *in situ* hybridization,[25,26] and arguably the most powerful molecular method to characterize diversity, bacterial tag-encoded FLX-titanium amplicon pyrosequencing (bTEFAP).[22,23]

In addition to types of bacteria levels of colony forming units (cfu) per gram of tissue, or mL of wound exudates, specifically above 10^5 cfu, are traditionally used as predictors of nonhealing/infection in a wound. This association between specific bacterial numbers and wound infection was based on a number of research studies published many years ago, particularly in diabetic foot ulcers and pressure ulcers.[10] In these and other studies, the authors' microbiological and sampling techniques utilized have now come under scrutiny. Because of this, many areas related to numbers of bacteria, types of bacteria, and sampling method employed in wound care have led to confusion.[27–29] It is, however, only relatively recently that the issue of bacterial numbers, the wound microbiology, and association with healing is being revisited by microbiologists.

It is also of interest to appreciate that some bacteria, found naturally on the skin, have been shown to enhance the healing of infected wounds.[30] Some strains of *Staphylococcus aureus*, when inoculated into rats, have been shown to accelerate wound healing.[31] Within this study, it was found that a inoculation of 10^5 to 10^8 cfu/g of *S. aureus* accelerated wound healing.

The significance of numbers and species diversity of microorganisms in a wound will be discussed below.

Numbers

Bendy and colleagues[27] in 1964 reported the use of topical antibiotics with non-antibiotic treatment regimes to wound healing. The authors concluded from their studies that wound healing progressed only when bacterial counts in wound fluid were below 10^6 cfu/mL. Above 10^6 cfu/mL the authors suggested that in pressure ulcers (decubiti) wound healing was inhibited. Robson and Heggers[29] also found that infection risk increased when the microbial load of a wound was $>10^5$ cfu/g of tissue. Based on the above research papers and additional studies,[29,31–33] it was concluded that quantitative microbiology, principally levels of bacteria above 10^5 cfu/g or mL, could be useful to help predict wound healing and risk of infection. Unfortunately, accurately determining microbiological values as a part of standard practice has many confounding variables and remains outside the ability of most

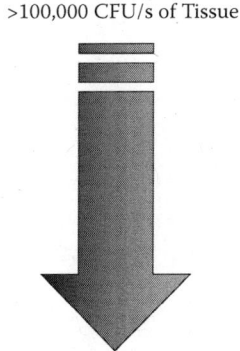

>100,000 CFU/s of Tissue

Infection and/or Delayed Healing

FIGURE 6.4   Numbers of microorganisms and delayed healing.

standard clinical microbiology laboratories. It is important to remember that it is not bacteria per se that cause infection but rather their products—namely, their toxins and enzymes. Especially confounding is the presence of biofilm phenotype bacteria known to be highly prevalent if not universal in chronic wounds.

The 10^5 cfu/g of tissue or mL of wound exudate rubric suggests that a chronic wound, no matter what its status, complexity, and microbial composition, has the potential to become infected if the microbial load is confined at or above this level (Figure 6.4). However, Pruitt and others[33] reported that quantitative microbiology alone could not differentiate between wound colonization and infection. The authors concluded that histological analysis was considered more effective for determining wound infection rather than microbiological numbers. Robson and colleagues[34] concluded that levels of bacteria have no clinical bearing. In addition, James[35] suggested that in pressure ulcers the microbial bioburden does not have an effect on healing. Clinically it is imperative to prevent a wound from achieving critical colonization or a mature biofilm. However, for chronic wounds this is often not achievable considering the rate at which bacteria multiply and develop into a biofilm in a wound.

In many diseases and infections of the human, the clinical role microorganisms play in healing is related to the microbial load—principally due to the fact that high levels of microorganisms will exacerbate a subclinical or asymptomatic condition.[36–38] For example, in acute wounds the probability of infections occurring as the microbial load (infectious dose) increases is likely and accepted, but one must also consider the virulence and pathogenicity of the bacteria within a wound bed and not be focused on numbers alone. The probability that a wound will progress and heal will, in addition to bacterial numbers and the virulence of the microbial ecosystem, also relate to the site where the wound is located, the care the patient is receiving, and the patient's underlying comorbidites. Such factors lead to a more complex wound.

There is a fine balance between numbers, virulence, and pathogenicity of bacteria and wound infection. If bacteria are able to overcome the armories of the host's immune system, invasion of bacteria into the host tissue can occur, leading to local and systemic responses.

SPECIES DIVERSITY IN ACUTE AND CHRONIC WOUNDS

One important question in wound microbiology that remains controversial today is whether delayed healing or infection in a wound is caused by one or more particular bacterial species or the end result of the metabolic activities of any combination of many bacteria within the wound. Bacteria that have been involved in wound infections have included *Mycobacterium fortuitum,*[39] *Acinetobacter* sp[40–43] and, of growing importance, *Candida albicans. C. albicans* is not generally screened for in wounds but represent potent highly virulent opportunistic pathogens that are known to interact synergistically with bacteria.[44,45] In addition to the microorganisms mentioned above, an array of other bacteria have been isolated from burns[46,47] and infected diabetic foot ulcers.[48-49] Without doubt, meticillin-resistant *S. aureus* (MRSA) are major concerns in both acute and chronic wounds, and as a result, screening in hospitals is becoming routine practice to prevent the spread of this bacteria.[50]

The most frequently isolated pathogens observed in surgical site infections based upon clinical culture methods, as proposed by Mangram and colleagues,[51] have included *Enterococcus* spp., *S.aureus,* coagulase-negative staphylococci (CNS), *Escherichia coli, Enterobacter* spp., and *P. aeruginosa.* In a recent study of 390 patients, comprising 280 surgical wounds, 92 ulcers, and 32 sinuses and lacerations, coliforms (63.5%) were found to be the most predominant microorganisms isolated followed by *Proteus* sp (37.2%). Multiple species infections have been reported in 22.7% of surgical wounds and 24.6% in ulcers.[52]

Other opportunistic pathogens have also been associated with delayed wound healing and infections and have included CNS such as *S. epidermidis* as well as *Enterococcus, Micrococcus,* and *Corynebacterium* spp. (often referred to as diphtheroides). *Streptococcus* sp are also cultured frequently in wounds.[53,54]

Gram-negative bacteria that have been proposed as a cause of infection in wounds have included *Pseudomonas* sp, *Acinetobacter* sp (a bacteria now becoming multidrug resistant),[55] and *Stenotrophomonas* spp. Other bacteria predominately found in wounds and significant to infections in both acute and chronic wounds have also included *E. coli, Enterobacter cloacae, Aeromonas, Klebsiella, Enterococcus,* and *Proteus* species.[56–59] In a study by Daltrey and colleagues,[60] it was reported that in 71% of nonhealing pressure ulcers, the bacteria most predominately cultured included *P. aeruginosa, Proteus* sp, and *Bacteroides* sp, *P. aeruginosa* are commonly selected for and therefore isolated from chronic wounds.[61,62]

Anaerobic bacteria such as Petostreptococcus are found in high numbers in chronic wounds.[63,64] The role these and other anaerobes play in wound infections is still poorly understood and as such more research is required in this area. Recent work using modern methods indicate that anaerobes such as *Fingoldia magna, Peptoniphilus* spp, *Anaerococcus,* and *Bacteroides* are highly ubiquitous in chronic wounds. In addition to these, *Prevotella* sp have been isolated frequently.[65]

It is also important to note that many surgical site and burn infections are reported as "culture negative," meaning nothing could be isolated.[66] This is an important concern particularly if guided therapies and appropriate use of antimicrobials is needed.

The effect of specific types of microorganisms on wound healing has been widely published. Yet because these studies are typically based upon culture methodologies, their results may exaggerate the contribution of these notable and easy to culture bacteria. Recent data using more powerful molecular methods somewhat contradict these findings, indicating there may be more involvement from previously uncharacterized anaerobes and many types of bacteria that are notoriously difficult to culture in the laboratory.

Based upon the literature, the main bacteria cited as causing a wound infection and delaying wound healing have been *S.aureus*, *P.aeruginosa*, and beta-hemolytic streptococci.[67–69] Such wound pathogens are known to produce a number of potentially destructive virulence factors, specifically enzymes and toxins.[70] In particular, *S. aureus*, predominately meticillan-resistant *S. aureus* (MRSA), is known to cause major concerns in heathcare settings.[71] Bacteria that have been linked to the "one bug one disease" concept (Koch's postulates) are discussed below.

STAPHYLOCOCCUS AUREUS

S. aureus is considered to be the most problematic bacterium in surgical and burn wound infections and has also been identified, using selectively biased culturable techniques, as the most predominant bacteria found in chronic wounds.[72–77] *S. aureus* has also been reported to be the single most commonly isolated bacteria responsible for a high percentage of cutaneous abscesses and necrotizing fasciitis.[37,75,78]

In numerous studies, *S. aureus* has been found in association with other bacteria such as coliforms, *Enterococcus* sp, *S. epidermidis*, *Streptococcus* spp, and *P. aeruginosa*.[79–81] Consequently, the role *S. aureus* plays as a causative agent for infections in wounds is open to debate because a clear and definitive correlation between the presence of *S. aureus* and wound infection has not been established. In addition, some recent studies have revealed the bias of previous reports showing that *S. aureus*, in a large number of wounds, is only present in minor populations.[22,23,82] Of increasing concern to wound care is the growing incidence and prevalence of vancomycin-resistant *S. aureus*.[83]

BETA-HEMOLYTIC STREPTOCOCCUS

Wright and colleagues[84] reported that a surgical wound could not be successfully closed if a hemolytic *S. pyogenes* strain was present. In addition, a study by Robson and Heggers[85] highlighted that beta-hemolytic streptococcus was the only bacterium that caused infections in tissue. Further studies have revealed that *S. pyogenes* is the sole pathogen in some cases of necrotizing fasciitis.[75] In bite wounds, for example, beta hemolytic streptococci have been proposed as the main cause of infection.[50,86,87]

Based on the evidence to date, beta-hemolytic streptococci are significant to wound healing and have been reported to often require guided therapies to eradicate it from wounds.

Pseudomonas aeruginosa

P. aeruginosa has been implicated as the primary cause of infection in a chronic wound.[88] They are opportunistic pathogens and associated with approximately 10% to 20% of all hospital-acquired infections.[89,90] *P. aeruginosa* have been isolated frequently from chronic wounds, based solely on culturable studies.[91–93] *P. aeruginosa* are intrinsically resistant to antimicrobial agents and are known to produce an array of virulence factors which include adhesions, alginate, pili, flagella, iron-binding proteins, leukocidins, elastase, proteases, phospholipase C, hydrogen cyanide, exotoxin A, and exoenzyme S.[94] Lipopolysaccharide (LPS) and other products derived from *P. aeruginosa* have been shown to have a dose-dependent inhibition of keratinocytes, which effects their migration. Preventing the migration of keratinocytes is considered significant to wound healing.[95] In addition, *P. aeruginosa* are avid biofilm formers. Removal of biofilms containing *P. aeruginosa* would be beneficial to wound healing. Essentially, all virulence factors produced by *P. aeruginosa* have a role to play in delaying wound healing.[96]

ANAEROBIC BACTERIA

The importance of anaerobic bacteria in nonhealing and infected chronic wounds is significant but poorly documented and studied. Anaerobic bacteria that have been isolated from chronic wounds have included *Prevotella*, *Finegoldia*, *Peptostreptococcus*, and *Petoniphilus*.[91]

Despite the prevalence of anaerobes in wounds, few studies have documented the incidence of anaerobic bacteria in different chronic wound types. A recent study by Dowd et al.[97] reported that in venous leg ulcers 1.5% of molecular sequences were found to be consistent with anaerobic bacteria. The study also confirmed that the presence of anaerobic bacteria accounted for 30% of the microbial population of diabetic foot ulcers and 62% in pressure ulcers. In diabetic wounds, commonly isolated anaerobic bacteria have included *Peptostreptococcus*, *Bacteroides*, and *Prevotella* spp.[98]

As anaerobic bacteria possess an array of virulence factors, they are known to impair wound healing. Such virulence factors have included capsular polysaccharides, fimbriae, hemagglutinin, tissue-damaging exoenzymes, toxins, and immunoglobulin inhibitors. Anaerobic bacteria are considered more significant to wound healing when they exist in an aerobic and anaerobic synergistic relationship.[99]

COMMUNITY HYPOTHESIS AND WOUND INFECTIONS/HEALING

No study has been able to demonstrate consistently and unequivocally a significant relationship between clinical outcome and the composition of the chronic wound microbiota. It is probable that no single etiological association exists.[100] Dowd et al.[97] described the concept of "functional equivalent pathogroups" to indicate that diverse communities of bacteria could develop in flexible ecological structures to achieve a pathogenic phenotype even if individual members of this community were not themselves inherently pathogenic. This concept works in concert with the "community hypothesis" discussed later in this chapter to explain how bacteria can work together to maintain the chronic nature of wounds and then achieve a state where they initiate infection.

S. aureus have been found based upon culture-methodologies to be the most prevalent bacteria in wounds. Even though new methods are indicating it is less important, its relationship between presence and clinical infection is still considered significant. In addition, bacteria such as beta-hemolytic streptococci, although highly virulent in a wound, are considered significant only to soft tissue wounds and acute wounds.[101] *P. aeruginosa* has been reported to be the causative bacteria of an infection in a chronic wound.[82] However, other studies have shown that *P. aeruginosa* are often found together with beta-hemolytic streptococci.[102]

Many papers have been published and data generated using selective culture media to isolate specific bacteria only in the majority of the wound studies undertaken. As a result, this has limited and biased growth of specific genera and species of bacteria. Consequently, the interpretation of research findings with biased culturable techniques has resulted in the *specific species wound hypothesis*.[103] However, numerous wound studies have shown that infections in wounds are not due to just one specific species but are in fact due to a community of microorganisms, hence the *community wound hypothesis* and the *functionally equivalent pathogroup hypothesis*.[104–105]

In a study investigating the bacteriology of chronic leg ulcers, Trengove[101] reported that no single microorganism or group of microorganisms was more detrimental to wound healing than any other. The researchers reported that there was a significantly lower probability of wound healing (the wound was more likely to remain chronic) if four or more bacterial groups were present in any ulcer. This and other studies indicate that a community of microorganisms that had undergone a series of microbial interactions may have an effect on the wound healing process. Such observations support an earlier view of Kingston and Seal,[103] who argued that all species of bacteria associated with a microbial disease should be considered significant to wound healing, as they are all able to interact synergistically with certain species of other bacteria. Synergy of bacteria is known to enhance virulence of bacteria.

Armstrong and colleagues[106] have proposed that evidence of specific pathogens does not necessarily suggest that particular bacteria are the agents of infection. Based on this and other evidence to date and the fact that most infections are caused by a combination of microorganisms, it seems plausible to suggest that wounds are polymicrobial and the community per se is responsible for the effect observed with delayed wound healing and infection. The community-based hypothesis has been found to occur in other infections including prostatitis, osetomyelitis, gum disease and peridontium.[107]

THE ROLE OF A COMMUNITY

As mentioned previously, infections in acute or chronic wounds are essentially complex in nature and only generally occur when there is an appropriate combination of bacteria. Synergy between bacteria, nutrient availability, and a susceptible wound environment aid bacterial proliferation. Synergy between bacteria in a wound is significant and should be considered to occur in all wound types that are colonized.

Microbial synergy is known to increase the "net pathogenic effect" of a community of microbes which will have a positive effect on the indigenous bacteria to the detriment of the wound. Synergy, particularly between aerobic and anaerobic bacteria, has been documented to increase the severity of an infection.

Many studies have shown that bacterial interactions play a critical role in the persistence of many species of bacteria, particularly in hostile environments. Within a chronic wound environment, particularly in the presence of devitalized sloughy tissue, anaerobes are considered the dominant microorganisms despite the close contact of air with the wound. Anaerobes are able to cope with the toxic effects of oxygen via interacting with bacteria that are able to grow in air. As the aerobic bacteria grow in the wound, they reduce the environmental levels of oxygen, enabling the anaerobic bacteria to proliferate. This has been demonstrated in laboratory studies involving communities of oral bacteria.[107-114] Such studies suggest that there is a very close association with aerotolerant, oxygen-resistant, and sensitive bacterial species that alter environments to enable obligate anaerobes to survive and grow.

Bacterial coaggregation (or co-adhesion) within the wound environment may be a key process in the formation of microbial communities in wounds. Coaggregation enables both intra- and intergeneric attachment of bacteria. Such interactions may ensure that the microorganisms needing to cooperate for nutritional or other environmental modifying purposes are then appropriately spatially organized in a wound biofilm. In addition, growing as an aggregate of cells or a biofilm is known to enhance communal protection from phagocytosis by polymorphonuclear leukocytes (PMNs). Furthermore, a community of bacteria acts synergistically to share essential nutrients, reduce intracellular killing by the immune system, enhance the production of essential bacterial growth factors, modify the local environment (e.g., redox potential), and enhance the protection of sensitive species within the community. This protection process has been named "indirect pathogenicity"—in certain situations some pathogens are found to be antibiotic sensitive but are rendered "resistant" by other members of the mixed infection and through formation of biofilm communities.

In order to proliferate at a site, microorganisms need to acquire their essential nutrients and modify the environment. This will make conditions more favorable for growth. These nutrients are broken down by a cooperative action of several groups of microorganisms with complementary enzyme profiles. Within a chronic wound environment, endogenous nutrients such as glycoproteins, sugars, and proteins will exert an effect in maintaining the diversity of the wound microbiota. In order to utilize these specific nutrients, numerous enzymes are required, ranging from proteases to glycosidases. Studies have shown that when bacteria are growing in a consortia, higher cell densities of bacteria are obtained than when bacteria are grown in a pure state.[115,116]

Microbial communities are able to defy the constraints imposed by the external macroenvironment by creating, through their metabolism, a mosaic of microenvironments that enable the survival and growth of niche species of bacteria. Additionally, microbial communities are significant to disease and have inherent characteristics similar to those of a multicellular organism. Microbial communities display characteristics of multicellularity in terms of spatial organization. This aims to enhance cellular division of labor, the use of communication systems and metabolic cooperation.

UNCULTURABLE BACTERIA AND WOUNDS

Ninety-nine percent of bacteria found in soil and the aquatic environment are not able to be cultured. This has often been referred to as the "great plate count anomaly."[117] In cases of acute otitis media, between 15% and 56% of bacteria cannot be cultured.[118,119] Additionally, it has been shown that a large percentage of wound bacteria cannot be cultured. It is probable that this lack of culturability in a wound may be due to nutritional constraints, requirements for synergistic interaction with other bacteria or the host, and the fastidious nature of the inherent microorganisms. The failure of bacteria isolated from wounds to grow on an artificial media means that the correct environment for growth has not yet been achieved. Artificial media often lack nutrients required by microorganisms or the incubation atmosphere or temperature and pH used for growth may be suboptimal. Essentially, a number of isolation media used in traditional microbiological laboratories may contain toxic substances for growth, or other bacteria within the sample may produce inhibitory substances that also will prevent growth.

Veeh and colleagues have demonstrated that conventional microbiology techniques failed to detect colonization of bacteria on some human tissues.[120] With new molecular biology techniques, such as polymerase chain reaction (PCR) and fluorescent in situ hybridization (FISH), and bTEFAP[23] more bacteria living in biofilms are being discovered. As technology advances, researchers may uncover additional bacteria living in biofilms that cause or contribute to diseases and infection. In a recent study of surgical site infections, Dowd et al.[97] originally identified using bTEFAP a previously uncharacterized Bacteroidetes, which occurred in 100% of surgical site infections they evaluated. In more than one instance, the percentage of this unknown bacteria in the surgical site infection was nearly 100%. In subsequent work, this bacteroide was unable to be cultured in isolate form using methods designed specifically for fastidious anaerobes, and only with a complex host-derived media combined with polymicrobial synergistic cultures (addition of other bacteria to the culture) could it be propagated. This provides a fundamental appreciation of how there are unknown bacteria associated with chronic wounds which cannot be evaluated or identified using traditional culture methodologies and how certain bacteria may only be cultured as part of a climax community or pathogroup.

Many bacteria have been identified from chronic wounds using molecular techniques when compared to culturable techniques.[121] Redkar et al.[122] examined the bacterial flora present in a chronic diabetic foot and the culture-based methods used identified single anaerobic species *Bacteroides fragilis*. Molecular-based methods employing 16S sequencing identified *B. fragilis* as a dominant organism and *Pseudomonas* (*Janthinobacterium*) *mephitica* as a minor component. Hill and colleagues[123] applied molecular and enhanced cultural techniques to determine the microbial composition of a chronic, nonhealing venous leg ulcer. Both tissues and swab samples taken were compared between the 16S sequence method and culture.

Culture analysis detected *Acinetobacter* spp in both types of samples. Swab samples yielded *Proteus* spp and *Candida tropicalis,* whereas tissue samples cultivated *S. epidermidis.* Molecular analysis identified clones that were closely related to these cultured organisms; however several clones were closely related to organisms that were not identified by culture techniques. These included *Morganella morganii, Bacteroides urelyticus, Enterococcus faecalis,* and *Peptostreptococcus octavius.* This study not only exemplified the ability of molecular techniques to identify a wider range of organisms in wounds, but also highlighted the effect of sampling techniques on the types of viable organisms that could be recovered. A similar study conducted by Davies et al.[124] also found a greater diversity of organisms when 16S ribosomal and denaturing gradient gel electrophoresis (DGGE) was employed to analyze the microfloras of healing and nonhealing chronic venous leg ulcers. Of the sequences obtained, 40% represented microorganisms that failed to be identified using culture. In addition, four organisms were identified by sequencing which have not previously been associated with chronic wounds—namely, *Paenibacillus* spp, *Gemella* spp, *Sphingomonas,* and *Afipia* spp. A direct comparison of the microfloras of the healing and nonhealing ulcers highlighted a significant difference in the carriage of *Micrococcus* and *Streptococcus* spp with a higher incidence in nonhealing wounds. A more recent study by James et al.[77] utilized molecular methods in both acute and chronic wounds and revealed a diverse polymicrobial community and the presence of bacteria, including strictly anaerobic bacteria, that were not identified using culturable techniques. Furthermore, Andersen and colleagues[125] verified DGGE as a powerful tool for elucidating the clinical microbiology of chronic diseases. The results of this study suggested that skin graft operations were a novel way of obtaining multiple samples for *in vivo* bacteriology and for establishing the spatial distribution of bacteria in the complex microenvironment of chronic wounds.

The use of the 16S rRNA gene has helped substantially to appreciate the true microbiology of the wound environment and overcome the problems associated with culturable techniques in the identification of bacteria. Most research on and within the human body which has utilized this technology has been principally in the oral[126] and the gastrointestinal (GI) tract areas. The skin, however, is an area that has not been studied in any great detail using molecular microbiology. This was discussed in Chapter 2.

Frank and colleagues[8] determined the microbiology of 19 chronic wounds using molecular analysis, rRNA, and culturing methods using swab samples. The authors concluded that 75% of the sequences belonged to Staphylococcus (25%), Corynebacterium, (20%), Clostridiales (18%), and Pseudomonas (12%). It was found that 0.5% of sequences were potentially new species of microorganisms. They found that in half of all the wound specimens, PCR and culturing methods gave different results. Similar results have also been observed in other wound studies.[127,128]

Historically, the microbiological analysis of chronic wounds, based on culturable techniques alone, indicated domination by Gram-positive bacteria. Despite a number of molecular studies that have been applied to study the microbiology of chronic wounds, it is true that molecular technologies will undoubtedly provide improved microbiological profiles of the wound environment than culture-based studies have generated.[126–129] Recently with the emergence of next-generation molecular techniques, Dowd et al.[97] in a series of in-depth survey studies have cataloged the microbial diversity of chronic wounds using culture-based, DGGE, cloning, and Sanger sequencing and bTEFAP. In their initial studies that evaluated the three primary categories of chronic wounds (venous leg ulcers, pressure ulcers, and diabetic extremity ulcers), they resolved that there were specific major populations of bacteria that were evident in the biofilms of all chronic wound types, including *Staphylococcus, Pseudomonas, Peptoniphilus, Enterobacter, Stenotrophomonas, Finegoldia,* and *Serratia* spp. Each of the wound types revealed marked differences in bacterial populations, such as pressure ulcers in which 62% of the populations were identified as obligate anaerobes. There were also populations of bacteria that were identified but not recognized as wound pathogens, such as *Abiotrophia paraadiacens* and *Rhodopseudomonas* spp. Results of molecular analyses were also compared to those obtained using traditional culture-based diagnostics. Only in one wound type did culture methods correctly identify the primary bacterial population, indicating the need for improved diagnostic methods. The unavoidable inference is, if clinicians can improve their understanding of wound microbiota, it will provide enhanced comprehension of the wound's ecology, thus promoting improved management and patient prognosis. Dowd and colleagues' research highlighted the necessity to begin evaluating, studying, and treating chronic wound pathogenic biofilms as multispecies entities in order to improve the outcomes of patients. This survey also fostered the pioneering and development of new molecular diagnostic tools that can be used to identify the community compositions of chronic wound pathogenic biofilms and other medical biofilm infections.[22,23,82] In subsequent work, this group individually analyzed each wound type to evaluate individual wound communities, providing a detailed evaluation of how more specific functional equivalent pathogroups may interact to promote infection and through the community hypothesis they are able to maintain the chronic nature of wounds through cooperative synergism.

As our understanding of chronic wound microbiology advances, new antimicrobial treatment strategies can be employed to better target the individual populations of as yet, unknown and uncultured microbes, bringing a more logical and personalized clinical approach to wound care. If treatment modalities can be identified that change and respond along with the microbial populations, then it is only logical that we can better control such highly responsive and adaptable microbial communities. The old adage of "know thy enemy" can be made to work to the advantage of the clinician and, subsequently, the patient.

Chronic wound medicine is historically slow at responding to change. Currently, wound microbiology will remain reliant on sampling techniques to provide informative microbiological reports.

SAMPLING OF A WOUND

Confusion exists as to whether microbiological sampling of a wound is clinically warranted. How should the sample be collected? When is the most appropriate time for sampling to take place? It is generally accepted that it is inappropriate for all wounds to be microbiologically sampled, but consistency in practice is variable. Analysis of a wound is only warranted if it is failing to make progress, deteriorating, increasing in size, being screened for antibiotic-resistant microorganisms, and clinically infected and has not responded to empirical antimicrobial therapy. Numerous hospitals will not undertake wound swabbing unless there is

1. Spreading cellulitis
2. Systemic symptomatology (i.e., fever and malaise, leukocytosis, etc.)
3. An immunosuppressed patient (diabetes, steroids, malignancy, malnutrition)
4. The necessity to screen for MRSA or other multiresistant microorganisms

Sadly, the motivation for sampling a wound may be driven by avoiding accusations of negligence rather than through clinical necessity.

In addition, it is important that any microbiological data generated be interpreted in conjunction with clinical observations. This will ensure that the most informed clinical decisions can be made so that appropriate and effective wound management strategies can be employed. A recent study was set up to investigate microbiological reporting policies and guidelines when submitting swabs from venous leg ulcers.[130] The authors concluded that clinicians need to include clinical information with the swab so that laboratories are better placed to interpret microbiological results accurately. This was suggested to help reduce the use of antibiotics in the management of wounds. Specific clinical information used in conjunction with microbiology data can be seen in Table 6.2.[131] Additional clinical data that are generally required include the type of wound, its location, any underlying pathology, signs of infection, and any antimicrobial strategies utilized.

Interpretation of microbiology data in isolation of relevant clinical data is considered inappropriate.

Another theory beginning to emerge takes into account the presence and nature of microbial communities. This theory involved predatory and nonpredatory populations,

TABLE 6.2
Clinical Indicators for Wound Infections

Heat

Redness

Swelling

Temperature increase

Exudate (purulent, serous, or serosanguinous)

Delayed healing

Malodor

Bleeding

Epithelial bridging

Tissue breakdown and necrosis

Poor development of granulation tissue

Systemic illness

Pain

Increased wound size

the rapid adaptability of microbial populations, and the pathogen evolution theory. The evolutionary goal of pathogens is to develop a relationship with the host where they develop a niche and maintain equilibrium with the host. Thus, in the case of a chronic wound, the microbial population would seek not to expand its area of influence but to maintain a commensalism and mutualistic relationship with the host.

SUPERFICIAL WOUND SWABS

Before a swab is taken, the wound should be sluiced with sterile saline or water. The aim of this is to make sure that the samples obtained do not represent surface contaminant bacteria but reflect the deeper seated microbiology of the wound bed. The Levine method of surface swabbing a wound is an accepted and easily performed sampling principle. The method involves rotating a swab (cotton-tipped, rayon-tipped, or alginate-tipped) over 1 cm^2 of wound tissue and applying gentle pressure.[132,133] A zig-zag sampling pattern is often employed, and areas containing slough or with a high level of discharge are often avoided. If wounds contain a lot of pus (this is often sterile) and discharge, aspirated samples may also be obtained. Cotton swabs have been shown to contain fatty acids that may suppress or increase the growth of certain bacteria. If the wound is dry, swabs are premoistened in the transport medium. The swab is then transported to a microbiology laboratory without undue delay where it is then analyzed.

The Levine technique has been reported to accurately measure wound bioburden,[133] but surface swabbing has been found to underestimate the microbial richness and diversity of a wound because this technique does not recover many "deep-seated" microorganisms found within the wound bed. For analysis of the microbiology deep within a tissue, more aggressive techniques are required. Surface swabbing essentially will recover only transient, loosely adhered, and planktonic bacteria, and such bacteria may not be those responsible for delaying healing or causing overt infection.[134,135] Viscoelastic biofilms are generally not sampled adequately during a swabbing technique, as they have the ability to significantly recoil—a property considered similar to the pulling and recoiling of an elastic band.

Even though superficial wound swabbing may be considered a poor method for obtaining microbiological specimens, the technique is inexpensive, noninvasive, and less distressing and traumatic to a patient when compared to other methods such as a tissue biopsy.

Furthermore, surface swabbing can be used on patients in both the community and the clinic, and the time involved is significantly less than taking other wound samples.[136] However, the lack of information regarding the collection of samples via swabbing in many studies, and also the different wound preparations and area sampled and transport time to the laboratory have resulted in poor interpretation of research findings and diagnostic validity.

ASPIRATE/TISSUE

An aspirate may provide an alternative sampling method to surface swabbing when seeking a microbiological profile of a wound. Additionally, such a sample is likely to incorporate detached wound biofilm. The aspirate sample harvested is similar to surface swabbing in that it is primarily composed of planktonic bacteria which in colonized or infected wounds poorly represent the total populations. However, a wound fluid aspirate sample is considered to provide a better sample for microbiological analysis, when compared to surface swabbing alone. Despite the benefits of an aspirate sample, on occasion this method is found not to be as good as surface swabbing.[137,138]

Tissue biopsy is one of the most invasive methods of sample collection, but when analyzed using molecular methods such a bTEFAP provides a highly accurate representation of the microbial populations. However, many clinicians have argued that tissue biopsies are not necessary to determine the wound microbiology, and impediments to acquiring biopsy samples exist. For example, in acute wounds a correlation has been found between quantitative tissue biopsy and semiquantitative superficial swab analyses.[139,140] Punch biopsy samples obtained from deep tissue are aseptically removed from the wound tissue, weighed, homogenized, and then diluted in an appropriate fluid. On occasion, accessibility in acquiring a biopsy deep tissue sample can prevent acquiring a representative sample. For this reason, superficial and devitalized tissue is often used as an alternative.

A number of authors, when analyzing pressure ulcers, have suggested that analysis of deep-tissue biopsy specimens should be used to assist in diagnostic criteria of infection as clinical histopathological data can be obtained.[141–144] However, Wheat and colleagues reported that the results generated from a surface swab correlated well to a tissue biopsy obtained from a diabetic foot infection.[145] A 75% correlation between a swab and a tissue biopsy sample was also reported by Sapico and colleagues.[24]

In order to gain a truly representative wound sample for microbiological analysis, the value of a single biopsy from a large wound is considered to not fully reflect the true microbial diversity of the wound. Hence, for correct interpretation of findings, a sample size greater than one would be required. A study by Woolfrey and colleagues[146] reported that in burn biopsies there was a 25% chance of not isolating a number of bacteria. The authors reported that this may be due to the distribution of bacteria on the wound surface.

A final method that arguably provides the best of all the sampling methods avoiding the invasiveness of tissue biopsy is sharp/surgical debridement. This method removes all devitalized tissue, biofilm, and microcolonies, exposing the healthy tissue underneath. The procedure is relatively simple and can be performed as part of standard of care for most chronic wounds provided the clinician has acquired the requisite skills. The surface of the wound is removed of surface contaminants, often with gauze and sterile saline to remove the types of populations and then rinsed again with saline. Following this, devitalized tissue is aseptically debrided and placed into sterile sample tubes for shipping to wound-specific molecular diagnostic laboratories where these laboratories utilize molecular diagnostic methods developed specifically for wounds. Such methods have been shown to provide more comprehensive evaluation of wound and biofilm microbial populations. Preliminary results are returned within several hours of the laboratories receiving the samples, and final comprehensive reports follow up after 48 hrs providing in-depth analysis of the microbial communities including any therapeutic susceptibilities as determined bioinformatically. The service of these laboratories provides a needed evolution of diagnostics to improve and ultimately reduce the cost of wound care.

Wound Sampling Overview

A tissue biopsy is viewed by many as the "gold standard" when seeking to obtain a sample that will provide accurate microbiological analysis and interpretation. However, sharp-debridement sampling or deep curetting of the wound bed provides a logical alternative and constant standard of care approach for obtaining microbiological samples. This technique can help clinicians to continually monitor the microbial communities when coordinated with next-generation diagnostic approaches. In addition to this, a study by Davies and colleagues[147] has shown that following analysis of the wound microbiota of 70 patients with chronic leg ulcers using both biopsy and swab samples, they concluded that "a significant association between healing and bacterial diversity in the wound as assessed by swab was demonstrated." In addition, the authors concluded that bacterial density was an independent predictor of wound healing and that microbiological analysis of biopsies provided no additional prognostic information when compared to the microflora obtained from surface swabs.

Microbiological Specimen Handling for Laboratory Culture-Based Diagnostics

Once sample(s) have been taken from a wound these are placed in appropriate transport medium and then conveyed to the laboratory in a timely manner—a delay in processing of samples will affect bacterial recover, viability and therefore growth on agar plates.[148] In most cases once a sample is collected it is placed in a recovery or enrichment broth which is designed to allow certain and more fastidious bacteria to recover from stressed states so that they can be grown in the clinical laboratory. Consequently during long-term transportation of samples the more robust bacteria will significantly outgrow the more fastidious bacteria suppressing their recovery and detection. This is an area often not considered or appreciated by many wound care practitioners Consequently, for microbiological specimens the transportation step is of paramount importance and should not be delayed. In the UK there are standard operating procedures for investigating both skin and superficial wound swabs (BSOP 11), abscesses, postoperative wounds and deep-seated infections (BSOP 14).[149]

In the microbiology laboratory microorganisms removed from a wound sample will be cultured and identified are then screened for antibiotic sensitivity against a library of antibiotics. Antibiograms are generated and this data is used to help guide topical or systemic antibiotic usage.

Anaerobic bacteria and yeasts/fungi are generally not routinely screened for in wound specimens and therefore are not often included in any microbiological profiles generated, unless requested for by the clinician. The reason for this resides around the fact anaerobes and fungi take a lot longer to grow and more selective agars are required.

Consequently, the identification of anaerobes and yeasts/fungi and the role they may play in wound infections have not been determined by the numerous clinical studies documented. As mentioned previously, the processing of microbiological specimens often involves the use of highly nutritious, selective and non-selective enrichment and diagnostic agars and as such the more fastidious bacteria, which are found in wounds, often go undetected even if they are the major populations in the wound. Consequently, only fast growing, robust and non-fastidious bacteria will be recovered.

Results obtained from all diagnostic microbiology laboratories can often be misleading as to which bacteria or community of microorganisms are responsible for infections and nonhealing infected wounds. As mentioned above the use of molecular microbiological techniques has significantly helped researchers to achieve a greater understanding of the microbial diversity and richness of acute and chronic wounds. However, molecular techniques are not employed routinely in diagnostic microbiology laboratories to investigate the microbiology of wound specimens. As such, the "viable but not culturable" (VBNC) bacteria are not documented in the majority of studies despite their significance to disease and infection.[150,151]

It may be that knowledge of VBNC bacteria will help to stimulate the development of improved management strategies for those wounds which are not progressing because they have become infected.

MICROBIOLOGICAL SPECIMEN HANDLING FOR LABORATORY MOLECULAR-BASED DIAGNOSTICS

Sample handling for molecular diagnostics is a simple procedure of placing the sample into a sterile transport vial and shipping it to the diagnostic facility. Samples assayed in molecular-based diagnostics do not require specialized enrichment media or preservatives, as this approach detects the bacterial genetic signature rather than relying on the bacteria's ability to grow on special nutrients within the laboratory environment. Because DNA is well protected inside of the bacterial cell, extremes of temperature will not significantly alter the profile of the microbial communities' DNA signature. It is rare that during shipment any given population of bacteria would grow and alter the signature due to the lack of enrichment or nutrients available after the sample is removed from the host environment. Once in the molecular diagnostic laboratory, the bacteria and biofilms are disrupted, and the DNA from the entire community is extracted and purified. Then molecular methods are able to specifically identify which bacteria are present. More advanced molecular methods are also able to determine the relative contribution of individual bacteria within a sample providing the clinician with vital information on which bacteria they might consider targeting with a more individualized approach to therapy. One of the criticisms of molecular methods is that information on antibiotic susceptibility is not generated. It is important to remember that standard clinical laboratories rely on the ability of bacteria to grow in a planktonic isolated single colony state before performing antibiotic sensitivity. Studies fail to provide clinically valid information especially if the bacteria they are analyzing are not the primary wound populations and are no longer in the same physiological state when they resided within the wound (e.g., biofilm phenotype). It is likely that most antibiotic susceptibility testing may soon be considered of little benefit when dealing with biofilm phenotype bacteria and bacteria that are part of complex communities. Modern molecular approaches combined with bioinformatics that resolve the complexity of known susceptibilities may soon provide the most logical approach to evaluating potential therapeutics by providing clinicians with more accurate data that will help to bestow targeted therapeutics.

CONCLUSION

Better knowledge and understanding about the numbers of microorganisms, evidence of biofilms, interactions between bacteria, and virulence of microorganisms as well as clinical signs are vital to help guide effective wound management strategies for both acute and chronic wounds. Knowledge of wound microbiology and the impacts bacteria have on wounds has not kept pace with developments in other areas of medicine and remains reliant on traditional approaches with their subsequent limitations.

Communities of microorganisms—the "microbial communities" and the "functional equivalent pathogroup"—are fundamental to the development of climax communities and infections and sustained nonhealing of chronic wounds. The community hypothesis was initially proposed in the dental arena. In wound care, this hypothesis proposes that the dynamic situation at the wound surface is in equilibrium with the host. Various components of the complex microflora can be disturbed by changes in the local wound environment. This will cause fundamental changes in the different species or subclones within species of microorganisms found on or in a wound bed so that different microorganisms will dominate the wound environment at different times. Similarly, the functionally equivalent pathogroup hypothesis suggests that communities of bacteria combine synergistically and as a whole possess the necessary genetic arsenal of generalized and virulence factors necessary to coerce heavily colonized tissue into infection. Together, the microbial community hypothesis and the functionally equivalent pathogroup hypothesis explain why chronic wounds remain chronic even when host comorbidities are controlled.

Once a wound infection is initiated, the progression of the infection occurs deeper into the tissue. This progression is accompanied by successional changes in the microflora of the wound. Limitations in respect of the knowledge surrounding bacterial wound infections exist, and there is a need for continuing research in this area to determine exactly what goes on at the wound biofilm interface during the early stages of infections in wounds. Such information will help facilitate the development of preventative measures. Development of protocols to prevent onset and progression of infection in wounds will follow improvements in our understanding of microbial initiation of infection. The quantitative and qualitative aspects of wound microbiology are thought to be fundamental to the development of infection. It is probable that risk of wound infection increases as the microbial load increases, up to a critical level where failure to heal as a result of infection is considered to be inevitable. This hypothesis seems justifiable as a result of the recent evidence linking biofilm formation to delayed healing, especially in chronic wounds, and that the composition of the polymicrobial wound flora is likely to be more important than the presence of specific pathogens.

However, a critical factor in wound healing and infection is the efficacy of the host immune response in dealing with the wound microflora. Local environmental factors such as tissue necrosis, hypoxia, ischemia, diabetes mellitus, chronic granulomatous disease, and other immune deficiencies will have a role to play in affecting the microbial community in a wound and in generating an accurate prognosis. Microorganisms help to compromise the immune response. By assessing the host and microbial factors collectively, the risk of a wound infection occurring can be addressed. The presence of biofilms in wounds will cause problems in relation to the management of a wound infection and the efficacy of the innate immune response. Biofilms are frequently observed in chronic wounds and will be discussed further throughut the book.[152]

The presence of the biofilm, rather than individual species found in a wound, will have an effect on wound healing. Consequently, in terms of control, we should be aiming to control the biofilm (i.e., *community hypothesis*) and to continually monitor and adapt our therapies to the changes in the functionally equivalent pathogroups and not individual species of bacteria (i.e., *specific species hypothesis*). This represents a significant and important evolution of medicine and a notable "anti-Koch approach" to infection and disease.

REFERENCES

1. Davis, M.H., P. Dunkley, R.M. Harden, K. Harding, J.M. Laidlaw, A.M. Morris, and R.A.B. Wood. (1992) Cause: Types of wound and ulcers, pp. 109–132. In *The wound programme*. Centre for Medical Education, Dundee, United Kingdom.
2. Anon. (2008) Wound infection in clinical practice. An international consensus. *Int Wound J* 5:3–11.
3. Moffatt, C.J. (2004) Perspectives on concordance in leg ulcer management. *J Wound Care* 13(6):243–248.
4. Mansbridge, J. (2009) Hypothesis for the formation and maintenance of chronic wounds. *Adv Skin Wound Care* 22(4):158–160.
5. Carr, S.C. (2008) Diagnosis and management of venous ulcers. *Perspect Vasc Surg Endovasc Ther* 20(1):82–85.
6. VanGilder, C., MacFarlane, G., Meyer, S., and Lachenbruch, C. (2009) Body mass index, weight, and pressure ulcer prevalence: An analysis of the 2006–2007 International Pressure Ulcer Prevalence Surveys. *J Nurs Care Qual* 24(2):127–135.
7. Posnett, J., and Franks, P.J. (2008) The burden of chronic wounds in the UK. *Nurs Times* 104(3):44–45.
8. Frank, D.N., Wysocki, A., Specht-Glick, D.D., Rooney, A., Feldman, R.A., St Amand, A.L., Pace, N.R., and Trent, J.D. (2009) Microbial diversity in chronic open wounds. *Wound Repair Regen* 17(2):163–172.
9. Howell-Jones, R.S., Baker, I.B., and McNulty, C.A. (2008) HPA GP Microbiology Laboratory Use Group. Microbial investigation of venous leg ulcers. *J Wound Care* 17(8):353–358.
10. Gardner, S.E., and Frantz, R.A. (2008) Wound bioburden and infection-related complications in diabetic foot ulcers. *Biol Res Nurs* 10(1):44–53.
11. Healy, B., and Freedman, A. (2006) Infections. *BMJ* 332(7545):838–841.
12. Kingsley, A. (2001) A proactive approach to wound infection. *Nurs Stand* 15(30): 50–58.

13. Schmidt, K., Debus, E.S., St, J., Ziegler, U., and Thiede, A. (2000) Bacterial population of chronic crural ulcers: Is there a difference between the diabetic, the venous, and the arterial ulcer? *Vasa* 29:62–70.
14. Moffatt, C.J. (2005) Identifying criteria for wound infection. In *Identifying criteria for wound infection EWMA position document.* London: MEP Ltd.
15. Heinzelmann, M., Scott, M., and Lam, T. (2002) Factors predisposing to bacterial invasion and infection. *Am J Surg* 183(2):179–190.
16. Cutting, K.F., White, R.J., Mahoney, P., and Harding, K.G. (2005) Clinical Identification of wound infection: Delphi approach. In *Identifying criteria for wound infection EWMA position document.* London:MEP Ltd.
17. James, G.A., Swogger, E., Wolcott, R., Pulcini, E., Secor, P., Sestrich, J., Costerton, J.W., and Stewart P.S. (2008) Biofilms in chronic wounds. *Wound Repair Regen* 2008 16(1):37–44.
18. Davies, C.E., Wilson, M.J., Hill, K.E., Stephens, P., Hill, C.M., Harding, K.G., and Thomas, D.W. (2001) Use of molecular techniques to study microbial diversity in the skin: Chronic wounds reevaluated. *Wound Repair Regen* 9: 332–340.
19. Hansson, C., Hoborn, J., Moller, A., and Swanbeck, G. (1995) The microbial flora in venous leg ulcers without clinical signs of infection. *Acta Derm Venereol* 75:24–30.
20. Halbert, A.R., Stacey, M.C., Rohr, J.B., and Jopp-McKay, A. (1992). The effect of bacterial colonisation on venous ulcer healing. *Australas J Dermatol* 33:75–80.
21. Murdoch, D.A., Mitchelmore, I.J., and Tabaqchali, S. (1994) The clinical importance of Gram-positive anaerobic cocci isolated at St. Bartholomews Hospital, London, in 1987. *J Med Microbiol* 41: 36–44.
22. Dowd, S.E., et al. (2008a) Survey of bacterial diversity in chronic wounds using pyrosequencing, DGGE, and full ribosome shotgun *sequencing. BMC Microbio*, 8:23.
23. Dowd, S.E. et al. (2008b) Polymicrobial nature of chronic diabetic foot ulcer biofilm infections determined using bacterial tag encoded FLX amplicon pyrosequencing (bTE-FAP). *PLoS ONE* 3: e3326.
24. Sapico, F.L., Ginunas, V.J., Thornhill-Joynes, M., Canawati, H.N., Capen, D.A., Klein, N.E., Khawam, S., and Mongomerie, J.Z. (1986) Quantitative microbiology of pressure sores *in different st*ages of healing. *Diagn Microbiol Infect Dis* 5:31–38.
25. Leake, J.L., Dowd, S.E., Wolcott, R.D., and Zischkau, A.M. (2009) Identification of yeast in chronic wounds using new pathogen-detection technologies. *J Wound Care* 18:103–104.
26. Malic, S., Hill, K.E., Hayes, A., Thomas, D.W., Percival, S.L., and Williams, D.W. (2009) Detection and identification of specific bacteria in wound biofilms using peptide nucleic acid (PNA) fluorescent *in situ* hybridisation (FISH). *Microbiology* 155:2603–2611.
27. Bendy, R.H., Nuccio, P.A., Wolfe, E., Collins, B., Tamburron, C., Glass, W., and Martin, C.M. (1964) Relationship of quantitative wound bacterial counts to healing of decubiti. Effect of topical gentamicin. *Antimicrob Agents Chemother* 10:147–155.
28. Heggers, J.P., Robson, M.C., and Doran, E.T. (1969) Quantitative assessment of bacterial contamination of open wounds by a slide technique. *Trans R Soc Trop Med Hyg* 63:532–534.
29. Robson, M.C., and Heggers, J.P. (1969) Bacterial quantification of open wounds. *Mil Med* 134:19–24.
30. Tenorio, A., Jundrak, K., Weiner, et al. (1976) Accelerated healing in infected wounds. *Surg Gynecol Obstet* 142:537–543.
31. Robson, M.C., Lea, C.E., Dalton, J.B., and Heggers, J.P. (1968). Quantitative bacteriology and delayed wound closure. *Surg Forum* 19:501–502.
32. Krizek, T.J., Robson, M.C., and Kho, E. (1967) Bacterial growth and skin graft survival. *Surg Forum* 18:518–519.

33. Pruitt, B.A., Jr., McManus, A.T., Kim, S.H., and Goodwin, C.W. (1998) Burn wound infections: Current status. *World J Surg* 22:135–145.
34. Robson, M., Duke, W., and Krizek, T. (1973) Rapid bacterial screening in the treatment of civilian wound. *J Surg Res* 14:420–430.
35. James, H. (1994) The microbial burden and healing rates of pressure sores. *J Wound Care* 3(6): 274–276.
36. Majewski, W., Cybulski, Z., Napierala, M., Pukacki, F., Staniszewski, R., Pietkiewicz, K., and Zapalski, S. (1995) The value of quantitative bacteriological investigations in the monitoring of treatment of ischaemic ulcerations of lower legs. *Int Angiol* 14(4):381–384.
37. Brook, I., and Finegold, S.M. (1981) Aerobic and anaerobic bacteriology of cutaneous abscesses in children. *Pediatrics* 67:891–895.
38. Brook, I., and Frazier, E.H. (1990). Aerobic and anaerobic bacteriology of wounds and cutaneous abscesses. *Arch Surg* 125:1445–1451.
39. Verghese, S., Madhusudhan, B., Senthil, M.S., Thabitha, C., Leelavathy, S., Padmaja, P., and Madhusudhan, K. (2007) Chronic postoperative wound infection caused by *Myocobacterium fortuitum* complex. *J Commun Dis* 39(4):257–259.
40. Sebeny, P.J., Riddle, M.S., and Petersen, K. (2008) *Acinetobacter baumannii* skin and soft-tissue infection associated with war trauma. *Clin Infect Dis* 47(4):444–449.
41. Lima, A.L., Oliveira, P.R., and Paula, A.P. (2008) *Acinetobacter* infection. *N Engl J Med* 358(26):2846–2847.
42. Giamarellou, H., Antoniadou, A., and Kanellakopoulou, K. (2008) *Acinetobacter baumannii*: A universal threat to public health? *Int J Antimicrob Agents* 32(2):106–119.
43. Elston, J.W., Bannan, C.L., Chih, D.T., and Boutlis, C.S. (2008) *Acinetobacter* spp. in gunshot injuries. *Emerg Infect Dis* 14(1):178–180.
44. Yener, S., Topcu, A., Manisali, M., Comlekci, A., and Yesil. S. (2009) *Candida albicans* osteomyelitis in a diabetic foot ulcer. *J Diabetes Complications* 23(2):137–138.
45. Ballard, J., Edelman, L., Saffle, J., Sheridan, R., Kagan, R., Bracco, D., Cancio, L., Cairns, B., Baker, R., Fillari, P., Wibbenmeyer, L., Voight, D., Palmieri, T., Greenhalgh, D., Kemalyan, N., Caruso, D.; Multicenter Trials Group, American Burn Association. (2008) Positive fungal cultures in burn patients: A multicenter review. *J Burn Care Res* 29(1):213–221.
46. Polavarapu, N., Ogilvie, M.P., and Panthaki, Z.J. (2008) Microbiology of burn wound infections. *J Craniofac Surg* 19(4):899–902.
47. Oncul, O., and Acar, A. (2008) Bacterial infections in burn patients. *Indian J Med Res* 127(4):415.
48. Bansal, E., Garg, A., Bhatia, S., Attri, A.K., and Chander, J. (2008) Spectrum of microbial flora in diabetic foot ulcers. *Indian J Pathol Microbiol* 51(2):204–208.
49. Hirsch, T., Spielmann, M., Zuhaili, B., Koehler, T., Fossum, M., Steinau, H.U., Yao, F., Steinstraesser, L., Onderdonk, A.B., and Eriksson, E. (2008) Enhanced susceptibility to infections in a diabetic wound healing model. *BMC Surg* 8:5.
50. Edris, B., and Reed, J.F. 3rd. (2008) MRSA infection in lower extremity wounds. *Int J Low Extrem Wounds* 7(1):28–31.
51. Mangram, A.J., Horan, T.C., Pearson, M.L., Silver, L.C., and Jarvis, W.R. (1999) Guideline for prevention of surgical site infection. *Am J Infect Control* 27:97–134.
52. Ozumba, U.C. (2007) Bacteriology of wound infections in the surgical wards of a teaching hospital in Enugu, Nigeria. *Afr J Med Sci* 36:341–344.
53. Schraibman, I.G. (1990) Significance of beta-haemolytic streptococci in chronic leg ulcers. *Annals of the Royal College of Surgeons of England* 72:123–124.
54. Madsen, S.M., Westh, H., Danielsen, L., and Rosdahl, V.T. (1996) Bacterial colonization and healing of venous leg ulcers. *APMIS* 104:895–899.

55. Brook, I. (1996). Aerobic and anaerobic microbiology of necrotising fasciitis in children. *Pediatr Dermatol* 13:281–284.
56. Schmidt, K., Debus, E.S., Jebberger, S., et al. (2000) Bacterial population of chronic crural ulcers: Is there a difference between the diabetic, the venous, and the arterial ulcer? *Vasa* 29:62–70.
57. Fisher, K., and Phillips, C. (2009) The ecology, epidemiology and virulence of Enterococcus. *Microbiology* 155(6):1749–1757.
58. Flattau, A., Schiffman, J., Lowy, F.D., and Brem, H. (2008) Antibiotic-resistant gram-negative bacteria in deep tissue cultures. *Int Wound J* 5:599–600.
59. Tena, D., Aspiroz, C., Figueras, M.J., Gonzalez-Praetorius, A., Aldea, M.J., Alperi, A., and Bisquert, J. (2009) Surgical site infection due to Aeromonas species: Report of nine cases and literature review. *Scand J Infect Dis* 41(3):164–170.
60. Daltrey, D.C., Rhodes, B., and Chattwood, J.G. (1981) Investigation into the microbial flora of healing and nonhealing decubitus ulcers. *J Clin Pathol* 34:701–705.
61. Hansson, C., Hoborn, J., Moller, A., and Swanbeck, G. (1995) The microbial flora in venous leg ulcers without clinical signs of infection. *Acta Derm Venereol* 75:24–30.
62. Schmidt, K., Debus, E.S., Jebberger, S., et al. (2000) Bacterial population of chronic crural ulcers: Is there a difference between the diabetic, the venous, and the arterial ulcer? *Vasa* 29:62–70.
63. Brook, I. (1996). Aerobic and anaerobic microbiology of necrotising fasciitis in children. *Pediatr Dermatol* 13:281–284.
64. Cruse, P.J.E., and Foord, R. (1980) The epidemiology of wound infection. A 10-year prospective study of 62,939 wounds. *Surg Clin North Am* 60:27–40.
65. Ge, Y., MacDonald, D., Hait, H., et al. (2002) Microbiological profile of infected diabetic foot ulcers. *Diabetic Medicine* 19:1032–1035.
66. Polavarapu, N., Ogilvie, M.P., and Panthaki, Z.J. (2008) Microbiology of burn wound infections. *J Craniofac Surg* 19:899–902.
67. Danielsen, L., Balslev, E., Döring, G., Høiby, N., Madsen, S.M., Ågren, M., Thomsen, H.K., Fos, H.H.S., and Westh, H. (1998) Ulcer bed infection. Report of a case of enlarging venous leg ulcer colonised by *Pseudomonas aeruginosa*. *APMIS* 106:721–726.
68. Murakawa, G.J. (2004) Common pathogens and differential diagnosis of skin and soft tissue infections. *Cutis* 73(5 Suppl):7–10.
69. Heggers, J.P. (1998) Defining infection in chronic wounds: Methodology. An historical review of the quantitative assessment of microbial flora in wounds. *J Wound Care* 7:452–456.
70. Chiller, K., Selkin, B.A., and Murakawa, G.J. (2001) Skin microflora and bacterial infections of the skin. *J Investig Dermatol Symp Proc* 6(3).170–174.
71. Ruef, C., Ruef, S.A., Senn, G., Cathomas, A., and Imhof, A. (2009) Decolonisation of patients with wounds colonised by MRSA. *J Hosp Infect* 72(1):88–90.
72. Brook, I., and S.M. Finegold. (1981) Aerobic and anaerobic bacteriology of cutaneous abscesses in children. *Pediatrics* 67:891–895.
73. Meislin, H.W., Lerner, S.A., Graves, M.H., McGehee, M.D., Kocka, F.E., Morello, J.A., and Rosen, P. (1977). Cutaneous abscesses. Anaerobic and aerobic bacteriology and outpatient management. *Ann Intern Med* 87:145–149.
74. Page, G., and Beattie, T. (1992) Infection in the accident and emergency department, pp. 123–132. In E.W. Taylor (ed.). *Infection in surgical practice*. Oxford: Oxford University Press.
75. Regev, A., Weinberger, M., Fishman, M., Samra, Z., and Pitlik, S.D. (1998) Necrotising fasciitis caused by *Staphylococcus aureus*. *Eur J Clin Microbiol Infect Dis* 17:101–103.

76. Frank, D.N., Wysocki, A., Specht-Glick, D.D., Rooney, A., Feldman, R.A., St Amand, A.L., Pace, N.R., and Trent, J.D. (2009) Microbial diversity in chronic open wounds. *Wound Repair Regen* 17(2):163–172.

77. Chen, A.E., Cantey, J.B., Carroll, K.C., Ross, T., Speser, S., and Siberry, G.K. (2009) Discordance between *Staphylococcus aureus* nasal colonization and skin infections in children. *Pediatr Infect Dis J* 28(3):244–246.

78. Sotto, A., Lina, G., Richard, J.L., Combescure, C., Bourg, G., Vidal, L., Jourdan, N., Etienne, J., and Lavigne, J.P. (2008) Virulence potential of *Staphylococcus aureus* strains isolated from diabetic foot ulcers: A new paradigm. *Diabetes Care* 31(12):2318–2324.

79. Armstrong, D.G., Liswood, P.J., and Todd, W.F. (1995) Prevalence of mixed infections in the diabetic pedal wound. A retrospective review of 112 infections. *J Am Podiatr Med Assoc* 85:533–537.

80. Pathare, N.A., Bal, A., Talvalkar, G.V., and Antani, D.U. (1998) Diabetic foot infections: A study of micro-organisms associated with the different Wagner grades. *Indian J Pathol Microbiol* 41:437–441.

81. Gardner, S.E., Frantz, R.A., Saltzman, C.L., and Dodgson, K.J. (2004) *Staphylococcus aureus* is associated with high microbial load in chronic wounds. *Wounds* 16:251–257.

82. Wolcott, R.D., and Dowd, S.E. (2008) A rapid molecular method for characterising bacterial bioburden in chronic wounds. *J Wound Care* 17:513–516.

83. Hong, K.H., Park, J.S., and Kim, E.C. (2008) Two cases of vancomycin-intermediate *Staphylococcus aureus* isolated from joint tissue or wound. *Korean J Lab Med* 28(6):444–448.

84. Wright, A. E., Fleming, A., and Colebrook, L. (1918) The conditions under which the sterilisation of wounds by physiological agency can be obtained. *Lancet* i:831–838.

85. Robson, M.C., and Heggers, J.P. (1970) Delayed wound closures based on bacterial counts. *J Surg Oncol* 2:379–383.

86. Dutta, J.K. (1998) Fulminant infection by uncommon organisms in animal bite wounds. *Postgrad Med J* 74:611–612.

87. Akgülle, H.A., Kocaoglu, B., Erol, B., and Tetik, C. (2009) Human hand bite causing soft tissue infection and finger amputation: A case report. *Ulus Travma Acil Cerrahi Derg* 15(2):201–204.

88. Madsen, S.M., Westh, H., Danielsen, L., and Rosdahl, V.T. (1996) Bacterial colonization and healing of venous leg ulcers. *APMIS* 104:895–899.

89. Ikeno, T., et al. (2007) Small and rough colony *Pseudomonas aeruginosa* with elevated biofilm formation ability isolated in hospitalized patients. *Microbiol Immunol* 51:929–938.

90. Trautmann, M., Halder, S., Lepper, P.M., and Exner, M. (2009) Reservoirs of *Pseudomonas aeruginosa* in the intensive care unit. The role of tap water as a source of infection. *Bundesgesundheitsblatt Gesundheitsforschung Gesundheitsschutz* 52(3):339–344.

91. Howell-Jones, R.S., Wilson, M.J., Hill, K.E., Howard, A.J., Price, P.E., and Thomas, D.W. (2005) A review of the microbiology, antibiotic usage and resistance in chronic skin wounds. *J Antimicrob Chemother* 55:143–149.

92. Schmidt, K., Debus, E.S., St, J., Ziegler, U., and Thiede, A. (2000) Bacterial population of chronic crural ulcers: Is there a difference between the diabetic, the venous, and the arterial ulcer? *Vasa* 29:62–70.

93. Aufiero, B., Duanmu, Z., Guo, M., Meduri, N.B., Murakawa, G.J., and Falkow, S. (2004) *Staphylococcus aureus* infection of human primary keratinocytes. *J Dermatol Sci* 36(3):173–175.

94. Tredget, E.E., Shankowsky, H.A., Rennie, R., Burrell, R.E., and Logsetty, S. (2004) Pseudomonas infections in the thermally injured patient. *Burns* 30:3–26.

95. Loryman, C., and Mansbridge, J. (2007) Inhibition of keratinocyte migration by lipopolysaccharide. *Wound Repair and Regeneration* 16:45–51.

96. Danielsen, L., Westh, H., Balselv, E., et al. (1996) *Pseudomonas aeruginosa* exotoxin A antibiodies in rapidly deteriorating chronic leg ulcers. *Lancet* 347:265.
97. Dowd, S.E., Wolcott, R.D., Sun, Y., McKeehan, T., Smith, E., and Rhoads, D. (2008) Polymicrobial nature of chronic diabetic foot ulcer biofilm infections determined using bacterial tag encoded FLX amplicon pyrosequencing (bTEFAP). *PLoS ONE* 3;3(10):e3326.
98. Diamantopoulos, E.J., Haritos, D., Yfandi, G., Grigoriadou, M., Margariti, G., Paniara, O., and Raptis, S.A. (1998) Management and outcome of severe diabetic foot infections. *Exp Clin Endocrinol Diabetes* 106:346–352.
99. Jiang, W., Abrar, S., Romagnoli, M., and Carroll, K.C. (2009) *Clostridium glycolicum* wound infections: Case reports and review of the literature. *J Clin Microbiol* 47(5):1599–1601.
100. Brook, I. (1987) Synergistic aerobic and anaerobic infections. *Clin Ther* 10:19–35.
101. Trengove, N.J., Stacey, M.C., McGechie, D.F., and Mata, S. (1996) Qualitative bacteriology and leg ulcer healing. *J Wound Care* 5:277–280.
102. Schraibman, I.G. (1990) The significance of beta-haemolytic streptococci in chronic leg ulcers. *Ann R Coll Surg Engl* 72:123–124.
103. Kingston, D., and Seal, D.V. (1990) Current hypothesis on synergistic microbial gangrene. *Br J Surg* 77:260–264.
104. Rotstein, O.D., Pruett, T.L., and Simmons., R.L. (1985) Mechanisms of microbial synergy in polymicrobial surgical infections. *Rev Infect Dis* 7:151–170.
105. Wolcott, R.D., Rhoads, D.D., and Dowd, S.E. (2008) Biofilms and chronic wound inflammation. *J Wound Care* 17:333–341.
106. Armstrong, D.G., Liswood, P.J., and Todd, W.F. (1995) Prevalence of mixed infections in the diabetic pedal wound. A retrospective review of 112 infections. *J Am Podiatr Med Asoc* 85:533–537.
107. Marsh, P.D., and Bradshaw, D.J. (1995) Dental plaque as a biofilm. *J Ind Microbiol* 15(3):169–175.
108. Bradshaw, D.J., Marsh, P.D., Hodgson, R.J., and Visser, J.M. (2002) Effects of glucose and fluoride on competition and metabolism within in vitro dental bacterial communities and biofilms. *Caries Res* 36(2):81–86.
109. Kuramitsu, H.K., He, X., Lux, R., Anderson, M.H., and Shi, W. (2007) Interspecies interactions within oral microbial communities. *Microbiol Mol Biol Rev* 71(4):653–670.
110. Haffajee, A.D., Socransky, S.S., Patel, M.R., and Song, X. (2008) Microbial complexes in supragingival plaque. *Oral Microbiol Immunol* 23(3):196–205.
111. Kolenbrander, P.E. (1988) Intergeneric coaggregation among human oral bacteria and ecology of dental plaque. *Annu Rev Microbiol* 42:627–656.
112. Kolenbrander, P.E., Ganeshkumar, N., Cassels, F.J,, and Hughes, C.V. (1993) Coaggregation; Specific adherence among human oral plaque bacteria. *FASEB J* 7(5):406–413.
113. Kolenbrander, P.E., and London, J. (1993) Adhere today, here tomorrow: Oral bacterial adherence. *J Bacteriol* 175(11):3247–3252.
114. Kolenbrander, P.E., Parrish, K.D., Andersen, R.N., and Greenberg, E.P. (1995) Intergeneric coaggregation of oral *Treponema* spp. with *Fusobacterium* spp. and intrageneric coaggregation among *Fusobacterium* spp. *Infect Immun* 63(12):4584–4588.
115. Shah, H.N., and Gharbia, S.E. (1989) Ecological events in subgingival dental plaque with reference to *Bacteroides* and *Fusobacterium* species. *Infection* 17(4):264–268.
116. Homer, K.A., and Beighton, D. (1992) Synergistic degradation of bovine serum albumin by mutans streptococci and other dental plaque bacteria. *FEMS Microbiol Lett.* 69(3):259–262.
117. Staley, J.T., and Konopka, A. (1985) Measurement of *in situ* activities of nonphotosynthetic microorganisms in aquatic and terrestrial habitats. *Annu Rev Microbiol* 39:32–46.

118. Ogra, P.L., Giesuik, G.S., and Bareukamp, S.J. (1989) Microbiological, immunology and vaccination. *Ann Otol Rhinol Laryngol* 139:S29–S49.
119. Amann, R., Ludwig, W., and Schleifer, K.H. (1995) Phylogenetic identification and *in situ* detection of individual microbial cells without cultivation. *Microbiol Rev* 59:143–169.
120. Veeh, R.H., Shirtliff, M.E., Petik, J.R., Flood, J.A., Davis, C.C., Seymour, J.L., Hansmann, M.A., Kerr, K.M., Pasmore, M.E., and Costerton, J.W. (2003) Detection of *Staphylococcus aureus* biofilm on tampons and menses components. *J Infect Dis* 188(4):519–530.
121. Freeman, K., Woods, E., Welsby, S., Percival, S.L., and Cochrane, C.A. (2009) Biofilm evidence and the microbial diversity of horse wounds. *Can J Microbiol* 55(2):197–202.
122. Redkar, R., Kalns, J., Butler ,W., et al. (2000) Identification of bacteria from a nonhealing diabetic foot wound by 16S rDNA sequencing. *Mol Cell Probes* 14: 163–169.
123. Hill, K.E., Davies, C.E., Wilson, M.J., Stephens, P., Harding, K.G., and Thomas, D.W. (2003) Molecular analysis of the microflora in chronic venous leg ulceration. *J Med Microbiol* 52(Pt 4):365–369.
124. Davies, C.E., Wilson, M.J., Hill, K.E., Stephens, P., Hill, C.M., Harding, K.G., and Thomas, D.W. (2001) Use of molecular techniques to study microbial diversity in the skin: Chronic wounds reevaluated. *Wound Repair Regen* 9: 332–340.
125. Andersen, A., Hill, K.E., Stephens., P., Thomas, D.W., Jorgensen, B., and Krogfelt, K.A. (2007) Bacterial profiling using skin grafting, standard culture and molecular bacteriological methods. *J Wound Care* 16(4):171–175.
126. McBain, A.J., Bartolo, R.G., Catrenich, C.E., Charbonneau, D., Ledder, R.G., and Gilbert, P. (2003) Growth and molecular characterization of dental plaque microcosms. *J Appl Microbiol* 94(4):655–664.
127. Fredricks, D.N. (2001) Microbial ecology of human skin in health and disease. *J Investig Dermatol Symp Proc* 6(3):167–169.
128. Cochrane, C.A., Freeman, K., Woods, E., Welsby, S., and Percival, S.L. (2009) Biofilm evidence and the microbial diversity of horse wounds. *Can J Microbiol* 55(2):197–202.
129. Wolcott, R.D., and Dowd, S.E. (2008) A rapid molecular method for characterising bacterial bioburden in chronic wounds. *J Wound Care* 17(12):513–516.
130. Howell-Jones, R.S., Baker, I.B., McNulty, C.A. HPA GP Microbiology Laboratory Use Group (2008) Microbial investigation of venous leg ulcers. *J Wound Care* 17:353–358.
131. Healy, B., and Freedman, A. (2006) Infections. *BMJ* 332(7545):838–841.
132. Dow, G., Browne, A., and Sibbald, R.G. (1999) Infection in chronic wounds: Controversies in diagnosis and treatment. *Ostomy Wound Manage* 45(8):23–40.
133. Levine, N.S., Lindberg, R.B., Mason, A.D., and Pruitt, B.A. (1976) The quantitative swab culture and smear: A quick simple method for determining the number of viable bacteria on open wounds. *J Trauma* 16:89–94.
134. Gardner, S.E., Frantz, R.A., Saltzman, C.L., Hillis, S.L., Park, H., and Scherubel, M. (2006) Diagnostic validity of three swab techniques for identifying chronic wound infection. *Wound Repair Regen* 14(5) 548–557.
135. Sapico, F.L., Canawati, H.N., Witte, J.L., Montgomerie, J.Z., Wagner, F.W., and Bessman, A.N. (1980) Quantitative aerobic and anaerobic bacteriology of infected diabetic feet. *J Clin Microbiol* 2: 413–420.
136. Perry, C.R., Pearson, R.L., and Miller, G.A. (1991) Accuracy of cultures of material from swabbing of the superficial aspect of the wound and needle biopsy in the preoperative assessment of osteomyelitis. *J Bone Joint Surg* 73A:745–749.
137. Thomson, P.D., and Smith, D.J. (1994) What is infection? *Am J Surg* 167(1A):7S–11S.
138. Johnson, S., Lebahn, F., Peterson, L.R., and Gerding, D.N. (1995) Use of an anaerobic collection and transport swab device to recover anaerobic bacteria from infected foot ulcers in diabetics. *Clin Infect Dis* 20:S289–S290.

139. Bill, T.J., Ratcliff, C.R., Donovan, A.M., Knox, L.K., Morgan, R.F., and Rodeheaver, G.T. (2001) Quantitative swab culture versus tissue biopsy: A comparison in chronic wounds. *Ostomy/Wound Management* 47:34–37.

140. Lawrence, J.C. (1985) The bacteriology of burns. *J Hosp Infect* 6(B):3–17.

141. Nicolle, L.E., Orr, P., Duckworth, H., et al. (1994) Prospective study of decubitus ulcers in two long term care facilities. *Can J Infect Control* 9(2):35–38.

142. Livesley, N.J., and Chow, A.W. (2002) Infected pressure ulcers in elderly individuals. *Clin Infect Dis* 35(11):1390–1396.

143. Fowler, E. (1998) Wound infection: A nurse's perspective. *Ostomy Wound Manage* 44:44–53.

144. Salcido, R. (2007) What is bioburden? The link to chronic wounds. *Adv Skin Wound Care* 20(7):368.

145. Wheat, L.J., Allen, S.D., Henry, M., et al. (1986) Diabetic foot infections: Bacteriologic analysis. *Arch Intern Med* 146:1935–1940.

146. Woolfrey, B., Fox, J., and Quall, C. (1981) An evaluation of burn wound quantitative microbiology. *Am Soc Clin Pathol* 7:532–537.

147. Davies, C.E., Hill, K.E., Newcombe, R.G., Stephens, P., Wilson, M.J., Harding, K.G., and Thomas, D.W. (2007) A prospective study of the microbiology of chronic venous leg ulcers to reevaluate the clinical predictive value of tissue biopsies and swabs. *Wound Repair Regen* 15:17–22.

148. Brook, I. (1982) Collection and transportation of specimens in anaerobic infections. *J Fam Pract* 15:775–779.

149. Health Protection Agency. (2003) National Standard Operating Procedures— Bacteriology. Available from www.hpa.org.uk.

150. Tomic-Canic, M., and Brem, H. (2004) Gene array technology and pathogenesis of chronic wounds. *Am J Surg* 188:67–72.

151. Hegarty, J.P., Pickup, R., and Percival, S.L. (2001) Detection of viable but non-culturable bacterial pathogens. In *Biofilm community interactions: Chance or necessity? Species consortia.* Edited by Gilbert, P.G., Allison, D., Walker, J.T., and Brading, M., pp. 39–51. Cardiff: Bioline.

152. Kirketerp-Moller, K., Jensen, P.O., Fazil, M., Madesen, K.G., Pedersen, J., Moser, C., Tolker-Nielsen, T., Hoiby, N., Givskov, M., and Bjarnsholt, T. (2008) The distribution, organisation and ecology of bacteria in chronic wounds. *J Clin Microbiol* 46(8):2717–2722.

7 Types of Wounds and Infections

Randall D. Wolcott, Keith F. Cutting,
Scot E. Dowd, and Steven L. Percival

CONTENTS

Introduction.. 219
Wound Types.. 220
   Diabetic Foot Ulcers 220
   Venous Leg Ulcers 222
   Pressure Ulcers.. 224
   Wound Characteristics... 225
Infection 226
References.. 230

INTRODUCTION

The clinical evaluation of cutaneous wounds has been problematic throughout the entire history of wound care. The very definition of a chronic wound, "a wound that does not heal in an orderly set of stages and in a predictable amount of time the way most wounds do,"[1] can only be determined retrospectively; it is not particularly helpful clinically. In addition, attempting to ascertain precisely what is causing the wound and preventing healing has remained elusive. This has led to a generally accepted system of categorizing wounds based on their etiologies, such as diabetic foot ulcer, venous leg ulcer, pressure ulcers (decubitus), and others. To make matters more complex, wounds are also evaluated based on wound characteristics such as slough, exudate, maceration, wound edges, "infection," as well as complicating host factors such as critical limb ischemia, host immune response, and diabetes. It is clear that the answer as to what "type" of wound an individual patient possesses is much more of an art than a scientific description.

This inexact and often overlapping classification system has by necessity produced a wound care discipline that is largely experiential and steeped in inconsistencies and sometimes contradictory teachings. Many of the traditional aspects of wound care result from the collation of repeated clinical observations, and others are the consequence of mere trial and error. Although some of these interventions may well be steeped in tradition, in many instances they have been of benefit to patients. More recently, our rapidly developing insight into the complexities of wound care together with the associated underpinning science are helping to replace those interventions that are based on custom.

The arbitrary grouping of wounds according to etiology is at best a functional approach, because etiology tends to group wounds with similar barriers to healing. For example, most wounds encountered on the foot of a patient with diabetes mellitus would register elevated blood sugar, endothelial cell dysfunction, microcirculatory abnormalities, a degree of immune incompetency, possibly neuropathy/ischemia, and alterations in tissue integrity or function as a result of pressure. Therefore, clinical assessment of patients with diabetic foot wounds may bear some similarities, as will many of the treatment regimens that in turn may be compared to treatment efficacy.

However, division of wounds based on etiologies is becoming recognized as incomplete and is very unsatisfying.[2] The reason may lie in that many of the barriers associated with a specific type of wound such as a diabetic foot ulcer are not equally expressed in all diabetic wounds. In fact, many of the barriers may be of reduced significance or absent in different patients. Other barriers associated with different types of wounds also frequently frustrate the simple classification scheme. For example, a severe diabetic with significant venous insufficiency may encounter a serious pressure event on the medial malleolus resulting in a cutaneous wound that would be very difficult to classify by an "etiology"-based system.

Recent findings that indicate that chronic wounds have so much in common militates against dividing wounds by their presumed etiologies. That is, all chronic wounds regardless of the supposed etiology have similar biochemistries of elevated proinflammatory cytokines, elevated matrix metalloproteases, and diminished growth factors.[3-6] Also, the cellularity of a chronic wound demonstrates an amazingly constant excess of neutrophils across all types of wounds.[7,8] To put it another way, wounds are stuck in a persistent inflammatory state,[9] which suggests a common "bond" between chronic wounds. The observation that biofilm is prevalent in chronic wounds and rare in acute wounds[10] challenges the position that biofilm is the common element related to the chronic inflammatory state.[2] From a practice perspective, wound care specialists may use many similar therapeutic interventions across the arbitrary etiologic divisions with similar results in efficacy. Yet for the time being and until all wounds can be unified by a generally accepted theory, it seems most prudent to continue the tradition of discussing wounds based on their etiologies and wound characteristics.

WOUND TYPES

DIABETIC FOOT ULCERS

Diabetic foot ulcers are a major worldwide healthcare problem that is increasing at an alarming rate, possibly because of the double-digit increase in diabetes each year, increased longevity, and patients having diabetes for longer periods of time. Approximately 12% to 25%[11] of the 24 million diabetics currently in the United States will suffer a diabetic foot ulcer during their lifetime. And 84% of major limb amputations are preceded by the occurrence of diabetic foot ulcers,[11] resulting in over $11 to $13 billion cost annually for major limb amputations in the United States alone.[12]

The cost to patients is much higher. A diabetic who undergoes a major limb amputation will lose the other leg up to 50% of the time, within 3 years[13] report a very dismal quality of life,[14] and will suffer a much higher mortality rate.[15,16] As noted in previous and forthcoming chapters, the management of chronic infections by removing body parts in general, and specifically removing legs to cure diabetic foot ulcers, is detrimental to the patient and must be considered a failed strategy. Unfortunately, amputation remains a common "treatment" for diabetic foot ulcers.

Diabetics suffer these ulcers due to minor traumas, pressure events, dry, cracked skin from autonomic dysfunction, and even self-induced from "bathroom" surgeries. The peripheral neuropathy associated with diabetes often allows for or exacerbates this destruction of the skin, which is the most important barrier to microbes in an impaired host. Paul Brand has demonstrated in leprosy that neuropathy allows otherwise minor traumas to lead to severe cutaneous wounds and loss of body parts.[17,18] The most important management strategy that cannot be overlooked by the wound care specialist is to keep the skin intact through education and appropriate footwear and skincare products.[19–22] Because the diabetic has so many barriers to healing, it is extremely important to prevent the initial event.

Diabetes imposes a number of important pathophysiological situations on the patient, with each most severe in the lower extremities. These pathophysiologies present very difficult barriers to wound healing. It is also important to note that many of these barriers are sufficient in their own right to prevent wound healing, and therefore all of these host abnormalities must be treated simultaneously.

Hyperglycemia can glycosylate white blood cells just as it does red blood cells (hemoglobin A1C), causing white blood cells to respond more slowly to molecular signals, interfering with white blood cell migration out of the capillaries into wounded areas (transmigration), and slowing the chemotaxis of white blood cells in the tissue.[23] White blood cell dysfunction impacts the host response to microbes as well as initiation and maintenance of healing pathways.[24] Endothelial cells also feel the effects of high blood sugar which produces glycosylation of important membrane-bound receptors (especially adhesion molecules) and rapid osmotic shifts which leads to endothelial cell dysfunction.[25] This leads to defects in translocation of white blood cells through the capillaries, nutrient, and oxygen transport.

Peripheral neuropathy associated with diabetes mellitus leads to the loss of sensory nerves and cytokines important to healing.[26] For example, the growth factor Substance P is in very high quantities in sensory nerves and is responsible for stimulating proliferation, migration, and synthesis of not only fibroblasts and keratinocytes but also endothelial cells.[27] Because the sensory nerves have died off in the area of the wound, these potent healing signals are lost. The loss of sensation also allows the skin to encounter repetitive trauma, which eventually becomes sufficient to produce a breach in the skin or exacerbate an existing wound. The autonomic nervous system also suffers in the peripheral neuropathic process producing poor and abnormal control over skin perfusion, dry fissured skin from lack of sweat and sebum production, and a sluggish initial inflammatory response. Peripheral neuropathy contributes to a slowed impaired host response that is essential for microbes to become established in a host environment.

The microbiology of diabetic foot ulcers is complex.[28] Cutaneous wounds are exposed not only to potential pathogens but also to host skin commensals as well as environmental bacteria, which results in an almost endless potential for bacterial species in the diabetic wound. Culture-based surveys of bacteria in diabetic foot ulcers have demonstrated that certain well-known pathogens such as *Staphylococcus aureus*, coagulase-negative staphylococcus, *Pseudomonas aeruginosa*, *Enterococcus* sp., *Streptococcus* sp., and others are prevalent in diabetic foot ulcers.[29] However, molecular-based methods such as polymerase chain reaction (PCR) and pyrosequencing have shown a much more complex microbial ecology of chronic wounds.[30]

The interaction of host abnormalities and the makeup of the microbial community that invades the wound have major effects as to the clinical presentation and course of the wound. Many diabetic foot ulcers present with extensive degraded tissue producing a soft fluctuance around the wound, yet remarkably are void of signs of inflammation. This produces a wound that is much more extensive than it appears on initial inspection. It is therefore imperative that all aspects of the base of the diabetic wound are explored to ensure there is no tunneling or areas of soft degraded tissue intruding deeper into the base. Diabetic foot ulcers tend to locate in areas of high pressure and are commonly seen on the plantar surface of the foot, under the first metatarsal or fifth metatarsal head, the base of the fifth metatarsal, toes, heel, and boney prominences.

Often these wounds appear pale or even red and do not show the common yellowish hue of slough because the microbial community on the surface of diabetic wounds may be significantly thinner than other wound types. Also, the exudate of diabetic foot wounds tends to be at a lower level when compared to other wound types, possibly as a result of impaired inflammation, poor perfusion, lack of substance P, and other unknown factors. Diabetic foot ulcers more than any other type tend to result in tissue death, especially in more distally located structures such as toes.

VENOUS LEG ULCERS

Venous insufficiency is becoming epidemic in industrialized nations, with up to 60% of all females and 56% of all males suffering from incompetent veins.[31] A recent survey in Germany suggests the prevalence more likely to be 35% of the general population.[32] It is generally agreed that venous insufficiency is the direct cause of venous hypertension, which eventually leads to the development of venous leg ulcers. However, the precise molecular mechanisms that lead to a cutaneous wound have not been characterized.

Venous hypertension produces sheer and stretch forces on venules in the subcutaneous space.[33,34] Either by the increased pressures produced by the venous hypertension or by inflammatory mechanisms produced by the injury to the endothelial cells, these venules become "leaky." This allows plasma exudate to exit the capillary bed in the subcutaneous and cutaneous regions of the lower leg. The plasma contents including albumin, fibrinogen, and other protein components then contribute to perivascular cuffing. This thick proteinaceous coating around the capillary bed effectively prevents diffusion of oxygen and other nutrients to the cells served by the capillaries. Wayward red blood cells also escape into the tissue, producing hemosiderosis as a result of iron pigment deposits in the tissues.[35]

This has led Falanga to postulate a "trap hypothesis" for the development of venous leg ulcers which suggests that the capillary contents, especially fibrin deposits, leak into the surrounding tissue-binding growth factors and matrix materials. These "trapped" cytokines and matrix proteins become bound, thus producing a physical barrier as well as not allowing these molecules to perform their proper tasks.[36] This hypothesis explains many molecular mechanisms producing traumatic host inflammation, yet neglects bacterial contributions to inflammation as well as other molecular abnormalities.

Studies evaluating venous leg ulcers have found many molecular abnormalities. Patients with venous leg ulcers have been found to have elevated Factor XIII antigen levels and Factor XIII V34L polymorphisms that are felt to play a critical role in developing venous leg ulcers at a molecular level.[37] Other studies have associated elevated gamma interferon, a proinflammatory cytokine, as the agent producing skin ulcers.[38] Even more recently, syndecan 4 (heparin sulfate proteoglycan), which is important in inflammation and tissue formation in normal wound healing, may be deficient in patients with venous leg ulcers.[39] Yet other studies have associated the high iron load in the tissue as important for bacterial colonization.[40–42]

All these molecular mechanisms are important, but none seem sufficient to explain the persistence of venous leg ulcers. In fact, when venous hypertension is corrected either through multilayer wrapping or venous ablation procedures, many venous leg ulcers continue to persist.[43,44] It seems self-evident that if the "cause" of the venous leg ulcer was primarily venous hypertension, correction of this host abnormality would resolve most of these wounds. It is clear that another barrier to healing must be present. Biofilm provides one of the best explanations for the nonhealing of venous leg ulcers once the venous hypertension is adequately managed. Biofilm appears to maintain its host niche by soluble bacterial-derived molecules that produce patterns of inflammation similar to the physical damage of venous hypertension.[45] This may explain why it has been so difficult to understand the role of surface-associated bacteria in the nonhealing of wounds because the host inflammatory response to infection overlaps the pattern of inflammation produced by tissue damage from venous hypertension.

The bacteria associated with the venous leg ulcers may be complex as it is with other chronic wounds. Venous leg ulcers present a special wound environment in that there is usually interstitial edema fluid that percolates through the wound, a high iron load from hemosiderosis as well as a well-perfused wound bed which results in extensive production of plasma-derived exudate. Venous leg ulcers tend to generate more exudate than poorly perfused wounds (diabetic foot ulcers), and the intact nervous system tends to result in a more painful wound. Because of this nutrient-rich environment, there appears to be a higher biofilm load and an increased diversity in the organisms present in the wound. Microbial ecologists, when examining continuous ecosystems such as a pond or a lake, were surprised to find how hetero-geneous the microbial populations were. It seemed reasonable that because the environment was continuous with similar parameters throughout, the microbes would be evenly dispersed. But this clearly was not the case. The same seems to be true for chronic wounds. Although most of the species are represented throughout the wound, there are multiple microscopic regions of the wound that have high prevalence of specific species of microbes. This fact must be kept in mind when sampling small areas of a large wound for culture information.

PRESSURE ULCERS

Wounds produced by pressure, shear, and friction forces are an enormous problem for patients and their caregivers and may often result in admission to a healthcare facility. Pressure that is sufficient to either directly necrose tissue (high pressure, short duration) or to interrupt blood supply for a duration sufficient to produce necrosis (low pressure, long duration) affects millions of people each year in the United States. The incidence of pressure ulcers is rising so rapidly that the Centers for Medicare Medicaid Services (CMS) has decided not to reimburse the treatment of pressure ulcers that occur in nursing home settings (F313 Tag) or in a hospital environment.

Although pressure/shear/friction is an etiologic factor for these ulcers, it does not appear to be the reason for their persistence. Multiple studies have shown that removal of pressure from the wound and protection from any repetitive trauma will improve many pressure ulcers, yet others will fail to heal.[46,47] Correcting host factors such as nutrition, systemic diseases, and so forth, does not always bring resolution of the problem, and this suggests an alternative impediment to healing.

Diegelman et al.[7] demonstrated that pressure ulcers have excessive neutrophils surrounding the wound bed. These neutrophils express proteolytic enzymes such as elastase and MMP8, which is a consistent biochemistry noted in all chronic wounds. A photomicrograph included in the report also demonstrates the presence of mature surface-associated bacteria tightly adhered to the wound bed surface. This suggests that wound biofilm and excessive neutrophils may be interrelated and may play an important role in the nonhealing of pressure ulcers.

The microbiology of pressure ulcers is intriguing, as most ulcers are in well-perfused proximal locations, often in the hip girdle area. Ulcers that are primarily the result of pressure usually manifest superior to an underlying bony prominence. This produces fairly deep cavernous niches in the host, which is a very different environment from the shallow wounds often encountered in diabetic foot ulcers and venous leg ulcers. This may contribute to the microbial diversity seen in pressure ulcers.

A high prevalence of anaerobes in well-perfused tissue seems contradictory at first. However, biofilm possesses community mechanisms to produce a large hypoxic core within the surface-associated community.[48] The increased representation of anaerobes may pertain to the depth and seclusion of the wound along with the overlapping surfaces present in pressure ulcers. A surgical wound that fails to heal within 30 days after surgery (or up to a year after implantation of a medical device) is termed a surgical site infection. This may seem to be somewhat of a misnomer, because polling of any group of surgeons will clearly demonstrate that the reason a postoperative wound dehisces is not infection but rather patient factors such as poor protoplasm, poor compliance, and poor nutrition. Many surgeons support this view by stating that a lot of cultures of dehisced wounds show "no growth." Wound dehiscence after surgical procedure is in fact an infection—a biofilm infection. The findings of a negative culture should not be surprising, because many biofilm phenotype bacteria are viable yet not culturable.[49]

Some patient factors impact on postoperative infections. Impaired host immunity, poor nutrition, damaged tissue, and the presence of remote chronic wounds at the time of surgery all increase the rate of surgical site infections. The pathophysiology appears to be that the surgical site gets seeded with a fragment of biofilm or even planktonic bacteria which attaches to the surgical site and cannot be cleared rapidly by the host immune system. Once the wound edges are reapproximated, two surfaces now cover the early microcolony, giving it a decided advantage over the host. The microcolony can rapidly expand along the reapproximated surgical surfaces, preventing angiogenesis and the healing of the surgical margins. This progresses until the surgical margins dehisce and drain. Because many biofilm phenotype bacteria are viable but not culturable, molecular methods may be necessary to diagnose surgical site infections. Burns are a very special subset of wounds, yet most of the morbidity and mortality stems from a chronic infection that develops after the thermal injury.[50] A mouse burn model shows just how rapidly biofilm can form and how deeply it can penetrate the host.[51] Confocal microscopy demonstrates biofilm forming around capillary structures within 1 hour after a severe burn. In this instance, Pseudomonal biofilm seems to preferentially form around capillaries extending deep into the host tissue. The speed at which the biofilm forms around the capillaries and the fact that it selects for capillary basement membrane is remarkable and suggests that biofilm structures hold little regard for host immunity, which is consistent with *in vitro* research.

Wound Characteristics

Many characteristics and phenomena associated with chronic wounds are easily explained by viewing chronic cutaneous wounds as chronic infections. By closely observing subtle changes in the wound and correlating these changes with diagnostic tools and responses to therapy, a faint picture begins to emerge of some of the processes taking place on the wound bed. Understanding some of these activities can help direct our wound management decisions.

The edges of chronic wounds seem to yield much information. Wound edges that gently slope down to the wound bed and "feather" several millimeters into the wound suggest progression to healing. Conversely, edges that are raised off the wound bed (punched-out) or with a rim of undermining at the edge strongly suggest bacterial involvement. Recent studies on keratinocyte migration suggest soluble substances secreted by bacteria may impair keratinocytes which may yield these types of edges.

The wound edge may also reveal additional clinical information related to moisture. If the wound biofilm becomes more active and upregulates host inflammatory response, this will result in an increase in exudate production and consequential maceration. Maceration of the wound edge along with pain, swelling, and deterioration of the wound suggests an active biofilm.

The hyperkeratosis (callus) seen on plantar wounds does not always indicate inflammation from walking and repetitive trauma. Inflammation from the chronic infection present in many chronic wounds can also manifest as callus. This seems to be confirmed by the fact that hyperkeratosis can be seen in a patient who was on a ventilator and clearly not walking. Also the clinical finding that suppression of wound biofilm without escalating offloading can significantly reduce callus formation.

The edge of the wound also can give clues as to the aggressiveness of the wound biofilm. A relatively bright red/pink border that separates the keratinocyte margin of the wound is a positive sign of the host control of wound biofilm. However, slough that laps over the edge of the wound onto the keratinocyte margin suggests poor host defenses and the poor healing outcome. Scalding, maceration, and nonphysiologic color changes of the wound margin all indicate the chronic infection is overwhelming the patient's immune system and wound treatment strategies.

The middle of the wound bed tends to be more fibrotic, less exudative, less tender, and to possess less slough. This is evidenced clinically by epithelial islands forming in the midportion of the wound when wound biofilm is rapidly suppressed. From a bacterial standpoint, surface bacteria in the midportion of the wound should be less challenged by host defenses because there are no keratinocytes present and a less robust capillary bed.

Tunneling and undermining are common terms within wound care. These phenomena occur in many wound types. We suggest the common denominator is biofilm. Tunneling may occur along soft tissue structures such as the subcutaneous layer, adipose tissue, and muscle. Tissue is slowly degraded through the structure and a tract forms. Tunneling is also commonly used to describe degradation along solid structures such as tendon or bone, again extending the wound deeper in the host tissues. Undermining usually refers to the erosion of the subcutaneous layer under the wound edge involving a significant portion of the circumference of the wound edge.

INFECTION

As mentioned in previous chapters, Robert Koch was one of the early pioneers in determining what exactly causes infection in human beings. Koch was able to isolate anthrax in pure culture, which led him to the conclusion that one species of bacteria was responsible for a given infection. Others have found it nearly impossible to duplicate his pure culture techniques. "No matter how ingenious the machinery, how careful the researchers, they kept ending up with beakers of mixed bacteria. The inability to get anything but mixed cultures led many scientist to believe that bacteria had to be in mixed groups in order to thrive." Koch's view, however, won out and 150 years later the predominant view in infectious disease is that one "germ" equals one infection. Today this postulate has no value, as the evidence is indicating that wound infections are polymicrobial in nature and that particularly in chronic wounds the biofilm phenotype predominates.

It is difficult to state exactly what constitutes an infected wound. Obviously, chronic wounds that demonstrate the signs of Celsus—rubor, tumor, calor, and dolor—are infected. It is universally accepted that all wounds have some bacteria present on their surface and even that small numbers of certain bacteria have been found to stimulate wound healing. But it is also widely accepted that even if there is an absence of erythema (rubor), swelling (tumor), heat (calor), or pain (dolor) that still many of these wounds are clearly infected. Therefore it was necessary to develop secondary signs of wound infection.

Secondary signs of infection vary but generally include lack of healthy granulation tissue, unhealthy color, friable granulation tissue, excessive exudate, degraded wound bed, and a "stalled" healing trajectory. It is interesting to note the definition of a chronic wound is "a wound that does not heal in an orderly set of stages" or in other words, a "stalled" wound. So are all chronic wounds infected?

A definition of wound infection is the detrimental colonization of a host organism by a foreign species. In an infection, the infecting organism seeks to utilize host resources to multiply (usually at the expense of the host). Acute wounds progress through the normal stages of healing and therefore show no detrimental effects from bacteria. Acute wounds, wounds that heal normally (even in patients with diabetes mellitus, venous insufficiency, and other severe host impairments), show some bacteria but very little organized biofilm has been documented to be present on their surface. Most chronic wounds, on the other hand, have been found to house a biofilm and each wound demonstrated the secondary sign of infection of failure to progress. The presence of biofilm on the surface of chronic wounds (and not on acute wounds) raises the question as to what role biofilm may play in the nonhealing of wounds.

A retrospective study was conducted to evaluate the healing of chronic wounds utilizing standard of care plus antibiofilm strategies.[52] By specifically targeting the biofilm and comparing the results to standard of care alone, the findings suggested that suppression of biofilm improves wound healing. This gives indirect evidence that the presence of biofilm on the surface of chronic wounds is detrimental to wound healing. Although the exact molecular mechanisms have not been fully elucidated, chronic wounds may indeed be chronic cutaneous infections.

The confusion of what actually constitutes a wound infection most likely lies in the different behaviors of bacteria when they exist in different phenotypes. Planktonic (single cell) bacteria behave much differently *in vitro* and *in vivo* compared to the social behavior of those same bacteria in a biofilm phenotype. These different behaviors are underappreciated in medicine.

Planktonic bacteria lack colony defenses and are susceptible to environmental changes, ultraviolet (UV) light, host immunity, certain antibiotics, and most biocides. Whereas biofilm phenotype bacteria, bacteria encased in a self-secreted polymeric matrix with multiple different phenotypes, are up to 1000 times more resistant to biocides and antibiotics and unperturbed by host immunity. These characteristics of the two different manifestations of the life cycle of bacteria produce markedly different types of infections.

The National Institutes of Health (NIH) states that 80% of human infections are caused by biofilm phenotype bacteria that produce chronic infections such as endocarditis, chronic rhinosinusitis, Crohn's disease, medical device infections, and chronic wounds. These wounds are characterized by their incomplete response to antibiotics as the infection quickly reemerges once antibiotics are withdrawn. Most chronic infections respond to steroids or other potent anti-inflammatories like tumor necrosing factor alpha inhibitors. Chronic infections wax and wane, leading to the degradation of the structure that is infected to the point where it is removed, such as replacement of the heart valve, stripping of the sinuses, resection of the small intestine, removal of the medical device, or amputation of the limb. Chronic infections tend to be managed by removal of body parts.

The remaining 20% of infections, acute infections like sepsis and cellulites, are caused by planktonic phenotype bacteria that pursue a decidedly different strategy. Planktonic phenotype bacteria tend to pursue a more predatory course that degrades and then destroys its host. These bacteria tend to secrete virulence factors and other proteins geared at invading host immunity, killing host cells, and then degrading and feeding on the dead host material. This predatory behavior is rapid, destructive, and most importantly is not sustainable as seen in acute infections such as sepsis and cellulitis. Because these are individual bacteria without colony defenses, they are susceptible to appropriate antibiotics yielding resolution of acute infections in relatively short periods of time or if the antibiotics are ineffective, death of the host. With either outcome, the results of an acute infection are usually played out within a few days to weeks.

Chronic infections are quite different from acute planktonic infections in that they can easily last for decades. Yet even though these two types of infections are so different on a molecular level, they are most often lumped into that same category of "bacterial infection." Biofilm phenotype bacteria produce chronic infections by a plethora of mechanisms. The three following examples illustrate how different biofilm infection is from planktonic infection.

Lipopolysaccharide (LPS) is one of the major components of the Gram-negative bacterial cell wall. LPS has been demonstrated to decrease neutrophil response to IL-8 and also fouls an important receptor composed of phosphatidyl serine (PS) found on macrophages. Macrophages utilize the PS receptor to recognize neutrophils that need to be cleared. If these neutrophils are not clear, their proteolytic contents (MMP 8 and elastase) are released at the site of infection. LPS also is a potent inducer of continued chemotaxis of neutrophils to migrate into the wound bed. Multiple studies have identified that a hallmark of all chronic wounds is excessive neutrophils in the wound bed. Biofilms are known to constantly release LPS and membrane vesicles made up of LPS (previously unknown function) into the surrounding environment.

A second specific example is the release of planktonic cells. Up to 30% of mature biofilm dedifferentiates back into planktonic phenotype cells each day which are released through a process termed "seeding dispersal." Biofilms continually release planktonic "seeds" of different bacterial species from the safety of the biofilm matrix. In the chronic wound, the cells can work to bait the immune system by pathogen-associated molecular patterns (PAMP) recognition mechanisms and to continually recruit the inflammatory response of the host. In this way, the sacrifice of a few individual bacteria promotes the survival of the community through sustained host inflammation producing exudate, and therefore providing continual nutrient acquisition to the biofilm.

In addition, biofilms have pathways by which bacterial DNA is released from cells and incorporated into the biofilm matrix. In the laboratory, the purpose of this extruded bacterial DNA is to provide cross-linking of polymeric sugars which strengthens the matrix. However, *in vivo,* the DNA stimulates inflammatory recruitment in the localized tissue which stimulates the innate immune response and augments the hyperinflammatory state of the wound bed. Thus, through the use of LPS, planktonic cells, and bacterial DNA, biofilms possess an arsenal of tools, not unique to a single pathogen, which through the biofilm community can act as a functional equivalent group to stimulate and maintain a hyperinflammatory milieu in the wound bed. This hyperinflammatory state in the wound bed may provide sustained nutrition while preventing closure of the host niche.

There is an emerging understanding that "bacterial pathogens expressed a large array of virulence factors that dampened and or reoriented both the innate and adaptive immune response."[53] Multiple studies demonstrate that bacteria are able to manipulate the host innate immune response to upregulate proinflammatory cytokines and the general proteolytic environment. For example, Shigella expresses a plasma virulence gene (msbB) that causes an increase in proinflammatory cytokine expression. Quorum sensing molecules from Pseudomonas originals act directly on host cells to induce expression of proinflammatory cytokines. *Staphylococcus aureus* (and other pathogens) expressed "modulations" and other superantigens (e.g., toxic shock syndrome toxin-1), which can induce massive and sustainable proinflammatory cytokines released. It is important not to focus on the individual virulence genes but rather on the common biofilm strategies that work to commandeer the host immune response.

An important property of biofilms that impacts its infective behavior, especially in cutaneous wounds, is that biofilms are polymicrobial. Through specialized cultivation methods, DGGE, PCR-based studies, molecular amplicon studies, and metagenomics, it has been shown that chronic cutaneous wounds demonstrate fungus, bacteria, and viruses in great diversity. Coupled with this amazing diversity of microorganisms is the phenotypic hypervariability produced within the biofilm society which allows for vast adaptation of the biofilm community to any single therapy.

This ability of the biofilm community to adapt to treatment strategies stands in stark contrast to the deep planktonic perspective that is currently entrenched in medical microbiology. The twin pillars of planktonic dogma that dominates the current management of infectious diseases was firmly established by Robert Koch over 150 years ago. When Koch was first trying to understand bacterial infections, within each host infection he found multiple organisms in many different states of growth which led him to term the whole process "chaos." To make sense out of the situation, he asserted the presupposition that one organism is responsible for a specific infection. This planktonic precept has been canonized into a dogma that only one organism is responsible for an infection and all other organisms found must therefore be "contaminants."

A corollary to this dogma soon arose; because there is only one organism, there should be only one treatment to manage the infection. This has led to the bias of using only one strategy to eradicate the infection. If that one strategy fails, it is abandoned, and then the next strategy (antibiotic) is tried in a sequential fashion. Again this is a decidedly antiplanktonic strategy that is wholly unsuited for biofilm infections. Biofilm is a more mutalistic type of infection produced by multiple organisms with vast phenotypic diversity that has the ability to adapt to *any* given stress. Biofilm possesses colony defenses that render most antibiotics only marginally effective, plus an antibiotic will only affect the susceptible members of the biofilm society. The unaffected members will quickly predominate and the biofilm will reemerge. This is often witnessed clinically as a wound responds for the first two to three weeks of treatment only to regress as the same treatment is continued. It is biofilm's incomparable abilities to resist treatments (radiation, antibiotic, biocides, host immunity, etc.) by changing dominant populations, composition of the secreted matrix, phenotype changes, horizontal gene transfer, and other mechanisms which makes chronic infections so recalcitrant and so variable. This leads to the need not only to have multiple strategies but to apply them simultaneously and to change them frequently.

Until recently, the application of treatments for chronic wounds (and other chronic infections) has been instigated in the absence of comprehensive diagnostic information. Wound care treatments have been largely based on trial and error. A treatment is arbitrarily initiated, and if the wound responds the treatment is continued; however, if the wound does not respond then the treatment is changed. The first principle of medicine is to diagnose and then to treat, and although biofilm societies are verging on being incomprehensibly complex, through modern molecular methods and bioinformatics we can, at least on a "neophytic" level, understand the identity of the organisms present, their gene expression, and their collective strategy for maintaining the chronic wound and preventing it from healing. The diagnosis of biofilm on the surface of chronic wounds will lead to new treatments and their combinations for the suppression of wound biofilm and the improvement of healing outcomes for chronic wounds.

REFERENCES

1. Wikipedia. (2009) Available at http://en.wikipedia.org/wiki/Chronic_wound. Accessed 1 July 2009.
2. Wolcott, R.D., D.D. Rhoads, and S.E. Dowd. (2008) Biofilms and chronic wound inflammation. *J Wound Care* 17:333–341.
3. Sibbald, R.G., et al. (2003) Preparing the wound bed 2003: Focus on infection and inflammation. *Ostomy Wound Manage* 49(11):23–51.
4. Armstrong, D.G., and E.B. Jude. (2002) The role of matrix metalloproteinases in wound healing. *J Am Podiatr Med Assoc* 92(1):12–18.
5. Yager, D.R., R.A. Kulina, and L.A. Gilman. (2007) Wound fluids: A window into the wound environment? *Int J Low ExtremWounds* 6(4):262–272.
6. Schultz, G., et al. (2005) Wound healing and TIME; new concepts and scientific applications. *Wound Repair Regen* 13(4):S1–11.
7. Diegelmann, R.F. (2003) Excessive neutrophils characterize chronic pressure ulcers. *Wound Repair Regen* 11(6):490–495.
8. Smith, P.C. (2006) The causes of skin damage and leg ulceration in chronic venous disease. *Int J Low Extrem Wounds* 5(3):160–168.
9. Trengove, N.J., et al. (1999) Analysis of the acute and chronic wound environments: The role of proteases and their inhibitors. *Wound Repair Regen.* 7(6):442–452.
10. James, G.A., R. Wolcott, E. Swogger, E. deLancey Pulcini, P. Secor, J. Sestrich, J.W. Costerton, and P.S. Stewart. (2007) Biofilms in chronic wounds. ASM Conference: *Biofilms* 36–37.

11. Brem, H., et al. (2006) Evidence-based protocol for diabetic foot ulcers. *Plast Reconstr Surg* 117(7):193S–209S.

12. Falanga, V., et al. (2008) Maintenance debridement in the treatment of difficult-to-heal chronic wounds. Recommendations of an expert panel. *Ostomy Wound Manage Suppl*: 2–13.

13. Apelqvist, J., J. Larsson, and C.D. Agardh. (1993) Long-term prognosis for diabetic patients with foot ulcers. *J Intern Med* 233:485–491.

14. Boutoille, D., et al. (2008) Quality of life with diabetes-associated foot complications: Comparison between lower-limb amputation and chronic foot ulceration. *Foot Ankle Int* 29(11):1074–1078.

15. Pohjolainen, T., and H. Alaranta. (1998) Ten-year survival of Finnish lower limb amputees. *Prosthet Orthot Int* 22(1):10–16.

16. Faglia, E., et al. (2006) Early and five-year amputation and survival rate of diabetic patients with critical limb ischemia: Data of a cohort study of 564 patients. *Eur J Vasc Endovasc Surg* 32(5):484–490.

17. Brand, P. (1995) Coping with a chronic disease: The role of the mind and spirit. *Patient Educ Couns* 26:107–112.

18. MacMoran, J.W., and P.W. Brand. (1987) Bone loss in limbs with decreased or absent sensation: Ten year follow-up of the hands in leprosy. *Skeletal Radiol* 16:452–459.

19. Armstrong, D.G., and L.A. Lavery. (1998) Diabetic foot ulcers: Prevention, diagnosis and classification. *Am Fam Physician* 57(6):1325–1328.

20. Lavery, L.A., E.J. Peters, and D.G. Armstrong. (2008) What are the most effective interventions in preventing diabetic foot ulcers? *Int Wound J* 5(3):425–433.

21. Singh, N., D.G. Armstrong, and B.A. Lipsky. (2005) Preventing foot ulcers in patients with diabetes. *JAMA* 293(2):217–228.

22. Wu, S.C., et al. (2007) Foot ulcers in the diabetic patient, prevention and treatment. *Vasc Health Risk Manag* 3(1):65–76.

23. Yonem, A., et al. (2001) Effects of granulocyte-colony stimulating factor in the treatment of diabetic foot infection. *Diabetes Obes Metab* 3(5):332–337.

24. Ochoa, O., F.M. Torres, and P.K. Shireman. (2007) Chemokines and diabetic wound healing. *Vascular* 15(6):350–355.

25. Szabo, C. (2009) Role of nitrosative stress in the pathogenesis of diabetic vascular dysfunction. *Br J Pharmacol.*

26. Gershater, M.A., et al. (2009) Complexity of factors related to outcome of neuropathic and neuroischaemic/ischaemic diabetic foot ulcers: A cohort study. *Diabetologia* 52(3):398–407.

27. Antezana, M., et al. (2002) Neutral endopeptidase activity is increased in the skin of subjects with diabetic ulcers *J Invest Dermatol* 119(6):1400–1404.

28. Dowd, S.E., et al. (2008) Survey of bacterial diversity in chronic wounds using Pyrosequencing, DGGE, and full ribosome shotgun sequencing. *BMC Microbiol* 8(1): 43.

29. Rennie, R.P., R.N. Jones, and A.H. Mutnick. (2003) Occurrence and antimicrobial susceptibility patterns of pathogens isolated from skin and soft tissue infections: Report from the SENTRY Antimicrobial Surveillance Program (United States and Canada, 2000). *Diagn Microbiol Infect Dis* 45(4):287–293.

30. Dowd, S.E., et al. (2008) Polymicrobial nature of chronic diabetic foot ulcer biofilm infections determined using bacterial tag encoded FLX amplicon pyrosequencing (bTEFAP). *PLoS ONE* 3(10):e3326.

31. Robertson, L., C. Evans, and F.G. Fowkes. (2008) Epidemiology of chronic venous disease. *Phlebology* 23(3):103–111.

32. Maurins, U., et al. (2008) Distribution and prevalence of reflux in the superficial and deep venous system in the general population—Results from the Bonn Vein Study, Germany. *J Vasc Surg* 48(3):680–687.

33. Raju, S., and R. Fredericks. (1991) Hemodynamic basis of stasis ulceration—A hypothesis. *J Vasc Surg* 13(4):491–495.
34. Welkie, J.F., et al. (1992) The hemodynamics of venous ulceration. *Ann Vasc Surg* 6(1): 1–4.
35. Zamboni, P., et al. (2006) The overlapping of local iron overload and HFE mutation in venous leg ulcer pathogenesis. *Free Radic Biol Med* 40(10):1869–1873.
36. Falanga, V., and W.H. Eaglstein. (1993) The "trap" hypothesis of venous ulceration. *Lancet* 341:1006–1008.
37. Gemmati, D., et al.(2004) Factor XIII V34L polymorphism modulates the risk of chronic venous leg ulcer progression and extension. *Wound Repair Regen* 12(5):512–517.
38. Simka, M. (2006) A potential role of interferon-gamma in the pathogenesis of venous leg ulcers. *Med Hypotheses* 67(3):639–644.
39. Nagy, N., et al. (2008) The altered expression of syndecan 4 in the uninvolved skin of venous leg ulcer patients may predispose to venous leg ulcer. *Wound Repair Regen* 16(4):495–502.
40. Simka, M., and Z. Rybak. (2008) Hypothetical molecular mechanisms by which local iron overload facilitates the development of venous leg ulcers and multiple sclerosis lesions. *Med Hypotheses* 71(2):293–297.
41. Zamboni, P., et al. (2008) Inflammation in venous disease. *Int Angiol* 27(5):361–369.
42. Zamboni, P., et al. (2005) Hemochromatosis C282Y gene mutation increases the risk of venous leg ulceration. *J Vasc Surg* 42:309–314.
43. Marston, W.A., et al. (2008) The importance of deep venous reflux velocity as a determinant of outcome in patients with combined superficial and deep venous reflux treated with endovenous saphenous ablation. *J Vasc Surg* 48(2):400–405.
44. Neglen, P., K.C. Hollis, and S. Raju. (2006) Combined saphenous ablation and iliac stent placement for complex severe chronic venous disease. *J Vasc Surg* 44(4):828–833.
45. Wolcott et al.
46. Ochs, R.F., et al. (2005) Comparison of air-fluidized therapy with other support surfaces used to treat pressure ulcers in nursing home residents. *Ostomy Wound Manage* 51(2):38–68.
47. McInnes, E., et al. (2008) Support surfaces for pressure ulcer prevention. *Cochrane Database Syst Rev* 4:CD001735.
48. Stoodley, P., et al. (2002) Biofilms as complex differentiated communities. *Annu Rev Microbiol* 56:187–209.
49. Fux, C.A., et al. (2005) Survival strategies of infectious biofilms. *Trends Microbiol* 13(1):34–40.
50. Jeschke, M.G., et al. (2008) Pathophysiologic response to severe burn injury. *Ann Surg* 248(3):387–401.
51. Schaber, J.A., et al. (2007) *Pseudomonas aeruginosa* forms biofilms in acute infection independent of cell-to-cell signaling. *Infect Immun*.
52. Wolcott, R.D. and Rhoads, D.D. (2008) A study of biofilm-based management in subjects with critical limb ischaemia. *J Wound Care* 17:145–154.
53. Dupont, N., Lacas-Gervais, S., Bertout, J. et al. (2009) Shigella phagocytic vacuolar membrane remnants participate in the cellular response to pathogen invasion and are regulated by autophagy. *Cell Host Microbe* 6:137–149.

8 Biofilms and Significance to Wound Healing

Keith F. Cutting, Randall D. Wolcott,
Scot E. Dowd, and Steven L. Percival

CONTENTS

Introduction 233
Brief Overview of the Biology of Wound Healing 233
Development of a Wound Biofilm 234
Extracellular Polymeric Substances (EPS) 235
Evidence of Biofilms in Chronic Wounds 236
Characterizing Biofilm Infection 237
Immune Stimulation and Retardation by the Biofilm 240
Biofilm Detachment and Dispersal in Wounds: Clinical Significance 243
Conclusions 244
References 245

INTRODUCTION

Despite many decades of wound healing research and related microbiology, meaningful insight into the impact that microorganisms have on wound healing remains poorly understood. A recurring notion throughout this book suggests that many of the theories and interpreted findings have been based principally on outmoded methods of inquiry. Wound care management aims to reduce or remove factors known to impede "normal" wound healing and where appropriate to manage the wound bioburden. Although in many wounds a reduction in the wound bioburden can be achieved with the use of debridement, appropriate wound dressings, topical antimicrobials, and systemic antibiotics, others will remain recalcitrant to antimicrobial agents. Wounds that are nonhealing and recalcitrant to antimicrobials constitute a significant problem to patients and the healthcare profession. It is in these wounds, which are considered to harbor pathogens, that recent thinking indicates that biofilm phenotype bacteria have a significant role to play in delaying or preventing a chronic wound from healing.

BRIEF OVERVIEW OF THE BIOLOGY OF WOUND HEALING

Skin is the most important line of defense in the human body and protects underlying tissue from external factors such as noxious chemicals and microorganisms.

233

Breaches in the skin's integrity stimulate the physiological and biochemical processes of the body to reestablish the barrier function as quickly as possible, helping to limit the loss of tissue and fluids (e.g., blood, serum). In addition to this, invasion of the body tissues by particulates and microorganisms is reduced by immunological response mechanisms.

The wound healing response is the sequence of events that occur as a consequence of a trauma that leads to the physical breakdown of the skin barrier. Healing response progresses through a series of phases, with the appearance and resolution of each phase being orchestrated by numerous signaling mechanisms until the wound is healed. As mentioned in previous chapters, the earliest phase in the healing cascade is the inflammatory response, and it is here that inflammatory cells, including neutrophils and macrophages, help to prevent adhesion and proliferation of invading microorganisms and remove damaged components of the skin. The result of this process prepares the wound bed for the formation of granulation tissue, which is composed of new blood vessels and collagen. Inflammatory cells, in addition to the ones mentioned above, express numerous signaling molecules, which promote the influx of other cells vital to the production of new tissue components (e.g., blood vessels and connective tissue).

Inflammation that occurs in a wound is known to play a major role in the formation and persistence of chronic wounds—in particular, ulcers. Tissue debris that forms at the wound site results in the release of protein-degrading enzymes (proteases). Proteases in particular are responsible for breaking down proteinacious debris, leading to its removal from the wound bed. The activity and quantities of such proteases in a wound need to be tightly controlled and monitored. If protease levels in a wound become elevated and uncontrolled, such an imbalance will itself lead to tissue destruction and promote a nonhealing state within the wound bed.

DEVELOPMENT OF A WOUND BIOFILM

The naturally occurring state of bacteria is to be associated with a surface. In a wound bed, there is an abundance of surfaces and in general no freestanding pools of liquid within which the bacterial populations might establish a planktonic lifestyle. Thus, within a wound, the bacteria by default will become surface associated and eventually attach to the surfaces. The life cycle of a wound biofilm begins when a planktonic microorganism attaches to a surface. There are many different strategies utilized by individual bacteria to achieve attachment. The generalized mechanisms are reviewed in Chapter 1. After attachment, microbial cells divide until a critical density (quorum) of bacteria is reached. At this point, quorum-sensing molecules (pheromones) and up-regulation of specific genes aid to model and then help to determine the architecture and formation of a biofilm[1] and the initiation of bacterial community activities.

Microbial attachment to a wound surface is reported to be highly dependent on the species of microorganisms, the colonizing microenvironment, nutrient availability, and the physical and biochemical composition of the colonizing surface. However, once a bacterium is attached to a surface, it has been shown to exhibit an array of behaviors, including twitching, rolling, creeping, and aggregate formation.[2] In addition to this, coadhesion and coaggregation of different types of bacteria may be important.

Numerous research groups working in the area of biofilmology have evaluated biofilms using microarray technology. Results from these gene expression studies have demonstrated that a wide range of genetic pathways are induced and repressed during bacterial adhesion and biofilm formation.[3,4,5] These changes in gene expression result in production of proteins, enzymes, and signaling molecules, which as the biofilm develops, help to augment the biofilm's adaptability and sustainability.

EXTRACELLULAR POLYMERIC SUBSTANCES (EPS)

Attachment of bacteria to surfaces is termed "adsorption" and is initially due to simple van der Waals forces and basic electrostatic attraction. These basic electrical attractions can facilitate a more stable interaction between bacteria and receptor sites within host tissue through stereospecific (lock and key) interactions, which create a more stable adsorption of the bacteria than ionic or electrostatic forces alone. As discussed in Chapter 1, if the association between the bacterium and host tissue becomes intimate enough and persist long enough to become stable, other types of chemical and physical structures are formed which transform the reversible adsorption to a permanent and essentially irreversible attachment. It is during this stage that the bacteria begin to form a biofilm that is characterized by formation and accumulation of extracellular polymeric substances (EPSs). Most of the EPSs of biofilms within host tissues are polymers containing proteins, nucleic acid, with some sugar constituents. The EPSs act as the building blocks of the complex matrix structure of the biofilm and are formed by the bacteria from both external and internal processes. The evolutionary role of the biofilm is to provide a stable environment that the bacteria utilize as protection and a way of interacting with their environment.

The matrix of EPS laid down by the bacteria is used as the focus for the attachment and growth of other organisms, increasing the biological diversity of the community. The goal of bacteria in forming a biofilm is to generate a protective structure that is the natural phenotype of most bacteria.

Following microbial attachment and EPS production, the bacteria attached to the wound surface become phenotypically different than their planktonic counterparts. Each attaching bacterium produces a large number of new proteins that are not found in bacteria within the free-floating state. Biofilm bacteria have only a 30% to 50% protein homology in the outer membrane proteins when compared to their identical planktonic form.[6,7]

EPS, as outlined in a previous chapter, is often referred to as exopolysaccharide, extracellular polysaccharide matrix, glycocalyx, slime, and matrix material. The use of multiple and incorrect terminology in reference to EPS has given rise to confusion within the wound care arena. In clinical, industrial, and environmental biofilms, EPS has been universally recognized for decades as extracellular polymeric substances (EPSs). It is composed not just of polysaccharides but also nucleic acids (DNA, RNA), glycoproteins, proteins, and polysaccharides.[8]

However, the composition of EPS is known to fluctuate and the variations have been shown to be dependent upon the inherent bacterial community, the availability of bathing nutrients, and environmental conditions within the vicinity of the biofilm. Structurally EPS is supported by cations, particularly Ca^{2+} and Mg^{2+} and side chains of the polymers.[9]

Many antibiotics have been shown to have the ability to penetrate the biofilm EPS.[10] However, efficacies of these antibiotics are significantly reduced once within the biofilm matrix. This is attributed to the phenotypic resistance inherent to the microbial population within the biofilm. For example, despite evidence of antibiotic diffusion into a biofilm, beta lactamase has been shown to accumulate in a biofilm matrix faster than antibiotics can diffuse through the EPS.[11,12] Such a biofilm defensive mechanism has been shown to be effective against an array of different antibiotics. In addition to this, the EPS is generally positively charged and so is able to adhere and sequester aminoglycosides. The result of antimicrobial sequestering assists in reducing the antibiotics' ability to kill bacterial cells.[13] Moreover, the EPS produced from coagulase-negative staphylococci has been shown to inhibit glycopeptide antibiotics.[14,15] Despite evidence that EPS is able to retard the efficacy of antibiotics, the presence of EPS alone does not necessarily guarantee protection, and other biofilm phenotypic variations aid to provide biofilm resistance.[16]

EVIDENCE OF BIOFILMS IN CHRONIC WOUNDS

Biofilms have been observed and reported in both acute partial-thickness wounds[17] and chronic human wounds.[18–21] A recent study by James et al.[21] pursued microscopic evaluation of acute and chronic wounds and revealed that the chronic wound samples contained multiple species of microorganisms and that the microcolonies of bacteria were surrounded by amorphous material.[21] Micrographs from the scanning electron microscope revealed evidence of biofilm and EPS in at least 60% of the chronic wounds samples.[21] Similar techniques and sampling methods used to visualize chronic wounds were employed to evaluate 16 acute wounds. Biofilms were observed in only one of the 16 wounds.[21] The acute wounds healed quickly, generally in 2 to 3 weeks. However, the chronic wounds in the same area were still open 2 to 3 months later. It is possible to speculate that based on this study alone, difference in healing between acute and chronic wounds may be the indirect evidence that biofilm is an important barrier to healing. Based on the use of the burn mouse model, Schaber et al. determined evidence of early biofilm behavior.[22] By utilizing the mouse model and subjecting the mouse to a 10% surface area burn, followed by an inoculation of 102 *Psuedomonas aeruginosa* (planktonic bacteria), it was reported that evidence of biofilm was noted in the adipose tissue within three hours. Biofilm was visualized to a depth of 1 to 2 mm through the adipose tissue. These studies[17–22] and others have indicated that biofilms are prevalent in wounds. Their role, however, in acute wounds, if in fact biofilms are able to proliferate in this environment, suggests biofilms may not play such a significant role on the surfaces of acute wounds. Nevertheless, their development and sustained presence in acute wounds is likely to lead to chronicity, but such speculation needs be substantiated by solid research and evidence.

Based on the studies undertaken and ongoing, it is clear that biofilms in wounds are polymicrobial and form quickly in an impaired host, penetrating the surface of the wound. Our current understanding of biofilms indicates that they clearly play a significant role in impairing classical wound healing.

CHARACTERIZING BIOFILM INFECTION

Biofilms that develop on a host offer the best explanation for many of the clinical findings observed in chronic wounds. In particular, biofilms help to explain clinical and biological findings that include elevated matrix metalloproteases, decreased tissue inhibitors of metalloproteases, impaired host defenses (white blood cells, antibodies), elevated proinflammatory cytokines, and resistance to antimicrobials.

Biofilm formation by bacteria such as *Pseudomonas aeruginosa* and *Burkholderia cepacia* in cystic fibrosis patients has been used as a model to try and explain the observations in chronic wounds (i.e., persistent infection and resistance to antimicrobial therapy).

It has been iterated many times in this book that chronic wounds contain excessive numbers of neutrophils,[23] increased levels of active matrix metalloprotease (MMP) 8, and elevated proinflammatory cytokines.[24] High levels of proteolytic enzymes have been found to have an effect in delaying wound healing. In addition to this, these wounds have been shown to have a decreased level of cytokine receptors, on cell membranes and decreased levels of tissue inhibitors of metalloproteases (TIMPs).[25] Excessive amounts of extracellular enzymes have been shown to lead to the development of a nonhomeostatic state in the wound. Heterogeneity within the wound can cause the wound to enter a state of "quiet" and then "chronic" inflammation. Established preformed and "mature" biofilms in the wound environment have been shown to be quite thick, generally between 60 and 200 microns thick. Architecturally "mature" biofilms possess an irregular (variegated) outer surface and are gelatinous in texture. However, characteristics of a wound biofilm vary and specifically depend on the surrounding environment of the host and the location of the biofilm. Mature or climaxed biofilms, described earlier in the book, are reported to be composed of pillars that attach to the wound surface with enlarged tops that look similar in shape to "mushrooms." This mushroom architecture is thought to allow for the development of water channels that flow into and out of the biofilm supplying essential nutrients and removing harmful toxins, akin to a circulatory system in humans. The arrangement of these water channels, mushrooms, and pillars are influenced by cell-to-cell communication systems that have been detected in both developing and developed biofilms.[26]

Wolcott suggested that the clinical implications of biofilm infections management in chronic wounds are parallel to that of oral hygiene.[27] For example, each day biofilm, otherwise referred to as plaque, is abraded from the surface of teeth and mucous membrane utilizing toothbrushes. By brushing the teeth, the plaque biofilm is essentially debrided and kept in an immature state. In addition to this, antibiofilm toothpaste is employed which helps to eradicate the biofilm as much as possible until the selective pressure subsides and the biofilm reconstitutes. Despite repeated brushing and flossing, the biofilm reestablishes its original architecture and microbial composition within 12 to 24 hours. Such a sequence of events is now being related to chronic wounds and is being shown to significantly suppress biofilms. This method is termed biofilm-based wound care.

Slough is commonly found on chronic wounds and is composed of dead host tissue with some proteinaceous exudate, white blood cells, and a microbial community.[28] Slough is very common in poorly perfused wounds and is tethered to the wound bed. Interestingly, slough has been found to possess many of the physical characteristics of biofilm observed in the *in vitro* environment. Specifically, when slough is removed by sharp debridement, it returns very quickly. The authors have observed that if slough is removed routinely on a weekly basis, the wound does not get deeper. It is probably that slough is not just dead tissue but a thriving bacterial ecosystem. This opinion is derived from the observation that microcolonies in the sputum/mucus of patients with cystic fibrosis are of a similar morphology and architecture to the microcolonies observed in slough. Slough is sometimes regarded as an innocuous hindrance to healing that should not cause undue concern. The impression often generated is that slough is secondary to whatever nonhealing barrier is present (e.g., chronic venous insufficiency) and is not a primary cause of nonhealing. This is a false assumption.

In a prospective study of 108 surgically treated wounds, it was found that infection was significantly more frequent when slough developed than in wounds that were without slough.[29] This provides a foundation for the proposed association of slough with infection. The following observations provide an association of the presence of slough with biofilm formation. Slough that laps over the edge of the wound onto the skin margin suggests poor host defenses and prognosis of a poor healing outcome (Figure 8.1). This overgrowth of slough is not caused by host factors and can only reasonably be explained by biofilm. Maceration and nonphysiological color changes of the wound margin all indicate that chronic infection is overwhelming the patient's immune system and wound treatment strategies and indicates the aggressiveness of the wound biofilm. Conversely, a bright red wound border that separates the keratinocyte margin of the wound is a positive sign of host control of wound biofilm (Figure 8.2).

The middle of the wound bed tends to be more fibrotic, less exudative, less tender, and sometimes possesses less slough. This is evidenced clinically by epithelial islands forming in the midportion of partial thickness wounds when wound biofilm is suppressed. From a bacterial standpoint, surface bacteria in the midportion of the wound should be less challenged by host defenses because there are no keratinocytes present and a less robust capillary bed than the intact skin margin. However, clinically the bacteria in the midportion of the wound appear to be more quiescent.

Tunneling and undermining are not uncommon and occur in many wound types. The common denominator may be biofilm. Tunneling may occur along tissue structures much in the same way as plaque erodes into the enamel of a tooth. Tissue is slowly degraded through the structure on which the biofilm was attached and the tract forms. Tunneling is also commonly used to describe degradation along solid structures such as tendon or bone, again extending the wound deeper in the host tissues. Undermining usually refers to the erosion of the subcutaneous layer under the wound edge involving a significant portion of the circumference of the wound edge. The process of bacterial cell attachment and its relationship to cellular proliferation

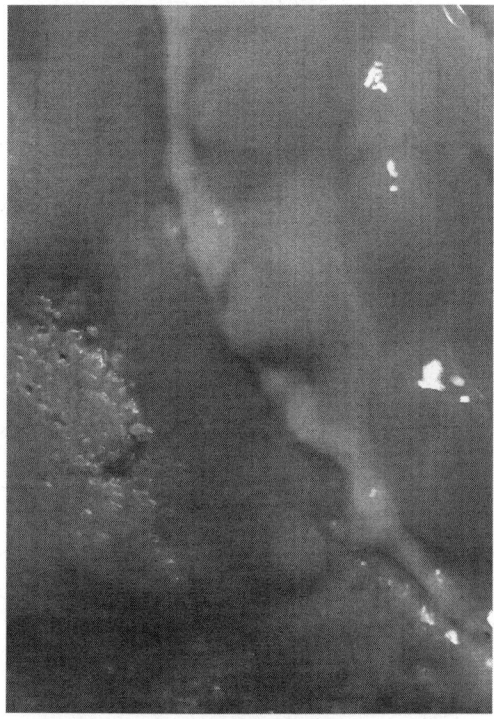

FIGURE 8.1 (Please see color insert following page 114.) Slough overlapping the wound edge.

and biofilm formation have been described elsewhere in this book. The importance of debriding undermined cavities and tunnels becomes clear when surface attachment and subsequent biofilm formation are considered. Wolcott and Rhodes[19] stated in their study of biofilm-based wound care and critical limb ischemia, that the first principle of debridement was to alter the anatomy of the wound by removing surfaces that touched each other and opened all tunnels and removed (laid open) undermined cavities. Debridement will therefore deny the biofilm the opportunity of attachment to two surfaces which would enhance further proliferation of the biofilm and in combination with bacterial population targeted therapies forms the basis of biofilm-based wound care.

The edges of chronic wounds can yield important information. In a healing wound, the wound edge gently slopes to the wound bed and will encroach several millimeters into the wound, whereas edges that are raised off the wound bed (rolled) or with a rim of undermining at the edge strongly suggest bacterial involvement. Recent studies on keratinocyte migration suggest soluble substances secreted by bacteria may impair keratinocytes, which may yield these types of edges.[30,31] If the wound biofilm becomes more active thereby upregulating host inflammation, this will result in an increase in exudate production, which if inadequately managed will cause periwound skin maceration. Maceration of the wound edge along with pain, swelling, and deterioration of the wound suggests an active biofilm.

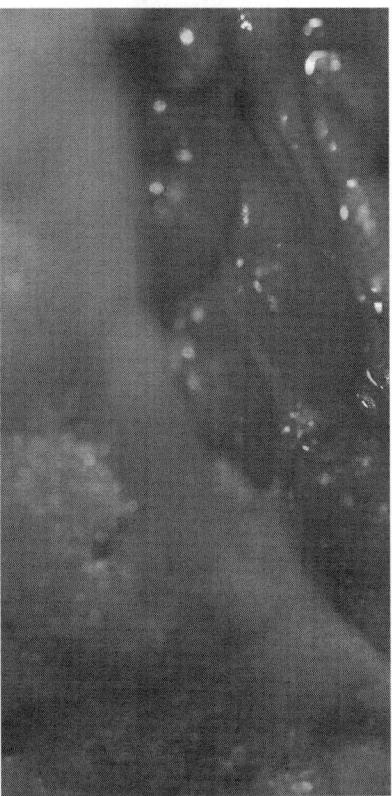

FIGURE 8.2 (Please see color insert following page 114.) Bright red wound border is a positive sign of host control of wound biofilm.

The hyperkeratosis (callus) seen on plantar wounds does not always indicate inflammation from walking and repetitive trauma. Inflammation from the chronic infection present in many chronic wounds can also manifest as callus.[32] This appears to be confirmed by the fact that hyperkeratosis can be seen in a patient who was ventilated and clearly not walking (Figure 8.3). In addition, clinical observation indicates that the suppression of wound biofilm without escalating offloading can significantly reduce callus formation.

IMMUNE STIMULATION AND RETARDATION BY THE BIOFILM

Many research papers have shown that biofilm bacteria are less susceptible to a host immune defense system and therapeutic agents when compared to their planktonic counterparts. As a consequence, this biofilm-associated wound infection will have an ability to persevere.

The biofilm matrix has been shown to be able to inhibit chemotaxis and degranulation by polymorphonucleocytes (PMNs) and macrophages and also depress the lymphoproliferative response of monocytes to polyclonal activators. A recent paper

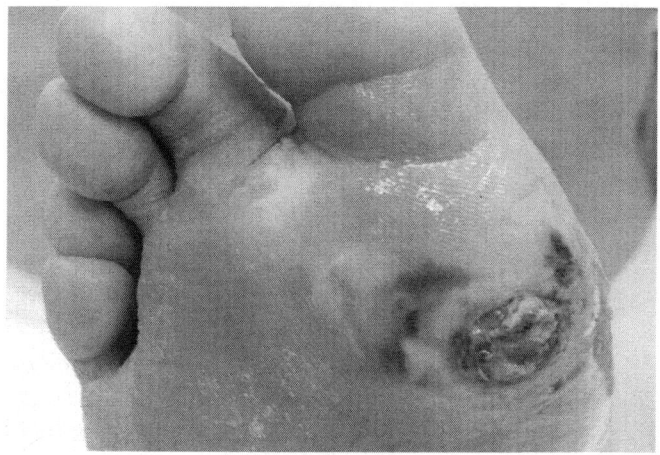

FIGURE 8.3   (Please see color insert following page 114.) Plantar hyperkeratosis in an immobile patient.

has shown that PMNs are ineffective at engulfing bacteria in biofilms causing them to release large amounts of proinflammatory enzymes and cytokines which long term will lead to the destruction of nearby tissues and chronic inflammation.[33] Neutrophils are very important in normal wound healing. However, they are also considered to delay the healing process of chronic wounds.[34,35] Armstrong and colleagues proposed that "it can be postulated with some confidence that neutrophil derived MMP 8 is the predominant collagenase present in normal human wounds and that over expression and activation of this collagenase may be involved in the pathogenesis of nonhealing chronic ulcers."[25] Neutrophils express MMP 8 as well as other enzymes. Overproduction of neutrophils can cause collateral damage to a nonhealing wound.[26]

In addition to neutrophils, macrophages are considered significant to wound healing. Macrophages are evident in smaller numbers in a chronic wound when compared with neutrophils. Macrophages have the ability to synthesize a large array of growth factors and other chemical messengers that not only direct host defenses but also move the wound from the inflammatory phase into the proliferative phase.

During trauma the host initiates a normal immune response to try and establish and return to some form of homoeostasis (healed wound). When a biofilm is detected on a wound surface, the host initiates additional innate immune responses to try and eliminate the colonizing bacteria. The biofilm in turn uses its virulence and generalized mechanisms to maintain its newly found niche. The biofilm seeks to remain firmly attached to the host tissue and propogate its community structure. As the biofilm develops, a climax community is achieved, implying stable associations and integrations of function between microbial populations and the wound bed. Thus the biofilm community is attempting to maintain its niche in the wound and reach a steady state with the host. This can almost be considered resetting the host's homeostasis settings. At this stage the microorganisms and biofilm, although interfering with the wound healing process, may not necessarily induce any clinical signs of infection.

Chronic wounds are both unpredictable and recalcitrant. Consequently chronic wound management and prevention of infection represents an area of concern for clinicians and physicians worldwide. As both the pathophysiology of wound healing and biofilm development are complex, such a combination in chronic wounds makes for a complex biological system that is dynamic and diverse.

The structure of biofilm is very important to its defenses and therefore the survival of the community of microorganisms. Overall, the biofilm is considered from *in vitro* work to consist of a number of important layers. These are the outer edge (which interfaces with the environment), the midportion of the biofilm, and base of the biofilm which is considered to be realtively dormant. The top layer of the biofilm is made up of metabolically active microbial cells. These bacteria can be easily shed or dispersed from the biofilm. This seeding or distribution of bacteria has been referred to as "seeding dispersal."[36]

Many characteristics and phenomena associated with chronic wounds are easily explained by viewing chronic cutaneous wounds as chronic infections. The relevance of observing subtle changes in wound bed characteristics and their relation to clinical diagnosis of infection has already been highlighted.[37,38] Closely observing subtle changes in the wound and then correlating these changes with diagnostic tools and responses to therapy may hint at some of the processes taking place on the wound bed at a cellular level. Gaining an understanding of these activities will enhance our proficiency in wound management.

Staphylococcus aureus is able to produce a polysaccharide intercellular adhesin (PIA). In strains of *S. aureus* that lack the PIA, polymorphonuclear white blood cells macrophages are able to reduce the numbers of bacterial cells from the biofilm.[39,40]

The biofilm matrix is known to afford protection from the host immune system. In fact it has been documented that EPS can completely inhibit macrophage activity[41] and antibodies.[42] The matrix of the biofilm is important for inhibition of host defenses and resistance to antimicrobials. EPS is an integral part of the biofilm, but it is not definitive for biofilm alone. For example, planktonic bacteria that have been exposed to subinhibitor concentrations of antibiotics have been documented to show up to a fifteenfold increase in the production of alginate.

The organizational structure of biofilm is critical to its defenses and therefore its survival. The three most important layers into which the bacteria will differentiate are the outer edge (which interfaces with the environment), the midportion of the biofilm, and the metabolically inactive base attached to the surface.

The environmental edge of the biofilm is composed of metabolically active bacterial cells imbedded in the biofilm matrix. All along the surface of the biofilm the bacteria of the environmental edge are in the process of developing small pods, which revert into planktonic cells and are dispersed out into the environment. This seeding dispersal appears to be a reproductive mechanism, as this portion of the biofilm is constantly shedding large fragments out into the environment, but it may also provide a defense mechanism much like our skin, which constantly sheds squamous cells from the stratum corneum.

Another defensive function of the environmental edge is to prevent penetration of toxic substances into the biofilm. This has been termed "reaction diffusion interaction." When biofilm was treated with high doses of hydrogen peroxide, a microelectrode deep inside the biofilm registered no sign of penetration of the peroxide for over 50 minutes.[43] The reaction at the environmental edge limits diffusion into the biofilm and the cells catalyze the hydrogen peroxide even after they were dead. This gives biofilm the property of being very resistant to hydrogen peroxide as well as other reactive substances.

In another biocide experiment, glutaraldehyde was tested against *P. aeruginosa* biofilm. Glutaraldehyde showed little or no penetration up to 200 minutes and it was only at 800 minutes that there was a 3-log killing of the bacteria in the biofilm. On the other hand, the planktonic form of the same strain of *P. aeruginosa* showed a 3-log killing in just a few minutes.[44] With prolonged exposure to monochloromine, biofilm resists penetration and produces a neutralizer.[44] Resistance to the penetration and, therefore, the killing by commonly used biocides such as glutaraldehyde, bleach, hydrogen peroxide, acetic acid, and others is a very important defense and is often used to define the presence of biofilm.

Bacteria that differentiate into the midportion phenotype in the biofilm have a different function. These bacteria remain metabolically active, but they are geared to produce protective responses, which are not well understood. Two-dimensional gel electrophoresis experiments show that the midportion of the biofilm will produce different, yet uncharacterized proteins in a very rapid fashion when the biofilm is subjected to stress. When antibiotics are applied, this region usually has subinhibitory concentrations of antibiotics.[45] This finding plus high cell density with mobile genetic material[46] may allow for the horizontal transfer of resistance. It is unclear what benefit the other proteins confer on the biofilm, but it is likely to be of a sophisticated nature.

A better characterized defense is the altered microenvironment in the base of the biofilm. The bacteria, which differentiate in the basilar region of the biofilm where it is attached to a surface, show no metabolic activity. These individual bacteria, whether due to cell-to-cell signaling, metabolic by-products, the extremely low oxygen tension in the region, or yet undetermined factors, have shut down all their DNA synthesis, protein synthesis, and other cell functions. This may lead to the production of cells that are viable but not culturable.[47]

BIOFILM DETACHMENT AND DISPERSAL IN WOUNDS: CLINICAL SIGNIFICANCE

Biofilm detachment from a surface is a complex process but is fundamentally significant to infections due to dissemination of biofilm bacteria. Biofilm sloughing is an area that is poorly reported on and presently there is a need to correlate this phenomenon with infection/disease risk. Biofilms are well known to be susceptible to detachment, and it has even been suggested that sloughing or dispersing of bacteria from a biofilm is preprogrammed.[48] Such an event is considered to be a reproductive process inherent in all biofilms. It is these detached biofilms that will be composed of very active microbial cells as sloughing and dispersal occurs generally from the outermost active regions of the biofilm first.

Detached sections of biofilm are free to enter the wound environment or exudate. Of clinical significance is the fact that these detached sections of biofilms have inherent characteristics of the "mother" biofilm (i.e., enhanced resistance to antimicrobials and the ability to attach and adapt quickly to virgin wound anchor points). Periodic shedding and dispersal of a biofilm is analogous to metastases in a tumor and aids to increase the turnover of the biofilm, enhancing its overall activity. Biofilm detachment is a process that is very important for the inherent defensive mechanisms to the biofilm.

Despite constant sloughing, the biofilm has the ability to reconstitute itself very quickly. Research has shown that biofilms have the ability to replace their entire mass in 24 hours.[48] Such a process is significant to wound care and the management of chronic and infected wounds.

The detached sections of a biofilm vary in size. These detached sections can be composed of just a few bacterial cells or contain millions of individual bacteria.[48] These detachment fragments are characteristic of biofilms and as such have the inherent properties and defenses of a mature biofilm.

Preprogrammed sloughing of detachment fragments has long been thought to be a reproductive mechanism. Because of the defenses that detachment fragments retain, this may make them more successful in reproducing the biofilm. A mature biofilm has the ability to replace its entire mass in approximately 24 hours through the sloughing of detachment fragments.[48] This theoretically means the entire area of a chronic wound may be reproduced and shed each day. These detachment fragments are shed into the environment and, in the case of a chronic wound, have the ability to "seed" the skin of the host. Because of the intact biofilm defenses of slow penetration, stress response, and metabolic inactivity, these detachment fragments are resistant to biocides and antibiotics. This means they may be resistant to many surgical antimicrobials and prophylactic antibiotics. Detachment fragments may explain why patients with chronic wounds tend to get more surgical site infections than patients who do not have chronic wounds.

The phenomena of detachment fragments may not only be thought of as a reproductive tool, but also a defensive mechanism much like our skin. The constant turnover of the epidermis prevents the adherence of pathogens and other detrimental environmental agents. Anything that does attach is quickly sloughed and new epidermis replaces the old. It is the constant turnover of the skin (and maybe the constant turnover of biofilm) that plays an important role in this defensive mechanism.

CONCLUSIONS

Open wounds are moist and as such are conducive to colonization from bacteria. Colonizing bacteria are able to attach to an array of biological material found in the wound bed such as fibronectin, keratinocytes, and fibroblasts. As the biofilm develops, it is then probable that the actual presence of the biofilm and the extracellular components released from the sessile bacteria will have the capacity to damage the surrounding tissue. By aiding bacterial attachment, the development of a biofilm will commence and be sustained in the wound. During biofilm development in the wound, it will progress to a "climax community" or maturity, and the extracellular components released from the sessile bacteria will have the capacity to damage the surrounding tissue. In addition to this, the phagocytic cells released from the host have been shown to not fully penetrate the biofilm so the enzymes released from the white blood cells will lead to extensive tissue damage.[46] In conjunction with this, if there is synergy between the human proteases released during wound healing and bacterial proteases, further healthy tissue breakdown would be inevitable and wound healing will be compromised.

We have suggested that the existence of biofilms in wounds is clinically evident, but their role in inhibiting the healing process is still being investigated. However, results obtained from biofilm-based management strategies have established positive clinical outcomes. All chronic wounds are colonized by bacteria and they become encased and develop into a biofilm which results in a prolonged inflammatory state in the host. In addition to this, proteases released by sessile bacteria, possibly working in synergy with human proteases, destroy human growth factors and tissue proteins that are essential to wound healing. A better understanding of the physiology and biochemistry of biofilms in wounds is needed, and this will aid in the development of more effective methods of biofilm treatment. This will ultimately result in better prognosis for patients and therefore improve wound healing.

REFERENCES

1. Stoodley, P., Sauer, K., Davies, D.G., and Costerton, J.W. (2002) Biofilms as complex differentiated communities. *Annu Rev Microbiol* 56:187–209.
2. Costerton, J.W., Lewandowski, Z., Caldwell, D.E., Korber, D.R., and Lappin-Scott, H.M. (1995) Microbial biofilms. *Annu Rev Microbiol* 49:711–745.
3. Brozel, V.S., Cloete, T.E.E., and Strydom, G.M. (1995) A method for the study of de novo protein synthesis in *Pseudomonas aeruginosa* after attachment. *Biofouling* 8:195.
4. Whiteley, M., et al. (2001) Gene expression in *Pseudomonas aeruginosa* biofilms. *Nature* 413:860–864.
5. Prigent-Combaret, C., Vidal, O., Dorel, C., and Lejeune, P. (1999) Abiotic surface sensing and biofilm-dependent regulation of gene expression in *Escherichia coli*. *J Bacteriol* 181:5993–6002.
6. Sauer, K., and Camper, A.K. (2001) Characterization of phenotypic changes in Pseudomonas putida in response to surface-associated growth. *J Bacteriol* 183: 6579.
7. Sauer, K., Camper, A.K., Ehrlich, G.D., Costerton, J.W., and Davies, D.G. (2002) *Pseudomonas aeruginosa* displays multiple phenotypes during development as a biofilm. *J Bacteriol* 184:1140.
8. Whitchurch, C.B., Tolker-Nielsen, T., Ragas, P.C., and Mattick, J.S. (2002) Extracellular DNA required for bacterial biofilm formation. *Science* 295:1487.
9. Stewart, P.S. (2006) Matrix mysteries hold keys to controlling biofilms. Web. 2-15-2006. Ref Type: Electronic Citation http://www.biofilmsonline.com/biofilmsonline/pdfs/biofilm_perspectives/Perspectives_Feb2006.pdf
10. Stewart, P.S. (1996) Theoretical aspects of antibiotic diffusion into microbial biofilms. *Antimicrob Agents Chemother* 40:2517–2522.
11. Bagge, N., et al. (2004) Dynamics and spatial distribution of beta-lactamase expression in *Pseudomonas aeruginosa* biofilms. *Antimicrob Agents Chemother* 48: 1168–1174.

12. Anderl, J.N., Zahller, J., Roe, F., and Stewart, P.S. (2003) Role of nutrient limitation and stationary-phase existence in *Klebsiella pneumoniae* biofilm resistance to ampicillin and ciprofloxacin. *Antimicrob Agents Chemother* 47:1251–1256.

13. Walters, M.C., III, Roe, F., Bugnicourt, A., Franklin, M.J., and Stewart, P.S. (2003) Contributions of antibiotic penetration, oxygen limitation, and low metabolic activity to tolerance of *Pseudomonas aeruginosa* biofilms to ciprofloxacin and tobramycin. *Antimicrob Agents Chemother* 47:317–323.

14. Konig, C., Schwank, S., and Blaser, J. (2001) Factors compromising antibiotic activity against biofilms of *Staphylococcus epidermidis*. *Eur J Clin Microbiol Infect Dis* 20:20–26.

15. Souli, M., and Giamarellou, H. (1998) Effects of slime produced by clinical isolates of coagulase-negative staphylococci on activities of various antimicrobial agents. *Antimicrob Agents Chemother* 42:939–941.

16. Fux, C.A., Costerton, J.W., Stewart, P.S., and Stoodley, P. (2005) Survival strategies of infectious biofilms. *Trends Microbiol* 13:34–40.

17. Serralta, V.W., Harrison-Balestra, C., Cazzaniga, A.L., Davis, S.C., and Mertz, P.M. (2001) Lifestyles of bacteria in wounds: Presence of biofilms? *WOUNDS* 13:29–34.

18. Bello, Y.M., Falabella, A.F., Cazzaniga, A.L., Harrison-Balestra, C., and Mertz, P.M. (2001) Are biofilm present in human chronic wounds? Presented at the Symposium on Advanced Wound Care and Medical Research Forum on Wound Repair in Las Vegas, NV, April 30–May 3, 2001.

19. Wolcott, R.D., and Rhoads, D.D. (2008) A study of biofilm-based wound management in subjects with critical limb ischaemia. *J Wound Care* 17:145–155.

20. Davis, S.C., Martinez, L., and Kirsner, R. (2006) The diabetic foot: The importance of biofilms and wound bed preparation. *Curr Diabetes Rep* 6:439–445.

21. James, G.A., Swogger, E., Wolcott, R., DeLancey, Pulcini, E., et al. (2008) Biofilms in chronic wounds. *Wound Repair Regen* 16:37–44.

22. Schaber, J.A., Triffo, W.J., Suh, S.J., et al. (2007) *Pseudomonas aeruginosa* forms biofilms in acute infection independent of cell-to-cell signaling. *Infect Immun* 75(8):3715–3721.

23. Diegelmann, R.F. (2003) Excessive neutrophils characterize chronic pressure ulcers. *Wound Repair Regen* 11:490–495.

24. Armstrong, D.G., and Jude, E.B. (2002) The role of matrix metalloproteinases in wound healing. *J Am Podiatr Med Assoc* 92:12–18.

25. Yager, D.R., and Nwomeh, B.C. (1999) The proteolytic environment of chronic wounds. *Wound Repair Regen* 7:433–441.

26. Stoodley, P., Sauer, K., Davies, D.G., and Costerton, J.W. (2002) Biofilms as complex differentiated communities. *Annu Rev Microbiol* 56:187–209.

27. Wound Care Center. Biofilm based wound management. Available at www.woundcarecenter.net/CHAPTERTEXT.pdf (2006) accessed 28 May 2009.

28. Williams, D., Enoch, S., Miller, D., Harris, K., Price, P., and Harding, K.G. (2005) Effect of sharp debridement using curette on recalcitrant nonhealing venous leg ulcers: A concurrently controlled, prospective cohort study. *Wound Repair Regen* 13:131–137.

29. Nylen, S., and Carlsson, B. (1980) Time factor, infection frequency and quantitative microbiology in hand injuries: A prospective study. *Scand J Plast Reconstr Surg* 14(2):185–189.

30. Tomic-Canic, M., Ayello, E.A., Stojadinovic, O., et al. (2008) Using gene transcription patterns (bar coding scans) to guide wound debridement and healing. *Adv Skin Wound Care* 10:487–492.

31. Pat Secor, personal communication with R. Wolcott.

32. Stojadinovic, O., Brem, H., Vouthounis, C., et al. (2005) Molecular pathogenesis of chronic wounds: The role of beta-catenin and c-myc in the inhibition of epithelialization and wound healing. *Am J Pathol* 167:59–69.

33. Williams, P. (1994) Host immune defences and biofilms. In: Wimpenny, J., Nichols, W., Stickler, D., and Lappin-Scott, H. (eds.). *Bacterial Biofilms and Their Control in Medicine and Industry*. Cardiff, UK: BioLine, 93–96.

34. Yager, D.R. et al. (1997) Ability of chronic wound fluids to degrade peptide growth factors is associated with increased levels of elastase activity and diminished levels of proteinase inhibitors. *Wound Repair Regen* 5(1):23–32.

35. Nwomeh, B.C., Yager, D.R., and Cohen, I.K. (1998) Physiology of the chronic wound. *Clin Plast Surg* 25:341–356.

36. Purevdorj-Gage, B., Costerton, W.J., and Stoodley, P. (2005) Phenotypic differentiation and seeding dispersal in non-mucoid and mucoid *Pseudomonas aeruginosa* biofilms. *Microbiology* 151:1569–1576.

37. Cutting, K.F., and Harding, K.G. (1994) Criteria for identifying wound infection. *J Wound Care* 3:198–201.

38. Cutting, K.F., White, R.J., Mahoney, P., and Harding, K. (2005) Clinical identification of wound infection: A Delphi approach. In: Identifying criteria for wound Infection EWMA Position Document MEP London. 2005.

39. Vuong, C., et al. (2004) Polysaccharide intercellular adhesin (PIA) protects Staphylococcus epidermidis against major components of the human innate immune system. *Cell Microbiol* 6:269–275.

40. Shiau, A.L., and Wu, C.L. (1998) The inhibitory effect of *Staphylococcus epidermidis* slime on the phagocytosis of murine peritoneal macrophages is interferon-independent. *Microbiol Immunol* 42:33–40.

41. Meluleni, G.J., Grout, M., Evans, D.J., and Pier, G.B. (1995) Mucoid *Pseudomonas aeruginosa* growing in a biofilm in vitro are killed by opsonic antibodies to the mucoid exopolysaccharide capsule but not by antibodies produced during chronic lung infection in cystic fibrosis patients. *J Immunol* 155:2029–2038.

42. Stewart, P.S., et al. (2000) Effect of catalase on hydrogen peroxide penetration into *Pseudomonas aeruginosa* biofilms. *Appl Environ Microbiol* 66:836–838.

43. Stewart, P.S., Grab, L., and Diemer, J.A. (1998) Analysis of biocide transport limitation in an artificial biofilm system. *J Appl Microbiol* 85:495–500.

44. Sanderson, S.S., and Stewart, P.S. (1997) Evidence of bacterial adaption to mono-chloramine in *Pseudomonas aeruginosa* biofilms and evaluation of biocide action model. *Biotechnol Bioeng* 56:201–209.

45. Rachid, S., Ohlsen, K., Witte, W., Hacker, J., and Ziebuhr, W. (2000) Effect of sub-inhibitory antibiotic concentrations on polysaccharide intercellular adhesin expression in biofilm-forming *Staphylococcus epidermidis*. *Antimicrob Agents Chemother* 44:3357–3363.

46. Fux, C.A., Costerton, J.W,, Stewart, P.S., and Stoodley, P. (2005) Survival strategies of infectious biofilms. *Trends Microbiol* 13:34–40.

47. Rayner, M.G., et al. (1998) Evidence of bacterial metabolic activity in culture-negative otitis media with effusion. *J Am Med Assoc* 279:296–299.

48. Stoodley, P., Wilson, S., Hall-Stoodey, L., Boyle, J.D., Lappin-Scott, H.M., and Costerton, J.W. (2001) Growth and detachment of cell clusters from mature mixed species biofilms. *Appl Environ Microbiol* 67(12):5608–5613.

9 Wounds, Enzymes, and Proteases

Steven L. Percival and Christine A. Cochrane

CONTENTS

Introduction 249
History and Structure of MMPs 250
Types, Mode of Action, and Sources of MMPs 250
    Collagenases 250
    Gelatinases 256
    Stromelysins 256
    Membrane-Type Metalloproteinases 257
Regulation of MMPs 257
    Transcriptional Level Regulation 257
    Zymogen Activation 257
    Tissue Inhibitors of Metalloproteinases (TIMPs) 258
Overall MMPs in Chronic Wounds 259
Factors That Stimulate MMPs 260
    Bacteria, Biofilms, and MMPs 260
    Aerobic Bacteria 262
    Anaerobic Bacteria 263
Management of MMPs 263
Conclusion 264
References 264

INTRODUCTION

The extracellular matrix (ECM) is composed of collagen, laminin, fibronectin, entactin, proteoglycans, as examples, and provides the ideal and optimum environment to support both cellular growth and proliferation necessary for effective wound healing. The degradation, proliferation, and remodeling of the ECM must be carefully controlled by matrix metalloproteinases (MMPs).

Within a chronic wound, the microenvironment resides within an imbalanced state culminating in the development of suboptimal physiological and biochemical conditions. The heterogeneity within the chronic wound is thought to induce an upregulation in MMP expression resulting in the production of excessive enzymatic activity and therefore tissue breakdown. Appropriate "optimal" levels of human MMPs are significant for healing, and to the remodeling of the wound.[1,2]

In addition to human MMPs, human neutrophils produce the serine protease elastase, which is often detected at elevated levels in the chronic wound exudate.[3–5] Furthermore, bacterial enzymes independently cause tissue breakdown that significantly delays wound healing and up-regulate human MMPs.

HISTORY AND STRUCTURE OF MMPS

MMPs were first reported in 1962 by Gross and Lapiere,[6] who documented enzymatic activity during the metamorphosis of the tail of a tadpole. The enzyme involved in this process was identified and named as interstitial collagenase (MMP-1). In 1968 the enzyme was reported in human skin.[7]

Presently, 23 MMPs have been identified in humans (Table 9.1).[8] MMPs are not stored in the body, and generally only expressed as inactive zymogens when they are required. The zymogen contains a domain that includes the propeptide, the catalytic domain, and also the hemopexin-like C-terminal domain (linked to the flexible hinge region). Within the propeptide domain region there is a "cysteine switch" composed of a cysteine residue that is able to interact with zinc. This switch is important as it prevents the binding and cleavage of a substrate to the active site of the enzyme. Consequently until zinc becomes exposed, the enzyme remains in the inactive form. Zinc is important for the activity of the enzyme, calcium is also significant. Calcium is important because it is required to maintain the structural integrity of MMPs.[9]

MMPs are expressed by keratinocytes, fibroblasts, macrophages, mast cells, neutrophils, and endothelial cells.[10] However, within uninjured skin, MMPs are not considered to be activated (see Figure 9.1).[11] The relevance and significance of MMPs in wound healing have been demonstrated in a number of different animal models and the data generated have been extrapolated to the activity and characteristics of human MMPs.[12–16]

In addition to MMPs having a significant role to play in remodeling[10,17–19] and degradation of the ECM, they can process a number of bioactive compounds. Such compounds have been shown to cleave surface receptors and activate/inactivate chemokines and cytokines.[20]

TYPES, MODE OF ACTION, AND SOURCES OF MMPS

In acute cutaneous wounds, the sources of MMPs are vast and have been documented by Toriseva and Kähäri (Figure 9.2 and Table 9.1).[21] Each MMP is categorized in relation to its primary structure, substrate specificity, and cellular location.

MMPs are divided into subgroups including gelatinases, collagenases, stromelysins, matrilysins, and those MMPs that are membrane bound.[22,23] There are a number of MMPs, however, that do not fit into any of the traditional groups.

COLLAGENASES

Collagenases are a group of MMPs that are able to degrade fibrillar collagen types I, II, III, V, and IX. They include collagenases 1, 2, 3 and MMPs 1, 8, and 13. Collagenase 1 has been documented to be expressed in most cells of the human body

TABLE 9.1
Classification of Matrix Metalloproteases (MMPs)

Enzyme	MMP	Location	ECM Substrate	Non-ECM Substrate	Activated by	Activator of
Collagenases						
Collagenase-1	MMP-1	Secreted	Collagens (I, II, III, VII, VIII, and X), gelatine, proteoglycan link protein, aggrecan, veriscan, tenacin, entactin	α_1-PI, ILb-1, pro-TNF, IGFBP-3, MMP-2, MMP9	MMP-3, -10 plasmin kallikrein, chymase	MMP-2
Collagenase-2	MMP-8	Secreted	Collagens (I, II, III, V, VII, VIII, and X) Gelatin, aggrecan	α_1-PI, α_2-antiplasmin, fibronectin	MMP3-10, plasmin	ND
Collagenase-3	MMP-13	Secreted	Collagens (I, II, III, IV, IX, X, XIV) gelatin, aggrecan, perlecan, large tenascin-C, fibronectin, osteonectin	MMP-9, plasminogen activator inhibitor-2	MMP-2, -3, -10, -14, -15, plasmin	MMP-2, -9
Collagenase-4	MMP-18	ND	ND	ND	ND	ND
Gelatinases						
Gelatinase A	MMP-2	Secreted	Collagens (I, IV, V, VII, X, XI, XIV) gelatine, elastin, fibronectin, laminin-1, laminin-5, Galectin-3, aggrecan, decorin, hyaluronidase-treated versican, proteoglycan link protein, osteonectin	IL-1b, α_1-PI, prolysyl oxidase fusion protein, MMP-1, MMP-9, MMP-13	MMP-1, -7, -13, -14, -15, -16, -17, -24, -25, tryptase?	MMP-1, -7, -13

(continued on next page)

TABLE 9.1 (continued)
Classification of Matrix Metalloproteases (MMPs)

Enzyme	MMP	Location	ECM Substrate	Non-ECM Substrate	Activated by	Activator of
Gelatinase B	MMP-9	Secreted	Collagens (IV, V, VII, X, XIV) gelatine, elastin, galectin-3, aggrecan, fibronectin, hyaluronidase-treated versican, proteoglycan link protein, entactin, osteonectin	α_1-PI, IL-1β, plasminogen	MMP-2, -3, -13, plasmin	ND
			Stomelysins			
Stromelysin-1	MMP-3	Secreted	Collagens (III, IV, V, and IX) gelatine, aggrecan, versican, hyaluronidase-treated versican, periecan, decorin, proteoglycan link protein, large tenacin-C, fibronectin, laminin, entactin, osteonectin	α_1-P, antithrombin-III, ovosstatin, substance P, IL-1β, serum amyloidal, IGFBP-3, fibrinogen, cross-linked fibrin, plasminogen, MMP-1 "superactivation," MMP-2/TIMP-2 complex, MMP-7, -8, -9, -13	Plasmin, kallikrein, chymase, tryptase	MMP-1, -7, -8, -9, -13
Stromelysin-2	MMP-10	Secreted	Collagens (III, IV, V) gelatin, casein, aggrecan, elastin, proteoglycan link protein aggrecan, elastin, proteoglycan link protein casein, laminin, fibronectin, gelatin, collagen IV and carboxymethylated transferring	α_1-P, casein, IGFBP-1	Furin	ND

Membrane-type MMPs

MT1-MMP	MMP-14	Membrane associated (type-I trans membrane MMP)	Collagens (I, II, and III) casein, elastin, fibronectin, gelatin, laminin, vitronectin, large tenacin-C, entactin, proteoglycans	α_1-P, MMP-2,-13	Plasmin, furin	MMP-2, -13
MT2-MMP	MMP-15	Membrane associated (type-I trans membrane MMP)	Large tenascin-C, fibronectin, laminin, entactin, aggrecan, perlecan	MMP-2	ND	MMP-2, -13
MT3-MMP	MMP-16	Membrane associated (type-I trans membrane MMP)	Collagen-III, gelatin, casein, fibronectin	MMP-2	ND	MMP-2
MT4-MMP	MMP-17	Membrane associated (glycosyl phospatidylinositol-attached)	ND	ND	ND	MMP-2
MT5-MMP	MMP-24	Membrane associated (Type I transmembrane)	ND	ND	ND	MMP-2
MT6-MMP	MMP-25	Membrane associated (Glycosyl phosphatidylinositol-attached)	ND	ND	ND	MMP-2

(continued on next page)

TABLE 9.1 (continued)
Classification of Matrix Metalloproteases (MMPs)

Enzyme	MMP	Location	ECM Substrate	Non-ECM Substrate	Activated by	Activator of
			Others			
Matrilysin	MMP-7	Secreted	Collagens IV and X, gelatine, aggrecan, decorin, proteoglycan link protein, fibronectin, laminin, insoluble fibronectin fibrils, entactin, large and small tenascin-C, osetonectin, β_4 integrin, elastin, casein, transferring	MMP-1, -2, -9 MMP-9/TIMP-1 complex, α_1-P, plasminogen	MMP-3, -10 plasmin	MMP-2
Matrilysin-2	MMP-26	ND	Collagen IV, gelatine, fibronectin	ProMMP-9, fibrinogen, α_1-P1	ND	ND
Metalloelastase (Macrophage)	MMP-12	Secreted	Collagen IV gelatin, elastin, casein, laminin, proteoglycan monomer, fibronectin, vitronectin, entactin	α_1-P1, fibrinogen, fibrin, plasminogen, myelin basic protein	ND	ND
No trivial name	MMP-19	ND	Gelatin	ND	Trypsin	ND
Enamelysin	MMP-20	Secreted	Amelogenin	ND	ND	ND
No trivial name	MMP-23	Membrane associated (Type-II transmembrane cysteine array)	Gelatin	ND	ND	ND
XMMP (Xenopus)	MMP-21	Secreted	ND	ND	ND	ND
CMMP (Chicken)	MMP-22	ND	ND	ND	ND	ND
Epilysin	MMP-28	Secreted				

Note: α_1-PI, α_1-proteinase inhibitor; IGFBP, insulin-like growth factor binding protein; IL-1, interleukin-1; TNF, tumor necrosis factor; ND, not determined.

Source: Adapted from Rawlings, N.D., Morton, F.R., and Barrett, A.J. (2006) *Nucleic Acids Res* 34: D270–272.

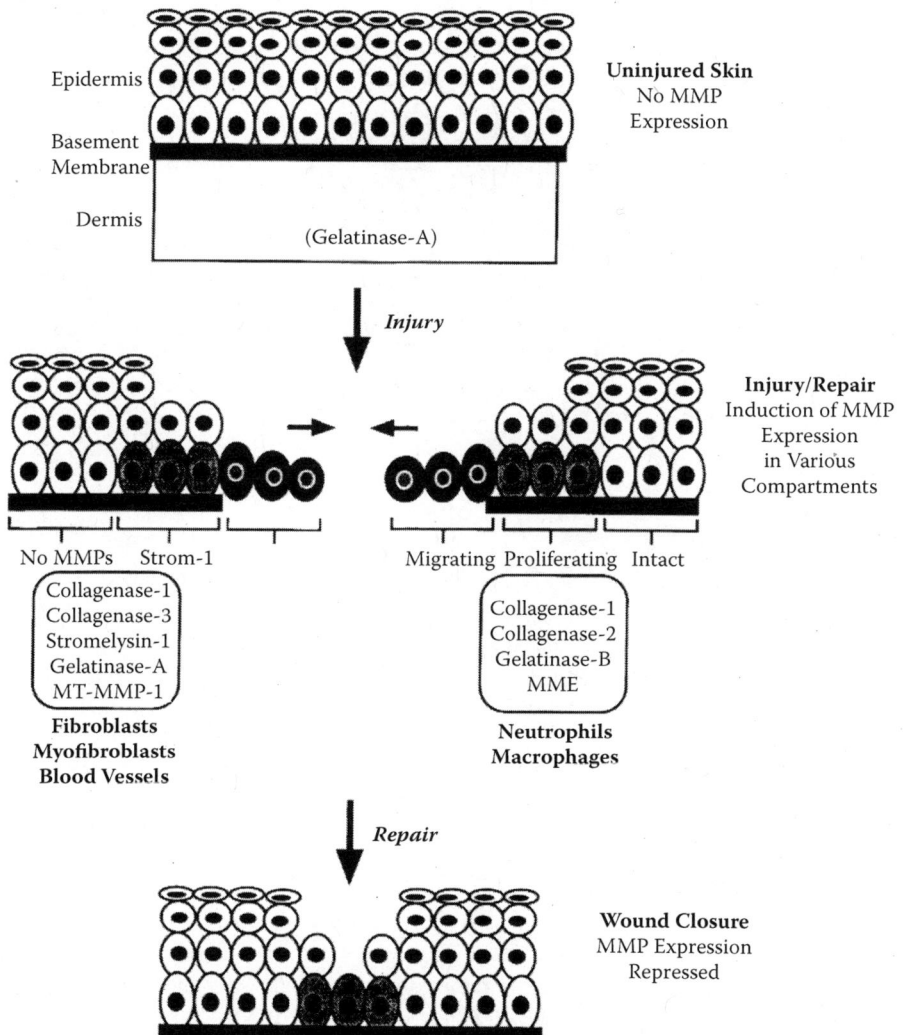

Epidermis

Basement
Membrane

Dermis

(Gelatinase-A)

Uninjured Skin
No MMP
Expression

Injury

Injury/Repair
Induction of MMP
Expression
in Various
Compartments

No MMPs   Strom-1   Migrating Proliferating Intact

Collagenase-1
Collagenase-3
Stromelysin-1
Gelatinase-A
MT-MMP-1

Fibroblasts
Myofibroblasts
Blood Vessels

Collagenase-1
Collagenase-2
Gelatinase-B
MME

Neutrophils
Macrophages

Repair

Wound Closure
MMP Expression
Repressed

FIGURE 9.1   Matrix metalloproteinase (MMP) expression in the epidermis and dermis during wound healing. (Parks, W.C. (1999) Matrix metalloproteinases in repair. *Wound Repair and Regeneration* 7(6):423–32. With permission)

including fibroblasts, keratinocytes, endothelial cells, monocytes, macrophages, and osteoblasts, and preferentially degrade type III collagen. Collagenase 2 is generally only produced by neutrophils, chondrocytes, fibroblasts, and endothelial cells, and preferentially degrades type I and type II collagens. Collagenase 3 is produced most predominately in developing bone and periodontitis and has been documented to be more effective for degrading type II collagen than any of the other collagenase.

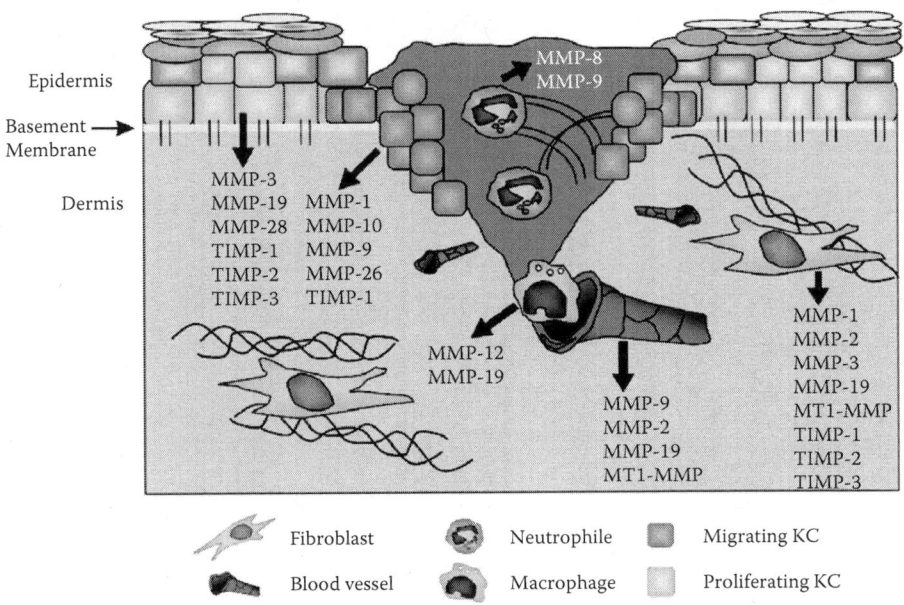

FIGURE 9.2   (Please see color insert following page 114.) The expression and cellular source of matrix metalloproteinases (MMP) and tissue inhibitors of MMPs (TIMP) in acute wounds. (Toriseva, M., and Kahari, V.M. (2009) Proteinases in cutaneous wound healing. *Cell Mol Life Sci* 66:203–224. With permission)

GELATINASES

Gelatinases A and B (MMP 2 and MMP 9) are able to cleave peptide bonds in collagen that has been denatured resulting in the formation of small chain length peptides. They are also able to breakdown collagen (IV, V, VII, and X) and elastin. Gelatinases A are produced from keratinocytes, chondrocytes, fibroblasts, monocytes, osteoblasts, and endothelial cells. However, gelatinases B are not produced from fibroblasts.

STROMELYSINS

The subgroup stromelysin is composed of stromelysins 1 (MMP 3), matrilysin (MMP 7), 2 (MMP 10), stromelysin 3 (MMP 11), and matrix metalloelastase (MMP 12) which are known to have a broad substrate range. In addition, they are known to activate other MMPs. Stromelysin 1 and 2 have been reported to be able to degade fibronectin, gelatin, elastin, proteoglycan, and collagens (type IV and V). Both stromelysins are produced by an array of different cells including keratinocytes, fibroblasts, chondrocytes, macrophages, and endothelial cells.

Membrane-Type Metalloproteinases

The membrane-type metalloproteinases (MT-MMPs 1, 2, 3, 4, 5 and MMPs 14, 15, 16, 17, 24) are bound to the cell membrane and have a role to play in activating other MMPs as well as degrading the ECM.

REGULATION OF MMPS

As mentioned previously, the overall synthesis and degradation of the components of ECM during wound healing is in balance. This balanced state has to be maintained, and subsequently regulation is necessary.

MMPs, including the three collagenases (MMP-1, MMP-8, and MMP-13) and MMP-2, -3, -9, -19, -26, and MMP14/MT1-MMP, have been shown to be expressed at many different wound sites during healing. By blocking the activity of a number of these enzymes, wound healing has been shown to be stimulated and also inhibited. This highlights the significance for their regulation.

MMP activity is regulated by enzymatic activation of zymogens, tissue inhibitors of metalloproteinases (TIMPs), and control of MMP genes.[21,23] In addition, MMPs have been shown to be regulated by binding onto plasma proteinase inhibitors, alpha 1-proteinase inhibitor or alpha 2 macroflobulins, and thrombospondin-1 and -2.[24] A number of other mechanisms have been reported.[25] One particular mechanism has included modification of MMPs by reactive oxygen species.[26] Furthermore, numerous papers have studied and reported on the regulation of MMPs at the genetic level.[23,27] MMP regulation is an active area of intensive research in wound care.

MMP regulatory mechanisms are discussed below.

Transcriptional Level Regulation

MMP gene expression is regulated at the transcription level in a cell by agents such as cytokines, chemokines, and growth factors. Tumor necrosis factor (TNF)-alpha, platelet-derived growth factor (PDGF), epidermal growth factor (EGF), and interleukin (IL)-1 are known to stimulate expression of numerous MMPs, but their expression is often dependent on the activating protein 1 (AP-1) binding site which constitutes a regulatory element of many of the MMPs. A number of MMPs are under secretory control.

Zymogen Activation

Generally, as mentioned previously, MMPs are released within their inactive forms called zymogens. Activation of these zymogens occurs outside of the cell they are produced from. Consequently, many MMPs have a lag or latency period prior to activation. They are dependent on the "cysteine switch" that occurs when cysteine interacts with zinc in the catalytic site which blocks the access of the substrate to its active site.[28]

The catalytic site of the enzyme is activated when the covalent bond between the cysteine and the catalytic zinc becomes dissociated. Numerous enzymes including plasmin, trypsin, and kallikrein convert the proenzyme into an intermediate active form that then autocatalytically cleaves itself to form an active form. The Furin-like prohormone convertage cleavage site, found in the MMP domain site, can be cleaved so that the enzyme can be activated.

TISSUE INHIBITORS OF METALLOPROTEINASES (TIMPs)

The levels and duration of activity of MMPs are generally carefully controlled by TIMPs. TIMPs together with plasmin activators help to control levels of MMPs.

TIMPs are a group of secretory proteins that inhibit MMPs by noncovalent binding onto the pre or active forms of MMP. All TIMPs are able to inhibit all MMPs except MT1-MMP. In vertebrates four TIMPs have been identified—namely, TIMP-1,-2,-3, and -4.

TIMPs contain chelating agents such as hydroxamic acid, a carboxylate, or thiol groups that enable the removal of the zinc ions from the MMPs, rendering them ineffective.

TIMPs inhibit MMPs in a 1:1 inhibitor-to-enzyme ratio through interaction of the N-terminal domain of the TIMP molecule with the active site of the MMP.[29,30] They coordinate the catalytic site Zn^{2+} and bind to the active site in a similar fashion to an MMP substrate.[29] A disintegrin and metalloproteinase (ADAM) are inhibited by TIMPs, although this inhibition is regulated primarily by TIMP-3.[30–32] TIMPs also function to regulate aspects of cell migration, apparently by restraining the activity of specific MMPs.

TIMP -1 is a glycoprotein that is very effective at degrading MMPs. In addition, they are known to promote cell proliferation in an array of different cell types and are thought to have anti-apoptotic qualities. Transcription of this gene occurs in response to many hormones and cytokines.[33] TIMP-1 is induced during reepithelialization, and excess levels of this inhibitor may contribute to the impaired epithelial cell migration observed in chronic wounds.[34] TIMP 1 is also known to be up-regulated by cytokines and many growth factors including IL-1beta and TGF-beta. TNF is also known to enhance the production of TIMP 1 when present at low concentrations. However, at high concentrations the opposite occurs.[35]

TIMP-2 is very effective in degrading MMPs but also has a role to play in suppressing the proliferation of endothelial cells. It is thought these TIMPs may have a specific role to play in returning the tissue ECM to a homeostatic state by suppressing proliferation of tissue growth factors.[33] TIMP-2 are found at lower levels than normal in ulcers that are non-healing.[35–37]

Both TIMPs 1 and 2 inhibit all MMPs but TIMP 1 specifically inhibits the activity of collagenase 1 and TIMP 2 specifically inhibits gelatinases A and B.

TIMP 3 is bound to the ECM and functions to inhibit TNF-α converting enzyme. It also functions to inhibit collagenases 1, gelatinases A and B, matrilysin, and stromelysin 1 but has a high specificity for gelatinase A and matrilysin. TIMP 3 is regulated by cytokines and a number of other mechanisms.

TIMP-4 is able to inhibit gelatinases A and B, stromelysin 1, matrilysin, and collagenase but has been reported to have the greatest efficiency against matrilysin and gelatinase A.

In addition to functioning as MMP regulators, TIMPs have also been reported to stimulate growth of keratinocytes and fibroblasts. Furthermore, TIMP-1 and -3 have been reported to inhibit angiogenesis.

OVERALL MMPS IN CHRONIC WOUNDS

Trengove and colleagues[38] have found that the activities of MMPs in chronic wound fluid are 30 times higher when compared to levels in acute wounds. The authors also have reported from their studies that levels of proteases decrease substantially in patients with venous ulcers that have shown signs of healing.[38] This has been supported by findings in a number of other studies.[39–42]

In normal intact skin, low levels of MMPs are expressed. MMP-7 and MMP-19 have been detected in both sweat and subaceous glands of intact skin.[43,44] However, if a skin injury occurs, multiple MMPs are produced. MMPs that have been detected have included collagenases (MMP-1 and MMP-8),[36,45] gelatinases (MMP-2 and MMP-9),[42] stromelysins (MMP-3 and MMP-10),[33,46] metalloelastase (MMP-12),[47] MT1-MMP,[42] MMP-19,[48] MMP-26,[49] and MMP-28.[50] In addition, TIMP-1, -2, and -3, but not -4, have been reported in acute skin wounds.[43]

In a number of studies, MMP-2 and -9 have been predominately reported to be the most significant to wound healing.[42,46,51] In diabetic foot ulcers, for example, high levels of these two gelatinases have been detected.[40] MMP-2 and MMP-9 are thought to be involved in ECM remodeling and are present for a long period of time during the wound healing process.[52] MMP-2 has been reported to activate TGF-β[48,49,53–55] and TNF-alpha. MMP-9 has also been shown to digest fibrin.[56] Because of its many attributes, MMP-9 is thought to be involved in the remodeling of the wound bed together with MT1-MMP and MMP-2 assist in angiogenesis.[57–59,60]

MMP-1 is produced during the early stages of wound healing, particularly when the basement membrane is affected. Levels of MMP-1 are known to subside after reepithelialization of skin but have an important role to play in wound reepithelialization, as overexpression has been found to result in delayed closure of a full thickness wound.[51] Together with MMP-13, MMP-1 is thought to help regulate the survival of fibroblasts in human fetal cutaneous wounds.[61] However, a number of other studies have shown that these MMPs do not alter the wound healing cascade.[62]

Collagenase 1 has been studied in great detail during wound repair. It is expressed by keratinocytes in many chronic wounds and is known to be stimulated by many growth factors. Many chronic wound fluids have been shown to have elevated levels of MMP-2, -9, -1, and -8 and are detected on a regular basis.[63–66]

Collagenase-2 (MMP-8) is produced specifically by neutrophils and secreted during their activation.[67] The biological functions of MMP-8 are vague and it is thought they are involved in tissue remodeling during the inflammatory stage of wound healing. In many excision wounds, MMP-8 has been found to be very abundant.[35] They are the main wound collagenases and can be found in both healing and nonhealing wounds. A number of studies have analyzed MMP expression patterns in many wounds and have shown an increased production of MMP-8. It is therefore plausible to suggest that MMP-8 participates in the process of wound healing. However, little information is available regarding the exact function of this enzyme in cutaneous wound repair.

MMP-3 and MMP-10 (stromelysins) are produced by keratinocytes and are evident during wound healing. MMP-3 is also produced by fibroblasts,[39,47] and its role is believed to be in the development of the basement membrane, cell migration, and proliferation of keratinocytes.

MMP-10 is found to be evident following a few days after the formation of a wound and its levels are reported to be regulated by cytokines. MMP-10 is thought to regulate both the organization and migration of keratinocyte, specifically during reepithelialization of skin.

MMP-19 has been found in wounds undergoing repair and found in epithelium cells, fibroblasts and macrophages and the metalloelastase.

MMP-26 (matrilsin-2, endometase) and MMP-28 (epilysin) are also found during wound healing. In normal wound healing, MMPs are controlled effectively and so damaged tissue is removed to allow for the development of new tissue. Correct regulation of MMPs is necessary for normal wound healing, so if the balance between MMPs and TIMPs is disrupted then a nonhealing wound may develop.[68–70]

In diabetic foot ulcers, for example, high levels of MMP-1, -2, -8, -9, -14 with low levels of TIMP-2 are common when compared to normal wounds. In particular, high levels of MMP-9 have been found to correlate to poor healing in venous wounds. In addition, numerous studies have shown that elevated levels of MMP-9, MMP-8, MMP-3, and MMP-1 are detected in chronic wounds and significant to chronic wound healing.

FACTORS THAT STIMULATE MMPS

Activated inflammatory cells stimulate MMP production. These have also been shown to suppress TIMPs. TNF-α and IL-1β in particular impair the healing process via increased degradation of the ECM components, growth factors, and receptors contributing to multiple negative feedback loops preventing wound closure.[71,72] In addition, a number of MMPs have been reported to activate each other and therefore stimulate overall enzymatic activity (see Figure 9.3).[73]

Bacteria have been shown to be very important stimulators of MMPs in wounds. For example, a wide range of proteases of the thermolysin family are secreted by *Pseudomonas aeruginosa* and *Vibrio cholera* which activate pro-MMP-1, -8, and -9.

BACTERIA, BIOFILMS, AND MMPS

Microbial proteases have been documented to be detrimental to wound healing. For example, bacerial enzymes have affected the host defenses, caused tissue breakdown

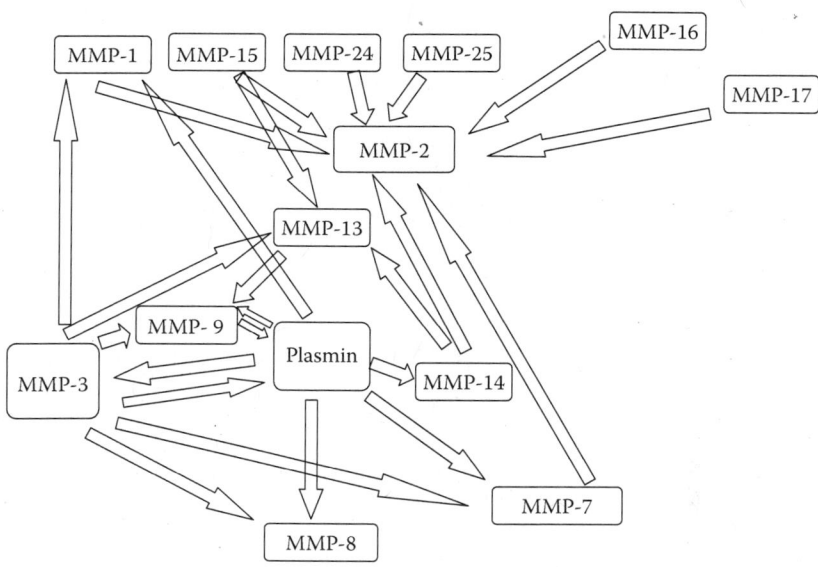

FIGURE 9.3 Mutual activation of matrix metalloproteinases (MMPs). (Coussens, L.M., and Werb, Z. (2002) Inflammation and cancer. *Nature* 420:860–867. With permission)

and destruction, and induced both inflammation and infection at the wound site (seen in Figure 9.4).[74] Proteases produced by bacteria are unaffected by many of the host's defenses as these enzymes effectively are able to degrade them. *Pseudomonas, Clostridium, Bacteroides,* and *Staphylococcus* spp. are also able to produce concentrations of enzymes at a tissue site which induces the production of phagocytic cells. The subsequent release of proteolytic enzymes produced from neutrophils which are used to kill bacteria causes tissue degradation which in turn is thought to up-regulate bacterial proliferation.

Microbial proteases have been associated with many infections, specifically periodontal disease.[75] For example, elastase produced from *P. aeruginosa* has been shown to be a cause of corneal ulcers and skin lesions. Hyaluronan, which is a component of the ECM, expressed by keratinocytes[76] and known to have a role in supporting cell proliferation and migration during inflammation,[77] can be broken down by bacteria that produce hyaluronidase. Hyaluronidase enables bacteria, and their toxins, to spread from a site of infection to other sites in the host, increasing the risk of a systemic infection. Numerous studies have shown that hyaluronidase is present at many sites of infection,[78–80] and constitutes a significant role in wound healing. Many strains of bacteria are known to produce hyaluronidase.

Many metallocysteine and serine proteases are produced by an array of common wound opportunistic pathogens and are capable of inflicting tissue destruction on the host by degrading collagen, elastin, and fibronectin resulting in tissue destruction. In addition, many bacterial proteolytic enzymes can cause a reduction in neutrophil recruitment to a site of infection.[81–86]

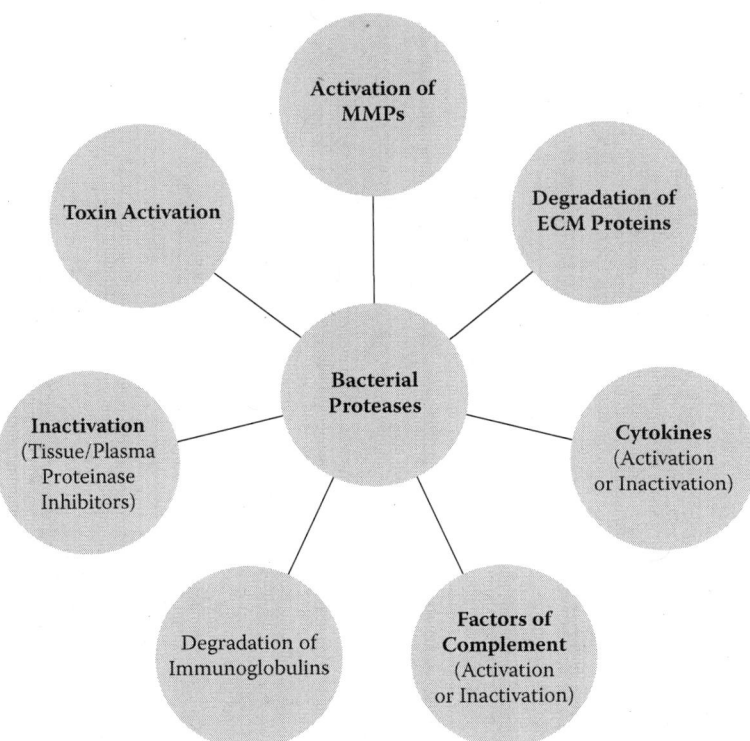

FIGURE 9.4 Functions of bacterial proteinases. (Adapted from Chakraborti et al. (2003) *Mol Cell Biochem* 253:269–285. With permission)

Aerobic Bacteria

TIMPs and other inhibitors of human MMPs are susceptible to inactivation by bacterial proteases.[86–88] Many proteinases produced by bacteria have, in general, a weak degradative activity to collagen;[86] however, many others are known to significantly cause destruction of the ECM of a wound.[87] However, the mechanisms by which bacteria cause destruction of the ECM is poorly understood.

Pseudomonas aeruginosa are significant in chronic wounds and are known to produce many proteolytic enzymes. These include elastase A (Las A), elastase B (LasB), protease IV, and alkaline protease.[89] Elevated levels of elastase have been shown to induce degradation of wound fluid and human skin proteins during infection and have been shown to degrade complement, fibroblast proteins, and proteoglycan decorin, and inhibit fibroblast growth.[89,90]

Alkaline proteases have been found in the sputum of cystic fibrosis patients, and these have been associated with poor clinical outcome.[91,92] Oldak and Trafney[93] concluded from their studies that *P. aeruginosa* are highly resistant to antimicrobials if they secrete proteases, and secretion of these enzymes is not attenuated even when exposed to antibiotics.

Although it is generally accepted that ECM destruction is brought about by human MMPs, it is very important to understand that indirect mechanisms by pathogen-elicited proteinases are significantly involved in causing damage to the ECM. This includes activating human MMPs, as shown with *P. aeruginosa*; plasminogen binding and activation, as shown with *Streptococcus* sp. and *Staphylococcus* sp., inactivation of human TIMPs causing overproduction of MMPs, and transient activation of chemotactic pathways recruiting phagocytic cells at the site of infection.[93–96] However, a number of *Clostridium* sp., *P. aeruginosa*, *Serratia marcescens*, *S. epidermidis* directly degrade connective tissue.[96,97] Bacterial proteinases participate in the ECM destruction by activating proMMPs. *Pseudomonas aeruginosa* elastase strongly activates proMMPs to generate active forms of MMPs.[87] Many bacteria produce metalloenzymes.[95] Some of these enzymes are membrane bound and not secreted. *Staphylococcus aureus* in chronic leg ulcers have been shown to up-regulate fibroblast expression of MMPs, in particular MMP-1, -2, -3, -7, -10, -11, and -13. They have also been shown to up-regulate TIMP-1 and -2.[98]

ANAEROBIC BACTERIA

Many anaerobic bacteria produce hydrolytic enzymes that have a role to play in tissue destruction. For example, bacteria such as *Bacteroides melaninogenicus*, *Fusobacterium fusiforme*, and *Clostridium histolyticum* produce many collagenases.[99–101]

Fibrinolysin is another proteolytic enzyme produced by anaerobic bacteria. This enzyme is significant during the formation of a fibrin clot at a wound surface. This clot can be broken down and may result in a delay in the wound healing process. In addition, anaerobic bacteria produce enzymes that degrade hyaluronic acid and chondriotin sulfate. These enzymes are very important to delaying wound healing as they help to enhance spread of bacteria through tissue. *Clostridium* sp produce elastase and hyaluronidase which are important enzymes causing tissue breakdown, and many species of *Bacteroides*, *B. fragilis,* and *B. vulgatus* produce the enzymes neuraminidase, proteases, and lipases.

Compounds that colonizing bacteria produce in a chronic wound are known to enhance the expression of MMPs from fibroblasts. Consequently, in wound care management, it is important to reduce inflammation caused by the bacteria or biofilm particularly during early chronic wound development. Oldak and Trafny[102] have shown that *P. aeruginosa* in biofilms secrete very active proteases which allows them to survive in human tissue during a chronic infection.

MANAGEMENT OF MMPS

As mentioned, MMPs are required for normal wound healing. However, within chronic wounds, levels of MMPs are often found to be very high and generally uncontrolled so that they prolong the wound healing process, or prevent wound closure. To date, the underlying reasons as to why in chronic wounds certain MMPs, in particular the gelatinases, are expressed in high levels is unclear. Despite this, the general view is that by regulating MMPs in wounds this will help enhance the wound healing process.

The high levels of MMPs evident in chronic wounds seem to be due to the increased infiltration of leukocytes known to secrete high levels of MMP-8 and gelatinases.[35] By utilizing agents or compounds that absorb excessive MMPs in a chronic wound, this has been shown to have positive effects on wound healing.[103–105] In addition, studies have looked at ways of neutralizing high levels of protease utilizing doxycycline.[105] It is probable that by understanding how to regulate protease expression, this may aid in the development of specific therapeutic methods.[107]

It is well documented that collagenases from bacteria debride necrotic tissue and break down living tissue.[107] Therefore, inactivation of microbial enzymes would help significantly to reduce bacterial pathogenicity and therefore reduce the risk of an infection in a wound. Consequently, downregulation of bacterial enzymes is very important, particularly as their activity can enhance human MMP production and synergistically enhance the detrimental effects of human proteinases on healthy human tissue. High levels of enzymes produced by neutrophils, need also to be addressed as these will enhance further tissue breakdown in a wound.

CONCLUSION

Generally there is a basic understanding of the role of human MMPs and other proteases play in wound healing. However, there is very little documented data on the role bacterial proteases play in chronic wound healing. Human MMPs and bacterial enzymes, which include MMPs, act synergistically and have an additive effect in tissue breakdown. Schmidtchen and colleagues[90] have shown that elastase produced by *P. aeruginosa* induces degradation of fibroblast proteins and proteoglycans and mimicing proteolytic activity of human MMPs. They have been shown to degrade skin proteins and inhibit fibroblast cell growth. It has also been found that 50% of all chronic ulcers colonized with *P. aeruginosa* have shown that these bacteria express elastase.[109] Based on this and other pieces of evidence, it is clear that bacterial proteinases have a significant role to play in both infected and nonhealing chronic wounds. Travis and colleagues[75] have reported that "there appears to be little, if any, structural relationship between prokaryotic and eukaryotic proteinase," including their reactive sites. As suggested by this group, inhibitors could be exploited to be more effective on bacterial proteinases compared to host proteinases for certain desired results in disease and infection. One of these areas could include chronic wounds.

REFERENCES

1. Lund, L.R., Green, K.A., Stoop, A.A., Ploug, M., Almholt, K., Lilla, J., Nielsen, B.S., Christensen, I.J., Craik, C.S., Werb, Z., Dano, K., and Romer, J. (2006) Plasminogen activation independent of uPA and tPA maintains wound healing in gene-deficient mice. *EMBO J* 25:2686–2697.
2. Lund, L.R., Romer, J., Bugge, T.H., Nielsen, B.S., Frandsen, T.L., Degen, J.L., Stephens, R.W., and Dano, K. (1999) Functional overlap between two classes of matrix-degrading proteases in wound healing. *EMBO J* 18:4645–4656.

3. Smith, J.A. (1994) Neutrophils, host defences, inflammation: A double-edged sword. *J Leukoc Biol* 56:672–686.

4. Garner, W.L.. Rodriguez, J.L. and Miller, C.G. et al. (1994) Acute skin injury releases neutophil chemoattractants. *Surgery* 116:42–48.

5. Dipietro, L.A. (1995) Wound healing: The role of the macrophages and other immune cells. *Shock* 4:233–240.

6. Gross, J., and Lapiere, C. (1962). Collagenolytic activity in amphibian tissues: A tissue culture assay. *Proc Natl Acad Sci USA* 48:1014–1022.

7. Eisen, A., Jeffrey, J., and Gross, J. (1968) Human skin collagenase. Isolation and mechanism of attack on the collagen molecule. *Biochim Biophys Acta* 151(3):637–645.

8. Rawlings, N.D., Morton, F.R., and Barrett, A.J. (2006) MEROPS: The peptidase database. *Nucleic Acids Res* 34: D270–272.

9. Snyder, R.J. (2005) A holistic approach to understanding and addressing the wound microenvironment to facilitate healing. *Wounds* 17:12–17.

10. Armstrong, D.G., and Jude, E.B. (2002) The role of matrix metalloproteinases in wound healing. *J Am Podiatr Med Assoc* 92(1):12–18.

11. Parks, W.C. (1999) Matrix metalloproteinases in repair. *Wound Repair Regen* 7(6):423–432.

12. Gutiérrez-Fernández, A., Inada, M., Balbín, M., Fueyo, A., Pitiot, A.S., Astudillo, A., Hirose, K., Hirata, M., Shapiro, S.D., Noël, A., Werb, Z., Krane, S.M., López-Otín, C., and Puente, X.S. (2007) Increased inflammation delays wound healing in mice deficient in collagenase-2 (MMP-8). *FASEB J* 21(10):2580–2591.

13. Lafuma, C., El Nabout, R., Crechet, F., Hovanian, A., and Martin, M. (1994) Expression of 72-kDa gelatinase (MMP-2), collagenase (MMP-1), and tissue metalloproteinase inhibitor (TIMP) in primary pig skin fibroblast cultures derived from radiation-induced skin fibrosis. *J Invest Derm* 102:945–950.

14. Mirastschijski, U., Haaksma, C.J., Tomasek, J.J., and Agren, M.S. (2004) Matrix metalloproteinase inhibitor GM 6001 attenuates keratinocyte migration, contraction and myofibroblast formation in skin wounds. *Exp Cell Res* 299(2):465–475.

15. Salonurmi, T., Parikka, M., Kontusaari, S., Pirila, E., Munaut, C., Salo, T., and Tryggvason, K. (2004) Overexpression of TIMP-1 under the MMP-9 promoter interferes with wound healing in transgenic mice. *Cell Tissue Res* 315:27–37.

16. Cochrane, C.A. (1997) Models in vivo of wound healing in the horse and the role of growth factors. *Vet Dermatol* (4):259–272.

17. Parks, W.C., Wilson, C.L., and Lopez-Boado, Y.S. (2004) Matrix metalloproteinases as modulators of inflammation and innate immunity. *Nat Rev Immunol* 4:617–629.

18. Overall, C.M., and López-Otín, C. (2002) Strategies for MMP inhibition in cancer: Innovations for the post-trial era. *Nat Rev Cancer* 2:657–672.

19. Folgueras, A.R., Pendás, A.M., Sánchez, L.M., and López-Otín, C. (2004) Matrix metalloproteinases in cancer: From new functions to improved inhibition strategies. *Int J Dev Biol* 48:411–424.

20. Van Lint, P., and Libert, C. (2007). Chemokine and cytokine processing by matrix metalloproteinases and its effect on leukocyte migration and inflammation. *J Leukoc Biol* 82(6):1375–1381.

21. Toriseva, M., and Kahari, V.M. (2009) Proteinases in cutaneous wound healing. *Cell Mol Life Sci* 66:203–224.

22. Parks, W.C. (1999) Matrix metalloproteinases in repair. *Wound Repair Regen* 7:423–432.

23. Nagase, H., Visse, R., and Murphy, G. (2006) Structure and function of matrix metalloprotienases and TIMPs. *Cardiovasc Res* 69:562–573.

24. Baker, A.H., Edwards, D.R., and Murphy, G. (2002) Metalloproteinase inhibitors: Biological actions and therapeutic opportunities. *J Cell Sci* 115:3719–3727.

25. Emonard, H., Bellon, G., de Diesbach, P., Mettlen, M., Hornbeck, W., and Courtoy, P.J. (2005) Regulation of matrix metalloproteinase (MMP) activity by the low-density lipoprotein receptor-related protein (LRP). A new function for an "old friend." *Biochimie* 87:369–376.

26. Fu, X., Parks, W.C., and Heinecke, J.W. (2008) Activation and silencing of matrix metalloproteinases. *Semin. Cell Dev Bio* 19:2–13.

27. Baker, E.A., and Leaper, D.J. (2000) Proteinases, their inhibitors, and cytokine profiles in acute wound fluid. *Wound Repair Regen* 8(5):392–398.

28. Van Wart, H., and Birkedal-Hansen, H. (1990). The cysteine switch: A principle of regulation of metalloproteinase activity with potential applicability to the entire matrix metalloproteinase gene family. *Proc Natl Acad Sci USA* 87(14):5578–5582.

29. Brew, K., Dinakarpandian, D., and Nagase, H. (2000) Tissue inhibitors of metalloproteinases: Evolution, structure and function. *Biochim Biophys Acta* 1477:267–283.

30. Amour, A., Knight, C.G., Webster, A., Slocombe, P.M., Stephens, P.E., Knauper, V., et al. (2000) The *in vitro* activity of ADAM-10 is inhibited by TIMP-1 and TIMP-3. *FEBS Lett* 473(3):275–279.

31. Nwomeh, B.C., Liang, H.X., Cohen, I.K., and Yager, D.R. (1999) MMP-8 is the predominant collagenase in healing wounds and nonhealing ulcers. *J Surg Res* 81(2):189–195.

32. Loechel, F., Fox, J.W., Murphy, G., Albrechtsen, R., and Wewer, U.M. (2000) ADAM 12-S cleaves IGFBP-3 and IGFBP-5 and is inhibited by TIMP-3. *Biochem Biophys Res Commun* 278(3):511–515.

33. Hornebeck, W. (2004). Down-regulation of tissue inhibitor of matrix metalloprotease-1 (TIMP-1) in aged human skin contributes to matrix degradation and impaired cell growth and survival. *Pathol Biol* 51(10):569–573.

34. Nwomeh, B.C., Liang, H.X., Cohen, I.K., and Yager, D.R. (1999) MMP-8 is the predominant collagenase in healing wounds and nonhealing ulcers. *J Surg Res* 81(2):189–195.

35. Ito, A., Sato, T., Iga, T., and Mori, Y. (1990) Tumor necrosis factor bifunctionally regulates matrix metalloproteinases and tissue inhibitor of metalloproteinases (TIMP) production by human fibroblasts. *FEBS Lett* 269:93–95.

36. Vaalamo, M., Leivo, T., and Saarialho-Kere, U. (1999) Differential expression of tissue inhibitors of metalloproteinases (TIMP-1, -2, -3, and -4) in normal and aberrant wound healing. *Hum Pathol* 30(7):795–802.

37. Bullen, E.C., Longaker, M.T., Updike, D.L., Benton, R., Ladin, D., Hou, Z., and Howard, E.W. (1995) Tissue inhibitor of metalloproteinases-1 is decreased and activated gelatinases are increased in chronic wounds. *J Invest Dermatol* 104(2):236–240.

38. Trengove, N.J., Stacey, M.C., MacAuley, S., Bennett, N., Gibson, J., Burslem, F., Murphy, G., and Schultz, G. (1999) Analysis of the acute and chronic wound environments: The role of proteases and their inhibitors. *Wound Repair Regen* 7(6):442–452.

39. Rogers, A.A., Burnett, S., Moore, J.C., Shakespeare, P.G., and Chen, W.Y.J. (1995) Involvement of proteolytic enzymes—plasminogen activators and matric metalloproteinases—in the pathophysiology of pressure ulcers. *Wound Rep Reg* 3:273–283.

40. Wysocki, A., Staiano-Coico, L., and Grinnell, F. (1993) Wound fluid from chronic leg ulcers contains elevated levels of metalloproteinases MMP-2 and MMP-9. *J Invest Derm* 101:64–68.

41. Yager, D.R., Zhang, L.Y., Liang, H.X., Diegelmann, R.F., and Cohen, I.K. (1996) Wound fluids from human pressure ulcers contain elevated matrix metalloproteinase levels and activity compared to surgical wound fluids. *J Invest Derm* 107:743–748.

42. Mirastschijski, U., Impola, U., Jahkola, T., Karlsmark, T., AGren, M.S., and Saarialho-Kere, U. (2002) Ectopic localization of matrix metalloproteinase-9 in chronic cutaneous wounds. *Hum Pathol* 33(3):355–364.

43. Sadowski, T., Dietrich, S., Muller, M., Havlickova, B., Schunck., M., Proksch, E., Muller, M.S., and Sedlacekm, R. (2003) Matrix metalloproteinase-19 expression in normal and diseased skin: Dysregulation by epidermal proliferation. *J Invest Dermatol* 121:989–996.

44. Saarialho-Kere, U.K., Vaalamo, M., Airola, K., Niemi, K.M., Oikarinen, A.I., and Parks, W.C. (1995) Interstitial collagenase is expressed by keratinocytes that are actively involved in reepithelialization in blistering skin disease. *J Invest Dermatol* 104(6):982–988.

45. Inoue, M., Kratz, G., Haegerstrand, A., and Stahle-Backdahl, M. (1995) Collagenase expression is rapidly induced in wound-edge keratinocytes after acute injury in human skin, persists during healing, and stops at re-epithelisation. *J Invest Dermatol* 104:479–483.

46. Vaalamo, M., Weckroth, M., Puolakkainen, P., Kere, J., Saarinen, P., Lauharata, J., and Saarialho-Kere, U.K. (1996) Patterns of matrix metalloproteinase and TIMP-1 expression in chronic and normally healing human cutaneous wounds. *Br J Dermatol* 135:52–59.

47. Madlener, M., Parks, W.C., and Werner, S. (1998) Matrix metalloproteinases (MMPs) and their physiological inhibitors (TIMPs) are differentially expressed during excisional skin wound repair. *Exp Cell Res* 242(1):201–210.

48. Hieta, N., Impola, U., Lopez-Otin, C., Saarialho-Kere, U., and Kahari, V.-M. (2003) Matrix metalloproteinase-19 expression in dermal wounds and by fibroblasts in culture. *J Invest Dermatol* 121:997–1004.

49. Ahokas, K., Skoog, T., Suomela, S., Jeskanen, L., Impola, U., Isaka, K., and Saarialho-Kere, U. (2005) Matrilysin-2 (matrix metalloproteinase-26) is upregulated in keratiniocytes during wound repair and early skin carcinogenesis. *J Invest Dermatol* 124:849–856.

50. Lohi., J., Wilson, C.L., Roby, J.D., and Parks, W.C. (2001) Epilysin, a novel human matrix metalloproteinase (MMP-28) expressed in testis and keratinocytes and in response to injury. *J Biol Chem* 276:10134–10144.

51. Rechardt, O., Elomaa, O., Vaalamo, M., Paakkonen, K., Jahkola, T., Hook-Nikanne, J., Hembry, R.M., Hakkinen, L., Kere, J., and Saarialho-Kere, U. (2000) Stromelysin-2 is upregulated during normal repair and is induced by cytokines. *J Invest Dermatol* 115:778–787.

52. Tammi, M.I., Day, A.J., and Turley, E.A. (2002) Hyaluronan and homeostasis: A balancing act. *J Biol Chem* 277:4581–4584.

53. Yu, Q., and Stamenkovic, I. (2000) Cell surface-localized matrix metalloproteinase-9 proteolytically activates TGF-beta and promotes tumor invasion and angiogenesis. *Genes Dev* 14(2):163–176,

54. Dallas, S.L., Rosser, J.L., Mundy, G.R., and Bonewald, L.F. (2002) Proteolysis of latent transforming growth factor-beta (TGF-beta)-binding protein-1 by osteoclasts. A cellular mechanism for release of TGF-beta from bone matrix. *J Biol Chem* 277(24):21352–21360.

55. Imai, K., Hiramatsu, A., Fukushima, D., Pierschbacher, M.D., and Okada, Y. (1997) Degradation of decorin by matrix metalloproteinases: Identification of the cleavage sites, kinetic analyses and transforming growth factor-beta1 release. *Biochem J* 322(3):809–814.

56. Lelongt, B., Bengatta, S., Delauche, M., Lund, L.R., Werb, Z., and Ronco, P.M. (2001) Matrix metalloproteinase 9 protects mice from anti-glomerular basement membrane nephritis through its fibrinolytic activity. *J Exp Med* 193(7):793–802.

57. Kato, T., Kure, T., Chang, J.H., Gabison, E.E., Itoh, T., Itohara, S., and Azar, D.T. (2001) Diminished corneal angiogenesis in gelatinase A-deficient mice. *FEBS Lett* 508(2):187–190.

58. Hiraoka, N., Allen, E., Apel, I.J., Gyetko, M.R., and Weiss, S.J. (1998) Matrix metalloproteinases regulate neovascularization by acting as pericellular fibrinolysins. *Cell* 95(3):365–377.

59. Chun, T.H., Sabeh, F., Ota, I., Murphy, H., McDonagh, K.T., Holmbeck, K., Birkedal-Hansen, H., Allen, E.D., and Weiss, S.J. (2004) MT1-MMP-dependent neovessel formation within the confines of the three-dimensional extracellular matrix. *J Cell Biol* 167(4):757–767.

60. Pins, G.D., Collins-Pavao, M.E., Van De Water, L., Yarmush, M.L., and Morgan, J.R. (2000) Plasmin triggers rapid contraction and degradation of fibroblast-populated collagen lattices. *J Invest Dermatol* 114(4):647–653.

61. Ravanti, L., Toriseva, M., Penttinen, R., Crombleholme, T., Foschi, M., Han, J., and Kähäri, V.M. (2001) Expression of human collagenase-3 (MMP-13) by fetal skin fibroblasts is induced by transforming growth factor beta via p38 mitogen-activated protein kinase. *FASEB J* 15(6):1098–1100.

62. Hartenstein, B., Dittrich, B.T., Stickens, D., Heyer, B., Vu, T.H., Teurich, S., Schorpp-Kistner, M., Werb, Z., and Angel, P. (2006) Epidermal development and wound healing in matrix metallo-proteinase 13-deficient mice. *J Invest Dermatol* 126:486–496.

63. Vaalamo, M., Mattila, L., Johansson, N., Kariniemi, A.L., Karjalainen-Lindsberg, M.L., Kähäri, V.M., and Saarialho-Kere, U. (1997) Distinct populations of stromal cells express collagenase-3 (MMP-13) and collagenase-1 (MMP-1) in chronic ulcers but not in normally healing wounds. *J Invest Dermatol* 109(1):96–101.

64. Saarialho-Kere, U.K., Vaalamo, M., Airola, K., Niemi, K.M., Oikarinen, A.I., and Parks, W.C. (1995) Interstitial collagenase is expressed by keratinocytes that are actively involved in reepithelialization in blistering skin disease. *J Invest Dermatol* 104(6):982–988.

65. Sadowski, T., Dietrich, S., Muller, M., Havlickova, B., Schunck., M., Proksch, E., Muller, M.S., and Sedlacekm, R. (2003) Matrix metalloproteinase-19 expression in normal and diseased skin: Dysregulation by epidermal proliferation. *J Invest Dermatol* 121:989–996.

66. Inoue, M., Kratz, G., Haegerstrand, A., and Stahle-Backdahl, M. (1995) Collagenase expression is rapidly induced in wound-edge keratinocytes after acute injury in human skin, persists during healing, and stops at re-epithelisation. *J Invest Dermatol* 104:479–483.

67. Hasty, K.A., Hibbs, M.S., Kang, A.H., and Mainardi, C.L. (1986) Secreted forms of human neutrophil collagenase. *J Biol Chem* 261(12):5645–5650.

68. McQuibban, G.A., Gong, J.H., Wong, J.P., Wallace, J.L., Clark-Lewis, I., and Overall, C.M. (2002) Matrix metalloproteinase processing of monocyte chemoattractant proteins generates CC chemokine receptor antagonists with anti-inflammatory properties in vivo. *Blood* 100(4):1160–1167.

69. Van den Steen, P.E., Proost, P., Wuyts, A., Van Damme, J., and Opdenakker, G. (2000) Neutrophil gelatinase B potentiates interleukin-8 tenfold by aminoterminal processing, whereas it degrades CTAP-III, PF-4, and GRO-alpha and leaves RANTES and MCP-2 intact. *Blood* 96(8):2673–2681.

70. Ladwig, G.P., Robson, M.C., Liu, R., Kuhn, M.A., Muir, D.F., and Schultz, G.S. (2002) Ratios of activated matrix metalloproteinase-9 to tissue inhibitor of matrix metalloproteinase-1 in wound fluids are inversely correlated with healing of pressure ulcers. *Wound Repair Regen* 10(1):26–37.

72. Puente, X.S., Sánchez, L.M., Overall, C.M., and López-Otín, C. (2003) Human and mouse proteases: A comparative genomic approach. *Nat Rev Genet* 4:544–558.

73. Coussens, L.M., and Werb, Z. (2002) Inflammation and cancer. *Nature* 420:860–867.

74. Chakraborti, S., Mandal, M., Das, S. et al. (2003) Regulation of matrix metalloproteinases: An overview. *Mol Cell Biochem* 253:269–285.
75. Travis, J., Potempa, J., and Maeda, H. (1995) Are bacterial proteinases pathogenic factors. *Trends Microbiol* 3:405–407.
76. Bretz, W.A., Lopatin, D.E., and Loesche, W.J. (1990) Benzoyl-arginine naphthylamide (BANA) hydrolysis by *Treponema denticola* and/or *Bacteroides gingivalis* in peridontal plaques. *Oral Microbiol Immunol* 5:275–279.
77. Agren, U.M., Tammi, M., Ryynänen, M., and Tammi, R. (1997) Developmentally programmed expression of hyaluronan in human skin and its appendages. *J Invest Dermatol* 109(2):219–224.
78. Tammi, M.I., Day, A.J., and Turley, E.A. (2002) Hyaluronan and homeostasis: A balancing act. *J Biol Chem* 277:4581–4584.
79. Hashioka, K., Suzuki, K., Yoshida, T., Nakane, A., Horiba, N., and Nakamura, H. (1994) Relationship between clinical symptoms and enzyme-producing bacteria isolated from infected root canals. *J Endod* 20:75–77.
80. Takao, A., Nagashima, H., Usui, H., Sasaki, F., Maeda, N., Ishibashi, K., and Fujita, H. (1997) Hyaluronidase activity in human pus from which *Streptococcus intermedius* was isolated. *Microbiol Immunol* 41(10):795–798.
81. Unsworth, P.F. (1989) Hyaluronidase production in *Streptococcus milleri* in relation to infection. *J Clin Pathol* 42:506–510.
82. Oda, T., Kojima, Y., Akaike, T., Ijiri, S., Molla, A., and Maeda, H. (1990) Inactivation of chemotactic activity of C5a by the serratial 56-kilodalton protease. *Infection Immun* 58:1269–1272.
83. Molla, A., Akaike, T., and Maeda, H. (1989) Inactivation of various proteinase inhibitors and the complement system in human plasma by the 56-kilodalton proteinase from *Serratia marcescens*. *Infect Immun* 57(6):1868–1871.
84. Molla, A., Matsumoto, K., Oyamada, I., Katsuki, T., and Maeda, H. (1986) Degradation of protease inhibitors, immunoglobulins, and other serum proteins by Serratia protease and its toxicity to fibroblast in culture. *Infect Immun* 53:522–529.
85. Bejarano, P.A., Langeveld, J.P., Hudson, B.G., and Noelken, M.E. (1989) Degradation of basement membranes by *Pseudomonas aeruginosa* elastase. *Infect Immun* 57: 3783–3787.
86. Sandholm, L. (1986) Proteases and their inhibitors in chronic inflammatory periodontal disease. *J Clin Periodontol* 13:19–26.
87. Weiss, S., and Regiani, S. (1984) Neutrophils degrade subendothelial matrices in the presence of alpha-1-proteinase inhibitor. Cooperative use of lysosomal proteinases and oxygen metabolites. *J Clin Invest* 73:1297–1303.
88. Okamoto, T., Akaike, T., Suga, M., Tanase, S., Horie, H., Miyaijima, S., Ando, M., Ichinose, Y., and Maeda, H. (1997) Activation of human matrix metalloproteinases by various bacterial proetinases. *J Biol Chem* 272:6059–6066.
89. Caballero, A.R., Moreau, J.M., Engel, L.S., Marquart, M.E., Hill, J.M., and O'Callaghan, R.J. (2001) *Pseudomonas aeruginosa* protease IV enzyme assays and comparison to other *Pseudomonas proteases*. *Anal Biochem* 290(2):330–337.
90. Schmidtchen, A., Holst, E., Tapper, H., and Bjorck, L. (2003) Elastase-producing degrade plasma proteins and extracellular products of human skin and fibroblasts, and inhibit fibroblast growth. *Microb Pathog* 34:47–55.
91. Döring, G., Obernesser, H.J., Botzenhart, K., Flehmig, B., Høiby, N., and Hofmann, A. (1983) Proteases of *Pseudomonas aeruginosa* in patients with cystic fibrosis. *J Infect Dis* 147(4):744–750.

92. Jaffar-Bandjee, M.C., Lazdunski, A., Bally, M., Carrere, J., Chazalette, P., and Galabert, C. (1995) Production of elastsase, exotoxin A, and alkaline protease in sputa during pulmonary exacerbation of cystic fibrosis in patients chronically infected by *Pseudomonas aeruginosa*. *J Clin Microbiol* 33:924–929.

93. Ołdak, E., and Trafny, E.A. (2005) Secretion of proteases by *Pseudomonas aeruginosa* biofilms exposed to ciprofloxacin. *Antimicrob Agents Chemother* 49(8):3281–3288.

94. Lottenberg, R., Minning-Wenz, D., and Boyle, M.D. (1994) Capturing host plasmin(ogen): A common mechanism for invasive pathogens? *Trends Microbiol* 2(1):20–24.

95. Travis, J., and Potempa, J. (2000) Bacterial proteinases as targets for the development of second-generation antibiotics. *Biochim Biophys Acta* 1477:35–50.

96. Harris, E.D. Jr., and Krane, S.M. (1974) Collagenases (first of three parts). *N Engl J Med* 291(11):557–563.

97. Häse, C.C., and Finkelstein, R.A. (1993) Bacterial extracellular zinc-containing metalloproteases. *Microbiol Rev* 57(4):823–837.

98. Kanangat, S., Postlethwaite, A., Hasty, K., Kang, A., Smeltzer, M., Appling, W., and Schaberg, D. (2006) Induction of multiple matrix metalloproteinases in human dermal and synovial fibroblasts by *Staphylococcus aureus*: Implications in the pathogenesis of septic arthritis and other soft tissue infections. *Arthritis Res Ther* 8(6):R176.

99. Gibbons, R.J., and Macdonald, J.B. (1961) Degradation of collagenous substrates by *Bacteroides melaninoge*. *J Bacteriol* 81:614–621.

100. Waldvogel, F.A., and Swartz, M.N. (1969) Collagenolytic activity of bacteria. *J Bacteriol* 98(2):662–667.

101. Kaufman, E.J., Mashimo, P.A., Hausmann, E., Hanks, C.T., and Ellison, S.A. (1972) Fusobacterial infection: Enhancement by cell free extracts of *Bacteroides melaninogenicus* possessing collagenolytic activity. *Arch Oral Biol* 17(3):577–580.

102. Oldak, E., and Trafny, E.A. (2005) Secretion of proteases by *Pseudomonas aeruginosa* biofilms exposed to ciprofloxacin. *Antimicrobial Agents Chemother* 49:3281–3288.

103. Veves, A., Sheehan, P., and Pham, H.T. (2002) A randomized, controlled trial of Promogran a collagen/oxidized regenerated cellulose dressing) vs standard treatment in the management of diabetic foot ulcers. *Arch Surg* 137(7):822–827.

104. Vin, F., Teot, L., and Meaume, S. (2002) The healing properties of Promogran in venous leg ulcers. *J Wound Care* 11(9):335–341.

105. Kakagia, D.D., Kazakos, K.J., Xarchas, K.C., Karanikas, M., Georgiadis, G.S., Tripsiannis, G., and Manolas, C. (2007) Synergistic action of protease-modulating matrix and autologous growth factors in healing of diabetic foot ulcers. A prospective randomized trial. *J Diabetes Complications* 21(6):387–391.

106. Chin, G.A., Thigpin, T.G., Perrin, K.J., Moldawer, L.L., and Schultz, G.S. (2003) Treatment of chronic ulcers in diabetic patients with a topical metalloproteinase inhibitor, doxycycline. *Wounds* 15: 315–323.

107. Ravanti, L., and Kähäri, V.M. (2000) Matrix metalloproteinases in wound repair (review). *Int J Mol Med* 6(4):391–407.

108. Mekkes, J.R., Zeegelaar, J.E., and Westerhof, W. (1998) Quantitative and objective evaluation of wound debriding properties of collagenase and fibrinolysin/deoxyribonulease in a necrotic ulcer animal model. *Arch Dermatol Res* 290:152–157.

109. Schmidtchen, A., Wolff, H., and Hansoon, C. (2001) Differential proteinase expression by *Pseudomonas aeruginosa* derived from chronic leg ulcers. *Acta Derm Venereol* 81:406–409.

10 Wound Healing Immunology and Biofilms

Emma J. Woods, Paul Davis, John Barnett, and Steven L. Percival

CONTENTS

Introduction...271
Overview of the Immune System...272
The Innate Immune System and Wound Healing 274
Identification of Microbial Intruders..276
Components of the Innate Immune System...277
Adaptive Immunity–Wound Interactions..281
Components of the Adaptive Immune System...282
Coordinated Responses of the Innate and Adaptive Immune Systems.................284
Chronic Wounds..284
Bacterial Immune Evasion 286
Focus on *S. aureus* and *P. aeruginosa* Immune Evasion......................286
Biofilm Formation..287
Conclusion 288
References..289

INTRODUCTION

Cutaneous injury resulting in breach of the skin barrier stimulates a cascade of events that ultimately can restore normal structure and function to the site of damage. While a wound is open, it is susceptible to contamination and potentially colonization by both exogenous and endogenous microorganisms. The human body provides an ideal incubated medium for the growth of microbes and, therefore, has evolved a highly orchestrated process with which to restore tissue integrity.[1] The overlapping responses that occur after damage has occurred encompass both innate and adaptive immunity to identify and eliminate microorganisms and other debris, together with migrational and proliferative cellular cues that culminate in the repair of the wounded tissue.

The acute wound healing response is divided into a number of stages, including the formation of a fibrin clot, an inflammatory phase, a proliferative phase, and a remodeling phase.[2,3] Each of these phases has an immunological element to ensure that wound debris is cleared and that wound infection is prevented. Restoration of skin integrity will restore the skin's physical barrier to colonization. Skin is continuously subjected to a number of potentially harmful insults ranging from wounds and microbial colonization to ultraviolet (UV) radiation. In order to respond to these stresses, the skin has a highly developed immune surveillance network distributed throughout the tissue. Langerhans cells, keratinocytes, cutaneous lymphocyte antigen (CLA) positive T lymphocytes and local lymph nodes are regarded collectively as skin-associated lymphoid tissue (SALT) and, because of its strategic location, it is this tissue that has the primary interaction with a skin wound.[4]

The overall immunological reaction that occurs in a human body following an invasion from microbial pathogens can be divided into the innate and adaptive immune response. The innate immune response represents the first line of defense against microorganisms and invading foreign particles. The cellular components constituting the innate immune system include barriers inherent to the epithelial cells, phagocytes (which include the neutrophils, macrophages, and dendritic cells), mast cells, natural killer (NK) cells, and the complement system.[5,6] Many of these cell types are involved in the critical inflammatory phase of wound healing and are responsible not only for removing microorganisms from the wound site but also for providing cues to surrounding tissue cells to promote proliferation and migration. The adaptive immune system mounts a more specific response toward foreign materials, especially bacteria, viruses, and eukaryotic parasites. Adaptive immunity is generally divided into the cellular immune response mediated by T lymphocytes (including production of cytokines) and humoral immunity, mediated by B lymphocytes and secreted antibodies,[7] although the two compartments are closely integrated. Through the use of long-lived memory cells, the cellular apparatus can rapidly launch itself into renewed vigorous action when the body reencounters a previously experienced infection, unleashing a highly specific and usually effective antibody attack.

This chapter will explain the stages of acute wound healing in relation to both innate and adaptive immune responses which are responsible for clearing infection from the wound site and in preventing subsequent infection. The immune processes that occur when systemic and local factors lead to the development of a chronic wound are also described as well as bacterial strategies and mechanisms for evading the immune system including the development of a biofilm that can further inhibit or evade the immunological response.

OVERVIEW OF THE IMMUNE SYSTEM

The immune system consists of a diverse group of specialized cells and proteins, which are active throughout the whole body.[8] Particular immunological cells are located in certain organs and tissue sites, including the spleen, lymph nodes, gut lamina propria, and skin, often with a particular structure and organization specific to that compartment. The common purpose of all these components is that of protecting the individual against invasion by foreign organisms, viruses, and toxins.

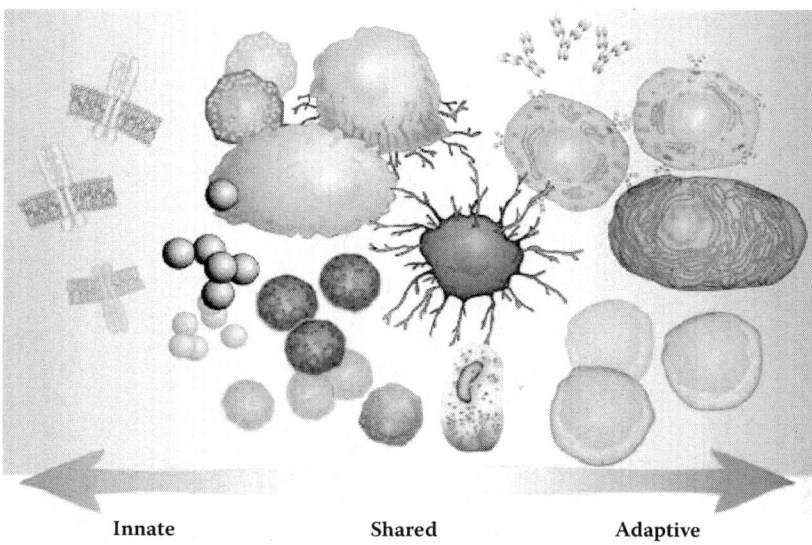

Innate   **Shared**   **Adaptive**

FIGURE 10.1   (Please see color insert following page 114.) Innate and adaptive compartments of the immune system.

In addition, the system undertakes a number of subsidiary, related roles, including removal of dead or spent cells, control of certain physiological processes, and the elimination of cancerous cells.[9] Key components of the immune system are shown in Figure 10.1, depicting the two arms or compartments of the system, called innate and adaptive immunity.

When a wound occurs, there is an urgent need for immediate defense mechanisms to be deployed in order to begin the initial removal of potential pathogens. However, defenses that can be deployed with extreme rapidity are, of necessity, based on mechanisms of action that are relatively nonspecific. Although they are capable of recognizing and attacking almost any kind of foreign invader of whatever molecular complexion, their action results in an amount of collateral damage to host tissue. The innate immune system is defined as all the ready-to-go defenses and the definition usually includes the physical and chemical barriers to entry at the various body surfaces (Figure 10.2).

In contrast to the innate immune system, the adaptive immune response, which includes antibodies and activated T lymphocytes, is able to home-in to their targets with great specificity and minimal collateral damage. Adaptive immune responses are much more powerful and effective than innate responses, but it can be several days before the first IgM antibodies are ready to be deployed, and it takes even longer before the really high-efficiency, affinity-matured IgG antibodies can be perfected.

Even though there are distinct activities and roles attributable to the innate and adaptive (often also called "acquired") compartments of the immune system, it is also important to appreciate that they are closely linked and interactive, if we are to understand immunity as a whole.[10,11]

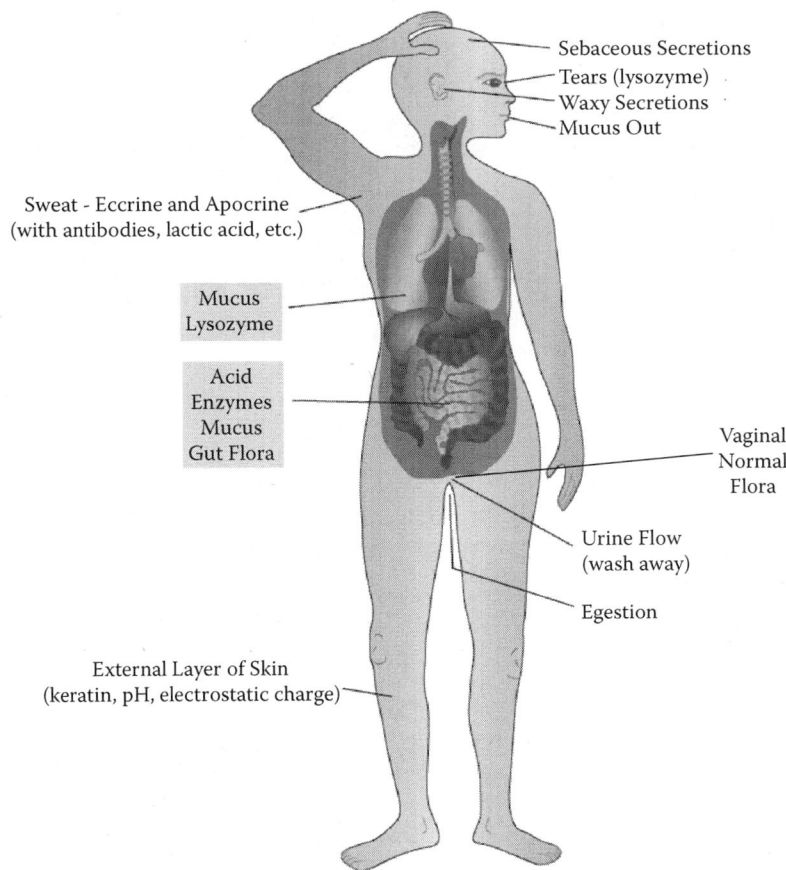

Sebaceous Secretions
Tears (lysozyme)
Waxy Secretions
Mucus Out

Sweat - Eccrine and Apocrine
(with antibodies, lactic acid, etc.)

Mucus
Lysozyme

Acid
Enzymes
Mucus
Gut Flora

Vaginal
Normal
Flora

Urine Flow
(wash away)

Egestion

External Layer of Skin
(keratin, pH, electrostatic charge)

FIGURE 10.2   Physical and chemical barriers to entry at the surfaces of the body.

THE INNATE IMMUNE SYSTEM AND WOUND HEALING

The role of the innate immune system in acute wound healing is far more pronounced than the more specific adaptive system.[12] Innate immunity is based on cells and molecules that can work together to mount a powerful first line of defense.[13] They are present from birth and are always poised to defend the host when invaded by potential pathogens (Figure 10.3). By their very nature, innate immune mechanisms are relatively nonspecific, which means that they are more limited in scope and have more potential to cause damage. In minor uncomplicated wound healing by primary intention, the restoration of barrier function of the skin through the migration of keratinocytes across the site of injury is normally complete 1 to 2 weeks post-injury, thus protecting the wound from further invasion by microorganisms. The innate response is most conspicuous in the inflammatory phase of wound healing, a stage that overlaps with the initial clotting phase and the later reepithelialization and, to some extent, remodeling phases of repair.

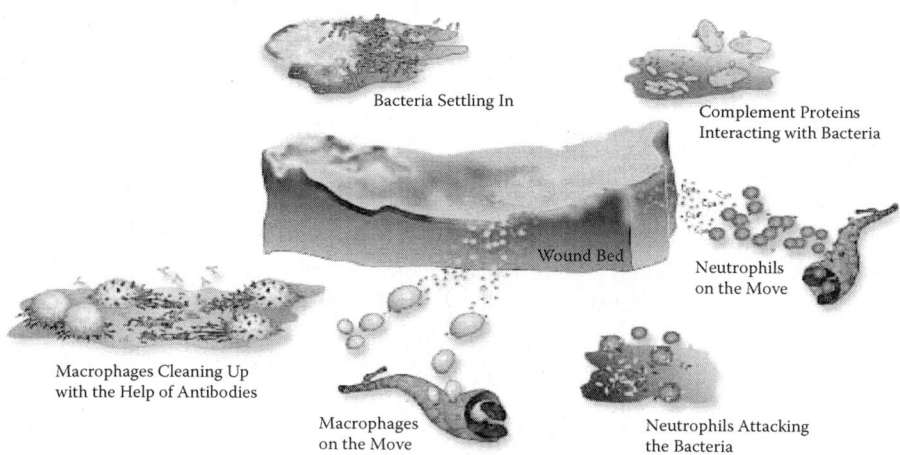

Bacteria Settling In

Complement Proteins
Interacting with Bacteria

Wound Bed

Neutrophils
on the Move

Macrophages Cleaning Up
with the Help of Antibodies

Macrophages
on the Move

Neutrophils Attacking
the Bacteria

FIGURE 10.3   (Please see color insert following page 114.) Battle plan: the immune defense of the wound.

While the wound is open, vulnerable subepidermal tissues are exposed and vulnerable to to bacterial colonization. Immediately after a wound has occurred, there is an influx of platelets that act to prevent blood loss through the formation of a clot. Once they arrive at the site of injury, platelets release chemokines and proinflammatory factors including epidermal growth factor (EGF), fibronectin, fibrinogen, histamine, platelet-derived growth factor (PDGF), and transforming growth factor β (TGFβ) which stabilize the wound via the formation of a fibrin clot. They also signal surrounding cells to increase their rate of proliferation. These signals released from blood platelets and also from nearby skin-resident mast cells, cause vasodilation and endothelial cell retraction in the nearest blood vessels, thus changing their flow dynamics and allowing fluid leakage to trigger local edema. Due to the number of neutrophils normaly present in the circulation, there is passive collection at the wound site in the blood clot. After this they migrate to the wound surface together with additional actively recruited neutrophils. Degranulation of platelets also promotes the inflammatory response especially by attracting additional circulating neutrophils toward the wound.

Within minutes to hours of an injury occurring, neutrophil numbers at the site begin to increase further to carry out their role in clearing microbial invaders and preventing invection. Keratinocytes also have a complex and crucial role during this stage of the wound healing response. These characteristics allow them to interact and activate immune cells arriving in the wound. Keratinocytes can synthesize a large number of cytokines and other inflammatory mediators in response to injury or UV radiation. These mediators induce the expression of adhesion molecules on nearby endothelial cells to recruit immune system cells into the wound as well as increasing vascular permeability.[14] Various adhesion molecules are expressed such as selectins-P and -E, which act as a molecular semaphore by interacting with integrins on the surface of neutrophils. Chemotactic fragments of complement proteins also provide molecular cues to guide neutrophils to the damaged site. The molecular signals displayed on the endothelial cells in the capillary lumen slow and arrest the required leukocytes, as well as stimulating them to transmigrate across the vessel walls toward the site of injury. To achieve this exit, they must cross through connective tissue barriers by releasing specialist proteases, especially neutrophil elastase and matrix metalloprotease 8 (MMP8, also known as neutrophil collagenase). Once out of the confines of the capillary vessel, neutrophils follow a chemical trail of chemotactic factors (e.g., chemokines) emanating from the injured site (chemokine-mediated chemotaxis) to begin clearing the wound.

Once at the site of damage, neutrophils encounter proliferating microbes that have taken advantage of the breakdown in the skin's barrier function. They recognize bacteria by means of specific receptors (known as pattern-recognition receptors, PRRs) to identify surface molecules on microbes.[15] They also detect and lock-on to foreign cells coated with activated complement molecular fragments. Recognition through either of these mechanisms causes the neutrophils to engulf and eliminate microbes, simultaneously discharging destructive enzymes and oxidizing agents including hydrogen peroxide, free oxygen radicals, and hypochlorite into the microenvironment—a process that can kill adjacent microbes but that can also cause damage to the tissues of the wound bed if allowed to progress uncontrolled. The inflammatory activity leads to increased levels of proinflammatory cytokines such as interferon γ (IFNγ) tumor necrosis factor (TNF-α), interleukin 1β (IL-1β), IL-6, and Il-8 which attracts further neutrophils and monocytes to the region of injury. Once exposed to the gradient of chemokines and proinflammatory cytokines, local monocytes mature into macrophages. These cells, as well as other leukocytes, are attracted to a wound by the endothelial semaphore system of selectins, integrins, and complement signals as neutrophils.[16] Macrophages migrate to the site when the neutrophils have completed their antimicrobial task, whereupon they clear up the wound by engulfing and digesting remains of neutrophils or bacteria that are left behind. Macrophages also help to direct the specific adaptive immune response by acting as antigen-presenting cells. The inflammatory phase of an acute wound is typically most active for 3 to 5 days after injury. At this point, wound reepithelization should be well advanced in the restoration of barrier function and in the protection of the host from any further bacterial invasion. Only in the presence of other local or systemic factors, which can include persistent infection, does the inflammatory process continue. Any extension of the nonspecific inflammatory response affects subsequent healing phases and can lead to the formation of a chronic wound (discussed below).

IDENTIFICATION OF MICROBIAL INTRUDERS

A key factor for the initiation of an innate response is the detection of cell surface components displayed by invading organisms that are not typically displayed by the host. These constitutive components include lipopolysaccharide (LPS), lipoproteins, peptidoglycan (PGN), and unmethylated DNA containing a CpG motif (CpG-DNA) and are collectively referred to as pathogen-associated molecular patterns (or PAMPs). PAMPs are recognized by receptors of the innate immune system, which are known as pattern-recognition receptors (PRR), such as mannose-binding lectin (MBL), ficolins, nucleotide-binding site leucine-rich repeat proteins (NOD1 and NOD2), and Toll-like receptors (TLR).[17]

Although the identification of microbial invaders by phagocytes is chiefly via PRRs, macrophages and neutrophils also recognize microbes by antibodies (when they become available) via their high-affinity IgG receptors.

COMPONENTS OF THE INNATE IMMUNE SYSTEM

Key cells involved in the innate arm of wound defense are phagocytes (derived from the Greek *phag*, meaning eat and *kytos*, meaning vessel or cell). These, as their name suggests, swallow up and then kill invading microbes through a well-defined process of phagocytosis (Figure 10.4). The two main phagocytic cell types are macrophages and the smaller, aggressive neutrophil leukocytes (neutrophils for short) also known as polymorphonuclear granulocytes (PMNs). The innate immune system also includes a nonphagocytic type of leukocyte called NK cells, with a role in killing virus-infected cells and cancer cells by a lethal contact binding event, often referred to as the "kiss of death." Mast cells are another innate immune cell type more commonly known for their role in allergic reactions but which also play a role in perpetuating the inflammatory response through the release of histamine and chemokine-containing granules.

Neutrophils are the most numerous type of white blood cell (or leukocyte) found in the circulation, yet they are completely absent from healthy skin tissue. Even so, they are the first defensive cells to reach the scene of any microbial infection or tissue injury, rushing into the affected site from nearby capillary blood vessels or postcapillary venules. Their arrival at the site of tissue damage is normally within 30 minutes of injury.[18] They possess an array of potent enzymes that include proteases, lipases, and amidases as well as antimicrobial peptides. It has been documented that these enzymes work in a synergistic manner to cause an effective

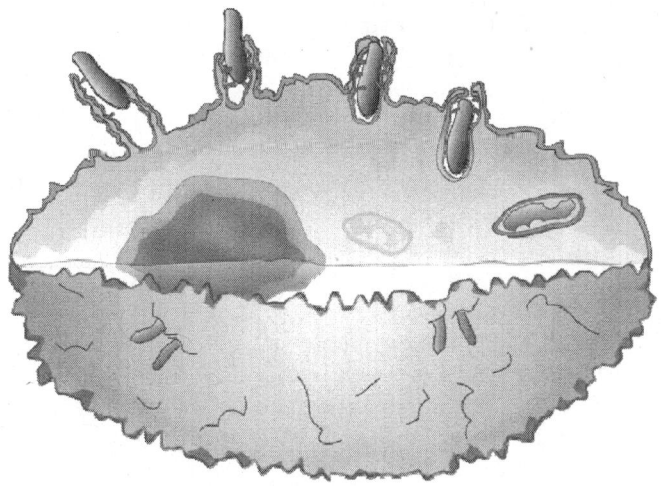

FIGURE 10.4   (Please see color insert following page 114.) The process of phagocytosis.

combination for bacterial destruction.[19] To underline their importance in wounds, patients with immunodeficiency involving neutrophils, such as chronic granuloma-tous disease (CGD), are especially susceptible to bacterial infections of the skin and mucosal membranes, most frequently with staphylococci.

Neutrophils are generated in the bone marrow and released into the blood, where they remain for several hours. In normal tissues they may last for a few days, but in a wound, after engulfing microbes, they die quickly. Various proteases (e.g., elastase, MMP8, MMP9), other enzymes (e.g., myeloperoxidase, lysozyme), and functional proteins (e.g., lactoferrin, phagocytin) which are carried in their granules are released as they work. Neutrophil-derived proteases and reactive oxygen species (e.g., hydro-gen peroxide) cause major local tissue.[20]

The attraction of neutrophils into infection sites or inflammatory lesions is caused by multiple signaling molecules[21] including platelet-derived growth factor (PDGF), the chemokine-connective tissue-activating peptide-III (CTAP-III), which is con-verted proteolytically into neutrophil-activating peptide-2 (NAP-2; CXCL7) by neu-trophils attached to the thrombus, growth-related oncogene-α (GRO-α; CXCL1), and ENA-78 (CXCL5). IL-8 also stimulates CXCR1 (a chemokine receptor on neutro-phils) to initiate a secondary response by inducing migration of more neutrophils to the wound bed. Bacterial products such as lipopolysaccharides, formyl-methionyl peptides, and N-acetylmuramyl-L-alanyl-D-isoglutamine can also accelerate the directed neutrophil locomotion. The complement fragment C5a also participates in neutrophil recruitment.

Tissues containing large numbers of neutrophils become severely hypoxic, which is a result of the high level of oxygen required to carry out phagocytosis and oxygen-dependent antimicrobial activities (respiratory burst). Once they have performed their phagocytic duties, neutrophils undergo apoptosis (programmed cell death) and are rapidly recognized and cleared from the wound by incoming macrophages.

Macrophages are also phagocytic cells, the "macro" prefix reflecting their greater size, in comparison with neutrophils (neutrophils are sometimes known as "microphages"). Macrophages are extremely versatile cells that differentiate from monocytes when exposed to cytokines at a wound site, for example. As stated above, the name macrophage derives from *phag*, to eat and *macro*, big, to give "big-eater"— a name that is appropriate but which only describes a very small part of the role they play. There is a network of macrophages throughout the body, known as the mono-nuclear phagocyte system with several crucial roles, including the removal of foreign cells and debris, acting as a primary defensive sentinel network, as well as stimu-lating and signaling cells of the adaptive immune system. Macrophages have far-reaching effects, as they orchestrate various aspects of complex immune responses and secrete numerous growth factors and cytokines involved in controlling tissue repair and regeneration.[9,18]

Monocyte chemotactic protein-1 (MCP-1; CCL2), which is synthesized by dermal mononuclear cells and basal keratinocytes at the wound edge, is a dominant mono-cyte chemoattractant during wound healing. MCP-1 recruits immune cells including monocytes, memory T cells, and dendritic cells to a site of tissue injury and infec-tion. Selective expression of MCP-1 mRNA during healing of incisional skin wounds is correlated with the profile of monocyte infiltration. The chemotactic cytokine macrophage inflammatory protein-1α (MIP-1α; CCL3) is expressed by macrophages after they are stimulated with bacterial endotoxins. Expression of this chemokine attracts and activates human neutrophils and also induces the synthesis and release of other proinflammatory cytokines such as interleukin 1 (IL-1), IL-6, and TNF-α from fibroblasts and macrophages.[22]

Monocytes stay in the blood for up to 100 hours and then migrate into tissues to become resident macrophages (histiocytes). In normal tissues they may remain for many months. When needed in increased numbers (e.g., during infection or certain phases of inflammation), reinforcements are called in from the blood (much as neutrophils are) and from surrounding tissues. Like neutrophils, their full antimicrobial potential is dependent on oxygen supply. Macrophages have a slightly different oxygen-dependent antimicrobial system. One major difference is that they use reactive nitrogen intermediates (RNIs), including nitric oxide, alongside their respiratory burst activities.[23] Within a wound, macrophages can engulf and kill microbes directly or clear up dead neutrophils and remaining bacteria after the battle. They secrete a wide variety of enzymes, signaling molecules and active substances, reflecting their multiple roles in controlling processes beyond inflammation and antimicrobial defense. Macrophages are a major link between the innate and adaptive immune system via their ability to produce chemokines and signaling molecules to influence surrounding cells and present processed antigens (as peptides) to appropriate T-lymphocytes as a key part of antibody induction and T-cell activation.

Mast cells are resident in connective tissue, especially in association with blood vessels and nerves. These cells have two major distinctive features—the presence of numerous conspicuous granules in their cytoplasm, and a high density of IgE antibodies carried on their outer surface. The granules are loaded with preformed inflammatory substances. On activation, the granules swell and move to the cell membrane where they empty histamine, heparin, and TNFα into the surrounding tissues in a well-defined process called degranulation. Several other mediators can also be synthesized and released following activation. Activation is achieved by several mechanisms, especially by antigen (often an allergen) binding to the IgE antibodies held on the surface by high-affinity IgE receptors through which the activation signal is mediated. Activation is also achieved by binding of complement fragments C3a, C5a, and C4a to special receptors. The main consequences of the released mediators are increased vascular permeability and vasodilation in local blood vessels and recruitment of neutrophils and macrophages, thus assisting in the perpetuation of the wound inflammatory response. Mast cells are best known for their prominent role in immediate-type allergic reactions, where allergen binding to their surface IgE antibodies cause massive release of their granular contents, so causing the familiar range of allergy symptoms, from a runny nose to full blown anaphylactic shock.

Natural killer (NK) cells are a type of cytotoxic lymphocyte. Although their function has previously been understood to focus on the eradication of tumor and virally infected cells, they have also been shown to participate in wound healing and inflammation.[24] NK cells migrate into a wound on a similar timescale to that of neutrophil infiltration. They are a source of the proinflammatory cytokine IFNγ, so an increased presence of these cells at the wound site during the initial phases of healing assists in the establishment of a chemoattractive gradient to summon further immune cells.

The *complement system* is a collective term for a group of more than 25 interacting serum proteins found throughout the body that participate in a sequential reaction cascade when the initiating members are stimulated by the presence of a microbe or other foreign structure. The complement system is able to recognize and attack foreign intruder cells either by direct binding of complement component C3 to the foreign cell surface or indirect binding of the component C1 complex to IgG or IgM antibodies specifically bound to antigens of the foreign cell. This activity occurs throughout the duration of the bacterial eradication reaction, from the moment bacterial presence is detected to the moment infection is cleared.

There are three main functions of complement. The first is to directly cause the lysis of foreign cells, especially bacteria and enveloped viruses, through the assembly of molecular complement complexes that create holes in microbial membranes. The second function is to coat foreign cells or particles with specific complement protein fragments that can be recognized by receptors on the cell membranes of neutrophils and macrophages, thus helping these phagocytes to engulf and kill them. This is termed *opsonization,* derived from the Greek "prepare to eat"—in other words, complement is acting as a kind of molecular relish, encouraging phagocytes to get on with the job of clearing infection as fast as possible. The third main function involves the active signaling molecular fragments, such as C5a and C3a fragments, in helping to orchestrate the recruitment of phagocytes from the blood. Activation of the cascade results in the generation of C3a and C5a signaling molecules, which diffuse away from the site toward the nearest microvascular blood vessels, together with other signal substances. When they reach the blood vessels, these molecules bind to specific receptors on the endothelial cells lining the inner vascular surfaces, so triggering them to express adhesion molecules that capture for passing neutrophils on the internal surface of capillaries and postcapillary venules. Weaker binding between the ligands and the receptors caused the neutrophil to roll, which effectively slows the neutrophil. Later stronger binding between the ligands and receptors causes them to stop.

Complement is such a powerful trigger of inflammation that it has to be equipped with a number of suppression mechanisms. Its potentially devastating power can be seen in the clinical condition of hereditary angioedema, in which there is a deficiency of the inhibitor of the first component of complement (C1q). Patients with this condition suffer from recurrent episodes of severe inflammatory edema (typically non itchy), usually localized to the skin, larynx, and/or gut. If there is a severe attack involving the larynx, the edema can block the airways leading to collapse or death. On the other hand, a deficiency that prevents complement from unleashing its full power against invading microbes can leave a patient unable to fight off infection (e.g., deficiency of C3), causing them to suffer recurrent and sometimes life-threatening infections.

Antimicrobial peptides form a key element of the host defense system against invading microbes. In the absence of injury, these peptides act at body surfaces that are normally exposed to high microbial loads, such as the epithelial surfaces (skin, the moist surfaces of the eyes, nose, airways, and the lungs, mouth, and the digestive tract, and the urinary and reproductive systems). These peptides are typically less than 100 amino acids in size and have been identified in a number of vertebrate, invertebrate, and plant species. They have a number of modes of action which can ultimately kill microbes and prevent infection. In keeping with their localized action, high concentrations of antimicrobial peptides are secreted by the epithelial cells resident at these sites. Phagocytic cells contain storage granules that hold these peptides as well as lytic enzymes such as proteases. These organelles deliver their contents into phagocytic vacuoles containing ingested microbes, thus exposing captured bacteria and other invaders to a lethal cocktail of destructive peptides and enzymes. Phagocytes also secrete peptides into the local extracellular fluid to aid in the eradication of infective agents.

At the right concentrations, antimicrobial peptides can cause lysis and then killing of bacteria within a short timescale (seconds to minutes). The majority of peptides with antimicrobial activity are cationic[25] and, in humans, can be divided into two broad categories: cathelicidins and defensins.[26] Their mode of action targets specific commonly encountered pathogenic invaders such as *Escherichia coli* (*E. coli*), *P. aeruginosa, S. aureus,* and *Enterococcus faecalis* (*E. faecalis*). The cationic nature of many of these antimicrobial peptides allows for strong electrostatic and hydrophobic interactions between themselves and bacterial cytoplasmic membranes.[27] There are a number of hypotheses to explain how antimicrobial peptides work. Insertion of peptides into the bacterial cell wall can create physical holes through which water can enter ultimately leading to cell lysis. Binding of peptides to critical intracellular components of invading cells can interfere with metabolic function. Alternatively, peptides may disrupt bacterial cell membranes and cause fatal depolarization and/or interfere with the usual distribution of lipids in the cell membranes.[25,27]

Some antimicrobial peptides (e.g., secretory leukocyte protease inhibitor of leukocytes [SLPI]) play an additional key role in modulating the immune response and inhibiting proteases.[28,29] Some are proinflammatory (by inducing proinflammatory cytokine and chemokine expression, activating adaptive immunity, promoting phagocytosis, promoting the expression of matrix metalloproteinases [MMPs] and triggering mast cell degranulation), while others are anti-inflammatory. There is evidence of a synergistic relationship between epithelial cells and neutrophils assisting in the modulation of wound healing and antimicrobial peptide responses via an autocrine signaling loop.[30]

ADAPTIVE IMMUNITY–WOUND INTERACTIONS

The main components of the adaptive immune system are the B and T lymphocytes, antibodies, and chemokines. In contrast to the innate compartment of the immune system, adaptive immune responses are initiated only in response to the presence of foreign material such as microorganisms. The adaptive immune response is able to discriminate between self and non-self matter with fine specificity, allowing it to provide a more specifically targeted response than the innate immune response. In addition, the adaptive system provides an immunological memory to protect the body against repeat encounters with pathogens.

When microbial invaders are engulfed by macrophages, the cells' digestive machinery degrades and processes the microbial cell into small peptide fragments.

This serves two purposes: disposal of the infectious agent and gathering intelligence on the antigenic nature of the invading microbe. The macrophage presents these incriminating pieces of molecular evidence to receptive T lymphocytes in order to direct the adaptive immune attack onto these specific targets. This task is also carried out by another type of cell found in skin, called the Langerhans cells. These are similar in many ways to macrophages, but they specialize in taking up, processing, and presenting any antigens they come across in that location. Langerhans cells are part of a group of antigen-presenting cells collectively called dendritic cells.

The process of antigen presentation and T cell activation is the starting point in the process of antibody induction, either within the wound or, more likely, in the nearest lymph node.[31] Bacteria and bacterial metabolic or degradation products that escape into tissues, blood circulation, or lymphatic system are captured by other antigen-presenting cells. This happens most efficiently in local lymph nodes, which are hotspots of the draining lymphatic system and, more distantly, within the spleen. These items of molecular debris are used by the immune system as stimuli to make specific antibodies, memory cells, and activated T lymphocytes.

The big advantage of the adaptive system is that it tailors its responses very specifically to the foreign invaders, be they toxins, viruses, bacteria, fungi, or eukaryotic parasites. By tailoring its responses, the adaptive system can deliver much more powerful attacks against its foes, partly because there is so much less collateral damage. There is a problem, however, for it takes several days for the specific, tailored responses to be assembled by the lymphocytes. The lymphocytes are the custodians and drivers of adaptive immunity. The active agents they turn out are, principally, antibodies (made by B lymphocytes) and cytotoxic cells of the T-lymphocyte lineage, complete with their own recognition molecules (T-cell receptors) and signaling molecules (e.g., interleukins). With a few exceptions, B cells can only make antibodies with the assistance of helper T cells that have the job of corecognizing foreign material presented by macrophages and dendritic cells through which they trigger a second signal that enables antibody production.

Powerful adaptive immunity is essential to life due to the ability of microbial attacks to evade and overcome the initial onslaught of the innate defenses. In wounds, however, adaptive immunity only seems to become a significant factor as a result of prolonged microbial presence at the site of the wound or if infections spread outside the wound margins and into surrounding tissue. Even so, this arm of the immune system is especially important in containing infections within the wound and stopping systemic spread.

COMPONENTS OF THE ADAPTIVE IMMUNE SYSTEM

In a normal individual there about a trillion lymphocytes (10^{12}) at any one time. All resting lymphocytes look virtually uniform, yet there are several different types with distinct roles. The T lymphocytes and the B lymphocytes are the most important to our understanding of adaptive immunology.[32]

T lymphocytes (or T cells) are schooled through the thymus (hence, the "T" prefix) and have the job of recognizing and responding to foreign proteins by means of their amazingly versatile molecular recognition receptors, prosaically named T-cell receptors (TCRs). In effect, they "see" the universe of foreign proteins through their TCRs, but TCRs can only "see" proteins that have been cleaved into small peptides and displayed on the surface of specialized antigen-presenting cells (usually dendritic cells and macrophages). TCRs are not secreted, so T cells can only attack or control target cells by close combat, or by influencing other immune cells (especially macrophages). Cutaneous lymphocyte antigen (CLA) is the main T-cell adhesion molecule for homing to the skin. This carbohydrate epitope assists in targeting memory T cells to sites of inflammation in the skin and is particularly relevant in cases of psoriasis, atopic and allergic contact dermatitis, and cutaneous T-cell lymphoma where T cells play a pathogenic role.[33]

Depending on the other surface receptors they express alongside the TCRs, and on the profile of the lymphokines they secrete, they adopt different roles. Some T cells help B cells to make antibodies and others become lethal cytotoxic cells, capable of killing virus-infected cells and pathogens hiding inside host cells. T cells can activate macrophages and help them to kill intracellular parasites. They have a central role in controlling and directing the adaptive immune responses and some subsets (e.g., T-helper or Th cells) can even direct responses toward allergic-type responses (Th2 type) or toward the cellular immunity responses (Th1 type).

$\gamma\partial$ T cells or dendritic epidermal T cells (DETC) are also associated with immune responses in the skin. They are found in the epidermis in association with damaged, stressed, or transformed keratinocytes. Studies in mouse wound models have shown that healing is delayed if DETC are deficient, and it is evident that they provide crucial signaling molecules controlling keratinocyte actions and macrophage infiltration.[34]

B lymphocytes (or B cells), are schooled through the bone marrow in adults (hence the "B" prefix), and are concerned with recognizing and defending against pathogens in the body fluids, rather than those inside cells. Their job is to make antibodies which they secrete in great quantities (around 2000 molecules/minute/cell) from fully matured, activated B cells (antibody-producing cells [APCs] or plasma cells). Initially, these cells are visually indistinguishable from T cells but undergo major changes once activated and as they mature toward their role as an antibody factory. Each B cell is capable of making only one specificity of antibody, which it permanently displays on its surface membrane. It can only become activated to start antibody production when it receives two signals: One is by the foreign substance binding to its antibody and the other is by particular cytokines released from adjacent helper T cells. In effect, B cells need permission from T-helper cells to make antibodies, because antibodies are so powerful that devastating damage would be caused if one was produced that attacked self-molecules (as can be seen in an autoimmune disease like rheumatoid arthritis). APCs usually congregate in lymph nodes or spleen to set up multicell antibody factories, causing the symptom of hard, swollen lymph nodes.

Antibodies are large (MW ~150,000), modular, multidomain globular proteins. They come in various forms and sizes, but the most abundant type is the IgG class, with a general architecture as depicted in Figure 10.5.

New antibodies circulating in the blood and lymph may find their way into the wound, where they attach to any recognizable bacteria, so helping neutrophils, macrophages, and complement to unleash further powerful attacks. More importantly,

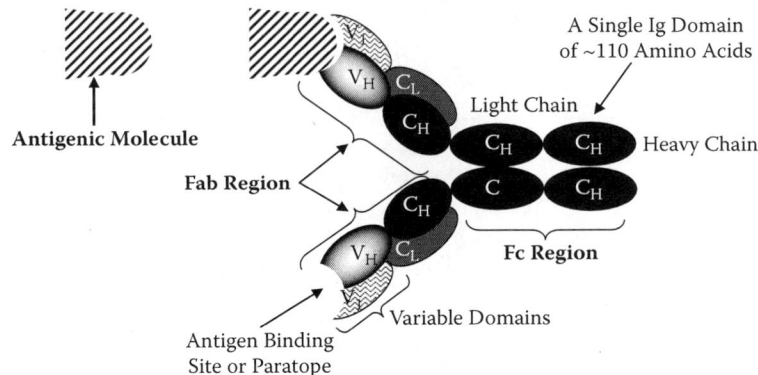

FIGURE 10.5   IgG antibodies.

they act systemically to eliminate any living bacteria that escape from the wound site, thus containing the infection and reducing the chances of systemic infection.

COORDINATED RESPONSES OF THE INNATE AND ADAPTIVE IMMUNE SYSTEMS

Despite the apparently clear distinction between innate and adaptive immunity, it soon becomes apparent to a student of immunology that many of the cells and molecules are closely involved with both forms of immunity.[35] For example, as described previously, neutrophil leukocytes are the first innate immune cells to arrive at the scene of bacterial invasion. Neutrophils possess receptors for antibodies, and yet antibodies can only turn up several days after the start of infection. Macrophages form a key link between the two arms of immunity. Although their primary role is to provide their phagocytic functions as members of the nonspecific innate system, they are also key providers of chemokine cues and cell surface signals that can initiate and direct the adaptive response.

As members of the innate arm, macrophages carry the special recognition molecules described above as PRRs. While these selectively bind to pathogenic microbes as a primary recognition agent, they are also central to the adaptive system where they participate in the overall process of antigen presentation and the commissioning of specific antibodies and T lymphocytes. Cells of the adaptive system are absolutely dependent on this sentinel and intelligence-gathering role of the innate cells (as well as other explicit antigen-presenting cells) for the task of generating a specific immune response toward infection. The enigmatic complement proteins are part of nonspecific innate immunity, yet they can be activated and deployed with fine specificity and power by the antibodies of the adaptive system.

CHRONIC WOUNDS

Continued neutrophil domination of a wound is a sign that a wound is becoming chronic in nature, and macrophage domination indicates that progress is being made and recovery is under way.[36] Although neutrophils are an essential and important component of defense, their prolonged presence at a wound site becomes a barrier to healing,[37] as they can cause a self-perpetuating state of chronic inflammation and hypoxia, dominated by excessive, destructive protease activity.[20] Persistent edema, inflammation, and tissue hypoxia in a wound that has shown little or no healing provide additional cues that a patient has systemic complications that are impeding the normal healing process.[38] The resultant lack of healing is often due to an overactivation of the wound healing inflammatory response—a condition that emphasizes the importance of keeping the rapid, nonspecific innate immune response under control.

Evidence-based medical innovation is the key to progress in wound management. The combined assessment of all available patient information is essential in determining the course of treatment. It is now recognized that assessment and treatment of the whole patient,[39] rather than only the wound itself, is required for effective wound care and to achieve healing outcomes. The overall immune status of the patient as well as the immunological status of the wound are crucially important factors, and there is a real need for new diagnostic tools capable of revealing these key parameters to the wound care clinician. The diagnosis of underlying pathologies often provides an understanding of how and why a wound has failed to heal in the normal way.

In addition to core immunological factors, there are several underlying systemic pathologies that can impair the wound healing process and which have links to an impaired immune response to injury. During the clinical assessment of a nonhealing wound, it is important to consider the overall health of the patient as the basis for a treatment regime. A patient with poorly managed diabetes, uncontrolled edema, an immunodeficiency disease, poor nutrition, or a dependency on cigarettes, alcohol, or drugs may present with a wound that exhibits slow or incomplete healing. Local factors can also cause healing to be inhibited. These can include the presence of a foreign body such as a retained suture or gauze material, untreated deep infection such as osteomyelitis, or the sheer size of an initial wound injury can prove too much for the body to repair.[39] An impaired blood supply to a wound site directly leads to lower tissue oxygenation and damaging hypoxia, in turn leading to localized necrosis. This is often the case in patients with unmanaged diabetes and edema affecting the extremities and lower limbs. Without an early diagnosis, this can ultimately lead to the breakdown of skin tissue and the formation of a wound. Tissue hypoxia is further compounded by excessive oxygen demand by neutrophil respiratory burst activities in response to the presence of infectious agents. Continued hypoxia maintains the neutrophil recruitment signals in the nearby microvasculature, so perpetuating a state of inflammation. The chronic state that ensues allows the wound to become stuck in a condition dominated by hypoxia, edema (contributing to ischemia), and proteolytic damage to the extracellular matrix and growth factor profile.[40] A reduced or insufficient supply of blood to the site tips the balance in favor of bacterial proliferation, thus leading to clinical infection. The host's response is to continue to try and heal the wound by recruiting further inflammatory cells to the site to clear infection. The combined effects of these local wound-inhibitory factors, and the continuing influx of destructive inflammatory cells in response to the foreign body or underlying infection make healing impossible. Moreover, prolonged, intense inflammation leads to further breakdown of surrounding healthy tissue due to the release of destructive proteases and reactive oxygen species from the invading neutrophils, causing wound enlargement.

The transition from colonization of a wound to an infected state is not only dependent on the host's immune status, but also on the bacterial count, the species present, the number of different species present, the virulence of the organisms, and the presence of synergistic interactions between the different species.[41] Initial infection of a wound is often caused by bacteria such as *Staphylococcus aureus, Staphylococcus epidermidis,* and *Pseudomonas aeruginosa* which reside on the skin surface, from where they can opportunistically colonize a wound when the skin barrier function is breached. In a healthy, healing, acute wound, these pioneer bacteria are destroyed by the innate response and the damaged skin is reepithelized and barrier function restored. However, the onset of a chronic wound state provides the ideal environment for colonization and progression toward infection. Infection by these pioneer bacteria then provide further opportunities for bacterial invasion from external sources and may include Gram-negative bacteria (in addition to *P. aeruginosa*) and, also, anaerobic species such as *Peptostreptococcus*.[41] As the immune system struggles to manage the increasing bacterial load, the invaders respond by utilizing means to evade detection and destruction by the host.

BACTERIAL IMMUNE EVASION

It may seem surprising that any microbes can survive the initial onslaught of responses from the innate system; however, bacteria possess a number of mechanisms by which they can evade immunity, at least to the extent that some of their number can survive. For a bacterial inoculum to cause infection, it must be able to evade the innate immune system sufficiently well to allow enough survivors to start taking control of the new environment. Bacteria such as *S. aureus* have been shown to be very effective in evading the effects of defensins and the oxygen dependent antimicrobial system of neutrophils. They also have mechanisms for evading attack by the complement system.[42] Current research has shown that some bacteria are directly able to initiate the coagulation of blood when they are clustered together and evade effects of the immune response.[43] Both *S. aureus* and *P. aeruginosa* (of particular relevance in wound infection), together with other related bacteria, possess specific mechanisms through which they are able to evade many elements of the immune response.

FOCUS ON *S. AUREUS* AND *P. AERUGINOSA* IMMUNE EVASION

The evasive actions taken by wound pathogens are determined by the immune responsiveness of the particular host they are invading. If the host has an intact immune system, the majority of invading cells are exterminated by the phagocytic cells of the innate immune system.[44] In this instance, bacterial cells are driven toward a defensive phenotype in order to persist, such as entering a biofilm state that provides physical protection from immune cells. In a patient with compromised defenses making them unable to mount a successful attack on bacterial invaders, or in a patient exposed to a massive bacterial assault, *P. aeruginosa* cells may survive as planktonic cells—a state from which they can proliferate, infect a wound site, and even go on to disseminate through the bloodstream to other tissues.[44]

S. aureus and *P. aeruginosa* can avoid eradication by the immune system by utilizing a key mechanism common to a number of wound pathogens, based on the production of large quantities of extracellular products that interfere with host defenses. The secretion of alkaline protease and elastase by *P. aeruginosa* causes the degradation of laminin, immunoglobulins, and a number of cytokines, all of which are critical elements of the wound repair process. Furthermore, these bacterial enzymes interfere with immune cell function by inhibiting chemotaxis and phagocytosis in macrophages and neutrophils, and reducing the activation of a CD4[+] T cell response.

Staphylococci possess a number of additional strategies through which they are able to inhibit wound healing and repress the immune response. The secretion of lytic enzymes such as hyaluronidase and lipase by Staphylococci assists in the dissemination of bacterial toxins and amplifies host tissue damage.[45] In addition to this, *S. aureus* cells also secrete a chemotaxis inhibitory protein of staphylococci (CHIPS) that can block complement-mediated signaling in turn inhibiting the migration of neutrophils from the blood to the wound site.[45] *S. aureus* secretes cytolytic toxins that are able to kill leukocytes that are attempting to phagocytose and kill them. These toxins target the host immune cells and cause cell death through the formation of pores in the cytoplasmic membrane.[45] Staphylococci are able to avoid and repress the phagocytic response after infection through the avoidance of opsonization, which occurs via the expression of anti-opsonic proteins and also via the expression of surface proteins integrated with an external microcapsular layer that interferes with neutrophil and monocyte recognition. Staphylococci can also resist antimicrobial peptides. Through this capability, engulfed bacteria can avoid death from neutrophil peptides even when the first steps of phagocytosis have been successful. This is predominantly due to cell wall modifications that reduce the affinity of cationic antimicrobial peptides and repel them from the bacterial cytoplasmic membrane.[45] Avoidance of cell death after phagocytosis has been described as a further virulence factor contributing to the persistence of *S. aureus* and other Staphylococcal infections.[46]

BIOFILM FORMATION

As mentioned throughout this book, biofilms are recognized as significant to many chronic infections, including infections of wounds. The resistance of a biofilm to the immune response is generating substantial interest in the scientific community.[47] In particular, it has been documented that biofilms possess mechanisms to avoid both killing and clearance of bacteria by the immune system. Such mechanisms include the following:[48] resistance to penetration of leukocytes into the biofilm—in fact, leukocytes have been shown to adhere to the surface of biofilms,[48,49] inactivation of leukocyte antimicrobial processes by the biofilm matrix itself, resistance to phagocytosis of leukocyte biofilm bacteria, evidence of regulators and quorum sensing molecules that enhance biofilm resistance to leukocytes, and (becoming more relevant) triggering of genetic switches that enhance cooperative resistance to components of the immune response.

P. aeruginosa has been shown to express a viscous exopolysaccharide called alginate which, in the context of a biofilm, is a significant virulence factor. Alginate is only produced by mucoid *P. aeruginosa* and has been shown to be very effective in decreasing the phagocytosis of planktonic mucoid *P. aeruginosa* by both neutrophils and macrophages.[50,51] It has been found that high levels of alginate can protect *P. aeruginosa* from antibiotics and peptides. A recent paper has shown that alginate may be important for the protection of mucoid *P. aeruginosa* biofilm bacteria from the immune system.[48]

Neutrophils appear to resist biofilms under some circumstances by preventing their formation,[52] yet, paradoxically, enhancing biofilm development under different conditions.[53] Some peptides have been shown to have effects on biofilms; for example, lactoferrin has been shown to prevent bacterial biofilms[52] as has LL-37.[54] Overhage and colleagues[54] were able to demonstrate that LL37 affected biofilm formation by decreasing the attachment of bacterial cells, affecting twitching motility and quorum sensing and down-regulating genes known to be essential for biofilm formation. Shapira and colleagues[49] have shown that neutrophils are able to reduce the number and viability of *Streptococcus mutans* when attached to beads which suggested that neutrophils were able to attach to biofilms and retain their antibacterial activity.

CONCLUSION

This chapter has described the immune response that ensues following the formation of a wound as part and parcel of a wound healing response. Under normal circumstances, the key players, the innate immune cells, can effectively clear a wound of pathogens and provide crucial local and systemic signals to other cells required for the immune and repair response to continue. In normal, acute wound healing in a healthy individual, these mechanisms are able to restore skin barrier function while preventing bacterial infection. However, where underlying systemic and local pathologies exist and a wound enters a chronic state, the very mechanisms that are key to the progression of injury repair are also those that are responsible for tissue damage and continued inflammation. It is this loss of balance between acute and chronic healing responses that must be addressed when treating problem wounds. The initial assessment of overall patient health, as well as the status of the wound itself, is critical in developing a wound treatment strategy.

Chronic wound healing is complex and variable. Knowledge of the essentials of wound immunology can help to ensure that clinicians make the right decisions at the right time, especially by helping them to accurately interpret the available clues and symptoms observed in the patients and in the wounds they treat. The additional evidence that biofilms can be present in these wounds also needs to be addressed, and new diagnostic tools and treatment strategies are required to overcome this and other bacterial survival mechanisms. These strategies may include means to destroy a biofilm as a whole, may target critical species within the biofilm, or perhaps interfere with quorum sensing molecules so as to allow anti-infective agents, together with the immune system, to have greater penetration and effect. It is likely that no single treatment will have the capacity to heal all classes of wound, but it will be the development of accurate diagnostic methods for patient pathology and critical wound healing factors such as infection that will allow tissue repair to be monitored and treatments adapted accordingly. Restoring the balance of the immune responses, which are either overactivated due to systemic and/or local factors or overwhelmed in the presence of infection, will then also allow healing to progress along its normal route and repair skin barrier function.

For the future, a greater understanding of the immunology of wound healing in both acute and chronic wounds will, undoubtedly, lead to improved therapies and new diagnostic tests. These advances must take into account the complexities of acute wound healing and wound immunology and identify which aspects are impaired in chronic wounds. The development of bespoke wound care protocols, depending on the patients' immune status and any other underlying pathology, is a practice currently being developed in a number of wound care clinics. Advances in treatments that harness and modulate the patients' innate ability to fight infection will be of great benefit. This might be achieved, for example, through antimicrobial peptide therapy to suppress infection and promote repair of injured tissue. Treatment with specific growth factors and cytokines is a current focus of many research groups, and others are pursuing the means to control redox signaling and the impact of tissue hypoxia on the activities of the immune system.

REFERENCES

1. Brown, S., K. Palmer, and M. Whiteley. (2008) Revisiting the host as a growth medium. *Nat Rev Microbiol* 6:657–666.
2. Abbas, A., and A. Lichtman. (2009) General properties of immune responses. In Cellular and Molecular Immunology, 5th ed. A. Abbas, A. Lichtman, and S. Pillai, eds. pp. 3–13.
3. Singer, A., and R. Clark. (1999) Cutaneous wound healing. *N Engl J Med* 341:738–746.
4. Streilein, J.W. Skin-associated lymphoid tissues (SALT): Origins and functions. *J Invest Dermatol* (1983) 80 Suppl:12s–16s.
5. Martin, P., and S. Leibovich. 2005. Inflammatory cells during wound repair: The good, the bad and the ugly. *Trends Cell Biol* 15:599–607.
6. Beutler, B. (2004) Innate immunity: An overview. *Mol Immunol* 40:845–859.
7. Medzhitov, R., and C. J. Janeway. (2000) Innate immunity. *N Engl J Med* 343:338–344.
8. Friedl, P., and B. Weigelin. (2008) Interstitial leukocyte migration and immune function. *Nat Immunol* 9:960–969.
9. Park, J., and A. Barbul. (2004) Understanding the role of immune regulation in wound healing. *Am J Surg* 187:11S–16S.
10. Iwasaki, A., and R. Medzhitov. (2004) Toll-like receptor control of the adaptive immune responses. *Nat Immunol* 5:987–995.
11. Hoebe, K., E. Janssen, and B. Beutler. (2004) The interface between innate and adaptive immunity. *Nat Immunol* 5:971–974.
12. Jones, S., R. Edwards, and D. Thomas. (2004) Inflammation and wound healing: The role of bacteria in the immuno-regulation of wound healing. *Int J Low Extrem Wounds* 3:201–208.

13. Eming, S., T. Krieg, and J. Davidson. (2007) Inflammation in wound repair: Molecular and cellular mechanisms. *J Invest Dermatol* 127:514–525.
14. McKay, I., and I. Leigh. (1991) Epidermal cytokines and their roles in cutaneous wound healing. *Br J Dermatol* 124:513–518.
15. Robinson, M., D. Sancho, E. Slack, S. LeibundGut-Landmann, and C. Reis e Sousa. (2006) Myeloid C-type lectins in innate immunity. *Nat Immunol* 7:1258–1265.
16. Hart, J. (2002) Inflammation. 1: Its role in the healing of acute wounds. *J Wound Care* 11:205–209.
17. Takeda, K., and S. Akira. (2005) Toll-like receptors in innate immunity. *Int Immunol* 17:1–14.
18. Parish, W.E. (1998). Inflammation. In *Textbook of Dermatology*. R.H. Champion, B.J.L., D.A. Burns, and S. Breathnach, eds. Blackwell Science, Oxford.
19. Mayer-Scholl, A., P. Averhoff, and A. Zychlinsky. (2004) How do neutrophils and pathogens interact? *Curr Opin Microbiol* 7:62–66.
20. Dovi, J., A. Szpaderska, and L. DiPietro. (2004) Neutrophil function in the healing wound: Adding insult to injury? *Thromb Haemost* 92:275–280.
21. Gillitzer, R., and M. Goebeler. (2001) Chemokines in cutaneous wound healing. *J Leukoc Biol* 69:513–521.
22. Engelhardt, E., A. Toksoy, M. Goebeler, S. Debus, E.B. Bröcker, and R. Gillitzer. (1998) Chemokines IL-8, GROalpha, MCP-1, IP-10, and Mig are sequentially and differentially expressed during phase-specific infiltration of leukocyte subsets in human wound healing. *Am J Pathol* 153:1849–1860.
23. Fang, F.C. (2004) Antimicrobial reactive oxygen and nitrogen species: Concepts and controversies. *Nat Rev Microbiol* 2:820–832.
24. Agaiby, A., and M. Dyson. (1999) Immuno-inflammatory cell dynamics during cutaneous wound healing. *J Anat* 195(Pt 4):531–542.
25. Ganz, T. (2003) Defensins: Antimicrobial peptides of innate immunity. *Nat Rev Immunol* 3:710–720.
26. Steinstraesser, L., T. Koehler, F. Jacobsen, A. Daigeler, O. Goertz, S. Langer, M. Kesting, H. Steinau, E. Eriksson, and T. Hirsch. (2008) Host defense peptides in wound healing. *Mol Med* 14:528–537.
27. Zasloff, M. (2002) Antimicrobial peptides of multicellular organisms. *Nature* 415:389–395.
28. Dürr, M., and A. Peschel. (2002) Chemokines meet defensins: The merging concepts of chemoattractants and antimicrobial peptides in host defense. *Infect Immun* 70:6515–6517.
29. Doumas, S., Kolokotronis, A., and Stefanopoulos, P. (2005) Anti-inflammatory and antimicrobial roles of secretory leukocyte protease inhibitor. *Infect Immun* 73:1271–1274.
30. Borregaard, N., K. Theilgaard-Mönch, J. Cowland, M. Ståhle, and O. Sørensen. (2005) Neutrophils and keratinocytes in innate immunity—Cooperative actions to provide antimicrobial defense at the right time and place. *J Leukoc Biol* 77:439–443.
31. Brady, R., J. Leid, A. Camper, J. Costerton, and M. Shirtliff. (2006) Identification of *Staphylococcus aureus* proteins recognized by the antibody-mediated immune response to a biofilm infection. *Infect Immun* 74:3415–3426.
32. Martin, C., and I. Muir. (1990) The role of lymphocytes in wound healing. *Br J Plast Surg* 43:655–662.
33. Fuhlbrigge, R.C., J.D. Kieffer, D. Armerding, and T.S. Kupper. (1997) Cutaneous lymphocyte antigen is a specialized form of PSGL-1 expressed on skin-homing T cells. *Nature* 30;389:978–981.
34. Akira, S., K. Takeda, and T. Kaisho. (2001) Toll-like receptors: Critical proteins linking innate and acquired immunity. *Nat Immunol* 2:675–680.

35. Jameson, J., and W. Havran. (2007) Skin gammadelta T-cell functions in homeostasis and wound healing. *Immunol Rev* 215:114–122.

36. Diegelmann, R., and M. Evans. (2004) Wound healing: An overview of acute, fibrotic and delayed healing. *Front Biosci* 9:283–289.

37. Sansonetti, P. (2006) The innate signaling of dangers and the dangers of innate signaling. *Nat Immunol* 7:1237–1242.

38. Grey, J., S. Enoch, and K. Harding. (2006) Wound assessment. *BMJ* 332:285–288.

39. Sibbald, R., K. Woo, and E. Ayello. (2006) Increased bacterial burden and infection: The story of NERDS and STONES. *Adv Skin Wound Care* 19:447–461; quiz 461–443.

40. Hopf, H., and M. Rollins. (2007) Wounds: An overview of the role of oxygen. *Antioxid Redox Signal* 9:1183–1192.

41. Edwards, R., and K. Harding. (2004) Bacteria and wound healing. *Curr Opin Infect Dis* 17:91–96.

42. Rooijakkers, S., K. van Kessel, and J. van Strijp. (2005) Staphylococcal innate immune evasion. *Trends Microbiol* 13:596–601.

43. Kastrup, C., J. Boedicker, A. Pomerantsev, M. Moayeri, Y. Bian, R. Pompano, T. Kline, P. Sylvestre, F. Shen, S. Leppla, W. Tang, and R. Ismagilov. (2008) Spatial localization of bacteria controls coagulation of human blood by "quorum acting." *Nat Chem Biol* 4:742–750.

44. Kharazmi, A. 1991. Mechanisms involved in the evasion of the host defence by *Pseudomonas aeruginosa*. *Immunol Lett* 30:201–205.

45. Foster, T. (2005) Immune evasion by staphylococci. *Nat Rev Microbiol* 3:948–958.

46. Gresham, H., J. Lowrance, T. Caver, B. Wilson, A. Cheung, and F. Lindberg (2000). Survival of *Staphylococcus aureus* inside neutrophils contributes to infection. *J Immunol* 164:3713–3722.

47. Wolcott, R., D. Rhoads, and S. Dowd. (2008) Biofilms and chronic wound inflammation. *J Wound Care* 17:333–341.

48. Leid, J., C. Willson, M. Shirtliff, D. Hassett, M. Parsek, and A. Jeffers. (2005) The exopolysaccharide alginate protects *Pseudomonas aeruginosa* biofilm bacteria from IFN-gamma-mediated macrophage killing. *J Immunol* 175:7512–7518.

49. Shapira, L., P. Tepper, and D. Steinberg. (2000) The interactions of human neutrophils with the constituents of an experimental dental biofilm. *J Dent Res* 79:1802–1807.

50. Cabral, D., B. Loh, and D. Speert. (1987) Mucoid *Pseudomonas aeruginosa* resists nonopsonic phagocytosis by human neutrophils and macrophages. *Pediatr Res* 22:429–431.

51. Pedersen, S., A. Kharazmi, F. Espersen, and N. Hiby. (1990) *Pseudomonas aeruginosa* alginate in cystic fibrosis sputum and the inflammatory response. *Infect Immun* 58:3363–3368.

52. Singh, P., M. Parsek, E. Greenberg, and M. Welsh (2002). A component of innate immunity prevents bacterial biofilm development. *Nature* 417:552–555.

53. Walker, T.S., K.L. Tomlin, G.S. Worthen, K.R. Poch, J.G. Lieber, M.T. Saavedra, M.B. Fessler, K.C. Malcolm, M.L. Vasil, and J.A. Nick. (2005) Enhanced *Pseudomonas aeruginosa* biofilm development mediated by human neutrophils. *Infect Immun* 73:3693–3701.

54. Overhage, J., A. Campisano, M. Bains, E. Torfs, B. Rehm, and R. Hancock (2008). Human host defense peptide LL-37 prevents bacterial biofilm formation. *Infect Immun* 76:4176–4182.

11 Antimicrobial Interventions for Wounds

*Steven L. Percival, Rose A. Cooper,
and Benjamin A. Lipsky*

CONTENTS

Introduction.. 294
Types of Antimicrobial Agents 295
The Rationale for Using Antimicrobial Interventions in Wounds 297
Physical/Mechanical Methods for Reduction/Removal of Bioburden and
Biofilms.. 299
    Debridement... 299
    Enzymic Debridement... 300
    Biological Debridement 300
Antibiotics... 301
Antiseptics... 303
Iodine 304
    Mode of Action 304
    Effectiveness on Biofilms... 305
Silver 305
    Mode of Action 306
    Efficacy on Biofilms... 306
Polyhexamethylene Biguanides (PHMB) 307
    Mode of Action 307
    Efficacy against Biofilms 307
Chlorhexidine.. 307
    Mode of Action 308
    Efficacy on Biofilms... 308
Acetic Acid.. 308
    Mode of Action 309
    Efficacy on Biofilms... 309
Honey 309
    Mode of Action 310
    Efficacy on Biofilms... 310
Hydrogen Peroxide 311
    Mode of Action 311
    Efficacy on Biofilms... 312
Lactoferrin... 312

Mode of Action 312
Efficacy on Biofilms... 312
Xylitol 312
Mode of Action 313
Efficacy on Biofilms... 313
Bacteriophages... 313
Mode of Action 313
Effectiveness on Biofilms.. 314
Ethylenediamine Tetraacetic Acid (EDTA)... 314
Mode of Action 315
Effectiveness of Biofilms 315
Quorum-Sensing Inhibitors... 316
Essential Oils 316
Conclusions.. 316
References... 317

INTRODUCTION

Coping with wounds has always been a serious problem for mankind. How ancient civilizations did so is difficult to imagine, as the only documentary evidence of their treatment strategies is provided by fragmentary records in the form of drawings, engravings, and scripts that can be traced back for less than 5000 years. It is only in the last 40 years that the detailed sequence of events in the wound healing process, and the factors that influence the various phases, have begun to be understood, allowing modern treatment strategies to be based on this knowledge. Primitive remedies must have been discovered gradually and serendipitously, and any effective therapeutic advances were presumably passed on verbally to successive generations. Early interventions included surgical procedures, topical treatments, and medicines administered orally as draughts.[1] Although concoctions devised by ancient peoples would have been restricted to locally available animal, vegetable, and mineral products, many different mixtures seem to have been developed.[1] Some of those remedies were likely effective, but others were based on religious and superstitious beliefs; not all "advances" represented therapeutic improvements.

Second only to achieving healing, the most consistent aim of wound management has been the treatment and prevention of infection. For those surviving the initial wound, infection was the leading cause of fatalities. Until the advent of antiseptics, and more recently antibiotics, controlling infection was difficult. Although the availability of antimicrobials revolutionized the approach to infection, it induced a false sense of security. Healthcare providers began to believe that we could control all infections with available, or newly developed, antimicrobials, leading the U.S. Surgeon General to have claimed that it was now possible "to close the book on infectious disease."[2] But the emergence and growing prevalence of antibiotic-resistant strains of microorganisms has confounded that statement. We now recognize that the selection of multidrug-resistant strains poses a real threat to our own and future generations. For various reasons, the development of new antibiotics has slowed and there are relatively few new agents in the developmental pipeline. This has led to efforts to find novel antimicrobial interventions and to the investigation of antimicrobial agents that were formerly abandoned. The increased awareness of the presence and importance of biofilms in wounds has also fueled the search for additional effective treatments against this phenomenon.

This chapter reviews the use of antimicrobial interventions in wounds, summarizes the mode of action of selected agents, and provides evidence for their efficacy on biofilms.

TYPES OF ANTIMICROBIAL AGENTS

The efficacy of an antimicrobial agent is variable on different types of microorganisms, and efficacy can even vary for different strains of the same species. Figure 11.1[3] presents an overview of the relative susceptibilities of various types of microorganisms to available antimicrobials. Antimicrobial agents, except for many antibiotics, usually target multiple sites in the microorganism, which results in generalized damage to microbial cells.

To appreciate the overall effects of an antimicrobial, it is important to understand both the structure of bacterial cells and the differences between the two major types

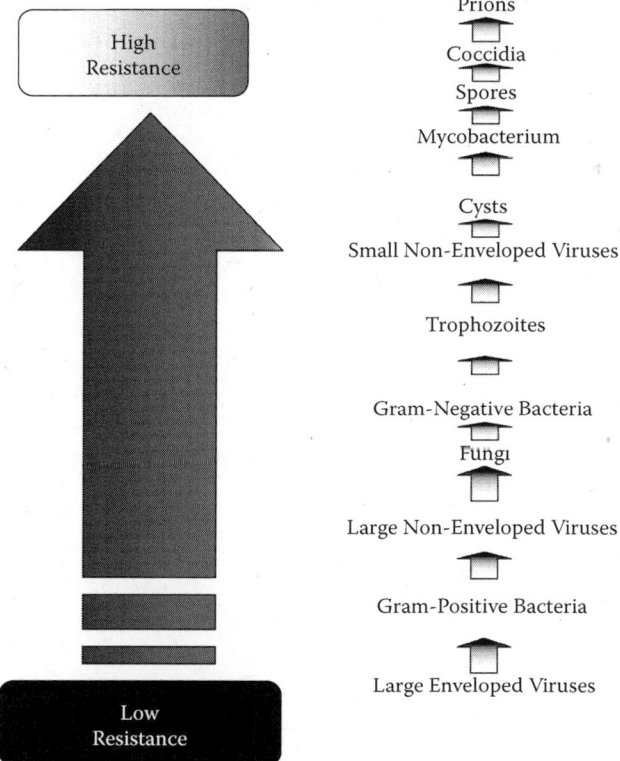

FIGURE 11.1 Sensitivity of various types of microorganisms to available antimicrobial agents. (Adapted from McDonnel, G., and Russell, D. (1999) *Clin Microbiol Rev* 12:147–179.)

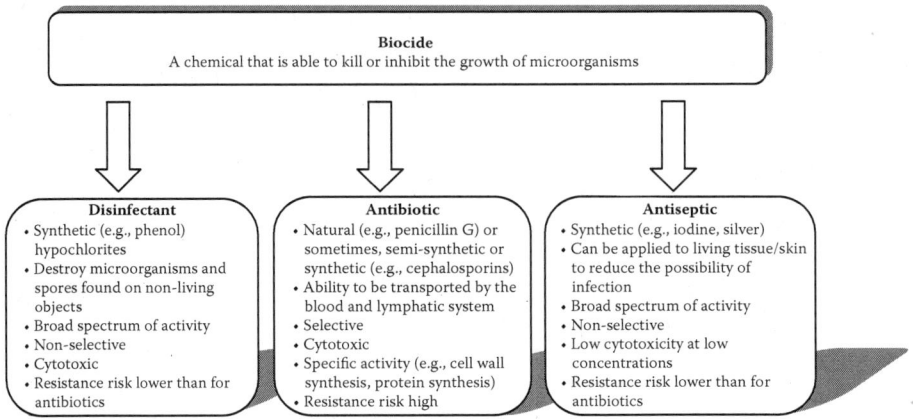

FIGURE 11.2   Biocide overview.

of bacterial cells (i.e., Gram positive and Gram negative). Gram-negative bacteria are generally more resistant to antimicrobial agents than are Gram-positive bacteria.

Growth of microorganisms can be limited in several ways, including restricting their access to essential nutrients, manipulating the physical environment away from the optimal conditions that support growth, and adding inhibitors that block their metabolic pathways. The resulting inhibition may be transient or permanent, depending on whether cellular viability is affected. When growth is inhibited without loss of viability, a bacterostatic effect is achieved; the loss of viability denotes a killing or bactericidal effect. The complete destruction or removal of viable cells is termed *sterilization*; *in vitro* this is relatively easily attained, but *in vivo* it is virtually unobtainable.

The use of antimicrobial agents to reduce the numbers of microorganisms (or microbial "load") has been given several labels, depending on what type of agent and in what manner it is used (Figure 11.2). *Disinfection* is the process of eliminating potentially pathogenic microbes from an inert surface without achieving sterilization; it usually involves employing chemical agents called disinfectants. Antiseptics are used to eliminate potentially pathogenic microbes from the surface of living tissues. The distinction between disinfection and antisepsis is not always clearly defined, and some texts refer to skin disinfection. Some definitions of relevant terms with suitable examples are provided in Table 11.1.

Reducing microbial load *in vivo* relies on using chemotherapeutic agents that selectively inhibit or kill microbial cells without damaging host cells. A diverse range of agents is available, but in clinical practice antiseptics and topical antibiotics have most frequently been used for this purpose. Agents used for clinical care should not adversely affect mammalian cells. The efficacy of all disinfectants and antiseptics is influenced by many factors, including their concentration, temperature, and contact time; the types and numbers of microbial cells present; and the presence of other organic matter. Treating invasive wound infections usually requires using systemic (oral or parenteral) antibiotics, most of which are different than those used topically. Some antibiotics are bacterostatic but most are bactericidal; the advantage of the latter is mainly limited to treating patients who are immunocompromised or in selected

TABLE 11.1
Conventional Antimicrobial Strategies

Term	Definition	Selected Examples	Antimicrobial Effect
Antiseptic	Agents that eliminate potentially pathogenic microbes from living tissue	Chlorhexidine Iodine Silver	Microbicidal Microbicidal Microbicidal
Antibiotic	A substance derived from a microorganism or produced synthetically that kills or inhibits the growth of other microbial species	Chloramphenicol Mupirocin Penicillin	Bacteriostatic Bactericidal Bactericidal
Biocide	An agent that literally kills *all* types of living cells, but implicitly it is intended to be selective	Silver	Biocidal (microbicidal)
Disinfection	The elimination of potentially pathogenic microbes from inanimate surfaces	Glutaraldehyde Phenol	Microbicidal Microbicidal
Sterilization	The complete removal or destruction of living cells from an object or an environment	Elevated temperatures (autoclaving) Irradiation with gamma rays	Microbicidal Microbicidal

severe infections such as endocarditis or meningitis. Unlike antibiotics used systemically, topical antibiotics are often rapidly inactivated within wounds. An antibiotic will act on a specific target site within specific microbial cells, and antiseptics, disinfectants, and sterilants typically induce nonspecific effects on multiple microbial cellular target sites and therefore possess broader specificity (Figure 11.3).

The emergence of microbes becoming resistant to antibiotics, and to a lesser extent to disinfectants, together with the persistence characteristic of biofilm-associated infections, has led to a broadening of approaches to microbial control. These include biological methods of controlling infections, such as maggots (larval biotherapy) and viruses (bacteriophages). No antimicrobial intervention is universally suitable for all wounds. For most antiseptics, such as iodine, there have been concerns about the possibility of toxicity to host cells.[5,6] These agents also have raised issues concerning their antimicrobial efficacy and chemical stability.[7]

THE RATIONALE FOR USING ANTIMICROBIAL INTERVENTIONS IN WOUNDS

As discussed in a previous chapter, the skin is the normal habitat for a variety of microbial species. Prior to a planned wound being initiated (e.g., a surgical incision), that microbial load is normally reduced by swabbing the skin with antiseptics. During nonelective injury, skin microorganisms can contaminate the wound and additional microbial cells may enter the wound from many sources (e.g., surrounding skin, other body sites, the hands of other people, including healthcare personnel, airborne organisms, animals, and local environmental surfaces).

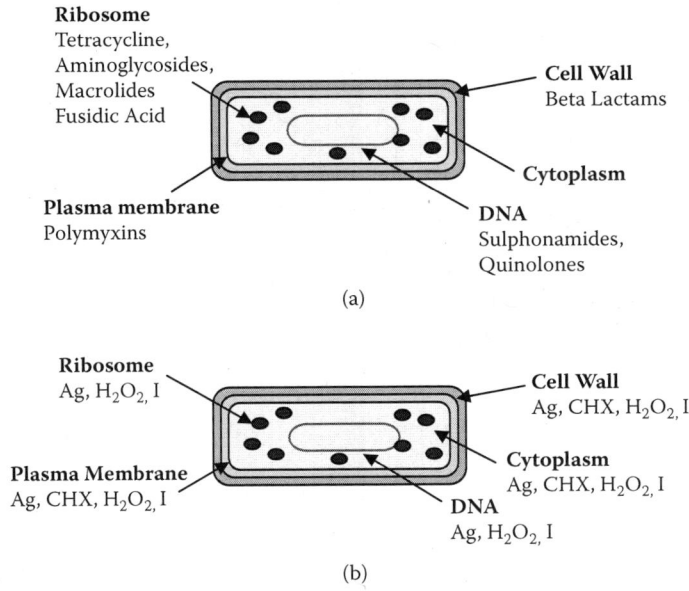

FIGURE 11.3 Target sites of antimicrobial agents in bacteria. (a) Examples of target sites for antibiotics. (b) Examples of target sites for antiseptics. Ag = silver, H_2O_2 = hydrogen peroxide, I = iodine, CHX = chlorhexidine.

Clearly uninfected wounds do not usually require any type of antimicrobial therapy. Prolonged observations and experimental investigations have helped clarify some aspects of the involvement of microorganisms in the pathogenesis of wounds, although the process is incompletely understood. What is known is that all clinically infected wounds require antimicrobial therapy, as acute wound infections interrupt healing.[8,9] For some wounds that are not clearly infected but may have a high colonizing bacterial load, there are data suggesting that reducing the bacterial bioburden (with topical agents) may improve rates of healing.[10] In these instances, it is necessary to weigh the risks of cytotoxicity against the potential benefits of the antimicrobial. There is also controversy about how best to define infection in a wound, both on clinical and microbiological grounds. A wound does not have to be cleared of all microbial cells for it to heal successfully, and antimicrobials are designed to cure infections (which usually takes days), not heal wounds (which may take months). There is no reason to continue antibiotics once the evidence of infection has resolved. Because of the continuing selection of microbial strains with antimicrobial resistance, routine antimicrobial intervention for all wounds is inappropriate.

The pathogens most frequently associated with wound infection are staphylococci, streptococci, pseudomonads, and obligate anaerobes. Whether a wound infection develops depends on a complex and often unpredictable series of interactions between pathogens and host.[11] Systemic antibiotics are used extensively for infected wounds, particularly in patients with other underlying pathologies. Antimicrobial interventions in this situation aim to limit the development of lymphangitis, bacteremia, septicemia, and death. A range of systemic antibiotics are used to treat infected wounds, including penicillins, cephalosporins, macrolides, tetracyclines, and fluoroquinolones.[12,13] Guidelines for the selection of systematic antibiotics and for deciding when to admit patients with serious wound infections to hospital have been formulated.[14]

Removing any necrotic tissue together with its associated microbial bioburden will assist acute and chronic wounds to heal in a timely manner. Not only will this help to limit the spread of infection, but it will reduce the quantity of organic debris, which could diminish the efficacy of any antimicrobial agents applied topically. This process (debridement) has been linked to enhanced rates of healing.[15] Moreover, failure to debride necrotic wounds has been associated with failure to heal and recurring infections.[16,17] The removal of necrotic tissue is best achieved surgically (with scalpel and forceps) but can sometimes be done chemically or mechanically (with various technologies).

Aside from debridement, the application of topical antimicrobials, either as antiseptics or as active or passive wound dressings, has been reported to contribute to the control of bioburden in wounds that are infected or at risk of infection.[17–19] Both infected and noninfected wounds can benefit from appropriate wound dressings. However, the type of dressing selected will depend on the type of wound and the severity of infection, or the risk of infection.[20,21]

In addition to deciding whether to use antibiotics, debridement techniques, or antiseptics, the wound practitioner should consider the possibility that chronicity is linked to the presence of a biofilm.[22] Evidence that biofilms actively impair healing[23] is rapidly accumulating, and wound management strategies will have to adapt to consider how to deal with this phenomenon. Biofilms are a community of bacteria at high density encased within an extracellular matrix of polymers, with reduced growth rates that make them less susceptible to inhibitors. Through spontaneous mutation, stress response, and bacterial synergies, the constituent bacteria may resist an attack by a single antimicrobial agent. Because bacteria in biofilms are notoriously resistant to both antimicrobial interventions and host immunological reactions, effective antibiofilm management strategies are essential. At present, such strategies are new to wound care and to medicine in general, but biofilms are controlled effectively in industrial situations, such as food processing, pulp and paper manufacture, shipping, and oil production.

Available antimicrobials and antibiofilm agents function by different mechanisms. Each may be effective in suppressing biofilm on the surface of a wound, but in routine clinical use they appear to apply only one selective pressure and within a short period of time biofilms may become less responsive to inhibitors. It has been suggested that by rotating different selective biocides, the capacity of a biofilm to adapt may be diminished and combination therapy will offer more effective control.[24]

PHYSICAL/MECHANICAL METHODS FOR REDUCTION/REMOVAL OF BIOBURDEN AND BIOFILMS

DEBRIDEMENT

Physical removal of the microbial bioburden, necrotic tissue, and biofilm is widely considered to be beneficial to wounds, although it is not advised for ischemic wounds.

In addition to biofilms,[23] the presence of devitalized tissue impedes wound healing. Based on one study, Steed and colleagues concluded that "debridement is a vital adjunct to the care of patients with chronic diabetic foot ulcers,"[25] yet no universal consensus exists on how to debride, when to debride, or how much tissue to remove.[26,27] Sharp removal of devitalized tissue may further help to enhance the wound healing process by reducing microbial toxins, proteases, and neutrophils, which are often present in wounds at elevated levels.[28] Presumably, debridement may prevent a biofilm progressing or "maturing" to a state where it becomes harder to eradicate. It has been shown that sharp debridement and topical treatment with lactoferrin and xylitol improved outcomes in patients with critical limb ischemia,[29] but further scientific validation is indicated before frequent debridement is accepted as a means to suppress biofilms. Clinicians will ultimately decide on the timing for surface management of biofilm, and this timing would be based on the progress of the wound and will be influenced by host factors. For example, biofilms may grow at a faster rate in immunocompromised patients than in immunocompetent individuals. Debridement can be discontinued once a self-sustaining wound is achieved. If a wound regains sufficient perfusion and effective host defenses are restored, it will be more difficult for biofilm to redevelop. This redevelopment of a biofilm may be preventable with the use of appropriate dressings and possibly with effective topical antiseptic products, but this remains to be proven. Because all open wounds are vulnerable to biofilm formation, appropriate wound management is required until complete wound closure is achieved.

ENZYMIC DEBRIDEMENT

Biofilms are, as mentioned above, embedded in a matrix composed of a heterogeneous mixture of extracellular polymeric substances (EPS) which includes polysaccharides, proteins, and nucleic acids. These viscous, slimy structures can occur in pipelines and on the surfaces of contact lens; they have been occasionally reduced significantly with preparations of enzymes. Despite the use of enzyme preparations such as collagenase, streptokinase, streptodornase, and krill as debriding agents in necrotic wounds, their potential in removing biofilms from wounds has not been fully investigated. Streptodornase (varidase) degrades DNA and has long been used in breaking down blood clots.

DNA is a significant component of the EPS of *Pseudomonas aeruginosa* biofilm, and it has been reported that application of varidase to silicone sheets with established *P. aeruginosa* biofilms effectively destroyed them.[30] Glucose oxidase and lactoperoxidase have been shown to be bactericidal for staphylococcal and pseudomonad biofilms, but polysaccharide hydrolyzing enzymes were needed to remove these biofilms from inert surfaces *in vitro*.[31] The complex nature of biofilm EPS suggests that mixtures of enzymes are more likely to be effective in eradicating biofilms from wounds than preparations containing single enzymes.

BIOLOGICAL DEBRIDEMENT

Another means of removing exogenous material from wounds is with maggots (i.e., larvae of flies). For many years, maggots have been known to improve wounds.

In modern times, the use of maggot debridement therapy for wounds was introduced in 1931 by an orthopedic surgeon.[32] Not only will this remove devitalized tissue and help to reduce bioburden, but larval secretions appear to be beneficial in disrupting biofilm,[33–35] and they help to counteract the inflammatory effects of neutrophils.[36] Compared to hydrogel, larval therapy has been shown to significantly reduce the time for debridement.[37] The exact mechanism of action of larval biotherapy is not yet clear.

ANTIBIOTICS

Clinically infected wounds, especially those with signs of spreading (e.g., cellulitis or erysepalis, lymphangitis) or deep-seated infection (abscesses or osteomyelitis) require systemic antibiotic therapy. The specific agents are chosen either empirically, based on clinical evidence, or specifically when the identity of the causative agent and its antibiotic susceptibilities have been determined. Unless the infection involves bones or joints, the duration of treatment is usually 7 to 14 days, when signs and symptoms are expected to resolve. Prolonged antibiotic administration (i.e., 4 to 6 weeks) is recommended for osteomyelitis. Prescribing details are available in pharmacology texts and guidelines,[14] but local authorities should be consulted. Details of modes of action of systemic antibiotics and some antiseptics are summarized in Table 11.2 and Table 11.3.

Topical antibiotics are usually administered as creams, gels, or ointments for periods of less than 2 weeks. They have been shown to be effective in treating superficial infections, such as impetigo They have also been used off-label without published evidence of efficacy for prophylaxis against acute infection, as well as to eradicate staphylococci (particularly methicillin-resistant *S. aureus* [MRSA]) from colonized wounds and to prevent cross-infection. The topical antibiotics most often used on skin and wound infections are fusidic acid (in countries where it is licensed), gentamicin, metronidazole, mupirocin, neomycin sulfate, and silver sulfadiazine. Many antifungal agents (e.g., imidazoles and allylamines) are also applied topically for superficial fungal infections (e.g., tinia). Topical metronidazole appears to be effective at reducing the odor of wounds colonized by obligately anaerobic bacteria.

Antibiotic usage, especially those administered topically, is a controversial area in chronic wound management. Antibiotic therapy is associated with potential adverse effects, with expense, but most importantly with selective pressure that increases the prevalence of strains of antibiotic-resistant microbes. Organisms with multiple antibiotic resistance determinants, and even antiseptic resistance, make effective antimicrobial treatment increasingly difficult. Antibiotic resistance concerns are also compounded by the presence of biofilms that have inherent resistance to antibiotics and antimicrobial agents in general. Physicians commonly utilize antibiotics to control bacteria in wounds as though they exist in a free-living (planktonic) state, where they are highly susceptible to appropriately selected agents. The realization that the use of antibiotics alone often produces a limited response in chronic wounds, partially because bacteria are likely in a biofilm, has led to recommendations that their usage be limited to short duration.[38] If future studies show that antimicrobials are effective for organisms in biofilms in chronic wounds, they should be considered

TABLE 11.2
Target Sites for Antiseptics

Target Site	Mechanism	Biocide
Cell wall	Prevents cross-linking	EDTA, chlorhexidine, chlorine, silver, phenols
Outer membrane (Gram-negative bacteria)	Increased permeability	EDTA, chlorhexidine, chlorine, silver, phenols, cationic compounds, lactoferrin, transferrin, polyphosphates, hypochlorites
Cytoplasmic (plasma) membrane	Increased permeability	Organic acids, alcohols, chlorhexidine, phenols, cationic compounds, lactoferrin, transferrin, polyphosphates, hypochlorites
	Membrane potential and electron transport chain	Organic acids, alcohols, chlohexidine, phenols, PHMB
	Inhibits adenosine triphosphate synthesis	Cationic compounds, PHMB
	Inhibition of enzyme activity	PHMB, chlorhexidine, phenols, silver, iodine
Cytoplasmic constituents	Generalized coagulation	Ionic silver, phenols, chlohexidine, PHMB
	Nucleic acids	Organic acids, chlorine, iodine, ionic silver, ozone
	Ribosomes	Hydrogen peroxide
Interactions with specific groups	Thiol groups	Hydrogen peroxide, chlorine, iodine, ionic silver
	Amino groups	Ionic silver, iodine, hydrogen peroxide, chlorine, ionic silver
	Sulfhydryl groups	Ionic silver, iodine, hydrogen peroxide, chlorine, ionic silver
Biocide-induced autocidal activity	Causes accumulation of free radicals	Hydrogen peroxide, agents that damage the cytoplasmic membrane

Source: Adapted from Maillard, J.-Y. (2002) *J Applied Microbiol Symp Supplement* 92:16S–27S.

as an adjunct to any management strategy. This has been highlighted by Fux and colleagues who suggested that for "Successful treatment in these cases (biofilm infections) depends on long-term, high dose antibiotic therapies and the removal of any foreign body material."[24] It is interesting to speculate that other infections that have been found to often require treatment with high doses of antibiotics for long durations (e.g., osteomyelitis, prostatitis) are likely biofilm-associated diseases. However, it is important to appreciate that they are heterogeneous structures and different regions of the biofilm respond in different ways to antibiotics. Some bacteria deep within regions of the biofilm may be less susceptible to antibiotics than the actively growing bacteria at the outer regions. Studies have shown that organisms in superficial regions of the biofilm are susceptible to high doses of antibiotics, whereas other regions are not.[39,40] Well-designed clinical trials are needed to determine the role of antibiotics and antiseptics, in combination with frequent debridement, antibiofilm

TABLE 11.3

Target Sites for Antibiotics Used Systemically in Treating Wound Infections

Type of Antibiotic	Examples	Target Site	Antimicrobial Effect
β lactams	Penicillin, amoxicillin, flucloxacillin, cephalosporins	Peptidoglycan biosynthesis (bacterial cell wall)	Bactericidal
Quinolones	Ciprofloxacin, levofloxacin	Bacterial DNA synthesis	Bactericidal
Tetracyclines	Oxytetracycline, doxycycline	Bacterial protein synthesis	Bacteriostatic
Aminoglycosides	Amikacin, gentamicin	Bacterial protein synthesis	Bactericidal
Macrolides	Erythromycin, azithromycin	Bacterial protein synthesis	Bactericidal
Lincosamides	Clindamycin	Bacterial protein synthesis	Bactericidal
Glycopeptides	Vancomycin, teicoplanin	Peptidoglycan biosynthesis (bacterial cell wall)	Bactericidal
Nitroimidazoles	Metronidazole	Metabolic functions	Bactericidal (for obligate anaerobes)
Carbapenems	Imipenem, meropenem, ertapenem	Bacterial protein synthesis	Bactericidal

agents (lactoferrin, xylitol), and appropriate wound dressings, in improving stalled wound healing.

ANTISEPTICS

Antiseptics are commonly used in chronic wound care management, but the clinical data supporting each of the available agents are suboptimal. They are classed as nonspecific inhibitory agents because their intracellular effects are exerted on multiple microbial target sites and functions. Many antiseptic agents are effective against a vast array of infectious agents, but adverse effects may also be seen on host cells, particularly fibroblasts and keratinocytes, but also polymorphonuclear leukocytes. Appropriate and judicious usage of antiseptics at optimum concentrations is, therefore, essential. Historically, antiseptics have been valuable treatments for treating and preventing wound infections. The introduction of carbolic acid into surgery during the 19th century by Joseph Lister changed surgical practice. However, attitudes toward antiseptics were altered by two studies published in 1985. Both demonstrated the occurrence of cytotoxicity when topical antiseptics were applied to animal models,[41,42] and this influenced healthcare practitioners to restrict their use. Such *in vivo* models (a rabbit ear chamber and cultured human fibroblasts) and later experiments with acute wounds in rats may not accurately simulate the conditions of a human patient, especially one with a chronic wound. Laboratory experiments necessarily precede the clinical use of a treatment and provide a means of discarding ineffective and toxic candidate agents. Ultimately, only clinical studies can provide suitable data on the efficacy of topical agents. Critical evaluation of available data indicates that cytotoxicity issues may have been exaggerated for some agents[43] but well-designed clinical studies are still needed to demonstrate efficacy.[44]

A range of antiseptics are still used in modern wound care. Information on some of the most commonly used antiseptics in wound care is outlined below.

IODINE

The element iodine (I) was first discovered in 1811 by Bernard Courtois. In the early 19th century, tinctures of iodine solution and potassium iodide in ethanol were used to treat wounds. However, it had notable limitations in that it caused pain, irritation, and staining; also it was found to be cytotoxic when used at high concentrations.

Iodine is known to have both bactericidal and bacteriostatic actions and has a broad spectrum of activity against virtually all microorganisms. Consequently, iodine has been used extensively in the treatment of wounds that are, or have the potential to become, infected. Despite extensive use of iodine, worldwide debate on its efficacy and cytotoxicity continues even though the reasons for its unpopularity are probably unfounded.[45] A recent review on the use of iodine in wound care demonstrated that formulations delivering optimized but sustained levels of iodine have the potential to be effective antimicrobial agents which are noncytotoxic and also stimulate wound healing.[6]

New-generation iodine products, such as povidone and cadexomer iodine, have addressed many of the undesirable effects of iodide. They have thereby significantly helped to maintain interest in iodine therapy. Povidone iodine is an iodophore composed of elemental iodine and a synthetic polymer (polyvinylpyrrolidone). Povidone iodine has been shown to reduce inflammation, bacterial numbers,[46] and protease activity.[47] The clinical evidence for the safety of povidone iodine is conflicting. Some studies have shown that it is not detrimental to wound healing,[48–51] but animal models have not always provided confirmation. Possible reasons for this include the concentrations of iodine used, and the study subjects and healing markers utilized.

Cadexomer iodine (Iodosorb) is composed of microspherical beads of hydrophilic biodegradable starch and 0.9% (w/v) iodine. This formulation is purported to give improved iodine release rate and reduced cytoxicity compared to povidone iodine, which has led many wound care practitioners to consider this to be a superior iodiine product to povidone iodine.

There are several reviews on iodine which the reader may find useful.[6,52–55]

Resistance to iodine was first documented in iodinated swimming pools.[56] Resistance has also been reported in strains of *Staphylococcus aureus*.[57] To date, bacterial resistance to iodine is considered rare and sporadic.

MODE OF ACTION

The mode of action of iodine depends on its ability to bind to thiol and sulfydryl groups. This binding leads to the denaturation of proteins and inactivation of enzymes via oxidation of the S-H bonds found in amino acids. Such structural changes in proteins and enzymes will have detrimental effects on the microbial cell walls, membranes, and cytoplasmic components.[58] Iodine is also able to bind to fatty acids and to nucleic acids.

Povidone iodine is a broad-spectrum antiseptic that has been shown to be effective on Gram-positive and Gram-negative bacteria, fungi, viruses, protozoa, and bacterial spores. Its mode of action is similar to the mode of action of molecular iodine, its active ingredient. The effects that povidone iodine elicit in microbial cells have been investigated using electron microscopy.[59]

Cadexomer iodine is a highly absorbent agent known to release low levels of iodine when added to a chronic wound. When the polymer is hydrated with wound exudate it swells, then iodine begins to leak out of the cadexomer-iodine complex. An optimized and sustained low-level delivery of iodine is then apparently directed into the wound. Clinical studies have highlighted the effectiveness of iodine in wounds, including burns,[60] chronic leg ulcers,[61] and decubitus ulcers.[62] Cadexomer iodine has been shown to enhance wound healing by stimulating vascular endothelial growth factors.[63] Based on the available data, cadexomer iodine's qualities, in particular its reduced cytotoxicity, make it preferable to povidone iodine.

A number of new iodine-based products are now commercially available. These include Repithel®, which is composed of povidone iodine encapsulated in liposomes, and Oxyzyme®, which consists of a hydrogel sheet containing glucose. This hydrogel sheet is placed on a wound surface and a smaller gel sheet consisting of glucose oxidase is added to it. In the presence of oxygen, the catalyst glucose oxidase converts glucose to gluconic acid and hydrogen peroxide. Glucose in the lower sheet then diffuses into the upper gel sheet and the hydrogen peroxide is then released back into the gel which diffuses through the dressing. This then oxidizes iodide ions to free iodine and the release of oxygen into solution. Oxyzyme has been shown to have beneficial effects on postsurgical wounds[64] and chronic wounds. Iodozyme® is a similar product that makes higher levels of iodine available at the wound surface.

EFFECTIVENESS ON BIOFILMS

The antibiofilm efficacy of iodine has not been well researched to date. A number of studies have been reported to show that when bacteria grow within a biofilm they become more resistant to iodine.[66,67] Kunisada and colleagues[68] reported that povidone iodine was efficacious on biofilms. Some recent work has shown that povidone iodine is less effective than hydrogen peroxide and alcohols in eradicating *Staphylococcus epidermis* within *in vitro* biofilms.[69] Cadexomer iodine has been reported to decrease slough in chronic wounds and disrupt a *S. aureus* biofilm.[70]

No studies have yet been published on the efficacy of Oxyzyme® or Iodozyme® against microorganisms within biofilms. However, a large clinical study has demonstrated improved healing in a large number of wounds.[71]

SILVER

The efficacy of silver in treating infected wounds has been known since ancient Roman times. In the 1800s, silver nitrate and other silver compounds were used with positive results in wound care. In 1964 silver sulfadiazine (SSD) began to be used for the management of infections in burn wounds, especially in an effort to prevent wound-related sepsis.

The active agent in silver preparations is ionic silver, which has been shown to be effective against microorganisms found in a wound bed.[72,73] Provided optimized preparations that allow sustained levels of ionic silver are used, silver has been shown to maintain a broad-spectrum antimicrobial effect in wounds.

Resistance to silver is both rare and sporadic but has been reported and documented in bacteria isolated from chronic wounds.[74–77] A recent study investigated the prevalence of silver resistance genes in 172 bacterial strains that had been isolated from both human and equine wounds. The authors demonstrated evidence of silver resistance genes in six strains of *Enterobacter cloacae* isolated from chronic wounds, yet when the silver resistant strains of bacteria were exposed to ionic silver, they were effectively inhibited.[78]

MODE OF ACTION

Ionic silver (Ag^+) at low concentration is effective in killing bacteria, whereas the elemental form of silver (Ag^0) has no antimicrobial effects. The efficacy of ionic silver is affected by a number of factors, such as the concomitant presence of anions, proteins, sulfides, chloride ions, and phosphates. The levels of ionic silver achieved are also important; levels of 5 to 50 µg/mL (ppm) are inhibitory against bacteria in wounds.[79] Microbial susceptibility also varies among bacterial species.

Silver inhibits bacteria by many mechanisms, including hindering bacterial respiration, disrupting the cell membrane, denaturing proteins, and altering nucleic acids. Its bactericidal efficacy is thought to be due to its ability to bind to disulfide and sulfydryl groups of proteins found in the cell walls of bacteria,[80,81] as well as its ability to join to DNA.[82] Because silver affects so many of the metabolic processes found in bacteria, it is rapidly bactericidal.[83]

EFFICACY ON BIOFILMS

Ionic silver is highly effective in inhibiting planktonic bacteria, particularly in drinking water.[84,85] However, at levels of 100 µg/L, ionic silver was shown in one study to be ineffective at destroying bacteria in a biofilm.[86] Another study, however, found that Elastoguard silver-releasing rubber reduced *Pseudomonas aeruginosa* biofilm formation in water.[87] Ionic silver has been reported to either delay or prevent biofilm formation on therapeutic catheters,[88] artificial heart valves,[89] and other medical prostheses.[90] Studies utilizing microbial genomics have demonstrated the effectiveness of silver in preventing the formation of biofilms.[91] Also, a study by Bjarnsholt et al.[92] showed that silver was effective against mature biofilms of *P. aeruginosa*. It further highlighted the importance of an adequate silver concentration for biofilm eradication and suggested that wound dressings might not provide sufficient levels for activity *in vivo*. In a study employing atomic force microscopy to measure the effect of ionic silver on *S. epidermidis* biofilm,[93] a concentration of 5 to 10 µg/mL silver sulfadiazine eradicated the biofilm, whereas a lower concentration (1 µg/mL) had no effect. The authors concluded that the concentration of silver required to eradicate a bacterial biofilm was 10 to 100 times higher than that used to eradicate planktonic bacteria. They suggested that the concentration of silver in currently available wound dressings is much too low for treatment of chronic wounds with biofilm.[93]

The effectiveness of silver-containing wound dressings on biofilms is not well documented. A recent study by Percival et al.[94] found that a silver-containing dressing achieved total bacterial killing on biofilms grown in a chambered slide model after 48 hours. This research provided valuable evidence that this dressing may have an effect on biofilms found in recalcitrant chronic wounds. Silver-containing phosphate-based glasses have been found to reduce the growth of *Pseudomonas aeruginosa* and *Staphylococcus aureus* biofilms, two common organisms in nosocomial infections.[95] Silver in combination with antibiotics (e.g., tobramycin) has been shown to have better efficacy on biofilms than either agent used individually.[96]

POLYHEXAMETHYLENE BIGUANIDES (PHMB)

PHMBs are cationic biocides. These compounds have a broad spectrum of activity and have been used commercially as swimming pool sanitizers (Baquacil), as general biocides, and as antiseptics.[97] PHMB has also been used as a contact lens cleaning formulation[98] and for the treatment of *Acanthamoeba* keratinitis,[99,100] in mouthwashes,[101] as an agent for bacterial vaginosis,[102] genital warts,[103] and to treat chronic wounds.[104,105] PHMB possesses broad-spectrum antibacterial activity, and it does not induce the development of bacterial resistance.[106–107] No reported incidences of acquired resistance against this agent in bacteria have been documented to date.

MODE OF ACTION

PHMB has been shown to bind to the cell walls and membranes of both Gram-negative and Gram-positive bacteria, specifically to the lipopolysaccharides and peptidoglycan. It has been reported to cause the cell membrane to become more permeable, resulting in its disruption and the cellular leakage of numerous cations.[108] It has also been documented to bind to bacterial DNA, alter its transcription, and cause lethal DNA damage.[109] PHMB has low toxicity to human cells, largely due to their more complex membranes.

EFFICACY AGAINST BIOFILMS

There is limited research activity in this area, but efficacy of PHMB has been detected on *S. aureus* biofilms.[110]

CHLORHEXIDINE

Chlorhexidine is a biguanide (Figure 11.4) that was chemically synthesized in 1946 and introduced into clinical practice in 1954. It has been used as an antiseptic and disinfectant in many situations, but particularly as a surgical scrub, for infection

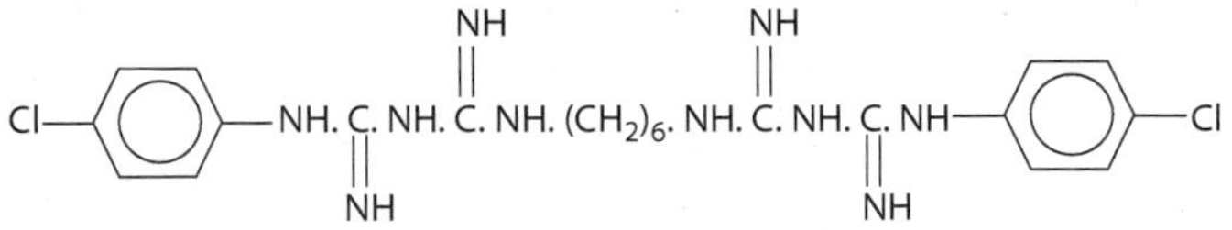

FIGURE 11.4   Structure of chlorhexidine.

prophylaxis in dentistry and urology procedures, and in the treatment of chronic wounds. In wound care it is used to irrigate and cleanse contaminated, traumatic wounds.[111]

Two systematic reviews have highlighted the paucity of evidence for the clinical efficacy of antiseptics in reducing the risk of postoperative infections. It is widely believed that antiseptics reduce skin microbiota, but the hypothesis that showering or bathing with an antiseptic before undergoing surgery prevents surgical site infection (SSI) has not been supported by objective evidence.[112] To date there are six available randomized controlled trials examining this issue with a total of 10,007 patients. Chlorhexidine gluconate was utilized in all of the studies, but it demonstrated no definitive advantage in diminishing SSIs compared to other wash products.[112] The evidence for the effectiveness of surgical hand antiseptics in reducing SSI is not very strong, either.[113] Data from four studies concerning aqueous scrubs indicated that those containing chlorhexidine gluconate were more effective in reducing the number of viable bacteria on hands than scrubs containing povidone iodine, but the reviewers considered that the evidence was limited. In dentistry studies, chlorhexidine rinses have been found to be effective in reducing the oral microbial load.[114]

MODE OF ACTION

The effect of chlorhexidine on bacterial cells depends on its concentration. At low concentrations it inhibits enzymes associated with bacterial membranes and causes leakage of cellular materials; at higher concentrations it causes denaturation of cytoplasmic components. A variety of bacteria are inhibited by chlorhexidine,[115] but resistance to this antiseptic has been reported.[116]

EFFICACY ON BIOFILMS

Unfortunately, much of the evidence concerning the efficacy of chlorhexidine on biofilms comes from *in vitro* rather than *in vivo* studies.[117,118] In the *in vitro* environment, studies of chlorhexidine on biofilm have shown it to be both effective[119] and noneffective,[120] leaving the question in doubt.

ACETIC ACID

Vinegar, or acetic acid, is reputed to have been discovered about 5000 years ago by the Babylonians when some of their alcoholic fermentations failed. It has since been utilized in various types of medicines and its first recorded use in wound care dates to Hippocrates c. 420 B.C. It is an agent that has a wide spectrum of activity against a broad range of microorganisms.

A 1% solution of acetic acid has been shown to eliminate *P. aeruginosa* (then called *Bacillus pyocyanea*) from military injury wounds within 2 weeks.[121,122] Acetic acid was used to successfully eliminate *P. aeruginosa* from the superficial wounds and burns in eight of nine patients within 3 weeks.[123] Similar results were found when concentrations of acetic acid between 0.5% and 5% were applied to burn and soft tissue wounds of 16 patients. *P. aeruginosa* was eliminated from 14 patients within 2 weeks and *in vitro* tests determined MIC values for all strains at 2% acetic acid.[124] This cheap and effective remedy for superficial *Pseudomonas* colonization is not used routinely today because a number of studies have shown that acetic acid is cytotoxic *in vitro*, although this does not seem to be the case *in vivo*.[125]

Mode of Action

Acetic acid has been shown to inhibit both Gram-positive and Gram-negative bacteria. Its mode of action is by preventing the uptake of essential substrates via active transport.

Efficacy on Biofilms

Few investigations into the efficacy of acetic acid on biofilms have been published. Treatments with low concentrations of acetic acid (1.25%, 15 min) have shown poor efficacy on biofilms.[126] However, Akiyama and colleagues have shown that acetic acid was effective on *S. aureus* biofilms.[127] Synergistic combination of 1% acetic acid with 50 or 100 µg/mL of monolaurin have eradicated planktonic cells of *Listeria monocytogenes*.[128] In the same study a population of 10^5 CFU/cm^2 of 1-day adherent cells was completely inactivated within 25 min by 1% acetic acid combined with 100 µg/mL of monolaurin.

HONEY

Of all of the antimicrobial agents used in wound care today, honey is probably the one with the longest history. Artefacts dating back more than 4500 years illustrate its use in many different cultures.[129] In British hospitals some patients' wounds were treated with commercially purchased (nonmedical) honey until the 1970s. This practice began to lose favor both because of concerns about spores that may contaminate nonmedical honey and the development of modern, occlusive and antibacterial dressings. The use of honey as topical treatment for wounds has recently been revived, however. The first modern wound care product was a blend of Australian and New Zealand honeys that was packaged into a tube, sterilized by gamma irradiation, and licensed by the Therapeutic Goods Agency for use in Australia and New Zealand in 1999. Now licensed wound care products containing medical grade honey are available throughout Europe, Hong Kong, Canada, and the United States. There is a large range of products, including tubes of sterile honey to gels containing honey and wax, ointments, hydrogels containing honey, mesh impregnated with honey, tulle impregnated with honey, calcium alginate impregnated with honey, and a flexible, nonsticky sheet or gel. The honeys utilized in these products vary; they may contain buckwheat honey, multifloral honey, manuka (with or without jellybush honey [*Leptospermum*]), or honey of unspecified floral origin. Except for remote communities in developing countries, commercially purchased honey is no longer used in conventional medicine because the antibacterial activity and microbiological quality cannot be assured.[130] Quality standards for honeys destined for wound care have been established[131] and medical-grade honey is distinct from that meant for human consumption.

There is extensive anecdotal evidence and increasing clinical evidence for the efficacy of honey in treating many different types of wounds.[132] A randomized controlled trial with venous ulcers recently showed that compared with a hydrogel, manuka honey increased the rate of healing and gave rise to a lower incidence of infection and more effective desloughing.[133] However, two recent randomized trials have demonstrated that the clinical benefits of honey were not statistically significantly better than standard care,[134,135] and three systematic reviews have criticized the quality of design of clinical studies using honey.[136–138] Notwithstanding the mixed evidence to date, the clinical use of honey continues to increase.

MODE OF ACTION

Honey is a supersaturated solution of sugars with an acidic pH, high osmolarity, and low water content. These characteristics alone restrict the growth of microorganisms.[139] Additional antimicrobial activity is generated on dilution by the activation of glucose oxidase (an enzyme incorporated into honey by bees) to produce hydrogen peroxide.[140] Plant-derived components also confer antimicrobial properties on *Leptospermum* honeys.[141,142]

Honey has a broad spectrum of antimicrobial activity and has been shown to inhibit more than 60 bacterial species,[143] as well as fungi, protozoa, and viruses.[144–147] Although the mode of action has not yet been fully elucidated, it has been shown that diluted honey inhibits bacteria at concentrations where the involvement of sugars is excluded.[148] Bactericidal activity against *P. aeruginosa* has been reported.[149]

EFFICACY ON BIOFILMS

All of the *in vitro* studies cited above have tested honey on planktonic cells, rather than biofilms. There is little experimental evidence available to evaluate the effect of honey on biofilms. In a preliminary laboratory study, manuka honey was shown to inhibit biofilms of *P. aeruginosa* at concentrations markedly higher than those required to inhibit suspensions of planktonic cells, and contact time was important.[150] Inhibition of biofilms formed by *S. aureus* and *P. aeruginosa* has also been demonstrated by manuka honey and a *Yemini* honey.[151] Two studies have suggested that honey prevents the attachment of bacteria to cell surfaces, which is a critical step in the initiation of biofilm formation.[152,153] It is possible that honey induces dispersal of established biofilms of *P. aeruginosa* by supplying excess nutrients,[154] but this hypothesis has not been tested.

Eradication of biofilm in wounds by honey has not yet been adequately demonstrated. One published example of biofilm control *in vivo* is a clinical trial that demonstrated that subjects who chewed a "honey leather" produced from manuka honey had significantly lower dental plaque scores than a control group.[155] Plaque is a biofilm found on teeth that has adverse effects on dental health.

HYDROGEN PEROXIDE

Hydrogen peroxide (H_2O_2) is a clear liquid that may be used as a 3% or 6% solution for cleansing traumatic, necrotic, or infected wounds. It has also been used as a disinfectant on inanimate surfaces within domestic and healthcare environments, especially contact lenses. Hydrogen peroxide has been found to be a good debriding agent by a number of researchers.[156] *In vitro* tests suggest that hydrogen peroxide is more effective against Gram-positive bacteria than Gram-negative ones, but it has cidal activity against a wide range of microorganisms, including bacterial spores. Hydrogen peroxide is generally recognized as a safe antimicrobial agent, by the U.S. Food and Drug Administration (FDA). The efficacy of hydrogen peroxide is reduced by catalase, which breaks it down to water and oxygen. Catalase is produced by both host cells in the wound tissue, and bacteria found in a wound.

The manufacture of hydrogen peroxide is made possible by the autoxidation of 2-ethyl-9,10-dihydroxyanthracene ($C_{16}H_{14}O_2$) to 2-ethylanthraquinone ($C_{16}H_{12}O_2$) and hydrogen peroxide using oxygen from the air. This whole process is known as the Riedl-Pfleiderer process (Figure 11.5).[157]

MODE OF ACTION

Hydrogen peroxide is relatively unstable in the presence of organic matter. It also decomposes (disproportionates) exothermically into water and oxygen gas spontaneously:

$$2\ H_2O_2 \rightarrow 2\ H_2O + O_2$$

The formation of free radicals during its oxidation, which cause irreparable breakage of strands in DNA molecules, probably accounts for its antibacterial activity. There have been safety concerns about hydrogen peroxide since reports of the formation of emboli,[158] and in most health care settings it is no longer used routinely on deep

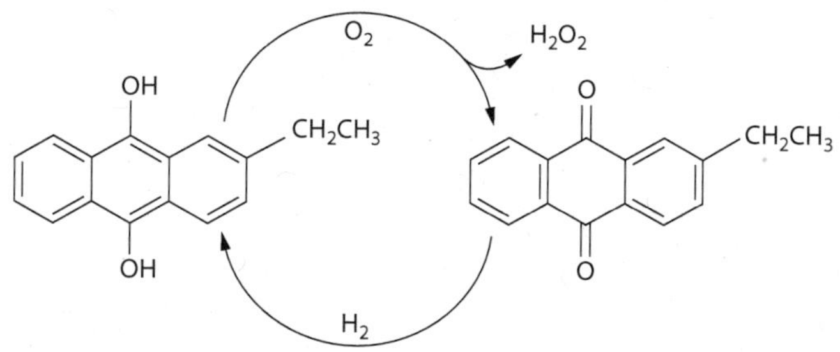

FIGURE 11.5 Riedl-Pfleiderer process.

wounds. New products, such as Oxyzyme, generate significantly lower concentrations of hydrogen peroxide than solutions of wound cleansers.

EFFICACY ON BIOFILMS

Hydrogen peroxide has been shown to be efficacious against *Streptococcus mutans* biofilm[159,160] and *S. epidermidis* biofilms.[69,161] Hydrogen peroxide has also been shown to be effective on biofilms in combination with other agents.[162]

LACTOFERRIN

Lactoferrin is part of the innate human immune system and is secreted in tears, saliva, mucous, and milk. Lactoferrin has a high affinity for iron, which is an essential trace metal for bacteria. Removing iron from their environment has been shown to have a bacteriostatic or a bactericidal effect on bacteria.[163]

MODE OF ACTION

Lactoferrin has an array of antibacterial effects, including enhancing the adhesion of neutrophils to endothelial cells,[164] and having high affinity for lipopolysaccharide (LPS) in the outer membrane of Gram-negative bacteria. It also binds to the lipid, which causes rigidity of acyl chains with resultant rapid release of lipopolysaccharides, causing enhanced cell membrane permeability with leakage of cell contents, followed by cell death.[165–167] Additionally, lactoferrin, as a serine protease, is able to cleave arginine, thereby inducing the degradation of secreted proteins essential for bacterial attachment.[168] Recent research has shown synergy between lactoferrin and polymorphonuclear leukocytes (PMNs), which may enhance their bactericidal potential.

EFFICACY ON BIOFILMS

The majority of research on the efficacy of lactoferrin has been with bacteria in the planktonic state, where it has bactericidal activity. For example, Singh and colleagues have demonstrated that lactoferrin prevented attachment of free-floating *P. aeruginosa* onto a surface, therefore interrupting biofilm formation.[169] A study by Weinberg found beneficial effects of lactoferrin on biofilms,[170] and another study[171] established that the combination of lactoferrin and xylitol decreased the viability of *P. aeruginosa*.

XYLITOL

Xylitol, widely used as an artificial sweetener, is a 5-carbon sugar alcohol that occurs naturally in small quantities in a number of plants, fruits, and vegetables.[172] It can be metabolized in the liver by the polyaldehydrogenase pathways. Studies have shown a beneficial effect of xylitol in managing paranasal sinusitis. Similarly, a study involving 857 children found that xylitol decreased the incidence of otitis media by more than 40% over an 8-week study period.[173] Xylitol solutions are commercially available as a nasal spray.

MODE OF ACTION

Xylitol has the ability to block the adhesion of Gram-positive bacteria by binding onto the outer surface of these bacteria, lowering the ability of bacteria to adhere to epithelial cell surfaces and tissue surfaces.[174] Xylitol is also able to complex with certain metal ions, such as calcium and iron, which are important for bacterial viability and virulence.

EFFICACY ON BIOFILMS

Xylitol has been found to be effective in preventing a number of biofilm-related diseases, including dental plaque, otitis media, and sinusitis. In synergy studies with farnesol, xylitol has been shown to inhibit biofilm formation on the skin.[175,176] Xylitol has also been found to inhibit the formation of EPS in S. aureus.[176]

BACTERIOPHAGES

Bacteriophages (or phages) are viruses that infect bacteria which were first reported by Twort and d'Herelle in 1915 and 1917, respectively.[177] Although their clinical potential for treating bacterial infections was recognized soon after their discovery, interest faded after the discovery of antibiotics. Phage therapy was initially developed and used in Georgia and parts of Russia, where it had some early setbacks. It has only relatively recently been taken up by Western countries, mostly in the setting of growing resistance of bacteria to available antibiotic agents.[178]

MODE OF ACTION

All viruses have an extremely narrow host range and bacteriophages are no exception. A virus can normally infect only its respective strain of host bacterial cell. The infection process begins when a bacterium binds to specific receptors on the surface of its respective host cell (Figure 11.6). Following viral enzymatic digestion of the host cell wall in the vicinity of attachment, phages inject their genetic material into the host cell. The phage-derived genetic material then directs the host's metabolic systems to reproduce phage components. The different parts of the phage are assembled into functional phages inside the host cell, and host enzymes cause the cell to lyse and release the replicated virus particles. Each of these viral particles can then infect additional host bacterial cells. Hence, one virus can potentially direct a bacterial cell to produce several hundred virus particles in less than an hour.

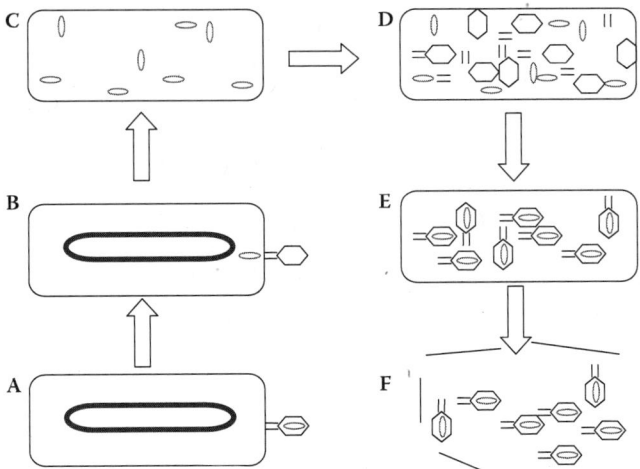

FIGURE 11.6   Replication of bacteriophage virus within a bacterial cell (viral infection of a bacterium).

EFFECTIVENESS ON BIOFILMS

Preliminary observations using mice suggested that phage therapy against *S. aureus*[179] and *P. aeruginosa* in a burn model[180] was effective in treating planktonic infections. Few studies have looked at the ability of bacteriophages to infect host bacteria in a biofilm phenotype. The available studies have shown that phage can absorb, replicate, and reinfect their host bacteria within a biofilm.[181–183] Some phage products have been shown to degrade bacterial EPS.[184–186] However, the EPS material produced by a particular bacterium depends on access to necessary substrates and interactions with other bacteria in the biofilm. So, it is unclear how effective phage therapy might be against biofilm in a clinical setting. At present the narrow host spectrum range of most bacteriophages tends to limit their clinical potential, because an array of different viral cultures would be required to combat the range of different species of bacteria likely to be found in any wound. Research into phage therapy is ongoing in a number of universities, and it offers some promise in the treatment of burn infections, infections with bacteria showing multiple antibiotic resistance, and in selected animal and human systemic infections.

ETHYLENEDIAMINE TETRAACETIC ACID (EDTA)

Ethylenediamine tetraacetic acid (EDTA) is widely used as a metal chelating agent in the food and water industry. The formula of EDTA is $[CH_2N(CH_2CO_2H)_2]_2$. This colorless and water-soluble chelating agent is able to "sequester" di- and tri-cationic metal ions, such as Ca^{2+} and Fe^{3+}. EDTA is able to inhibit bacteria in its own right,[187–189] as well as enhance the antibacterial activity of other agents. EDTA is used to remove excess iron from the body and for chelation therapy for mercury and lead poisoning. Research into the effect of EDTA on bacteria began in the 1960s.[190,191]

In theory, EDTA should be able to abstract and chelate calcium and magnesium cations, thereby causing the breakdown the EPS matrix of the biofilm. It follows that this would then expose the microorganisms to the effects of applied antimicrobials. Recently, a number of groups have examined this effect of EDTA, both alone and in conjunction with antimicrobial agents, on biofilms.

MODE OF ACTION

Because EDTA avidly binds metal ions such as magnesium and iron, it has multiple antimicrobial effects. The outer membrane of Gram-negative bacteria becomes more permeable when exposed to EDTA, which causes a loss of integrity and allows the ingress of other agents.[192] Gram-positive bacteria are affected to a lesser degree because they do not possess an outer membrane. Bacterial growth can be restricted by the chelation of essential metal ions such as iron, and biofilm initiation can be prevented by lack of cations, which are required for microbial adherence. The role of chelators in the prevention of biofilm formation and catheter-related infections has been reviewed by Raad and colleagues.[193]

EFFECTIVENESS OF BIOFILMS

Many studies have investigated the antibiofilm properties of EDTA.[194] Disodium EDTA (at 20 mg/mL) has been shown to have some bactericidal effect against clinical isolates of *S. epidermidis* in catheters *in vitro*.[195] The synergistic effect of minocycline and 30 mg/mL EDTA has also been demonstrated[196,197] against bacteria within a biofilm. This combination was also shown to prevent colonization, bacteremia, septic phlebitis, and endocarditis in a animal catheter model.[198] In addition to minocycline, vancomycin and gentamicin have been reported to act synergistically with EDTA against *S. epidermis* in an *in vitro* biofilm model.[199,200] Disodium EDTA was also found to be effective against fungal (*C. albicans*) biofilms *in vitro*.[201]

Research on catheter biofilms using the tetrasodium salt of EDTA has shown that 40 mg/mL has a broad-spectrum activity against *in vivo*–generated biofilms.[202,203] Tetrasodium EDTA has also been shown to remove biofilms of oral microorganisms grown on denture base discs or toothbrushes.[204]

Several commercially available products incorporating EDTA are available, including RescuDerm.® This water-soluble burn gel contains 0.1% EDTA along with vinegar, citric acid, and Carbopol. RescuDerm has been shown to have antibiofilm activity against *P. aeruginosa* and *S. epidermis in vitro*[205] and to prevent further *P. aeruginosa* growth in moderately infected rat full-thickness wound models.[206]

One study determined the effect of magnesium, calcium, and EDTA on slime production by 15 slime-positive and 13 slime-negative *Staphylococcus epidermidis* strains isolated from various clinical specimens. The slime production on tryptic soy broth was significantly enhanced after addition of 128 µmol/L Mg^{2+}. Similarly, the addition of Ca^{2+} caused a significant increase in slime production of all tested strains when concentration of Ca^{2+} exceeded 64 µmol/L. In contrast, in the presence of EDTA, the slime production by all strains was significantly reduced. Hence Ca^{2+} and Mg^{2+}

increase slime production of *S. epidermidis*. This finding is important in the context of the pathogenesis of biomedical implant infections caused by *S. epidermidis*.

QUORUM-SENSING INHIBITORS

An interesting antibiofilm strategy concerns the specific signaling agents that are involved in the development of biofilm and determinants of microbial virulence. There are microbial signaling substances called quorum-sensing molecules. They are produced by, and detected by, individual cells and allow each to gauge the size of the population of cells in which they are immersed. When the bacterial population recognizes that a critical density of bacterial cells (i.e., a quorum) has been reached, differential gene expression is induced. Essentially, a threshold number of quorum-sensing molecules must be present to cause the up-regulation or down-regulation of operons and regulons, which in turn control a suite of genes responsible for biofilm maturation. In theory, if communication between bacteria could be blocked (inhibited), this would affect biofilm formation and virulence of bacteria. Extracts of garlic and RNA molecules are currently being investigated for their ability to interfere with quorum sensing.

ESSENTIAL OILS

Because essential oils possess a broad spectrum of antimicrobial activity, they have been suggested as possible treatments for wounds. In *in vitro* tests with biofilms, those induced by MRSA and meticillin-sensitive *S. aureus* were eradicated by exposure to 5% tea tree oil for 1 hour, but those induced by coagulase-negative staphylococci were less susceptible.[207] A major component of essential oils is terpenes. Recently, ten terpenes were tested for antibiofilm activity against three species of *Candida,* and three (carvacol, geraniol, and thymol) were found to have exceptional activity.[208]

CONCLUSIONS

Humans have always suffered from skin and soft tissue wounds, and many of these become infected. It was only after the discovery of bacteria in the 19th century that we understood their crucial role in causing infection. Defining infection in acute wounds is relatively easy; it is manifested by either purulent secretions or at least some of the cardinal clinical signs and symptoms of inflammation (i.e., redness, warmth, tenderness, pain, induration, or loss of function). In chronic wounds, especially in patients with comorbidities such as peripheral neuropathy or vasculopathy, these signs and symptoms are less helpful. Thus some have used secondary or more subtle signs (e.g., friable, discolored, or pocketed granulation tissue or foul odor) to define infection. Others have advocated defining infection by microbiological standards, usually meaning the presence of greater than 10^5 organisms per gram of tissue. Defining when infection is present is crucial, as this determines which wounds may require antimicrobial therapy. Clinically uninfected wounds typically do not need to be treated with antimicrobials, but clearly, infected wounds usually require systemic antibiotic therapy. It is mostly the wounds that are not responding to otherwise appropriate therapy that may have subtle evidence of infection for which topical antimicrobials may be most useful.

Although antimicrobial agents have been used since ancient times, their use in modern medicine can only be clearly traced from the 19th century. Few would dispute the conviction that reducing the microbial load of wounds ought to reduce the risk of infection, yet objective evidence to support the practice is limited.[43,44,136–138,209–211] High-quality clinical evidence is desperately needed. Bacterial colonization, infection, and biofilms seem to be significant factors that compromise wound healing, yet "best practice" guidelines for managing these issues are not well defined or universally accepted. The widespread use of systemic and topical antibiotics is known to encourage the emergence of microbial resistance, such as methicillin-resistant *Staphylococcus aureus,* and antiseptics might well assist this process. Even though the use of certain antiseptic agents may be a valuable option for controlling infection, promoting healing, and killing bacteria, their potential in coping with biofilms is not proven. With the recent confirmation of the presence of biofilms in wounds and their role in delaying wound healing, it is important that the next generation of antimicrobials be effective against biofilms, because our traditional antimicrobial agents are not proving to be entirely satisfactory in facilitating rapid healing in a large number of recalcitrant wounds. Clinicians must, however, guard against unproven claims for the effectiveness of newly introduced wound antimicrobials. Properly conducted trials are crucial in determining which new agents are both safe and effective.

REFERENCES

1. Forrest, R.D. (1982) Early history of wound healing. *J Roy Soc Med* 75:198–205.
2. Nelson, R. (2003) Antibiotic development pipeline runs dry—New drugs to fight resistant organisms are not being developed, experts say. *Lancet* 362:1726–1727.
3. McDonnel, G., and Russell, D. (1999) Antiseptics and disinfectants: Activity, action and resistance. *Clin Microbiol Rev* 12:147–179.
4. Harding, K.G. (1996). Managing wound infection. *J Wound Care* 8:391–392.
5. Gilchrist, B. (1997). Should iodine be reconsidered in wound management? *J Wound Care* 6:148–150.
6. Cooper, R.A (2007). Iodine revisited. *Int Wound J* 4:124–137.
7. Gulliver, G. (1999) Wound care: Arguments over iodine. *Nursing Times* 95:68–70.
8. Bucknall, T.E. (1980) The effect of local infection upon wound healing: An experimental study. *Br J Surg* 67:851–855.
9. Robson, M.C., Stenberg, B.D., and Heggars, L. (1990) Wound healing alterations caused by infection. *Clin Plast Surg* 17(3):485–492.
10. Lyman, I.R., Tenery, J.H., and Basson, R.P. (1970) Correlation between decrease in bacterial load and rate of wound healing. *Surg Gynecol Obstet* 130(4):616–621.
11. Vowden, P., Apelqvist, J., and Moffat, C. (2008) Wound complexity and healing. In *European Wound Management Association (EWMA) Position Document: Hard-to-heal wounds: A holistic approach.* pp. 2–9. London. MEP.
12. Basu, S., Ramchuran Panray, T., Bali Singh, T., Gulati, A.K., and Shukla, V.K. (2009) A prospective, descriptive study to identify the microbiological profile of chronic wounds in outpatients. *Ostomy Wound Manage* 55(1):14–20.

13. Howell-Jones, R.S., Wilson, M.J., Hill, K.E., Howard, A.J., Price, P.E., and Thomas, D.W. (2005) A review of the microbiology, antibiotic usage and resistance in chronic skin wounds. *J Antimicrob Chemother* 55(2):143–149.

14. Eron, L.J., Lipsky, B.A., Low, D.E., Nathwani, D., Tice, A.D., and Vilturo, G.A. (2003) Managing skin and soft tissue infections: Expert panel recommendations on key decision points. *J Antimicrob Chemother* 52(1):13–17.

15. Dinh, T.L., and Veves, A. (2006) Treatment of diabetic ulcers. *Dermatol Ther* 19(6):348–355.

16. Roghmann, M.C., Siddiqui, A., Plaisance, K., and Standiford, H. (2001) MRSA colonization and the risk of MRSA bacteraemia in hospitalized patients with chronic ulcers. *J Hosp Infect* 47(2):98–103.

17. McGuckin, M., Goldman, R., Bolton, L., and Salcido, R. (2003) The clinical relevance of microbiology in acute and chronic wounds. *Adv Skin Wound Care* 16(1):12–23.

18. Dow, G., Browne, A., and Sibbald, R.G. (1999) Infection in chronic wounds: Controversies in diagnosis and treatment. *Ostomy Wound Manage* 45(8):23–7, 29–40.

19. White, R.J., Cutting, K., and Kingsley, A. (2006) Topical antimicrobials in the control of wound bioburden. *Ostomy Wound Manage* 52(8):26–58.

20. Chaby, G., Senet, P., Vaneau, M., Martel, P., Guillaume, J.C., Meaume, S., Téot, L., Debure, C., Dompmartin, A., Bachelet, H., Carsin, H., Matz, V., Richard, J.L., Rochet, J.M., Sales-Aussias, N., Zagnoli, A., Denis, C., Guillot, B., and Chosidow, O. (2007) Dressings for acute and chronic wounds: A systematic review. *Arch Dermatol* 143(10):1297–1304.

21. Thomas, S. (2006) Cost of managing chronic wounds in the UK, with particular emphasis on maggot debridement therapy. *J Wound Care* 15(10):465–469.

22. James, G.A., Swogger, E., Wolcott, R., Pulcini, E., Secor, P., Sestrich, J., Costerton, J.W., and Stewart P.S. (2008) Biofilms in chronic wounds. *Wound Rep Regen* 16(1):37–44.

23. Bjarnsholt, T., Kirketerp-Møller, K., Jensen, P.Ø., Madsen, K.G., Phipps, R., Krogfelt, K., Høiby, N., and Givskov, M. (2008) Why chronic wounds fail to heal: A new hypothesis. *Wound Rep Regen* 16(1):2–10.

24. Fux, C.A., Costerton, J.W., Stewart, P.S., and Stoodley, P. (2005) Survival strategies of infectious biofilms. *Trends Microbiol* 13:34–40.

25. Steed, D.L., Donohoe, D., Webster, M.W., and Lindsley, L. (1996) Effect of extensive debridement and treatment on the healing of diabetic foot ulcers. Diabetic Ulcer Study Group. *J Am Coll Surg* 183:61–64.

26. Steed, D.L. (2004) Debridement. *Am J Surg* 187:71S–74S.

27. Sibbald, R.G. et al. (2000) Preparing the wound bed—Debridement, bacterial balance, and moisture balance. *Ostomy Wound Manage* 46:14–18, 30.

28. Tarnuzzer, R.W., and Schultz, G.S. (1996) Biochemical analysis of acute and chronic wound environments. *Wound Rep Regen.* 4(3):321–325.

29. Rhoads, D.D., Wolcott, R.D., and Percival, S.L. (2008) Biofilms in wounds: Management strategies. *J Wound Care* 17(11):502–507.

30. Nemoto, K., Hirots, K., Murakami, K., Taniguti, K., Murata, H., Viducic, D., and Miyake, Y. (2003) Effect of varidase (streptodornase) on biofilm formed by *Pseudomonas aeruginosa*. *Chemotherapy* 49:121–125.

31. Johansen, C., Falholt, P., and Gram, L. (1997) Enzymatic removal and disinfection of bacterial biofilms. *Appl Environ Microbiol* 63(9):3724–3728.

32. Baer, W.S. (1931) The treatment of chronic osteomyelitis with the maggot larvae of the blowfly. *J Bone Joint Surg* 13:438–475.

33. Chambers, L., Woodrow, S., Brown, A.P., Harris, P.D., Phillips, D., Hall, M., Church, J.C., and Pritchard, D.I. (2003) Degradation of extracellular matrix components by defined proteinases from the greenbottle larva *Lucilia sericata* used for the clinical debridement of non-healing wounds. *Br J Dermatol* 148:14–23.

34. Van der Plas, Jukema, G.N., Wai, S.W., Dogterom-ballering, H.C., Lagendijk, E.L., van Gulpen, C., van Dissel, J.T., Bloemberg, G.Y., and Nibbering, P.H. (2008) Maggot excretions/secretions are differentially effective against biofilms of *Staphylococcus aureus* and *Pseudomonas aeruginosa*. *J Antimicrob Chemother* 61:117–122.

35. Cazander, G., van Veen, K.E., Bouwman, L.H., Bernards, A.T., and Jukema, G.N. (2009) The influence of maggot excretions on PAO1 biofilm formation on different biomaterials. *Clin Orthop Relat Res* Feb;467(2):536–545.

36. Van der Plas, M.J., van der Does, A.M., Baldry, M., Dogterom-Ballering, H.C., van Gulpen, C., van Dissel, J.T., Nibbering, P.H., Jukema, G.N. (2007) Maggot excretions/secretions inhibit multiple neutrophil proinflammatory responses. *Microbes Infect* 9:507–514.

37. Dumville, J.C., Worthy, G., Bland, J.M., Cullum, N., Dowson, C., Iglesias, C., Mitchell, J.L., Nelson, E.A., Soares, M.O., Torgerson, D.J., and Venus II team. (2009) Larval therapy for leg ulcers (VenUS II): randomized controlled trial. *BMJ* 338:b773.

38. Lipsky, B.A. et al. (2004) Diagnosis and treatment of diabetic foot infections. *Clin Infect Dis* 39:885–910.

39. Anderl, J.N., Zahller, J., Roe, F., and Stewart, P.S. (2003) Role of nutrient limitation and stationary-phase existence in *Klebsiella pneumoniae* biofilm resistance to ampicillin and ciprofloxacin. *Antimicrob Agents Chemother* 47:1251–1256.

40. Walters, M.C., III, Roe, F., Bugnicourt, A., Franklin, M.J., and Stewart, P.S. (2003) Contributions of antibiotic penetration, oxygen limitation, and low metabolic activity to tolerance of *Pseudomonas aeruginosa* biofilms to ciprofloxacin and tobramycin. *Antimicrob Agents Chemother* 47:317–323.

41. Brennan, S.S., and Leaper, D.J. (1985) The effect of antiseptics on the healing wound: A study using the rabbit ear chamber. *Br J Surg* 72(10):780–782.

42. Lineweaver, W., Howard, R., Soucy, D., McMorris, S., Freeman, J., Cairn, C., Roberston, J., and Rumley, T. (1985) Topical antimicrobial toxicity. *Arch Surg* 120:267–270.

43. Drosou, A., Falabella, A., and Kirsner, R.S. (2003) Antiseptics in wounds: A area of controversy. *Wounds* 15(5):149–166.

44. O'Meara, S.M., Cullum, N.A., Majid, M., and Sheldon, T.A. (2001) Systematic review of antimicrobial agents used for chronic wounds *Br J Surg* 88(1):4–21.

45. Gilchrist, B. (1997) Should iodine be reconsidered in wound management? *J Wound Care* 6:148–150.

46. Pierard-Franchimont, C., Paquet P., Arrese, J.E., and Pierard, G.E. (1997) Healing rate and bacterial necrotizing vasculitis in venous leg ulcers. *Dermatology* 194:383–387.

47. Lineaweaver, W., Soucy, D., and Howard, R. (1985). Cellular and bacterial toxicities of topical antimicrobials. *Plast Reconstr Surg* 75:394–396.

48. Mayer, D.A., and Tsapogas, M.J. (1993) Povidone iodine and wound healing: A critical review. *Wounds* 5:14–23.

49. Eming, S.A., Smola-Hess, S., Kurschat, P., Hirche, D., Krieg, T., and Smola, H. (2006) A novel property of povidone-iodine: Inhibition of excessive protease levels in chronic non-healing wounds. *J Invest Dermatol* 126: 2731–2733.

50. Gruber, R.P., Vistnes, L., and Pardoe, R. (1975). The effect of commonly used antiseptics on wound healing. *Plast Reconstr Surg* 55:472–476.

51. Drosou, A., Falabella, A., and Kirsner, R.S. (2003). Antiseptics in wounds: An area of controversy. *Wounds* 15(5):149–166.

52. Mayer, D.A., and Tsapogas, M.J. (1993). Povidone-iodine and wound healing: A critical review. *Wounds* 5:14–23.

53. Durani, P., and Leaper, D. (2008) Povidone-iodine: Use in hand disinfection, skin preparation and antiseptic irrigation. *Int Wound J* 5:376–387.

54. Fleischer, W., and Reimer, K. (1997) Povidone-iodine in antisepsis: State of the art. *Dermatology* 195:3–9.

55. Leaper, D.J., and Durani, P. (2008) Topical antimicrobial therapy of chronic wounds healing by secondary intention using iodine products. *Int Wound J* 5:361–368.

56. Favero, M.S., and Drake, C.H. (1966) Factors influencing the occurrence of high numbers of iodine-resistant bacteria in iodinated swimming pools. *Appl Microbiol* 14(4):627–635.

57. Mycock, G. (1985) Methicillin/antiseptic-resistant *Staphylococcus aureus*. *Lancet* 2:949–950.

58. Harvey, S.C. (1985) Antiseptics and disinfectants. In Gilman, A.G., and Goodman L.S., et al. (Eds.) *The Pharmacological Basis of Therapeutics* (7th ed.). New York: Macmillan.

59. Schreier, H., Erdos, G., Reimer, K., König, B., König, W., and Fleischer, W. (1997) *Dermatology* 195(2):111–116.

60. Han, K.H., and Maitra, A.K. (1989) Management of partial skin thickness burn wounds with Inadine dressings. *Burns* 15:399–402.

61. Audanaska, H., and Gustavson, B. (1988) In patient treatment of chronic varicose venous ulcers; A randomised trial of cadexomer iodine versus standard dressings. *J Int Med Res* 16:428–435.

62. Moberg, S., Hoffman, L., Grennert, M.L., and Holst, A. (1983) A randomised trial of cadexomer iodine in decubitus ulcers. *J Am Geriatr Soc* 8:462–465.

63. Tomoyuki, O., Mizuashi, M., Ito, Y., and Aiba, S. (2007) Cadexomer as well as cadexomer iodine induces the production of proinflammatory cytokines and vascular endothelial growth factor by human macrophages. *Exp Dermatol* 16:318–323.

64. Ivins, N., Simmonds, W., Turner, A., and Harding, K. (2007) The use of an oxygenating hydrogel dressing in VLU. *Wounds UK* 2007 3:1–5.

65. Maillard, J.-Y. (2002) Bacterial target for biocide action. *J Applied Microbiol Symp Supp* 92:16S–27S.

66. Pyle, B.H., and McFeters, G.A. (1990) Iodine susceptibility of pseudomonads grown attached to stainless steel surfaces. *Biofouling* 2:113–120.

67. Cargill, K.L., Pyle, B.H., Sauer, R.L., and McFeters G.A. (1992) Effects of culture conditions and biofilm formation on the iodine susceptibility of *Legionella pnueumophila*. *Can J Microbiol* 38:423–429.

68. Kunisada, T., Yamada, K., Oda, S., and Hara, O. (1997) Investigation on the efficacy of povidone-iodine against antiseptic-resistant species. *Dermatology* 195:14–18.

69. Presterl, E., Suchomel, M., Eder, M., Reichmann, S., Lassnigg, A., Graninger, W., and Rotter, M. (2007) Effects of alcohols, povidone-iodine and hydrogen peroxide on biofilms of *Staphylococcus epidermidis*. *J Antimicrob Chemother* 60:417–420.

70. Akiyama, H., Oono, T., Saito, M., and Iwatsuki, K. (2004) Assessment of cadexomer iodine against *Staphylococcus aureus* biofilm *in vivo* and *in vitro* using confocal laser scanning microscopy. *J Dermatol* 31: 529–534.

71. Ivins, N., Simmonds, W., Turner, A., and Harding, K. (2007) The use of an oxygenating hydrogel dressing in VLU. *Wounds UK* 2007. 3:1–5.

72. Leaper, D.J. (2006) Silver dressings: Their role in wound management. *Int Wound J* 3(4):282–294.

73. Rai, M., Yadav, A., and Gade, A. (2009) Silver nanoparticles as a new generation of antimicrobials. *Biotechnol Adv* 27(1):76–83.

74. Percival, S.L., Bowler, P., and Russell, D. (2005) Bacterial resistance to silver and its impact to woundcare, *J Hosp Infect* 60:1–7.

75. Woods, E.J., Cochrane, C.A., and Percival, S.L. (2009) Prevalence of silver resistance genes in bacteria isolated from human and horse wounds. *Vet Microbiol* Mar 24.

76. Landsdown, A.B., and Williams, A. (2007) Bacterial resistance to silver in wound care and medical devices. *J Wound Care* 16(1):15–19.

77. Percival, S.L., Cochrane, C., Nutekor, M., and Silver, S. (2008) Prevalence of silver resistance in bacteria isolated from diabetic foot ulcers and efficacy of silver-containing wound dressings. *Ostomy Wound Manage* 54(3):30–40.

78. Woods, E.J., Cochrane, C.A., and Percival, S.L. (2009) Prevalence of silver resistance genes in bacteria isolated from human and horse wounds. *Vet Microbiol* 138:325–329.

79. Yin, H.Q., Langford, R., and Burrell, R. (1999) Comparative evaluation of the antimicrobial activity of Acticoat antimicrobial barrier dressing. *J Burn Care Rehab* 20:195–200.

80. Slawson, R.M., Lee, H., and Trevors, J.T. (1990) Bacterial interactions with silver. *Biol Met* 3(3–4):151–154.

81. Slawson, R.M., Van Dyke, M.I., Lee, H., and Trevors, J.T. (1992) Germanium and silver resistance, accumulation, and toxicity in microorganisms. *Plasmid* 27(1):72–79.

82. Thurman, R.B., and Gerba, C.O. (1989) The molecular mechanisms of copper and silver ion disinfection of bacteria and viruses. *Crit Rev Environ Control* 18:295–315.

83. Liau, S.Y., Read, D.C., Pugh, W.J., Furr, J.R., and Russell, A.D. (1997) Interaction of silver nitrate with readily identifiable groups: Relationship to the antibacterial action of silver ions. *Lett Appl Microbiol* 25(4):279–283.

84. Silvestry-Rodriguez, N., Bright, K.R., Uhlmann, D.R., Slack, D.C., and Gerba, C.P. (2007) Inactivation of *Pseudomonas aeruginosa* and *Aeromonas hydrophila* by silver in tap water. *J Environ Sci Health A Tox Hazard Subst Environ Eng* 42(11):1579–1584.

85. Silvestry-Rodriguez, N., Bright, K.R., Slack, D.C., Uhlmann, D.R., and Gerba, C.P. (2008) Silver as a residual disinfectant to prevent biofilm formation in water distribution systems. *Appl Environ Microbiol* 74(5):1639–1641.

86. Silvestry-Rodriguez, N., Sicairos-Ruelas, E.E., Gerba, C.P., and Bright, K.R. (2007) Silver as a disinfectant. *Rev Environ Contam Toxicol* 191:23–45.

87. De Prijck, K., Nelis, H., and Coenye, T. (2007) Efficacy of silver-releasing rubber for the prevention of *Pseudomonas aeruginosa* biofilm formation in water. *Biofouling* 23(5–6):405–411.

88. Gentry, H., and Cope, S. (2005) Using silver to reduce catheter-associated urinary tract infections. *Nurs Stand* 19:51–54.

89. Ionescu, A., Payne, N., Fraser, A.G., Giddings, J., Grunkemeier, G.L., and Butchart, E.G. (2003) Incidence of embolism and paravalvar leak after St Jude Silzone valve implantation: Experience from the Cardiff Embolic Risk Factor Study. *Heart* Sep;89(9):1055–1061.

90. Darouiche, R.O. (1999) Anti-infective efficacy of silver-coated medical prostheses. *Clin Infect Dis* 29(6):1371–1377.

91. Wu, M.Y., Suryanarayanan, K., van Ooij, W.J., and Oerther, D.B. (2007) Using microbial genomics to evaluate the effectiveness of silver to prevent biofilm formation. *Water Sci Technol* 55(8–9):413–419.

92. Bjarnsholt, T., Kirketerp-Møller, K., Kristiansen, S., Phipps, R., Nielsen, A.K., Jensen, P.Ø., Høiby, N., and Givskov, M. (2007) Silver against *Pseudomonas aeruginosa* biofilms. *APMIS* 115(8):921–928.

93. Chaw, K.C., Manimaran, M., and Tay, F.E. (2005) Role of silver ions in destabilization of intermolecular adhesion forces measured by atomic force microscopy in *Staphylococcus epidermidis* biofolms. *Antimicrob Agents Chemother* 49 (12):4853–4859.

94. Percival, S.L., Bowler, P., and Woods, E.J. (2008) Assessing the effect of an antimicrobial wound dressing on biofilms. *Wound Repair Regen* 16(1):52–57.

95. Valappil, S.P., Knowles, J.C., and Wilson, M. (2008) Effect of silver-doped phosphate-based *glasses* on bacterial biofilm growth. *Appl Environ Microbiol* 74(16):5228–5230.

96. Kim, J., Pitts, B., Stewart, P.S., Camper, A., and Yoon, J. (2008) Comparison of the anti-microbial effects of chlorine, silver ion, and tobramycin on biofilm. *Antimicrob Agents Chemother* 52(4):1446–1453.

97. Fabry, W., Trampenau, C., Bettag, C., Handschin, A.E., Lettgen, B., Huber, F.X., Hillmeier, J., and Kock, H.J. (2006) Bacterial decontamination of surgical wounds treated with Lavasept. *Int J Hyg Environ Health* 209(6):567–573.

98. Santodomingo-Rubido, J. (2007) The comparative clinical performance of a new poly-hexamethylene biguanide- versus a polyquad-based contact lens care regime with two silicone hydrogel contact lenses. *Ophthalmic Physiol Opt* 27(2):168–173.

99. Larkin, D.F.P., Kilvington, S., and Dart, J.K.G. (1992) Treatment of *Acanthamoeba keratitis* with polyhexamethylene biguanide. *Ophthalmology* 99:185–191.

100. Lim, N., Goh, D., Bunce, C., Xing, W., Fraenkel, G., Poole, T.R., and Ficker, L. (2008) Comparison of polyhexamethylene biguanide and chlorhexidine as mono-therapy agents in the treatment of *Acanthamoeba keratitis*. *Am J Ophthalmol* 145(1):130–135.

101. Nascimento, A.P., Tanomaru, J.M., Matoba-Júnior, F., Watanabe, E., Tanomaru-Filho, M., and Ito, I.Y. (2008) Maximum inhibitory dilution of mouthwashes containing chlor-hexidine and polyhexamethylene biguanide against salivary *Staphylococcus aureus*. *J Appl Oral Sci* 16(5):336–339.

102. Minozzi, M., Gerli, S., Di Renzo, G.C., Papaleo, E., and Ferrari, A. (2008) The effi-cacy and safety of a single dose of polyhexamethylene biguanide gynaecologic solution versus a seven-dose regimen of vaginal clindamycin cream in patients with bacterial vaginosis. *Eur Rev Med Pharmacol Sci* 12(1):59–65.

103. Marelli, G., Papaleo, E., Origoni, M., Caputo, L., and Ferrari A. (2005) Polyhexamethylene biguanide for treatment of external genital warts: A prospective, double-blind, random-ized study. *Eur Rev Med Pharmacol Sci.* 9(6):369–372.

104. Motta, G.J., Milne, C.T., and Corbett, L.Q. (2004) Impact of antimicrobial gauze on bac-terial colonies in wounds that require packing. *Ostomy Wound Manage* 50(8):48–62.

105. Salas Campos, L., Gómez Ferrero, O., Estudillo Pérez, V., and Fernández Mansilla, M. (2006) Preventing nosocomial infections. Dressings soaked in polyhexamethylene biguanide (PHMB). *Rev Enferm* 29(6):43–48.

106. Nascimento, A.P., Tanomaru, J.M., Matoba-Júnior, F., Watanabe, E., Tanomaru-Filho, M., and Ito, I.Y. (2008) Maximum inhibitory dilution of mouthwashes containing chlor-hexidine and polyhexamethylene biguanide against salivary *Staphylococcus aureus*. *J Appl Oral Sci* 16(5):336–339.

107. Vowden, K., and Vowden, P. (2003) Understanding exudate management and the role of exudate in the healing process. *Br J Community Nurs* 8(11 Suppl):4–13.

108. Broxton, P., Woodcock, P.M., and Gilbert, P. (1983) A study of the antibacterial activity of some polyhexamethylene biguanides towards *Escherichia coli* ATCC 8739. *J Appl Bact*, 5445–5453.

109. Allen, M.J., White, G.F., and Morby, A.P. (2006) The response of *Escherichia coli* to exposure to the biocide polyhexamethylene biguanide.*Microbiology*152(4):989–1000.

110. Ueda, S., and Kuwabara, Y. (2007) Susceptibility of biofilm *Escherichia coli*, *Salmonella enteritidis* and *Staphylococcus aureus* to detergents and sanitizers. *Biocontrol Sci* 12(4):149–153.

111. Russell, A.D. (2002) Introduction of biocides into clinical practice and the impact on antibiotic-resistant bacteria. *J Appl Microbiol* 92: 121S–135S.

112. Webster, J., and Osborne, S. (2007) Preoperative bathing or showering with skin anti-septics to prevent surgical site infection. *Cochrane Database Syst* Rev 2: CD004985

113. Tanner, J., Swarbrook, S., and Stuart, J. (2008) Surgical hand antisepsis to reduce surgi-cal site infection. *Cochrane Database Syst* Rev 1: CD004288

114. Lambert, P.M., Moris, H.F., and Ochi, S. (1997) The influence of 0.12% chlorhexidine gluconate rinses on the incidence of infectious complications and implant success. *J Oral Maxillofac Surg* 55:25–30.
115. Payne, D.N., Babb, J.R., and Bradley, C.R. (1999) An evaluation of the suitability of the European suspension test to reflect in vitro activity of antiseptics against clinically significant organisms. *Lett Appl Microbiol* 28(1):7–12.
116. Cookson, B.D., Bolton, M.C., and Platt, J.H. (1991) Chlorhexidine resistance in methicillin-resistant *Staphylococcus aureus* or just an elevated MIC? An *in vitro* and *in vivo* assessment. *Antimicrob Agents Chemother* 35(10):1997–2002.
117. Millward, T.A., and Wilson, M. (1989) The effect of chlorhexidine on *Streptococcus sanguis* biofilms. *Microbios* 58:155–164.
118. Pratten, J., Barnett, P., and Wilson, M. (1998) Composition and susceptibility to chlorhexidine of multi-species biofilms of oral bacteria. *Appl Environ Microbiol* 64:3515–3519.
119. Sena, N.T., Gomes, B.P., Vianna, M.E., Berber, V.B., Zaia, A.A., Ferraz, C.C., and Souza-Filho, F.J. (2006) *In vitro* antimicrobial activity of sodium hypochlorite and chlorhexidine against selected single-species biofilms. *Int Endod J* 39(11):878–885.
120. Vitkov, L., Hermann, A., Krautgartner, W.D., Herrmann, M., Fuchs, K., Klappacher, M., and Hannig, M. (2005) Chlorhexidine-induced ultrastructural alterations in oral biofilm. *Microsc Res Tech* 68(2):85–89.
121. Taylor, K. (1916) Tissue Fragments and wound infection. *Ann Surg* 64(6):641–644.
122. Phillips, A.W. (1968) Burn therapy: V. Disaster management—to treat or not to treat? Who should receive intravenous fluids? *Ann Surg* 168(6):986–996.
123. Milner, S.M. (1992) Acetic acid to treat *Pseudomonas aeruginosa* in superficial wounds and burns. *Lancet* 340(8810):61.
124. Sloss, J.M., Cumberland, N., and Milner, S.M. (1993) Acetic acid used for the elimination of *Pseudomonas aeruginosa* from burn and soft tissue wounds. *J R Army Med Corps* 139(2):49–51.
125. Cooper, M.L., Laxer, J.A., and Hansbrough, J.F. (1991) The cytotoxic effects of commonly used topical antimicrobial agents on human fibroblasts and keratinocytes. *J Trauma* 31(6):775–784.
126. Peeters, E., Nelis, H.J., and Coenye, T. (2008) Evaluation of the efficacy of disinfection procedures against *Burkholderia cenocepacia* biofilms. *J Hosp Infect* 70(4):361–368.
127. Akiyama, H., Yamasaki, O., Tada, J., and Arata, J. (1999) Effects of acetic acid on biofilms formed by *Staphylococcus aureus*. *Arch Dermatol Res* 291(10):570–573.
128. Oh, D.H., and Marshall, D.L. (1996) Monolaurin and acetic acid inactivation of *Listeria monocytogenes* attached to stainless steel. *J Food Prot* 59(3):249–252.
129. Crane, E. (1999). *The World History of Beekeeping and Honey Hunting*. London: Duckworth.
130. Cooper, R.A., and Jenkins, L. (2009) A comparison between medical grade honey and table honeys in relation to antimicrobial efficacy. *Wounds* 21(2):29–36.
131. Molan, P.C., and Hill, C. (2009) Quality standards for medical grade honey. In: Cooper RA, Molan, P.C. White, R. (Eds.) *Honey in Modern Wound Management*. HealthComm, United Kingdom.
132. Molan, P.C. (2006). The evidence supporting the use of honey as a wound dressing. *Int J Lower Extremity Wounds* 5:40–54.
133. Gethin, G., and Cowman, S. (2009) Manuka honey vs. hydrogel: A prospective, open label, multicentre, randomised controlled trial to compare desloughing efficacy and healing outcomes in venous ulcers. *J Clin Nurs* 18(3):466–474.
134. Jull, A., Walker, N., Parag, V., Molan, P., and Rodgers, A. (2008) Honey as adjuvant leg ulcer therapy trial collaborators. Randomised clinical trial of honey-imprgenated dressings for venous leg ulcers. *Br J Surg* 95(2):175–182.

135. Robson, V., Dodd, S., and Thomas, S. (2009) Standardized antibacterial honey (Medihoney) with standard therapy in wound care: randomized clinical trial. *J Adv Nurs* 65(3):565–575.

136. Moore, O.A., Smith, L.A., Campbell, F., Seers, K., McQuay, H.J., and Moore, R.A. (2001) Systematic review of the use of honey as a wound dressing. *BMC Complement Altern Med* 1:2.

137. Jull, A.B., Rodgers, A., and Walker, N. (2008) Honey as a topical treatment for wounds. *Cochrane Database Syst Rev* (4):CD005083.

138. Brady, J., Slevin, N.J., Mais, K.L., and Molassiotis, A. (2008) A systematic review of honeybuses and its potential value within oncology. *J Clin Nurs* 17:2604–2623.

139. Molan, P.C. (1992) The antibacterial nature of honey: 1. The nature of the antibacterial activity. *Bee World* 73(1):5–28.

140. Bang, L.M., Buntting, C., and Molan, P.C. (2003) The effects of dilution rate on hydrogen peroxide production in honey and its implications for wound healing. *J Altern Compl Med* 9(21):267–273.

141. Mavric, E., Wittmann, S., Barth, G., and Henle, T. (2008) Identification and quantification of methylglyoxal as the dominant antibacterial constituent of Manuka (*Leptospermum scoparium*) honeys from New Zealand. *Mol Nutr Foods Res* 52:483–489.

142. Adams, C.J., Boult, C.H., Deadman, B.J., Farr, J.M., Grainger, M.N.C., Manley-Harris, M., and Snow, M.J. (2008) Isolation by HPLC and characterisation of the bioactive fraction of New Zealand manuka (*Leptospermum scoparium*) honey. *Carbohydr Res* 343:651–659.

143. Blair, S.E., and Carter, D.A. (2005). The potential for honey in the management of wounds and infections. *J Aust Infect Control* 10(1):24–31.

144. Irish, J. et al. (2006) Honey has an antifungal effect against *Candida* species. *Med Mycol* 44:289–291.

145. Brady, N.F., Molan, P.C., and Harfoot, C.G. (1996). The sensitivity of dermatophytes to the antimicrobial activity of manuka honey and other honey. *Pharm Sci* 2:1–3.

146. Zeina, B., Zohra, B.I., and Al-Assad, S. (1997). The effects of honey on Leishmania parasites: An *in vitro* study. *Tropical Doctor suppl* 1:36–38.

147. Al-Waili, N.S. (2004). Topical honey application vs. acyclovir for the treatment of recurrent herpes simplex lesions. *Med Sci Monit* 10 (8): 94–98.

148. Cooper, R.A., Molan, P.C., and Harding, K.G. (2002) The sensitivity to honey of Gram-positive cocci of clinical significance isolated from wounds. *J Appl Microbiol* 93(5):857–863.

149. Cooper, R.A., Halas, E., and Molan, P.C. (2002) The efficacy of honey in inhibiting strains of *Pseudomonas aeruginosa* from infected burns. *J. Burns Care Rehab* 23(6):366– 370.

150. Okhiria, O.A., Henriques, A.F.M., Burton, N.F., Peters, A., and Cooper, R.A. (2009) Honey modulates biofilms of *Pseudomonas aeruginosa* in a time and dose dependent manner. *J. ApiProduct ApiMedical Science* 1(1):6–10.

151. Alandejani, T., Marsan, J.G., Ferris, W., and Chan, F. (2008) Effectiveness of honey on *S. aureus* and *P. aeruginosa* biofilms. *Otolaryngol Head Neck Surg* 139(1):107.

152. Lerrer, B., Zinger-Yosovich, K.D., Avrahami, B., and Gilboa-Garber, N. (2007) Honey and royal jelly, like human milk, abrogate lectin-dependent infection-preceding *Pseudomonas aeruginosa* adhesion. *ISME J,* 1(2):149–155.

153. Alnaqdy, A., Al-Jabri, A., Al Mahrooqi, Z., Nzeako, B., and Nsanze, H. (2005) Inhibition effect of honey on the adherence of *Salmonella* to intestinal epithelial cells *in vitro. Int J Food Microbiol* 103:347–351.

154. Sauer, K., Cullen, M.C.C., Rickard, A.H., Zeef, A.H., Davies, D.G., and Gilbert, P. (2004) Characterization of nutrient-induced dispersion in *Pseudomonas aeruginosa* PAO1 biofilm. *J Bact* 186(21):7312–7326.

155. English, H.K.P., Pack, A.R.C., and Molan, P.C. (2004) The effects of manuka honey on plaque and gingivitis: A pilot study. *J Int Acad Periodontol* 6:63–67.

156. Rodeheaver, G.T. (1997) Wound cleansing, wound irrigation, wound disinfection. In Krasner, D., and Kane, D. (Eds.), *Chronic Wound Care: A Clinical Source Book for Healthcare Professionals* (2nd ed.), Wayne, PA: Health Management Publications, pp. 97–108.

157. Campos-Martin, J.M., Blanco-Brieva, G., and Fierro J.L.G. (2006) Hydrogen peroxide synthesis: An outlook beyond the anthraquinone process. *Angewandte Chemie Int Ed* 45(42):6962–6984.

158. Sleigh, J.W., and Linter, S.P. (1985) Hazards of hydrogen peroxide. *Br Med J (Clin Res Ed)* 291(6510):1706.

159. DeQueiroz, G.A., and Day, D.F. (2007) Antimicrobial activity and effectiveness of a combination of sodium hypochlorite and hydrogen peroxide in killing and removing *Pseudomonas aeruginosa* biofilms from surfaces. *J Appl Microbiol* 103(4):794–802.

160. Baldeck, J.D., and Marquis, R.E. (2008) Targets for hydrogen-peroxide-induced damage to suspension and biofilm cells of *Streptococcus mutans*. *Can J Microbiol* Oct; 54(10):868–875.

161. Glynn, A.A., O'Donnell, S.T., Molony, D.C., Sheehan, E., McCormack, D.J., and O'Gara, J.P. (2009) Hydrogen peroxide induced repression of icaADBC transcription and biofilm development in *Staphylococcus epidermidis*. *J Orthop Res* 27(5):627–630.

162. Surdeau, N., Laurent-Maquin, D., Bouthors, S., and Gellé, M.P. (2006) Sensitivity of bacterial biofilms and planktonic cells to a new antimicrobial agent, Oxsil 320N. *J Hosp Infect* 62(4):487–493.

163. Bullen, J.J. (1975) Iron-binding proteins in milk and resistance to *Escherichia coli* infection in infants. *Postgrad Med J* 51(3):67–70.

164. Oseas, R., Yang, H.H., Baehner, R.L., and Boxer, L.A. (1981) Lactoferrin: A promoter of polymorphonuclear leukocyte adhesiveness. *Blood* 57:939–945.

165. Brandenburg, K., Jurgens, G., Muller, M., Fukuoka, S., and Koch, M.H. (2001) Biophysical characterization of lipopolysaccharide and lipid A inactivation by lactoferrin. *Biol Chem* 382:1215–1225.

166. Appelmelk, B. J. et al. (1994) Lactoferrin is a lipid A-binding protein. *Infect Immun* 62:2628–2632.

167. Ellison, R.T., III, Giehl, T.J., and LaForce, F.M. (1988) Damage of the outer membrane of enteric Gram-negative bacteria by lactoferrin and transferrin. *Infect Immun* 56:2774–2781.

168. Ochoa, T.J. et al. (2003) Lactoferrin impairs type III secretory system function in enteropathogenic *Escherichia coli*. *Infect Immun* 71:5149 5155.

169. Singh, P.K., Parsek, M.R., Greenberg, E.P., and Welsh, M.J. (2002) A component of innate immunity prevents bacterial biofilm development. *Nature* 417:552–555.

170. Weinberg, E.D. (2004) Suppression of bacterial biofilm formation by iron limitation. *Med Hypotheses* 63(5):863–865.

171. Ammons, M.C.B., Ward, L.S., Fisher, S.T., Wolcott, R.D., and James, G.A. (2008) *In vitro* susceptibility of established biofilms composed of a clinical wound isolate of *Pseudomonas aeruginosa* treated with lactoferrin and xylitol. *Int J Antimicrob Agents*.

172. Granstrom, T.B., Izumori, K., and Leisola, M. (2007) A rare sugar xylitol. Part II: Biotechnological production and future applications of xylitol. *Appl Microbiol Biotechnol* 74:273–276.

173. Uhari, M., Tapiainen, T., and Kontiokari, T. (2000) Xylitol in preventing acute otitis media. *Vaccine* 19(1):S144–S147.

174. Tapiainen, T. et al. (2004) Ultrastructure of *Streptococcus pneumoniae* after exposure to xylitol. *J Antimicrob Chemother* 54:225–228.

175. Katsuyama, M., Ichikawa, H., Ogawa, S., and Ikezawa, Z. (2005) A novel method to control the balance of skin microflora. Part 1. Attack on biofilm of *Staphylococcus aureus* without antibiotics. *J Dermatol Sci* 38:197–205.

176. Ammons, M.C., Ward, L.S., Fisher, S.T., Wolcott, R.D., and James, G.A. (2009) *In vitro* susceptibility of established biofilms composed of a clinical wound isolate of *Pseudomonas aeruginosa* treated with lactoferrin and xylitol. *Int J Antimicrob Agents* 33(3):230–236.

177. Kutter, E., and Sulakvelidze, A. (2005) *Bacteriophages: Biology and Applications*. Boca Raton, FL: CRC Press.

178. Parisien, A., Allain, B., Zhang, J., Mandeville, R., and Lan, C.Q. (2008) Novel alternatives to antibiotics: Bacteriophages, bacterial cell wall hydrolases, and antimicrobial peptides. *J Appl Microbiol* 104(1):1–13.

179. Capparelli, R., Parlato, M., Borriello, G., Salvatore, P., and Iannelli, D. (2007) Experimental phage therapy against *Staphylococcus aureus* in mice. *Antimicrob Agents Chemother* 51(8):2765–2773.

180. McVay, C.S., Velásquez, M., and Fralick, J.A. (2007) Phage therapy of *Pseudomonas aeruginosa* infection in a mouse burn wound model. *Antimicrob Agents Chemother* 51(6):1934–1938.

181. Adams, M.H., and Park, B.H. (1956) An enzyme produced by a phage-host cell system. II. The properties of the polysaccharide depolymerase. *Virology* 2:719–736.

182. Donlan, R.M. (2009) Preventing biofilms of clinically relevant organisms using bacteriophage. *Trends Microbiol* 17(2):66–72.

183. Doolittle, M.M., Cooney, J.J., and Caldwell, D.E. (1995) Lytic infection of *Escherichia coli* biofilms by bacteriophage T4. *Can J Microbiol* 41:12–18.

184. Hughes, K.A., Sutherland, I.W., Clark, J., and Jones, M.V. (1998) Bacteriophage and associated polysaccharide depolymerases—Novel tools for study of bacterial biofilms. *J Appl Microbiol* 85:583–590.

185. Hughes, K.A., Sutherland, I.W., and Jones, M.V. (1998) Biofilm susceptibility to bacteriophage attack: The role of phage-borne polysaccharide depolymerase. *Microbiology* 144(11):3039–3047.

186. Sutherland, I.W., Hughes, K.A., Skillman, L.C., and Tait, K. (2004) The interaction of phage and biofilms. *FEMS Microbiol Lett* 232(1):1–6.

187. Kite, P., Eastwood, K., Sugden, S., and Percival, S.L. (2004) Use of *in-vivo* generated biofilms from haemodialysis catheters to test the efficacy of a novel antimicrobial catheter lock for biofilm eradication *in-vitro*. *J Clin Microbiol* 42:3073–3076.

188. Percival, S.L., Kite, P., and Donlan, R. (2005) Assessing the effectiveness of tetrasodium ethylenediaminetetraacetic acid as a novel central venous catheter (CVC) lock solution against biofilms using a laboratory model system. *Infect Control Hosp Epidemiol* 26(6):515–519.

189. Eastwood, K., Kite, P., and Percival, S.L. (2005) The effectiveness of TEDTA on biofilm eradication. Biofilm Club, *Biofilm Club* September 6–8, Bioline, Gregynog, Cardiff.

190. Brown, M.R., and Richards, R.M. (1965). Effect of ethylenediamine tetraacetate on the resistance of *Pseudomonas aeruginosa* to antibacterial agents. *Nature* 207:1391–1393.

191. Gray, G.W., and Wilkinson, S.G. (1965). The effect of ethylenediaminetetraacetic acid on the cell walls of some Gram-negative bacteria. *J Gen Microbiol* 39:385–399.

192. Leive, L. (1965). Release of lipopolysaccharide by EDTA treatment of *E. coli*. *Biochem Biophys Res Commun* 21:290–296.

193. Raad, I.I., Fang, X., Keutgen, X.M., Jiang, Y., Sherertz, R., and Hachem, R. (2008) The role of chelators in preventing biofilm formation and catheter-related bloodstream infections. *Curr Opin Infect Dis* 21(4):385–392.

194. Percival, S.L., Kite., P., and Stickler, D. (2009) The use of urinary catheters and control of biofilms using TEDTA. *Urological Research* 37:205–209.

195. Root, J.L., McIntyre, R.O., Jacobs, N.J., and Daghlian, C.P. (1988). Inhibitory effect of disodium EDTA upon the growth of *Staphylococcus epidermidis in vitro*: Relation to infection prophylaxis of Hickman catheters. *Antimicrob Agents Chemother* 32:1627–1631.

196. Raad, I., Chatzinikolaou, I., Chaiban, G., Hanna, H., Hachem, R., Dvorak, T., Cook, G., and Costerton, W. (2003). *In vitro* and *ex vivo* activities of minocycline and EDTA against microorganisms embedded in biofilm on catheter surfaces. *Antimicrob Agents Chemother* 47:3580–3585.

197. Raad, I., Hanna, H., Dvorak, T., Chaiban, G., and Hachem, R. (2007). Optimal antimicrobial catheter lock solution, using different combinations of minocycline, EDTA, and 25-percent ethanol, rapidly eradicates organisms embedded in biofilm. *Antimicrob Agents Chemother* 51:78–83.

198. Raad, I., Hachem, R., Tcholakian, R.K., and Sherertz, R. (2002). Efficacy of minocycline and EDTA lock solution in preventing catheter-related bacteremia, septic phlebitis, and endocarditis in rabbits. *Antimicrob Agents Chemother* 46:327–332.

199. Kim, H.-J., Dorn, V.L., and VanBriesen, J.M. (2004). The efficacy of ethylenediaminetetraacetic acid (EDTA) against biofilm bacteria. *Biomedical Engineering Society Annual Meeting*, Fall 2004.

200. Banin, E., Brady, K.M., and Greenberg, E.P. (2006). Chelator-induced dispersal and killing of *Pseudomonas aeruginosa* cells in a biofilm. *Appl Environ Microbiol* 72:2064–2069.

201. Ramage, G., Wickes, B.L., and López-Ribot, J.L. (2007). Inhibition on *Candida albicans* biofilm formation using divalent cation chelators (EDTA). *Mycopathologia* 164:301–306.

202. Kite, P., Eastwood, K., Sugden, S., and Percival, S.L. (2004) Use of *in vivo*-generated biofilms from hemodialysis catheters to test the efficacy of a novel antimicrobial catheter lock for biofilm eradication *in vitro*. *J Clin Microbiol* 42:3073–3076.

203. Percival, S.L., Kite, P., Eastwood, K., Murga, R., Carr, J., Arduino, M.J., and Donlan, R.M. (2005). Tetrasodium EDTA as a novel central venous catheter lock solution against biofilm. *Infect Control Hosp Epidemiol* 25:515–519.

204. Devine, D.A., Percival, R.S., Wood, D.J., Tuthill, T.J., Kite, P., Killington, R.A., and Marsh, P.D. (2007). Inhibition of biofilms associated with dentures and toothbrushes by tetrasodium EDTA. *J Appl Microbiol* 103:2516–2424.

205. Martineau, L., and Dosch, H.-M. (2007). Biofilm reduction by a new burn gel that targets nociception. *J Appl Microbiol* 103:297–304.

206. Martineau, L., and Dosch, H.-M. (2007). Management of bioburden with a burn gel that targets nociceptors. *J Wound Care* 16:157–164.

207. Brady, A., Loughlin, R., Gilpin, D., Kearney, P, and Tunney, M. (2006) *In vitro* activity of tea-tree oil against clinical isolates of meticillin-resistant and -sensitive *Staphylococcus aureus* and coagulase-negative staphylococci growing planktonically and as biofilms. *J Med Microbiol* 41(1):52–55.

208. Dalleau, S., Cateau, E., Berges, T., Berjeaud, J-M., and Imbert, C. (2008) *In vitro* activity of terpenes against Candida biofilms. *Int J Antimicrob Agents* 31:572–576.

209. Tanner, J., Swarbrook, S., and Stuart, J. (2008) Surgical hand antisepsis to reduce surgical site infection. *Cochrane Database Syst. Rev.* (1):CD004288.

210. Webster, J., and Osborne, S. (2007) Preoperative bathing or showering with skin antiseptics to prevent surgical site infection. *Cochrane Database Syst. Rev.* (2):CD004985.

211. Vermeulen, H., van Hattem, J.M., Storm-Versloot, M.N., and Ubbink, D.T. (2007) Topical silver for treating infected wounds. *Cochrane Database Syst Rev.* (1):CD005486.

12 Wound Dressings and Other Topical Treatment Modalities in Bioburden Control

Richard White

CONTENTS

Introduction 329
Wound Dressings 330
Generic Dressing Classes 331
    Hydrocolloids 331
    Alginate Dressings 331
Negative Pressure Wound Therapy 332
Biosurgery 332
Honey 333
Hyperbaric Oxygen 334
Sequestration 334
Control of Bacterial Virulence Determinants 335
Dressing Leakage and Strike-Through 336
The Microbiological Barrier Properties of Dressings 336
The Value of Close Association of Dressing with the Wound Bed 336
Fluid Handling Properties: Absorption/Retention, Lateral Wicking,
Sequestration 337
Conclusions 338
References 338

INTRODUCTION

With any wound—defined as a breach in the integrity of the skin—the objectives of clinical management should include healing wherever possible, as it is reepithelialization that restores full barrier function and the exclusion of potential pathogens. In chronic wounds, so called because of factors that delay or compromise healing, a necessary part of management is the control of bioburden. All wounds healing by secondary intent will be contaminated or colonized with microorganisms. This does not necessarily lead to infection, and does not prevent healing. The clinical skill of management is to ensure that the bioburden does not overwhelm host defenses and delay healing or cause infection. Although clinicians will, by necessity, have to resort to antimicrobial agents to control bioburden in some cases, not all chronic wounds will require such interventions. This chapter outlines how modern dressings and other topical treatments can help in this respect—without recourse to antimicrobial therapy. Where antimicrobial dressings are mentioned, it is in the context of the performance characteristics of the base dressing, or "carrier," as this technology is recognized as an important adjunct to the overall clinical performance.[1]

The received wisdom of wound management currently dictates that antimicrobial agents such as antibiotics and antiseptics be used whenever the bioburden requires reduction to facilitate healing. However, this need not necessarily always be the case. Wound bioburden may be "controlled" through bacteriostatic or bactericidal means, but this is not always the entire solution—some virulence determinants can now be controlled by other means. This chapter outlines current knowledge in this area, and so serves as a knowledge base for clinical and scientific developments over the coming years.

Wounds are typically covered or "packed" with dressings; this approach has served well for many years. Developments in wound dressings and other treatment modalities have enabled practitioners to take a variety of approaches to bioburden control and wound management. The recent return of "ancient" remedies, such as honey or maggots, together with some elegant scientific research, have provided alternatives to the traditional dressings approach. Such interventions offer real advantages; for example, where no chemical or biological antimicrobial agent is used, there can be no problems with resistance, irritancy, contact allergy, or toxicity. This latter attribute can be of great value, for example, in the management of wounds in pediatric patients. It also offers effective therapy which spares antibiotics.

WOUND DRESSINGS

Over the past 30 years, wound dressings have been transformed from purely absorptive materials, such as Gamgee[2] and gauzes, into a series of modern, interactive, multifunctional products. This has been driven by the definition and adoption of the moist wound healing (MWH) concept.[3] The development of a variety of modern wound dressings has enabled the practitioner to match the dressing to the wound, rather than use gauze for all wound types.[4] The clinical and scientific evidence for MWH has been accumulating steadily since the 1960s and is now substantial.[5-9]

Recent *in vitro* studies on dressings have elucidated a number of mechanisms whereby the wound bioburden can be reduced without recourse to antimicrobial agents. Some of these are still to be clinically validated, and others have given wound dressings a new dimension.

The basic principles of wound management, as encapsulated in the concepts of tissue, inflammation/infection, moisture, edge (TIME)[10] and applied wound management,[11] provide the basis for selection and use of the dressings described below. This is not intended as a systematic review, but merely to illustrate the ancillary modes of action of dressings and other modalities.

GENERIC DRESSING CLASSES

HYDROCOLLOIDS

For many, the hydrocolloids have been synonymous with moist wound healing, having been among the very first of the modern dressings. The hydrocolloids are probably the most researched class of dressings to date. They were introduced into the market in 1960s, first as an oral dressing for apthous ulcers, and later in 1972 as skin-protecting dressing for stoma patients.[12] The clinicians involved in wound management saw an opportunity to adapt these products for their patients, a development that was to change wound care for the better as many companies developed and marketed hydrocolloid dressings for specific wound applications.[13] The hydrocolloids are composed of carboxymethylcellulose, gelatine, pectin, polyisobutylene, or similar polymers and are occlusive, gelling, moist wound dressings.[14,15] They have very wide application in chronic wound management, being found to be generally safe and effective.[16–19]

Scientific and clinical research has shown these dressings to promote debridement through autolysis and reepithelialization.[20,21] There is evidence that one hydrocolloid (HCD) creates an environment that tends to restrict microbial growth.[22] According to Williams,[23] most hydrocolloids produce an acidic environment, owing to acid polymers such as pectin. This, together with an hypoxic environment,[24,25] is likely to restrict the growth of some microorganisms. The growth of *P. aeruginosa* was inhibited under hydrocolloid.[22] As occlusive dressings, hydrocolloids have been suspected of enhancing microbial growth;[26] however, laboratory findings, together with clinical data,[27] and with a meta-analysis of clinical trial data,[28,29] have shown this to be unfounded. In fact, as occlusive dressings, some have been demonstrated to have bacterial (and viral) barrier properties, rendering them suitable for "containing" wound microorganisms by an important infection control function.[30,31]

ALGINATE DRESSINGS

Calcium alginate, extracted from seaweed, has been used in wound management since the 1940s.[32,33] Since then, a number of alginate-based products have been developed and evaluated clinically and scientifically.[34] This class of dressings is known to create a moist environment by gelling in the presence of exudate[35]—they are, to a varying degree, hemostatic, and promote healing.[36–38] Extensive clinical research has shown them to be cost-effective in a variety of chronic wounds,[39–42] including infected ulcers.[43] The basic difference between the various alginates is in their chemical composition, whether the carbohydrate polymer is mannuronic- or guluronic acid based, and as a calcium or sodium salt. These differences endow the dressing with their distinctive performance characteristics. The alginates have been attributed with bacterial sequestration properties both *in vitro* and *in vivo* (see below), a feature that might have some clinical significance in the management of chronic wounds. The alginates evaluated in these studies included Sorbsan (a high mannuronate sodium alginate), Kaltostat (a high guluronate calcium-sodium alginate), and Algosteril (a high mannuronate calcium alginate).

NEGATIVE PRESSURE WOUND THERAPY

Also known as NPWT, this is often mistakenly regarded as a totally new development in wound care. A precursor, suction drainage of wounds, is well documented, particularly in surgery[44] where surgeons have used it to remove fluids from the wound. Multiple approaches to delivering NPWT are now available including the Argenta Morykwas technique[45] (uses foam dressings), the Kremlin technique[46] (uses rigid domes), hybrid techniques (uses modifiable, semirigid domes), and the Chariker-Jeter technique[47] (uses gauze). Each offers continuous, controlled negative, or subatmospheric, pressures in the order of –125 mm Hg (16.7 kPa).

NPWT is believed to work via a number of mechanisms, including the following:

1. Reduced surface pressure has been believed to induce increased perfusion of the wound area. However, very recent evidence suggests that this is not the case.[48,49]
2. Promotion of granulation tissue formation. In theory this will tend to reduce the risks of infection, as healing wounds are less likely to become infected. Clinical research has shown NPWT to promote the formation of granulation tissue and to decrease fibrinous slough in leg ulcers.[50] Slough is known to harbor microorganisms, and it has been postulated that it is a biofilm.[51]
3. Removal of exudate. This would logically be a measure that is likely to reduce bioburden, at least for those planktonic bacteria that are suspended in exudate.
4. Reduction of bioburden. Early research on pig wounds inoculated with *Staphylococcus* spp. revealed that TNP-treated wounds had a 3-log reduction in bacterial counts in 5 days.[52] Since then, a number of clinical studies have reported on wound microbiology.[53–55] However, Khashram et al.[55] found an increase in flora in chronic, noninfected venous leg ulcers and concluded from their study that NPWT "does not exert its effects by reducing the numbers of colonising bacteria or pathogens." This apparent conundrum could be ascribed to the presence of biofilm.[56] Further research is clearly needed.
5. Reduction of edema. The quantitative measurement of edema is difficult. NPWT is reported to reduce edema and to remove large quantities of fluid. However, there is no evidence to support a reduction in interstitial fluid.

Although there is a wealth of published clinical evidence on NPWT, a recent review claims that studies on chronic wounds have methodological flaws, and consequently more and better-quality research is needed.[57]

BIOSURGERY

A resurgence of interest in the use of maggots, initiated by Ron Sherman in the United States and John Church in the United Kingdom, has led to the commercial development of sterile larvae for wound care.[58] Products are available and are licensed for medical use in wounds in the United States, United Kingdom, and many other countries. Maggots infesting wounds is "myiasis" if spontaneous contamination occurs, or, if *Lucillia sericata* sterile larva are used, this is now known as "biosurgery."[59] The treatment has been subject to intensive research over the past decade. Major clinical[60] and cost-effectiveness studies have been reported[61,62] in front-line medical journals. To back up clinical findings, research studies have indicated anti-inflammatory mechanisms[63] and selective antimicrobial activity[64] without any detectable effects on phagocytosis or subsequent apoptosis.

In vitro studies on the secreta and excreta of *L. sericata* on wound pathogens showed a bactericidal effect on *S. aureus* and temporary inhibition of *E. coli* and *P. aeruginosa*.[64] The same study also involved 30 patients with chronic wounds. Results after maggot applications were positive for healing and antimicrobial effects. All 30 patients either healed or had a resolution of infection. The change in pathogenic bacteria from pre- to post-treatment with maggots showed elimination of *Bacteroides* sp., *Peptococcus* sp., *Serratia* sp., and groups C and G *Streptococcus* sp. There was a reduction in *P. aeruginosa*, *S. aureus*, coagulase-negative staphylocooci, and increases in diptheroids and enterococci. The authors concluded that this selective action is one of the factors that affect the success of larvae, but special precautions are needed when infections are due to *Proteus* spp. These findings are supported by other reports of successful management of necrotic and infected wounds.[65,66] The active antimicrobial entities have been investigated with varying results. Cazander et al.[67] found "no direct antibacterial effect of maggots *in vitro*," and Bexfield et al.[68] were able to detect two antibacterial factors and partially characterize them. These findings are corroborated by van der Plas et al.[69] in studies showing the efficacy of maggot excretions and secretions against biofilms of *Staphylococcus aureus* and *Pseudomonas aeruginosa*.

HONEY

The emphasis on honey research over the past 15 years has led to many clinical and scientific advances in our understanding, as well as a number of regulated wound treatments. Not least are the microbiology studies—a recent PubMed search of articles on honey and wound microbiology identified over 50 articles since 1992. Significant advances have been made by Cooper, Molan, and co-workers.[70 75] The very recent report of the action of honey on biofilms is notable, as this is likely to be of great importance in chronic wound management.[76]

The actions of honey in wound healing are varied.[77] Apart from inhibition of pathogens, it is known to promote healing, to promote autolytic debridement (through the creation of a moist environment), and to regulate pH within the wound.[78–80] These, combined, are claimed to be "a complete wound bed preparation product."[81]

Research has also shown honey to have anti-inflammatory properties, exerted through the stimulation of TNF-α via TLR-4,[82] and reduction of reactive oxygen species.[83]

These findings, together with the growing body of clinical evidence in wound care, makes the modern honey products valuable components of the clinician's armamentarium.[84–86]

HYPERBARIC OXYGEN

Oxygen plays an important role in the physiology of wound healing.[87–90] Hyperbaric oxygen (HBO) is the delivery of 100% oxygen at pressures greater than atmospheric; it has become widely used to treat a variety of nonhealing wounds. Of the so-called chronic wounds, only diabetic foot ulcers have been subjected to extensive HBO treatment.[91] Studies on other chronic wound types are required.[92] Multiple anecdotal reports and retrospective studies in HBO therapy in diabetic patients suggest that HBO can be an effective adjunct in the management of diabetic wounds. Prospective studies also show the beneficial effects of HBO, but because published studies suffer from methodological problems, there is an urgent need for a randomized prospective clinical trial for the application of HBO in diabetic foot lesions before it can be recommended as standard therapy in patients with foot ulcers. Although the mechanisms of action are still being elucidated, it is postulated that HBO can raise tissue oxygen tensions to levels where wound healing can be expected. Hyperbaric oxygen therapy induces vascular endothelial growth factor (VEGF), attenuates apoptosis, increases killing ability of leucocytes, is lethal for certain anaerobic bacteria, and inhibits toxin formation in other anaerobes.[93,94] From a practical perspective, HBO therapy is not without side effects, it is restricted by the availability of specialized equipment and experienced clinicians; there are potentially viable alternatives in development, whereby "oxygen balance" in the wound can be more easily achieved.[95]

In addition to the ancillary attributes listed above, there are a number of mechanisms whereby some dressings can reduce wound bioburden. These are discussed below.

SEQUESTRATION

It has been claimed that the capacity of a dressing to absorb and retain (i.e., sequester) bacteria is an important function, particularly in chronic wounds.[96] *In vitro* and animal *in vivo* microbiological studies have illustrated the extent of this effect in hydrofiber, alginate, cotton, and superabsorbent dressings.[97–102] The hydrofiber dressing Aquacel has been shown to immobilize potential wound pathogens such as Staphylococci and Pseudomonads within the gelled fibers. The true clinical significance of this feature is yet to be demonstrated conclusively, but it is likely to be of value in reducing bioburden in colonized wounds where antimicrobials are not indicated (e.g., routine use in chronic wounds and other wounds healing by secondary intent). This would have the distinct advantage of not selecting for resistance, as the function is purely physical. A similar function has been described for the binding of bacterial toxins.[103] In this context, a silver dressing containing activated charcoal has been shown to adsorb endotoxins from *E. coli* and *P. aeruginosa* in a standard *in vitro* assay. A similar principle applies to a ceramic microsphere dressing (Cerdak). This has been shown to adsorb endotoxins.[104] Although this, too, has yet to be demonstrated as clinically relevant, it is an important mechanism for neutralizing an important virulence determinant.

The initial events involved in a wound infection include the adhesion of a pathogen to the surface. This process is mediated by either electrostatic forces or specific hydrophobic interaction (e.g., through adhesins). The latter may be between the bacterial cell surface and extracellular matrix components.[105] The physical principle of hydrophobic interaction has been utilized to sequester bacteria through the addition of a hydrophobic coating containing a fatty acid derivative (dialkylcarbamoyl chloride, DACC) to the dressing fibers.[106] Bacteria and other microorganisms are "bound" to the dressing when in contact with a moist environment. The microorganisms are then removed when the dressing is changed. In a 116-patient multicenter study with a mean treatment period of 37 days, 81% of wounds showing signs of infection at the start of treatment healed. Twenty-one percent of patients' wounds healed, with a further 72% showing improvement in wound healing.[107] Hydrophobic interaction would appear to offer a "natural" approach to wound healing.[108] There are no chemically active antimicrobial agents and no known side effects or risk of bacterial/fungal resistance.

CONTROL OF BACTERIAL VIRULENCE DETERMINANTS

Bacteria exert their pathogenic effects via a number of factors, known as virulence determinants, which regulate their growth, and adaptive measures to enhance their survival.[109–112] These factors include:

- Biofilm formation
- Quorum sensing
- Synergy
- Immuno-evasive measures
- Adhesins
- Invasins, including bacterial (i.e., exogenous) proteases
- Pili
- Toxins: endo- and exotoxins

The current literature contains a number of initiatives designed to control pathogenicity through interruption of these factors. For the purposes of this chapter, only the invasions and toxins are known to be influenced by topical wound treatments that do not contain antimicrobials. Specific protease-modulating matrices (e.g., Promogran and Tegaderm Matrix) and carboxymethylated cellulose[97,113] are known to quench protease enzymes; charcoal (as included in Actisorb silver 220)[103] and ceramic granules[104] (Cerdak) have the capacity to absorb toxins.

The principle of protease modulation as a means of overcoming delayed healing in chronic wounds is based upon the discovery of raised activity of matrix metalloproteases (MMPs), particularly MMPs-2 and -9 in chronic wound fluid.[114] Their role in terms of healing and prognosis is accepted.[115] In effect, these enzymes can act as "chronicity factors," both in skin wounds and in gingiva, unless their action is modulated by the natural inhibitors—or TIMPs. Although many maintain that these enzymes are of predominantly neutrophil origin (i.e., intrinsic), there is little doubt that bacterial elastases also contribute.[116]

The first product with MMP-neutralizing (or inactivating) activity to be marketed was Promogran—a mixture of oxidized regenerated cellulose (ORC) and collagen.[117]

The mode of action is probably nonspecific, being an inert binding[97] and sacrificial protein substrate mechanism.[118]

This technical approach to overcoming these putative wound chronicity factors has become an intensive area of research and product development.[113,119,120]

DRESSING LEAKAGE AND STRIKE-THROUGH

The capacity of a dressing or wound management system to contain exudate is an important feature in the prevention of infection. The exudate from chronic wounds will invariably be contaminated with microorganisms, and as such presents a cross-infection hazard. The leakage of exudate from nonadherent or poorly adherent dressings, together with strike-through from saturated dressings/bandages provides a portal from the entry or escape of these microorganisms. Thus for adhesive dressings, a secure adhesion to the periwound skin over the wear time of the dressing is important. So, too, is the avoidance of strike-through. In most, if not all, cases, these can be avoided through diligent management. First, the wear time must be judged cautiously—to push this too far will often result in leakage. This also applies to strike-through. For optimal adhesion, particularly where the periwound skin is compromised, a skin barrier preparation will help.[121]

THE MICROBIOLOGICAL BARRIER PROPERTIES OF DRESSINGS

This applies to both the escape and spread of pathogens from the wound, and to the contamination of the wound from pathogens from other body sites or the environment.[29] The concept is not new, as clinicians have long been aware of the risks of cross-infection.[122] There are a number of important aspects to this aspect of wound care:

- Some dressings have been shown to reduce or eliminate the risks of bacterial dispersal by aerosol formation on dressing removal.[123]
- The capacity to exclude pathogens from the wound or to contain wound pathogens has been a feature of many modern dressings.[30,31,97]

THE VALUE OF CLOSE ASSOCIATION OF DRESSING WITH THE WOUND BED

According to Aristotle, "nature abhors a vacuum." This idiom is used to express the idea that empty or unfilled spaces are unnatural, as they go against the laws of nature. This applies to wounds insofar as such spaces adversely affect healing. Snyder[124] has recorded that the presence of "dead space" may act as a nidus for infection and so contribute to delayed healing. Robson et al., cited by Edberg,[125] stated that dead space lends itself to infection because it does not possess a defense mechanism.

These statements clearly indicate there is a need to avoid the creation of dead space (void within a viscous or between dressing and wound bed), as there is an apparent association of dead space with risk of infection.

In order to circumvent this situation, when applying a wound dressing the clinician should ensure that the dressing has the capacity to maintain a close association with the wound bed. It is also reasonable to assume that in those dressings with an absorptive capacity, a close association of the dressing with the wound bed will help promote absorption of exudates.[1] Vanscheidt et al.[126] have made the empirical observation that the gel matrix formed by the hydrofiber dressing used in their study on 18 patients with chronic leg ulcers molded itself over the wound surface and eliminated dead space. This observation was subsequently confirmed by Jones et al.,[127] who investigated the conformability *in vitro* of two silver dressings to human wound tissue, dried dermal membrane, and indented agar plates that had been seeded with MRSA or *P. aeruginosa*. The results showed that there was conformability of the dressing to the dermal tissue (wound bed), but this was less evident with Ag polyethylene mesh dressing.

The intimate association of a dressing with the irregular undulating topography of the wound bed would appear to offer advantages when considering the avoidance of the creation of dead space and absorption of exudate.

FLUID HANDLING PROPERTIES: ABSORPTION/RETENTION, LATERAL WICKING, SEQUESTRATION

Before the advent of products that incorporated antimicrobials, dressings were used principally from the perspective of material performance *in situ*. Until the 1960s, dressings were composed mainly of woven textiles with a primarily covering/protective function and were not regarded as agents capable of enhancing healing. Following the work of Winter,[128] dressing design took into account the contribution that the dressing material could make to the reparative process. However, traditional dressing material such as gauze continues to be used despite recognition that it does not comply with optimal management requirements.[129] Modern wound dressings have been developed primarily to support a moist wound environment, while at the same time providing an absorptive capacity. If problems associated with excess moisture at the dressing–wound interface are not managed correctly, then optimal healing will be compromised. An excessively moist environment predisposes to the growth of certain bacteria; for example, *Staphylococcus aureus* and *Pseudomonas aeruginosa* favor a high water activity (denoted a_W) for growth and toxin production.[130] Dressings and topical wound treatments which reduced will consequently inhibit the growth of these organisms; this might be a factor in the performance of some absorptive. Where absorption of exudate is required, the dressing should also be capable of absorbing and retaining the fluid to achieve an optimum moist environment,[131] ensuring at the same time that the periwound skin is not subjected to maceration.[132] This important performance parameter must apply to those dressings intended for use under compression bandaging. Further breaches in the skin barrier increase infection risks. Parsons et al.,[133] in an *in vitro* study, investigated the performance of seven proprietary silver-containing dressings, including fluid-handling properties and dressing pH. The authors state: "This study suggests that dressing selection should be based on the overall properties of the dressing clinically relevant to the wound type and condition."

It is a criticism often leveled at foam-based dressings that the fluid uptake is compromised by compression. Where data exists, it is *in vitro* and consequently subject to interpretation.

CONCLUSIONS

All wounds healing by secondary intent, whether they are acute or chronic, are colonized with bacteria. In most instances this bioburden will not require treatment with antimicrobials, either topical or systemic. As long as the host defenses can maintain immunological control of the wound environment, the bioburden should be effectively managed. Clinicians have at their disposal a variety of means whereby the bioburden can be reduced, without recourse to antimicrobials. This approach has the distinct advantages of not selecting for resistance and of being nontoxic. Such apparently "inert" means have been in use for many years without ill effect—indeed, with positive attributes, and often without clinicians being aware of the fact that the bioburden is being reduced by such means. This approach to wound care is to be commended. However, much more research is required in this area before we can fully understand the implications of these mechanisms.

REFERENCES

1. Cutting KF, White RJ, Hoekstra H. (2009) Topical silver-impregnated dressings and the importance of the dressing technology. *Int Wound J*, submitted.
2. Kapadia HM. (2002) Sampson Gamgee. *J Roy Soc Med* 95:96–100.
3. Moues CM, Heule F, Legerstee R, Hovius SE. (2009) Five millennia of wound care products—What is new? *Ostomy Wound Management* 55(3):16–32.
4. Alvarez O. (1988) Moist environment for healing: Matching the dressing to the wound. *Ostomy Wound Manage* 21:64–83.
5. Hammond MA. (1979) Moist wound healing: Breaking down the barrier. *Nurs Mirror* 149(18):38–40.
6. Alper JC, Welch EA, Ginsberg M, et al. (1983) Moist wound healing under a vapour permeable membrane. *J Am Acad Dermatol* 8(3):347–353.
7. Dyson M, Young S, Pendle CL, et al. (1988). Comparison of the effects of moist and dry conditions on dermal repair. *J Invest Dermatol* 91(5):434–439.
8. Chang H, Wind S, Kerstein M. (1996) Moist wound healing. *Dermatol Nurs* 8(3):174–176.
9. Bryan J. (2004) Moist wound healing: A concept that changed our practice. *J Wound Care* 13(6):227–228.
10. Schultz GS, Barillo DJ, Mozingo DW, et al. (2004) Wound bed preparation and a brief history of TIME. *Int Wound J* 1(1):19–32.
11. Gray DG, White RJ, Cooper P, Kingsley AR. (2006) Applied Wound Management. Chapter 4. In *Essential Wound Management,* Eds. D Gray, P Cooper, J Timmons. *Wounds*–UK Publications, Aberdeen, pp. 73–106.
12. Black P. (1994) Choosing the correct appliance. *Br J Nurs* 3(11):545–550.
13. Queen D, Orsted H, Sanada H, Sussman G. (2004) A dressing history. *Int Wound J* 1(1):59–77.

14. Ichioka S, Harii K, Nakahara M, Sato Y. (1998) An experimental comparison of hydrocolloid and alginate dressings, and the effect of calcium ions on the behaviour of alginate gel. *Scand J Plast Reconstr Surg Hand Surg* 32(3):311–316.
15. Fletcher J. (2003) The benefits of hydrocolloids. *Nurs Times* 99(21):57.
16. Tracy GD, Lord RS, Kibel C, et al. (1977) Varihesive sealed dressing for indolent leg ulcers. *Med J Aust* 1(21):777–780.
17. van Rijswijk L, Brown D, Friedman S, et al. (1985) Multicentre clinical evaluation of a hydrocolloid dressing for leg ulcers. *Cutis* 35(2):173–176.
18. Finnie A. (2002) Hydrocolloids in wound management: Pros and cons. *Brit J Comm Nurs* 7(7):340–342.
19. Heyneman A, Beele H, Vanderwee K, Defloor T. (2008) A systematic review of the use of hydrocolloids in the treatment of pressure ulcers. *J Clin Nurs* 17(9):1164–1173.
20. Bale S, Harding KG. (1990) Using modern dressings to effect debridement. *Prof Nurs* 5(5):244–245.
21. Porter JM. (1991) A comparative investigation of re-epithelialisation of split skin graft donor areas after application of a hydrocolloid and alginate dressings. *Br J Plast Surg* 44(5):333–337.
22. Gilchrist B, Reed C. (1989) The bacteriology of chronic venous ulcers treated with occlusive dressings. *Brit J Dermatol* 121(3):337–344.
23. Williams C. (1994) Granuflex. *Br J Nurs* 10(3):730–733.
24. Varghese MC, Balin AK, Carter DM, Caldwell D. (1986) Local environment of chronic wounds under synthetic dressings. *Arch Dermatol* 122(1):52–57.
25. Thomas S. (1990) Making sense of hydrocolloid dressings. *Nurs Times* 86(45):36–38.
26. Marshall DA, Mertz PM, Eaglstein WH. (1990) Occlusive dressings. Does dressing type influence the growth of common bacterial pathogens? *Arch Surg* 125(9):1136–1139.
27. Annoni F, Rosina M, Chiurazzi D, Ceva M. (1989) The effects of a hydrocolloid dressing on bacterial growth and the healing process of leg ulcers. *Int Angiol* 8(4):224–228.
28. Hutchinson JJ, Lawrence JC. 1991. Wound infection under occlusive dressings. *J Hosp Infect* 17(2):83–94.
29. Lawrence JC. (1994) Dressings and wound infection. *Am J Surg* 167(1A):21S–24S.
30. Mertz PM, Marshall DA, Eaglstein WH. (1985) Occlusive wound dressings to prevent bacterial invasion and wound infection. *J Am Acad Dermatol* 12(4):662–668.
31. Ameen H, Moore K, Lawrence JC, Harding KG. (2000) Investigating the bacterial barrier properties of four contemporary wound dressings. *J Wound Care* 9(8):385–388.
32. Bray C, Blaine G, Hudson P. (1948) New treatments for burns, wounds and haemorrhage. *Nurs Mirror* 86:239–242.
33. Morgan D. (1996) Alginate dressings. *J Tiss Viab* 7(1):4–8.
34. Paddle-Ledinek JE, Nasa Z, Cieland HJ. (2006) Effect of different wound dressings on cell viability and proliferation. *Plast Reconstr Surg* 117(7):110S–118S.
35. Thomas S, Loveless P. (1992) Observations on the fluid handling of alginate dressings. *Pharm J* 248:850–851.
36. Thomas S. (2000) Alginate dressings in surgery and wound management: Part 1. *J Wound Care* 9(2):56–60.
37. Thomas S. (2000) Alginate dressings in surgery and wound management: Part 2. *J Wound Care* 9(3):115–119.
38. Thomas S. (2000). Alginate dressings in surgery and wound management: Part 3. *J Wound Care* 9(4):163–166.
39. Motta GJ. (1989) Calcium alginate topical wound dressing: A new dimension in the cost-effective treatment for exudating dermal wounds and pressure sores. *Ostomy Wound Manag* 25:52–56.
40. Fanucci D, Seese J. (1991) Multi-faceted use of calcium alginates. *Ostomy Wound Manag* 37:16–2242.

41. McMullen D. (1991) Clinical experience with a calcium alginate dressing. *Dermatol Nurs* 34:216–219, 270–271.
42. Bale S, Baker N, Crook H et al. (2001) Exploring the use of an alginate dressing for diabetic foot ulcers. *J Wound Care* 10(3):81–84.
43. Young MJ. 1993. The use of alginates in the management of exudating, infected wounds: Case studies. *Dermatol Nurs* 5,5:359–363.
44. Garrie S. (1965) Negative pressure suction device. Surgery 58(6):979–980.
45. Argenta LC, Morykwas MJ. (1997) Vacuum-assisted closure: A new method for wound control and treatment: Clinical experience. *Ann Plast Surg* 38:563–577.
46. Kostiuchenok BM, Kolker II, Karlov VA, et al. (1986) Vacuum treatment in the surgical management of suppurative wounds. *Vestn Khir Im I I Grek* 137(9):18–21.
47. Chariker ME, Jeter KF, et al. (1989) Effective management of incisional and cutaneous fistulae with closed suction wound drainage. *Contemp Surg* 34:59–63.
48. Kairinos N, Voogd AM, Botha PH, et al. (2009) Negative pressure wound therapy II: Negative-pressure wound therapy and increased perfusion. Just an illusion? *Plast Reconstr Surg* 123(2):601–612.
49. Kairinos N, Solomons M, Hudson DA. (2009). Negative pressure wound therapy I: The paradox of negative-pressure wound therapy. *Plast Reconstr Surg* 123(2):598–600.
50. Lorée S, Dompmartin A, Penven K, et al. (2004). Is vacuum assisted closure a valid technique for debriding chronic leg ulcers? *J Wound Care* 13(6):249–252.
51. Wolcott R, Cutting KF, Contreras Ruiz J. (2008) Biofilms and delayed healing. In *Advancing your practice: Understanding wound infection an the role of biofilms*. Ed. KF Cutting, pp. 12–17. Malvern PA, Association for the Advancement of Wound Care.
52. Morykwas MJ, Argenta LC, Shelton-Brown EI, McGuirt W. (1997) Vacuum-assisted closure; a new method for wound control and treatment: Animal studies and basic foundation. *Ann Plast Surg* 38:553–562.
53. Moues CM, Vos MC, van den Bemd G, et al. (2004) Bacterial load in relation to vacuum-assisted closure wound therapy: A prospective randomized trial. *Wound Rep Regen* 12:11–17.
54. Weed T, Ratliff C, Drake DB. (2004) Quantifying bacteria during negative pressure wound therapy: Does the wound VAC enhance bacterial clearance? *Ann Plast Surg* 52(3):276–279.
55. Khashram M, Huggan P, Ikram R, Chambers S, Roake JA, Lewis DR. 2009. Effect of TNP on the microbiology of venous leg ulcers: A pilot study. *J Wound Care* 18,4:164–167.
56. Wolcott RD, Rhoads DD, Dowd SE. (2008) Biofilms and chronic wound inflammation. *J Wound Care* 17,6:331–341.
57. Ubbink DT, Westerbos SJ, Evans D, et al. (2008) Topical negative pressure for treating chronic wounds. Cochrane Database Syst Rev 16, 3, CD001898.
58. Whitaker IS, Twine C, Whitaker MJ, et al. (2006) Larval therapy from antiquity to the present day: Mechanisms of action, clinical applications and future potential. *Postgrad Med J* 83(980):409–413.
59. Church J. (1996) Biosurgery in wound healing. *J Wound Care* 5(2):60–69.
60. Dumville JC, Worthy G, Bland JM, et al. (2009). Larval therapy for leg ulcers (VenUS II): Randomised controlled trial. *BMJ* 388:b773. Doi:10.1136/bmj.b773.
61. Thomas S. (2006) Cost of managing chronic wounds in the UK with particular emphasis on maggot debridement therapy. *J Wound Care* 15(10):465–469.
62. Soares MO, Iglesias CP, Bland JM, et al. (2009) Cost effectiveness of larval therapy for leg ulcers. *BMJ* 388:b825. doi: 101136/bmj.b825.
63. Percivova J, Macickova T, Takac P, et al. (2008) Effect of the extract from salivary glands of *Lucilia sericata* on human neutrophils. *Neuro Endochrinol Lett* 29(5):794–797.
64. Jaklic D, Lapanje A, Zupancic K, et al. (2008) Selective antimicrobial activity of maggots against pathogenic bacteria. *J Med Microbiol* 57(5):617–625.

65. Steenvoorde P, Jacobi CE, Van Doorn L, Oskam J. (2007) Maggot debridement therapy in the palliative setting. *Am J Hosp Palliat Care* 24(4):308–310.

66. Steenvoorde P, Jacobi CE, Van Doorn L, Oskam J. (2007) Maggot debridement therapy of infected ulcers. *Ann Roy Coll Surg Engl* 89(6):596–602.

67. Cazander G, van Veen KE, Bernards AT, Jukema GN. (2009) Do maggots have an influence on bacterial growth? A study on the susceptibility of strains of six different bacterial species to maggots of *Lucilia sericata* and their excretions/secretions. *J Tissue Viability* [Epub ahead of print], 10 April 2009.

68. Bexfield A, Nigam Y, Thomas S, Ratcliffe NA. (2004) Detection and partial characterisation of two antibacterial factors from the medicinal maggot *Lucilia sericata*. *Microbes Infect* 6(14):1297–1304.

69. van der Plas MJ, Jukema GN, Wai SW, et al. (2008) Maggot excretions/secretions are differently effective against biofilms of *Staphylococcus aureus* and *Pseudomonas aeruginosa*. *J Antimicrob Chemother* 61(1):117–122.

70. Willix DJ, Molan PC, Harfoot CG. (1992) A comparison of the sensitivity of wound-infecting species of bacteria to the antibacterial activity of manuka honey and other honey. *J Appl Bacteriol* 73(5):388–394.

71. Cooper RA, Molan PC. (1999) The use of honey as an antiseptic in managing Pseudomonas infection. *J Wound Care* 8(4):161–164.

72. Cooper RA, Halas E, Molan PC. (2002) The efficacy of honey in inhibiting strains of *Pseudomonas aeruginosa* from infected burns. *J Burn Care Rehabil* 23(6):366–370.

73. Cooper RA, Molan PC, Harding KG. (2002) The sensitivity to honey of Gram-positive cocci of clinical significance isolated from wounds. *J Appl Microbiol* 93(5):857–863.

74. Cooper RA. (2008) Using honey to inhibit wound pathogens. *Nurs Times* 104(3):46–49.

75. Cooper RA, Blair S. (2009) Challenges in modern wound microbiology and the role for honey. Chapter 4. In: *Honey in Modern Wound Management*. Eds. R Cooper, P Molan, RJ White. Wounds-UK, Aberdeen, pp. 47–62.

76. Merckoll P, Jonassen T, Yad ME, et al. (2009) Bacteria, biofilm and honey. A study of the effects of honey on planktonic and biofilm-embedded chronic wound bacteria. *Scand J Infect Dis* 41(5):341–347.

77. Molan P. (2005) Mode of action. In: *Honey in Modern Wound Management*. Eds. R White, R Cooper, P Molan. Wounds UK, Aberdeen, pp. 1–23.

78. Gethin GT, Cowman S, Conroy RM. (2008) The impact of manuka honey dressings on the surface pH of chronic wounds. *Int Wound J* 5(2):185–194.

79. Simon A, Traynor K, Santos K, et al. (2009) Medical honey for wound care—Still the "latest resort"? *Evid Based Complement Alternat Med* 6(2):165–173.

80. Robson V, Cooper RA. (2009) Using leptospermum honey to manage wounds impaired by radiotherapy: A case series. *Ostomy Wound Manage* 55(1):38–47.

81. Acton C. (2008) Medihoney: A complete wound bed preparation product. *Br J Nurs* 17(11):S44–S48.

82. Tonks AJ, Dudley E, Porter NG, et al. (2007) A 58-kDa component of manuka honey stimulates immune cells via TLR 4. *J Leukoc Biol* 82(5):1147–1155.

83. van den Berg AJ, van den Worm E, van Ufford HC, et al. (2008) An *in vitro* examination of the antioxidant and anti-inflammatory properties of buckwheat honey. *J Wound Care* 17(4):172–178.

84. White RJ, Molan P. (2005) A summary of published clinical research on honey in wound management. In: *Honey in Modern Wound Management*. Eds. R White, R Cooper, P Molan. Wounds UK, Aberdeen, pp. 130–142.

85. Jull AB, Rodgers A, Walker N. (2008) Honey as a topical treatment for wounds. *Cochrane Database Syst Rev* 8(4); CD005083.

86. Robson V, Dodd S, Thomas S. (2009) Standardised antibacterial honey (Medihoney) with standard therapy in wound care: Randomised clinical trial. *J Adv Nurs* 65(3):565–575.
87. Hopf H, Rollins MD. (2007) Wounds: An overview of the role of oxygen. *Antioxid Redox Signal* 9:1183–1192.
88. Bishop A. (2008) Role of oxygen in wound healing. *J Wound Care* 17(9):399–402.
89. Sen CK. (2009) Wound healing essentials: Let there be oxygen. *Wound Rep Regen* 17:1–18.
90. Thom SR. (2009) Oxidative stress is fundamental to hyperbaric oxygen therapy. *J Appl Physiol* 106(3):988–995.
91. Niinikoski JH. (2003) Hyperbaric oxygen therapy of diabetic foot ulcers and transcutaneous oxymetry in clinical decision making. *Wound Rep Regen* 11:458–461.
92. Thackham JA, McElwain DL, Long RJ. (2007). The use of hyperbaric oxygen therapy to treat chronic wounds: A review. *Wound Rep Regen* 16:321–330.
93. Gordillo GM, Roy S, Khanna S, et al. (2008) Topical oxygen therapy induces VEGF expression and improves closure of clinically presented chronic wounds. *Clin Exp Pharmacol Physiol* 35:957–964.
94. Zhang Q, Chang Q, Cox RA, et al. (2008) Hyperbaric oxygen attenuates apoptosis and decreases inflammation in an ischaemic wound model. *J Invest Dermatol* 128(8):2102–2112.
95. Davis PJ. (2007) How might we achieve oxygen balance in wounds? *Int Wound J* 4(3):18–24.
96. Wysocki AB. (2002) Evaluating and managing open skin wounds: Colonization versus infection. *AACN Clinical Issues* 13:382–397.
97. Edwards JV, Howley PS. (2007) Human neutrophil elastase and collagenase sequestration with phosphorylated cotton wound dressings. *J Biomed Mater Res A* 83(2):446–454.
98. Walker M, Hobot JA, Newman GR, Bowler PG. (2002) Scanning electron microscopic examination of bacterial immobilisation in a carboxymethyl cellulose (AQUACEL) and alginate dressings. *Biomaterials* 24,5:883–890.
99. Edwards JV, Yager DR, Cohen IK, Diegelmann RF, Montante S, Bertoniere N, Bopp AF. (2001) Modified cotton gauze dressings that selectively absorb neutrophil elastase activity in solution. *Wound Repair Regen* 9(1):50–58.
100. Newman GR, Hobot JA, Walker M Bowler PG. (2006) Visualisation of bacterial sequestration and bactericidal activity within hydrating hydrofibre dressings. *Biomaterials* 27(7):1129–1139.
101. Bruggisser R. (2005). Bacterial and fungal absorption properties of a hydrogel dressing with a superabsorbent polymer core. *J Wound Care* 14(9):438–442.
102. Eming S, Smola H, Hartmann B, et al. (2008) The inhibition of matrix metalloprotease activity in chronic wounds by a polyacrylate superabsorber. *Biomaterials* 29(19):2932–2940.
103. Müller G, Winkler Y, Kramer A. (2003) Antibacterial activity and endotoxin-binding capacity of Actisorb Silver 220. *J Hosp Infect* 53(3):211–214.
104. Opoku AR, Sithole SS, Mthimkhulu NP, Nel W. (2007) The endotoxin binding and antioxidative properties of ceramic granules. *J Wound Care* 16(6):271–274.
105. Ljungh A, Wadstrom T. (1995) Binding of extracellular matrix proteins by microbes. In: *Methods in Enzymology: Microbial Adhesion*. Eds. RJ Doyle, I Ofek. New York: Academic Press, 253:501–514.
106. Ljungh A, Yanagisawa N, Wadström T. (2006) Using the principle of hydrophobic interaction to bind and remove wound bacteria. *J Wound Care* 15(4):175–180.
107. Kammerlander G, Locher E, Suess-Burghart A, von Hallern B, Wipplinger P. (2008) An investigation of Cutimed® Sorbact® as an antimicrobial alternative in wound management. *Wounds-UK* 4(2):10–20.

108. Hampton S. (2007). An evaluation of the efficacy of Cutimed Sorbact in different types of non-healing wounds. *Wounds-UK* 3(4):113–119.

109. Heinzelmann M, Scott M, Lan T. (2002). Factors predisposing to bacterial invasion and infection. *Am J Surg* 183:179–190.

110. La MV, Raoult D, Renesto P. (2008) Regulation of whole bacterial pathogen transcription within infected hosts. *FEMS Microbiol Rev* 32(3):440–460.

111. Marshall JC. (2008). Sepsis: rethinking the approach to clinical research. *J Leukoc Biol* 83(3):471–482.

112. van der Poll, Opal SM. (2008) Host-pathogen interactions in sepsis. *Lancet Infect Dis* 8(1):32–43.

113. Edwards JV, Yager DR, Cohen IK, et al. (2001) Modified cotton gauze dressings that selectively absorb neutrophil elastase activity in solution. *Wound Rep Regen* 9(1):50–58.

114. Wysocki AB, Staiano-Coico L, Grinnell F. (1993) Wound fluid from chronic leg ulcers contains elevated levels of metalloproteases MMP-2 and MMP-9. *J Invest Dermatol* 101(1):64–68.

115. Rayment EA, Upton Z. (2009) Finding the culprit: A review of the influences of proteases on the chronic wound environment. *Int J Low Extrem Wounds* 8(1):19–27.

116. Werthén M, Davoudi M, Sonesson A, Nitsche DP, Mörgelin M, Blom K, Schmidtchen A. (2004) *Pseudomonas aeruginosa*-induced infection and degradation of human wound fluid and skin proteins ex vivo are eradicated by a synthetic cationic polymer. *J Antimicrobial Chemother* 54(4):772–779.

117. Cullen B, Smith R, McCulloch E, et al. (2002) Mechanism of action of Promogran, a protease modulating matrix, for the treatment of diabetic foot ulcers. *Wound Rep Regen* 10(1):16–25.

118. Yang SQ, Wang CT, Gillmor SA, et al. (1998) Ecotin: A serine protease inhibitor. *J Mol Biol* 279(4):945–957.

119. Edwards JV, Bopp AF, Batiste S, et al. (1999) Inhibition of elastase by a synthetic cotton-bound serine protease inhibitor: *In vitro* kinetics and inhibitor release. *Wound Rep Regen* 7(2):106–118.

120. Edwards JV, Bopp AF, Batiste S, et al. (2003) Human neutrophil elastase inhibition with a novel cotton-alginate wound dressing formulation. *J Biomed Mater Res A* 66(3):433–440.

121. Schuren J, Becker A, Sibbald RG. (2005) A liquid-forming acrylate for peri-wound protection: A systematic review and meta-analysis. *Int Wound J* 2(3):230–238.

122. Piskozub ZT. (1968) The efficiency of wound dressing materials as a barrier to secondary bacterial contamination. *Br J Plast Surg* 21(4):387–401.

123. Lawrence JC, Lilly HA, Kidson A (1992) Wound dressings and airborne dispersal of bacteria. *Lancet* 339:8796:807.

124. Snyder RJ. (2005) Managing dead space: An overview. *Podiatric Mgmt* 24(8):171–174.

125. Edberg SC. (1981) Methods of quantitative microbiological analyses that support the diagnosis, treatment, and prognosis of human infection. *CRC Crit Rev Microbiol* 8(4):339–397.

126. Vanscheidt W, Lazareth I, Routkovsky-Norval C. (2003) Safety evaluation of a new ionic silver dressing in the management of chronic ulcers. *Wounds* 15(11):371–378.

127. Jones S, Bowler PG, Walker M. (2005) Antimicrobial activity of silver-containing dressings is influenced by dressing conformability with a wound surface. *Wounds* 17(9):263–270.

128. Winter G. (1962). Formation of the scab and rate of epithelialisation of superficial wounds in the skin of a young domestic pig. *Nature* 193;293–294.

129. Ovington LG. (2002) Hanging wet-to-dry dressings out to dry. *Adv Skin Wound Care* 15(2):84–86.

130. Qi Y, Miller KJ. (2000) Effect of low water activity on staphylococcal enterotoxins bio-synthesis. *J Food Prot* 63(4):473–478.
131. Bishop SM, Walker M, Rogers AA, Chen WYJ. (2003) Importance of moisture balance at the wound–dressing interface. *J Wound Care* 124:125–128.
132. White RJ, Cutting KF. (2003) Interventions to avoid maceration of the skin and wound bed. *Br J Nurs* 12(20):1186–1201.
133. Parsons D, Bowler PG, Myles V, Jones S. (2005) Silver antimicrobial dressings in wound management: A comparison of antibacterial, physical and chemical characteristics. *Wounds* 17:222–232.

13 Factors Affecting the Healing of Chronic Wounds: An Iconoclastic View

Marissa J. Carter and Caroline E. Fife

CONTENTS

Introduction...345
    Pressure Ulcers as an Example of Chronic Wounds 346
Colonization, Infections, and Wound Healing...347
The Chronic Wound: What to Do?...348
    Basic Wound Care...350
Compliance with Basic Wound Care 350
    Basic Wound Care Clinical Practice Guidelines 352
The Technology Ceiling...355
    A Short History of High-Tech Wound Healing Treatments...........................355
    The Verdict on Advanced Wound-Healing Technologies.................................358
    Efficacy versus Effectiveness 359
    Ischemia and Cellular Hypoxia...361
    Where Do We Go from Here?...362
Silver-Impregnated Wound Dressings as an Example of Advanced Technology....363
    The Concept 363
    Clinical Use of Silver-Impregnated Dressings...363
    Systematic Review and Meta-Analysis...364
Conclusions...366
References...367

INTRODUCTION

The Greeks were probably the first to recognize the difference between an acute and chronic wound (i.e., fresh versus non-healing), and Galen of Pergamum, who served the gladiators performing in Rome (circa 120–201 A.D.) also recognized the importance of wound site moisture for successful closure. However, it was not until the 16th century that Ambroise Paré, the French military barber/surgeon, conducted what might be considered the first controlled clinical trial regarding wound healing.[1]

Because he ran out of boiling oil to "neutralize the toxins" from residual gunpowder in wounds, he was forced to resort to a poultice of turpentine, egg whites, and oil of roses. He compared the two cohorts thus:

> In the night I could not sleep in quiet, fearing some default in not cauterising, that I should find the wounded to whom I had not used the said oil dead from the poison of their wounds; which made me rise very early to visit them, where beyond my expectation I found that those to whom I had applied my digestive medicament had but little pain, and their wounds without inflammation or swelling, having rested fairly well that night; the others, to whom the boiling oil was used, I found feverish, with great pain and swelling about the edges of their wounds. Then I resolved never more to burn thus cruelly poor men with gunshot wounds. (Paré, 1537)

The next major advance came from Joseph Lister, who identified airborne bacteria as the source of infection both in the wound, and in the operating room. The patient's skin or wound was first washed with 5% carbolic acid and the operation performed in an atmosphere of carbolic acid mist produced by a jet of steam mingling with a 5% solution of carbolic acid solution. Furthermore, all the sponges and instruments used during the operation were soaked in the same solution and bleeding points secured with carbolized gut or silk. Following wound closure, a silk dressing was applied over which was secured carbolized gauze.[2] By bathing the room in carbolic acid mist, postoperative mortality rates, which had been running as high as 90%, began to drop significantly. Dressings, which heretofore were often made from the sweepings collected from textile mill floors, now became ready-made, sterile, practical surgical dressings wrapped and sealed in individual packages thanks to Robert Wood Johnson.

Throughout the 20th century, hundreds of new dressing materials appeared, which balanced moisture, reduced bacterial load, and performed other specific functions to assist in wound healing. In addition, following the discovery of penicillin by Alexander Fleming, the advent of antibiotics allowed infection to be better controlled. The combination of these advances (and others) can be illustrated by examining the mortality rate resulting from battlefield wounds in the 19th century that resulted in amputation and then looking at modern day comparisons. For example, during the American Civil War, the mean mortality rate from nearly 30,000 amputations was 24.3%.[3] A more recent study of New York subway accident amputees from 1989 to 2003 demonstrated that the mortality rate was 5%, although the infection rate was still 32%.[4]

While the healing of acute wounds still garners attention today, the focus of wound healing has shifted to the healing of chronic wounds and subsequent improvement in the quality of life in populations aged ≥ 60 years, or those who have chronic diseases that interfere with wound healing, such as diabetes. The problem is that the incidence of chronic wounds has not changed much in the last 20 years.

Pressure Ulcers as an Example of Chronic Wounds

A pressure ulcer (PU) has historically been defined as an injury caused by unrelieved pressure on tissues presumed to result from local tensions exceeding the capillary pressure. It can originate in the subcutaneous tissue, particularly when muscle is subjected to external pressure and forced against a bony prominence, or on the skin itself if it is subjected to a high enough pressure. However, histological research conducted by Witkowski and Parish[5] demonstrated that the epidermis is more robust than muscle in regard to compression injury[6,7] and capable of withstanding ischemia for longer periods of time. Thus, subcutaneous and muscle tissue necrosis is much more likely to occur before any damage to the epidermal layer is visible. Unfortunately, this concept of deep tissue injury is poorly represented by the current PU staging system. Although the term "staging" in cancer terminology implies a progression from one stage to another, the current system used in the United States endorsed by the National Pressure Ulcer Advisory Panel (NPUAP) is meant to describe only the depth of tissue visible at the base of the wound. Thus, stage 2 ulcers do not progress to stage 3 ulcers, and so forth, nor can ulcers be "reversed staged" to describe the healing process[8] despite a recent NPUAP update.[9]

The prevalence of pressure ulcers has not decreased in the last 15–20 years. For example, one national survey conducted in 265 acute-care hospitals in 1995 determined that the prevalence of PUs was 10.1%, of which 74% were stage 1 and 2.[10] These results mirrored the results of four previous surveys. Coleman et al.[11] also carried out two cross-sectional surveys in 1992–1994 and 1997–1998 in which the crude prevalence of PUs was 8.5%, and a review of 15,121 nursing home residents in Ohio carried out by Spector and Fortinsky[12] in 1994 showed a prevalence of 12%. Comparable results have been found in Europe. An extremely large national survey conducted in Germany during 2002–2003 found a PU prevalence of 24.6% in hospitals and 19.3% in nursing homes,[13] and the most recent French study that involved 37,307 inpatients in 1149 hospitals indicated a prevalence of 8.9%, which the authors commented was exactly the same result compared to a previous French study 10 years prior.[14]

We can also compare the incidence of PUs to that of cancer, which also shows no major shift downward. For example, if the incidence of breast cancer is compared between the years 1980 and 2006, there is no overall change.[15] Despite advances in these respective fields, we have not made progress in mortality or healing rates. The point is that both these conditions are treated empirically without truly understanding their etiology; if their mechanisms of development were completely understood, one could argue that treatments, even if not completely efficacious, ought to have made a difference. We argue that we have reached a similar impasse when it comes to wounds and infection.

COLONIZATION, INFECTIONS, AND WOUND HEALING

The current model of infection encompasses colonization of a wound by bacteria or fungi, in which the body's defense mechanisms fail to halt growth, resulting in infection, or maintain an uneasy status quo with infectious agents in which colonization persists. For clinical purposes the gold standard used to define infection is $>10^5$ colony-forming units per gram of tissue with the exception of beta-hemolytic *Streptococci*.[16,17] However, because the taking of quantitative tissue biopsies to ascertain the status of a wound is a non-routine procedure, clinicians usually rely on diagnostic signs of infection. A groundbreaking study conducted by Gardner, Frantz, and Doebbeling[18] evaluated wound bed infection and classic signs of infection in several types of wounds and found little correlation, which indicates that visual signs of infection have little predictive value. In addition, a systematic review of the literature for diabetic foot ulcers showed that infection could not be reliably identified using clinical assessment.[19] Moreover, in chronic venous ulcers, signs of inflammation (induration and warmth of the skin, pain, swelling, tenderness to touch, copious drainage, and foul odor) are frequently confused with infection.[20] In an attempt to be more precise, some clinicians resort to taking swab cultures of wounds. But the results of these procedures can be even more misleading, because there is no universal standard technique: Does one take the swab before or after removing necrotic or devitalized tissue, and from what part of the wound does one take the sample? Thus, taking a swab of the wound's surface will reflect surface contamination or surface colonization, not the nature of the agent that might be causing an infection deep within the wound. Gardner et al.[21] attempted to answer this question and found that out of three techniques, swabs from wound exudates, the Z technique, and the Levine technique, the Levine technique was superior, with a sensitivity of 90%, a specificity of 57%, and a mean concordance of 78% between swab specimens when a critical threshold of 37,000 organisms per swab was utilized. (The Levine technique entails rotating the swab over a 1-cm^2 area in the cleansed wound center with sufficient pressure on the wound surface to extract fluid from inside the wound[22]). The summation of poor infection identification practices leads to two common occurrences in wound care, neither of which are beneficial: (1) overuse of topical microbial agents and systemic antibiotics, and (2) undertreatment of wounds that really are infected.[20,23–25] The consequences of excessive antibiotic administration leads to bacterial resistance, while overuse of topical agents can be injurious to tissue. On the other hand, neglect of infected wounds can lead to systemic complications or even risk of death.

THE CHRONIC WOUND: WHAT TO DO?

The vast majority of acute wounds in healthy individuals heal relatively quickly. However, when comorbidities, age >60 years, and wound colonization are factored in, healing can quickly slow or stall out despite appropriate treatment, which presents the clinician with the classic chronic non-healing wound.

What should be done next? This is a challenging question, not least because many therapeutic options are available. In our opinion, this involves identifying the rate-determining step (RDS) in each case, and the factors that are causing the "bottleneck" in the healing steps. In biochemistry, complex processes consist of several sequential steps, with each separate reaction catalyzed by a different enzyme that has a given kinetic rate constant (Figure 13.1). However, the step that has the lowest kinetic rate is the step that controls the overall speed of the entire process. In Figure 13.1A, it can be seen that the step from B to C is the fastest, while the slowest step is from C to D. The last step, C to D is the RDS that controls the speed of the entire

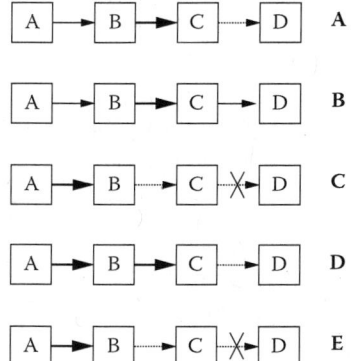

FIGURE 13.1   Biochemical reaction sequence: relative rates of reaction (Panels A and B). In Panels C through E, box A represents hemostasis, B inflammation, C proliferation, and D maturation.

reaction sequence. In this hypothetical sequence perhaps the enzyme that controls the last step is being inhibited, so if that inhibition is removed, the rate of that step increases, and now the reaction A to B becomes the RDS (Figure 13.1B). In similar fashion there are three stages of wound healing following hemostasis (inflammation, granulation [proliferation], and maturation [remodeling]),[26,27] although they are not so tightly coupled as biochemical reactions and overlap to some degree. We can also make the case that epithelialization represents a separate stage that overlaps with granulation, rather than considering it part of the proliferation stage. This is because there are wounds that granulate but do not create skin; moreover, there are products specifically designed to enhance granulation and others to specifically enhance epithelialization. Nevertheless, the stage concept is useful and Figure 13.1C–E shows the results of major problems with inflammation, proliferation, or a concurrent condition, such as corticosteroid administration, that affects all phases. Thus, it is critical to assess the chronic wound in terms of the stage in which stalling has occurred, and the possible factors that are inhibiting the process.

What does the current evidence tell us about the state of the chronic wound that will not heal? Several aspects have been pursued in the last decade that can be loosely grouped in to the first two phases of healing: (1) inflammation, which includes wound biofilms, the adverse effect of matrix metalloproteinases, and (2) proliferation, which includes growth factors, such as epidermal growth factor and insulin growth factor, tissue ischemia and its effect on angiogenesis mediated in part by the VEGF family (vascular endothelial growth factor), nitric oxide production, fibroblast senescence, and (3) older age, which likely interferes with many stages of healing, and covers nutritional deficiencies that might be the cause of insufficient protein expression, biochemical factors and a poor cellular redox status, as well as metabolic dysfunctions that appear to be part of the aging process.

We will discuss some of these factors in more detail, but first we have to ask an overriding question: Are all patients receiving wound basic care?

BASIC WOUND CARE

The definition of basic wound care varies according to the type of chronic wound. In its most fundamental form, the wound is cleansed, debrided, and a moist dressing applied, of which there are many choices. Those factors that contributed to the wound formation in the first place must also be managed. For pressure and diabetic ulcers offloading is essential, and for venous ulcers (VUs), compression is applied.[28] Basic wound care is effective in healing wounds in the majority of cases. For example, one of us (CF) reviewed data (1990–1998) from 300 consecutive VU patients at the Memorial Hermann Hospital in Houston, Texas, who received a hydrocolloid dressing and an Unna's boot covered with an elastic wrap. Overall, 85% of the patients healed with an average healing time of 22 weeks. The recalcitrant wounds were associated with patients who had diabetes mellitus, rheumatoid arthritis, or were on prednisone. Margolis, Berlin, and Strom[29] concluded from a review of the literature of the 1980s and the early 1990s that compression resulted in healing rates of 30% to 60% at 24 weeks, and 70% to 85% at 1 year, and a more recent study of wound healing trajectories in 232 patients (8 trials over 10 years) conducted by Steed et al.[30] demonstrated that 60% of patients on average healed at 20 weeks. So, our results are in line with those of other investigators and it can be summarized that in time, some 85% of patients with venous ulcers will be healed with basic wound care emphasizing adequate compression to overcome venous stasis. The flip side of this result, of course, is that 15% of patients will not heal with basic care, and that this number has not changed in 20 years.

Diabetic foot ulcers (DFUs) are harder to heal with basic wound care. A cohort study of 27,193 individuals with diabetic neuropathic foot ulcers who received "standard care" showed that between 1988 and 1990, 66% of patients had not healed at 20 weeks.[31] A more recent longitudinal UK study of 370 patients with DFUs performed over 31 months showed that 33% never became ulcer-free.[32] Of the population who became ulcer-free at some stage, (i.e., were healed), over 40% developed a new or recurring ulcer. Again, basic wound care emphasizing offloading will heal some two-thirds of ulcers, but the remaining third are refractory, and the percentages have not shifted in 20 years. These outcomes impart two axioms. First, if all patients receive basic care, a majority will heal. Second, there is a substantial remainder whose wounds will never heal with basic care. A discerning student of wound care might therefore ask if it is possible to identify those patients that might not heal with basic wound care—patients who should receive more advanced therapy options as soon as practicable. The short answers to this very complex question are probably and perhaps. It is certainly possible to identify high-risk patients; the dilemma lies between first fixing the colonized wound if it exists, determining a more advanced suitable treatment, or trying to ameliorate the underlying cormorbidities.

COMPLIANCE WITH BASIC WOUND CARE

It is naively assumed that all wounds receive basic care (or more advanced options) and that patients adhere to the prescribed treatment. Unfortunately, neither is true. Since the inauguration of the evidence-based medicine era there has been a move toward evaluating wound care practices and treatments through systematic reviews and meta-analysis, with the development of clinical practice guidelines (CPGs) founded on the conclusions of the systematic reviews. On the other hand, because wound care is often regarded as a stepchild rather than a true specialty in its own right, education of nurses and physicians regarding wound care practice is still lacking. For example, in a study of the professional and nonprofessional staff members of a community urban hospital it was noted that participants were poorly informed in several basic aspects of PU care while generally desiring more wound care education.[33] A more recent study of 79 physicians and 63 nurses from the surgical wards of a large academic center also found little agreement regarding their choice of gauze-based and occlusive wound dressings for 18 representative photos of open wounds, a task that had also been evaluated by a panel of wound experts.[34] Interobserver agreement (kappa values) was 0.07 and 0.23, respectively. Agreement vis-à-vis the expert panel was also low: 0.14, and 0.32, respectively. One must conclude that these values are indicative of a major lack of knowledge of wound dressings and their application, a finding also noted in a previous audit of wound dressing selection and use.[35] A Canadian study to determine the knowledge of effective treatments for venous ulcers that involved 107 physicians caring for 226 patients with leg ulcers also elicited considerable concern. Only 16% were confident about managing leg ulcers and 61% reported not knowing enough about wound care products. Most disturbing was the fact that more than 50% of the physicians were unaware that compression is the standard treatment for venous ulcers.[36] Some of the reasons cited for this finding were: lack of evidence-based CPGs (82%); absence of evidence-based protocols in home-care agencies (72%); lack of access to wound care products (69%) and wound care centers (66%); and poor communication between health-care workers (60%). Education of patients was also an issue in a UK investigation, which determined that health promotion was perceived as ineffective and leg ulcer aftercare services fragmented.[37] Another conclusion suggested that improved caregivers' and patients' understanding of factors influencing leg ulcer recurrence was much needed.

One of us (CF) is also affiliated to Intellicure, Inc., a company specializing in utilizing Level 4 electronic medical record (EMR) systems specifically designed to meet the unique documentation needs of wound center patients. A broad range of wound parameters can be tracked, including wound measurements, photos, dressings, co-morbid conditions, laboratory tests, studies and medications. Users of the software system agree to participate in the "Intellicure Research Consortium," which allows us access to data in a "deidentified" fashion in accordance with Health Insurance Portability and Accountability Act (HIPAA) privacy requirements. Drawing from the data of 28 facilities distributed across 18 states with clinics in both urban and rural areas, we investigated patients who had diabetic foot ulcers to determine what percentage received total contact casting (TCC), now considered the "gold standard" of care in terms of off-loading.[38] The preliminary results, drawn from data for the period January 1, 2006 through December 31, 2007, showed that only 17 out of 264 patients with diabetic foot ulcers (6%) received this treatment. The cost of care for the TCC patients was $11,946 versus $22,494 for the other patients—a savings of over 50%. Over a similar time frame, it was also noted that patients who received the advanced treatment Apligraf® for venous ulcers only received compression concurrently 20% of the time. Another analysis of the U.S.-based Intellicure Clinical Documentation and Facility Management Software Intellicure database, which contained 622 VU patients with 1377 VUs showed that approximately one third of identifiable VUs in the database did not receive compression of any kind.[39]

All of this evidence suggests that basic wound care is not being received by a large percentage of patients, even when they are seen at centers specializing in wound care. Moreover, a recent review of interventions designed to enhance patient adherence with leg venous ulcer treatment reported that patient adherence (the extent to which the compression system was worn or the treatment regimen followed) with class III stockings was enhanced compared to short-stretch compression bandages, but that intermittent pneumatic compression systems did not improve compliance.[40] However, there was no well-documented evidence that any health-care system intervention or educational programs improved compliance with compression. Other studies that have investigated patient adherence in regard to compression therapy that was instituted by clinicians found ranges of 39% to 75%.[41,42]

Basic Wound Care Clinical Practice Guidelines

Several clinical practice guidelines have been developed in the last several years for basic wound care. For example, Bolton et al.[39] developed a very comprehensive set of guidelines for venous ulcers that were validated by a multidisciplinary, all-volunteer task force that included 11 advanced practice nurses or wound ostomy continence nurses, five physicians, four physical therapists, two researchers (PhD level), one doctor of podiatric medicine, and a registered pharmacist. The guidelines consist of three elements of diagnosis (patient history, differential diagnosis, physical exams), specific management options, including patient education, lower leg elevation, ambulation or exercise, compression options, management of the periwound area, cleansing, debridement, filling deep wounds, managing excess exudate, maintaining a moist wound environment, use of an antimicrobial system, biologic dressing, surgical options, or other modalities if no healing occurs within 30 days.[43] Each guideline is detailed, given strength of evidence rating (A to C) and incorporates references used to support the evidence. A well-written series of guidelines for the management of venous leg ulcers is also available from the Royal College of Nursing in the UK,[44] as well as an implementation guide that utilizes a team-based approach and which describes in great detail how the CPGs should be implemented in practice. A search of the National Guideline Clearinghouse Web site in the United States (www.guideline.gov) also reveals several CPGs for different types of wounds in regard to basic and more sophisticated wound care. In addition, the National Health System (NHS) National Institute of Clinical Excellence (NICE), as well as several other national governments also maintain Web sites from which a wound care clinician can locate and download various guidelines for wound care aspects. Thus, it is clear that high-quality wound care CPGs are available, so the questions arise, what are the problems involved in implementation in health-care systems, and what issues are preventing clinicians from utilizing CPGs?

In the UK, a survey conducted in 2002 (476 respondents from the Tissue Viability Society) regarding implementation of clinical guidelines, particularly those involved with aspects of wound care, found that while the use of guidelines was common (64%), only 18% reported that guidelines were fully implemented.[45] Lack of resources, lack of awareness of guideline content, and lack of acceptance of the guideline recommendations were the most commonly cited reasons for failure to implement, which suggests that the problem is multifactorial. A German investigation of pressure ulcer treatment in 51 acute care hospitals and 15 nursing homes (long-term care) also found a major disconnect between clinical evidence-based practice guidelines and treatment; for example less than half of grade II through IV PUs received moist wound healing.[46] The authors suggested that both a lack of knowledge and economic factors were responsible. A recent Belgian study of pressure ulcer prevalence in connection with adherence to a national prevention of pressure ulcer guideline found an appalling lack of care: only 4% received preventive measures, 65% received measures that did not adhere to the guideline, and 31% were at risk for developing PUs in which prevention measures were lacking.[47] Again, there seemed to be a lack of awareness of the guideline and understanding of the evidence behind the guidelines. Similar findings were noted in a U.S. study of 400 patients with different types of ulcers based on chart reviews and data abstraction of patient histories.[48] One other reason for the failure of health-care centers to properly follow guidelines was explored in a Canadian pilot study, which clearly showed that even in the presence of a culture of organizational learning and transformational leadership that there was considerable variation in best practice implementation.[49]

Salcido[50] has made some excellent points regarding the implementation of pressure ulcer guidelines originally developed by the Agency for Health Care Policy and Research (AHCPR) in the early 1990s that have a bearing on this discussion. As he points out, part of the AHCPR's original mission was to plan periodic reviews and updates with appropriate meetings and solicitations of input from stakeholders. The new AHRQ (Agency for Healthcare Research and Quality, the organization that assumed the ACHPR's functions, has decided that they will not update CPGs; rather, they will maintain guideline summaries and links to the sponsoring (originating) organizations which now have the duty of maintaining periodic updates. The problem is that newer technologies and discoveries make CPGs obsolete, so unless timely updates are performed, a clinician not keeping up with developments in the field could find him- or herself prescribing treatments that are outdated, obsolete, or not supported by the evidence. In 2001, Shekelle et al.[51] attempted to "audit" the original 17 CPGs developed by the AHCPR (two were withdrawn) and determined that more than three-quarters were obsolete. Although unable to find any existing standard method by which the validity of existing guidelines could be assessed, the study authors did recommend that CPGs be updated every 3 years. However, we do not know if all wound care guidelines published to date are being periodically updated as they should be.

In the United States, the Centers for Medicare and Medicaid (CMS) published the new pressure ulcer guidelines (F314), which were so detailed that they amounted to a "clinical practice guideline" by themselves.[52] As we indicated earlier, although understanding of pressure ulcer development is still in its infancy, it is slowly being recognized that the current staging system is flawed and continued usage leads to variety of problems. Although in many ways the new CMS guidelines were an advance in that they spelled out a level of care in regard to dressings, physician involvement, risk assessment, monitoring, and nutrition assessment as well as many other items, the problems inherent in the staging system and a general lack of controlled trial data for the majority of the recommendations detract from current evidence-based practice guidelines.

So, what can be done in the overall situation so that every patient receives basic wound care? First, if your organization is responsible for developing a CPG, ensure that it is periodically reviewed and updated. Second, a "systems" approach to implementation of relevant CPGs must be adopted by every health-care organization, or entity, and that includes primary care physician practices,[50,53] an approach favored by the Joint Commission on Accreditation of Healthcare Organizations (JCAHO) in the United States to health-care practice in general. This approach entails education and training, the development of compliance standards, compliance monitoring or auditing, clear communication, designation of appropriate persons to carry out these functions, and finally development of a procedure for violations and disciplinary action. Third, create a culture of organizational learning and transformational leadership that promotes an atmosphere in which implementation can take place. Fourth, employ an electronic medical record system that thoroughly documents patient care. The ability to prompt for care while the clinician documents a patient's record and the use of clinical data support systems can considerably improve compliance with guidelines. Finally, if you are involved in payer policies, "do the right thing" and help the patient with basic wound care. A particularly egregious example of not doing the right thing applies to Noridian in the United States, which prohibits payment for compression bandaging, determining that such bandages ought to be self-applied. This policy is still in place despite the results of our recent study, which demonstrated that 55% of VU patients (N = 547) required assistance with activities of daily living (mainly issues with dressing and toileting), and therefore would be unlikely to achieve adequate positioning to perform self-bandaging.[54]

CMS is experimenting with "pay for performance" (P4P, www.cms.hhs.gov/apps/media/press/release.asp?counter=1343) and although some of the pilot project results look promising in terms of incentives paid and money saved, many physicians have been unable to view their reports and so are not sure if they are completing the necessary requirements to receive their bonuses. Administrative snafus aside, these kind of "one-size-fits-all" programs are likely to stay, but whether they will improve basic wound care remains to be seen. One particularly interesting aspect is whether wound care providers will now be forced into providing more basic care for all patients instead of employing more advanced technologies that are expensive and of questionable cost-effectiveness. Most importantly, it will be interesting to see if payers become more educated in regard to wound healing concepts and evidence-based medicine as health care budgets are slashed.

The take-home message for basic wound care is that it seems simple, but is in fact complex, a situation we term "simplexity."[55]

THE TECHNOLOGY CEILING

A certain percentage of wounds fail to heal with basic wound care depending on the type and become chronic, nonhealed wounds. If life were simple, it would be a matter of identifying the problem factor and either replacing it in some fashion or providing a treatment that would overcome the problem. Much of the "high-technology" wound care that has been developed in the last 10 to 15 years is based on this principle. Unfortunately, wound healing is a complex process controlled by many different types of interacting factors, and so the results of the high-tech "simple" solutions have not been as successful as was once hoped.

Factors that interfere with normal wound healing are often associated with one particular phase of healing, one or more phases, or can be an issue at any stage of healing (Table 13.1). This is far from an exhaustive list, and typically, in a chronic nonhealing wound, several factors are present that have to be addressed either sequentially or simultaneously.

A SHORT HISTORY OF HIGH-TECH WOUND HEALING TREATMENTS

Through increased understanding of the biochemical mechanisms that are responsible for problems associated with wound healing factors, various bioengineered or recombinant products have been developed in the laboratory, tested in animal models, and finally undergone phase I, II, and III clinical trials. For example, Promogran® is a topical dressing that uses a ratio of collagen to oxidized regenerated collagen of approximately 55:45 and is applied to a chronic wound stuck in the inflammatory phase. The rationale behind its development is that it reduces protease activity and protects growth factors from degradation, as well as acting as a mechanical matrix to stimulate fibroblast migration and enhance the metabolic activity of granulation tissue and massively bound fibronectin.[56–59] Promogran (Johnson & Johnson) received its FDA premarketing approval in 2002 for treatment of all types of ulcers and several randomized controlled trials (RCTs) have been reported that involved testing the efficacy of Promogran against standard or other advanced treatments.[59–63] Most trials showed promising results, with the RCT conducted by Kakagia et al.[64] in which Promogran, autologous growth factors delivered by the Gravitational Platelet Separation System, and a combination of both were delivered to three groups demonstrating that the combination was superior in healing diabetic ulcers.

A number of bioengineered skin products also started appearing in the marketplace in the late 1990s. The intent behind all of these products was to kick start the healing process by attaching a replacement skin structure that provided a number of growth and other factors that would bring the chaotic inflammation situation under control and then encourage the proliferation phase of healing (see also Table 13.1).[65,66] Gene-cloned human PDGF (recombinant platelet-derived growth factor, commercially available from Johnson & Johnson, becaplermin, Regranex®), is also another high-technology wound care product designed to replace lowered levels of endogenous PDGF often found in chronic wounds.[67] PDGF helps promote the proliferation of cells involved in wound repair and enhance the formation of granulation tissue. NPWT (negative pressure wound therapy), as exemplified by the VAC (KCI Concepts)

TABLE 13.1

Factors That Affect the Stages of Healing and Treatments Designed to Overcome the Problems Associated with These Factors

Factor	Underlying Mechanisms	Treatment	Comment
		Stage 1: Inflammation	
Infection	Bacterial wound counts exceed critical mass	Cleansing, debridement, and exudate management; systemic broad-spectrum antibiotics directed by microfloral analyses and topical antimicrobials	
Wound colonization	Colonization of wound by bacteria	Cleansing, debridement, and exudate management; topical antiseptics for difficult wounds; topical antibiotics; silver-impregnated dressings	Important to distinguish between normal and critical colonization
Cytokine levels	Excessive and/or prolonged levels of inflammatory cytokines cause inflammation stasis	Cytokine antibodies, e.g., infliximab for TNF-α	Little is known about the individual role of various cytokines in nonhealing wounds
Necrotic tissue	Defective matrix and cell debris impair healing	Periodic debridement	
Growth factors	Decreased levels or inappropriate localization	VEGF (vascular endothelial growth factor); PGDF (platelet-derived growth factor, e.g., becaplermin);	Phase I clinical trials of VEGF are still ongoing
Proteases (also important in proliferation stage)	Excessive levels of serine, cysteine, aspartic, or metalloproteases (MMPs) from neutrophil or monocyte abundance	Protease inhibitory modulating matrix (oxidized regenerated cellulose/collagen Promogran dressings) to absorb excessive amounts of proteases	Balance of MMPs and TIMPS (tissue inhibitors of MMPs) important; MMP levels peak at different times during normal wound healing; increased levels of MMP2 are beneficial at later stage; high levels of MMP9 at early stage are detrimental
Protease inhibitors	Reduced levels contribute to excessive levels of proteases	Amnion-derived cellular cytokine solutions to rebalance proteases and their inhibitors	Leads to accelerated wound closure and strength

TABLE 13.1 (continued)
Factors That Affect the Stages of Healing and Treatments Designed to Overcome the Problems Associated with These Factors

Factor	Underlying Mechanisms	Treatment	Comment
Moisture imbalance	Moisture slows epithelial cell migration; excessive fluid causes wound margin maceration	Moisture-balanced dressings; compression and/or NPWT fluid removal	

Stage 2: Proliferation

Factor	Underlying Mechanisms	Treatment	Comment
Cell adhesion, migration, and proliferation issues (non-advancing wound)	High levels of matrikines from high protease levels; problems with fibroblasts	NPWT (negative wound pressure therapy, e.g., VAC); human skin allografts (GammaGraft™; AlloDerm®; Apligraf); GM-CSF (granulocyte monocyte colony stimulating factor); keratocyte growth factor	
Ischemia	Leads to severe wound hypoxia; affects angiogenesis, growth factor stimulation, and resistance to infection	HBOT (hyperbaric oxygen therapy); HBOT and adjunct use of α-lipoic acid; revascularization in the case of peripheral arterial disease (PAD)	HBOT not recommended for routine wound care; revascularization may be a better long-term solution in the case of PAD
Fibroblast senescence	Impaired ability of fibroblasts to replicate	Aggressive debridement; repopulate using tissue-engineered skin or autologous keratinocytes	Common in the elderly suggesting age is a primary factor
Diabetes	Advanced glycation and lipoxidation of ECM proteins disrupts function	Keep diabetes controlled; angioplasty or arterial revascularization for ischemia; statins/ antiplatelet therapy for neuroischemic ulcers	

All Stages

Factor	Underlying Mechanisms	Treatment	Comment
Ischemia-reperfusion injury	Causes leukocyte/ complement activation, oxidative stress, and microvasculature dysfunction	Ischemic conditioning, anti-inflammatory agents, antioxidants, and complement therapy, although efficacy in humans is lacking	Cyclic ischemia reperfusion theories are still under vigorous debate

(continued on next page)

TABLE 13.1 (continued)
Factors That Affect the Stages of Healing and Treatments Designed
to Overcome the Problems Associated with These Factors

Factor	Underlying Mechanisms	Treatment	Comment
Oxidative stress, including hyaluronan modification	Low cellular and circulating levels of antioxidants	Esterified hyaluronic acid; antioxidant therapies	
Poor nutrition	Leads to protein metabolism dysfunction and deficits of many cofactors	Proper nutrition; well-balanced diet; explore malabsorption syndromes	Serum albumin is not a satisfactory biomarker of poor nutrition
Corticosteroid therapy	Interferes with inflammation, fibroblast proliferation, collagen synthesis and degradation, angiogenesis, wound contraction, and reepithelialization.	Vitamin A therapy; measures to prevent infection	

is another different kind of high-technology approach to dealing with the cytokine imbalance found in chronic wounds.[68,69] This methodology expands exudate control handled by dressings to one of draining wound fluid to accelerate closure. However, it appears that the VAC is much more than a mere suction device; the process also stimulates neovascularization and deposition of granulation tissue and reduces bacterial count in the wound. The likely mechanisms by which these processes occur are via microstrain on cellular mitogenesis, angiogenesis, and elaboration of growth factors, and enhancement of microcirculation dynamics through active evacuation of excess interstitial fluid in the form of edema.[70]

THE VERDICT ON ADVANCED WOUND-HEALING TECHNOLOGIES

Although a review of advanced technologies is beyond the scope of this chapter, most, with the exception of the VAC, and hyperbaric oxygen therapy (HBOT), which is a much older technology, have not lived up to clinicians' expectations. In an unpublished study of 19,236 wounds from 8628 patients, we determined that the time to heal wounds when advanced technology (AT) was employed was not different compared to moist wound care with the exception of amputations, arterial ulcers and surgical wounds, which healed more slowly, but this might have been an effect of patient selection. It is clear that advanced technology can be used to good effect in select populations. For example, one of the most surprising findings of the study was the disparate size of wounds treated by AT versus basic wound care: the mean size of the wounds treated by basic wound care was 5.5 cm^3, whereas it was 25.8 cm^3 for AT. Furthermore, healing times of wounds using AT were excellent when compared to basic wound care (203 versus 148 days) given that the average wound size was almost five times greater. In addition, the overall healing rate of wounds treated with AT was 85.7%—close to the 91.2% healing rate achieved by basic wound care. One surprising finding was that on average, advanced technologies were not initiated until after the average wound would have healed, that is, until more than 60 days had elapsed. This makes the subsequent improvement in the wound after the institution of advanced technology even more impressive. Thus, "real-world" data suggest that advanced technologies may be able to drive some large, difficult-to-heal wounds to closure.

RW Johnson (a Johnson & Johnson company) spent $30 million to bring Regranex® to market. However, the product has not been a big success financially and clinical success has also been limited. Dermagraft® failed to receive FDA approval for venous ulcers, was withdrawn and is now recently being marketed again for diabetic ulcers. Its competitor, Apligraf, continues to be relatively successful for DFUs and VUs but will not replace split thickness skin grafting. Advanced technology products continue to arrive in the market place but we have not seen dramatic decreases in the prevalence of chronic wounds nor overall healing rates. To frame an explanation for these observations, we must first take a short detour to discuss how new wound care products are clinically evaluated.

EFFICACY VERSUS EFFECTIVENESS

When new wound care products are tested in humans, at some point their efficacy has to be evaluated in a phase II clinical trial. For most treatments, the RCT continues to be the gold standard by which that is accomplished. However, as Carter and Warriner[71] have detailed, there are a myriad of issues concerning trial design and outcome measures. First, many wound care RCTs are small and therefore underpowered in the context of trying to measure a relatively small treatment effect. Second, they utilize one treatment instead of the multiple treatments that are typically given to wound care patients. Third, they have relatively short study periods, in part because most wound care treatments are not designed to be given over the whole healing period of the wound. Fourth, most wound care patients either have multiple comorbidities or are elderly, which compromises healing ability, and these kinds of patients are typically excluded from RCTs. Finally, while the FDA (the U.S. Food and Drug Administration) only considers complete wound healing as the only acceptable outcome, many studies rely on partial wound healing outcomes, or surrogate and composite end points. It is possible that scientifically validated partial healing end points are possible and could well be an answer to a tricky evaluation dilemma, but this would have to take account of the heterogeneity of wound trajectories.

Generalizability of RCT results to real-world wound care populations is a particularly thorny issue, and many systematic review algorithms attempt to estimate this problem in variety of ways. In essence, if the majority of eligible patients in a wound care RCT are excluded for various reasons, is it logical to assume that the trial results are likely to be applicable to general populations? According to Britton et al.[72] and other investigators, there must be serious doubts.

Because there were no data in the literature on generalizability of wound care RCTs to real-world populations, we recently undertook a study to estimate the percentage of patients in 17 AT wound care RCTs (DFUs, VUs, and PUs) that would have been excluded based upon the inclusion and exclusion criteria of each trial.[73] The results were disturbing. More than 50% of the study population would have been excluded in 15 out of the 17 RCTs. When less clinically relevant exclusion criteria were removed (such as infection or wounds less than 1 month old), 14 out of 17 RCTs would still have excluded between 25% and 50% of the study population. Among 8,611 wound center outpatients, 70 % would have been excluded from the wound-related RCTs at the "first pass" (i.e., exclusions on the basis of comorbid conditions, previous surgeries or medications, even before further tests are performed). That first pass exclusion included 8.4% of patients on steroids; 5% on renal dialysis (many others had renal insufficiency); 10% of patients with various forms of PVD (peripheral vascular disease; low estimate based on reported prior vascular surgery, transcutaneous oximetry or ABI [ankle-brachial index] results in their charts); and 26% of wounds that were not diabetic foot ulcers were found in patients who had diabetes. The situation is in fact worse than the scenario we have portrayed because the protocols in most RCTs require detailed diagnostic studies prior to trial inclusion after initial screening, such as transcutaneous oximetry, bone scans, hemoglobin A1C, or creatinine. Perhaps as many as 10 to 30% of patients who pass initial screening will fail these subsequent tests due to further scrutiny before enrolment. Thus, we identified only those patients who would have been summarily excluded from a clinical trial; it is clear that the percentage of patients actually excluded would have been much higher. Consequently, the vast majority of wound center patients would have been excluded from the clinical trials performed on Regranex, Apligraf, and Dermagraft because of their comorbid conditions.

Efficacy refers to whether the intervention *can* be successful when it is properly implemented under controlled conditions, and is best determined by prospective, controlled, randomized clinical trials. On the other hand, *effectiveness* is the *capability* of producing an effect, and refers to the impact in real world situations (i.e., whether the intervention typically is successful in actual practice). Thus the effectiveness of some wound care products currently on the market may be less than hoped due to the difference between controlled trial conditions and real world practice.

Although evidence from RCTs tends to be graded higher in most EBM hierarchical schemes, several investigations have shown that well-designed observational studies (cohort and case-control designs) can provide similar results in terms of the direction and magnitude of the effect size of the treatment.[74–78] That is not to say that in some aspects of medical fields or particular populations significantly different results and large heterogeneity between RCTs and observational studies have not been observed. On the contrary, this has occurred. However, the conclusion reached when assessing all these comparisons in the context of wound care suggests that observational trials can provide good estimates of treatment effects with more realistic wound care populations.

Ischemia and Cellular Hypoxia

There are also more fundamental physiological reasons why AT wound care might not be as beneficial as we had first thought in many patients, or why we might be reaching a "technology ceiling." In healthy patients, normobaric tissue oxygen levels when measured in the foot by transcutaneous oximetry ($PtcO_2$) are >50 mm Hg with typical values at sea level of 60 to 70 mm Hg.[79-82] However, many patients with chronic nonhealing wounds have hypoxia when breathing normobaric air, which is sufficient to impair or prevent wound healing and is defined as a $P_{tc}O_2$ < 40 mm Hg.[83] The two major diseases that cause wound hypoxia in leg ulcers are peripheral arterial disease (PAD) and venous insufficiency.

PAD affects some 8 millions Americans and is a progressive disease that primarily affects the arterial circulation of the lower extremities. Its prevalence increases with age, especially in those ≥ 60 years old, and appears to be a huge factor in a large percentage of leg ulcers. At least 10% of patients with venous ulcers have arterial disease,[84,85] but about half of those with diabetic ulcers have arterial disease,[86-88] and then there are those who present with pure arterial ulcers. That means that in any given wound center, as many as 30 to 50% of leg ulcer patients have arterial disease. However, relying on the ABI (a common measure of arterial insufficiency in the leg) as an indication for revascularization is fraught with difficulty. Other forms of screening are superior (e.g., transcutaneous oximetry or skin perfusion pressure) and patients who fail screening studies are referred for definitive vascular evaluation (angiography).

Chronic venous insufficiency (CVI), which is characterized by retrograde blood flow in the lower extremity venous system, and is associated with venous hypertension, edema, and inflammation of the venous walls and valves, is the prime cause of venous stasis ulcers. Some patients have more than one disease. The prevalence of CVI in our diabetic ulcer patients was approximately 11.5%, and 34.2% of our venous ulcer patients also had diabetes.[73]

One of the most critical issues in treating venous stasis ulcers is the detection of concomitant arterial disease because compression applied for treatment of the ulcer under such circumstances will impede or even cut off arterial inflow. Patients with venous disease who have concomitant arterial disease may require revascularization. The options for revascularization consist of endovascular techniques such as balloon angioplasty with or without stent emplacement, percutaneous catheter-directed thrombolytic therapy, or surgical bypass.[89] In our opinion, if the occlusion is relatively small, endovascular techniques are preferred because of the minimal complications associated with procedure even though the vessel may only be patent for a few years. A repeat procedure than can then be applied if necessary. Surgical bypass procedures are much more invasive and carry relatively high risks of mortality, although they may be the only method to address extensive occlusions. For elderly patients in particular, where life expectancies of the order of 5 to 10 years are common, endovascular techniques are preferred.

Mixed arterial/venous disease can be particularly difficult to treat and despite revascularization, limbs may require amputation.[90] Similarly, for diabetic foot ulcers, revascularization is not always successful in restoring a functional circulation in lower limbs.[91] HBOT has been used to treat ischemic ulcers in such situations, and is moderately successful, especially in regard to limb salvage. Data suggest that HBOT can reduce the need for major amputations and increase quality of life years in diabetics. However, it is an interim and not long-term treatment because it does not fix the causative factors. What is needed is a treatment that can restore the damaged microcirculatory system, presumably angiogenesis-based—but to date, very few such clinical studies have been attempted. Angiogenesis is a complicated process whose success depends on the interaction of several factors in the right concentrations levels in the right sequence (order).[92] One such study conducted by Nikol et al.[93] involved the intramuscular administration of NV1FGF, a plasmid-based angiogenic gene delivery system for local expression of fibroblast growth factor. The results of this double-blind randomized controlled trial were disappointing (at least in our opinion): Twenty-five weeks after treatment of the affected limb, complete healing in the treatment and control groups had occurred at rates of 19.6% and 14.3%, respectively ($p = 0.514$). Nevertheless, the treatment did reduce the risk of major amputation by nearly two-thirds, which is an important result. The key element for healing of hypoxic wounds is the correction of tissue hypoxia. In general, there are only two options: revascularization, which improves perfusion, and hyperbaric oxygen therapy, which increases tissue PO_2. Among hypoxic wounds, healing will not occur unless some minimum tissue PO_2 is achieved, regardless of what other interventions are offered.

WHERE DO WE GO FROM HERE?

We have discussed that chronic nonhealing wounds are frequently overwhelmed with unchecked hypercytokinemia and proteases, lack a variety of wound healing factors, and are often colonized by bacteria to an extent that interferes with wound healing. Worse, many such wounds are hypoxic and lack sufficient functioning vasculature to deliver proper nutrients to cells or have a problem with waste disposal because the lymphatic system is dysfunctional. Therefore, it should be no surprise that clinical trials of advanced technology applications are not always brilliant successes. To summarize: The majority of chronic wound patients can be healed by addressing their underlying basic etiologic factors, instituting edema control and off-loading for pressure and diabetic ulcers, and managing the bioburden and other local and systemic factors. The small subset of patients who need *and can also benefit from* advanced technology (e.g., Regranex, Apligraf, or Dermagraft) are those who have adequate arterial supply.

The "technology threshold" experienced by most of the "advanced therapeutics" currently on the market is due to their inability to effect a significant increase in angiogenesis. HBOT remains unique in the field of wound healing for its ability to enhance angiogenesis and for its demonstrated effectiveness in Wagner III wounds, but at the same time it continues to be negatively affected by the laws of physics, distribution (the nature of its administration) and organization. In addition, HBOT suffers from lack of research funding, is not properly recognized due to less than stellar evidence from an EBM point of view, and thus its use is hindered by the list of approved indications for it as well as reimbursement for actual treatment regimens. Finally, we lack good models for "ischemic wounds." At present we have no single treatment model for chronic wounds; rather, success depends on a series of consecutive treatments, each designed to overcome one or more barriers in phases of wound healing.

SILVER-IMPREGNATED WOUND DRESSINGS AS AN EXAMPLE OF ADVANCED TECHNOLOGY

Because this book focuses on the microbiology of chronic wounds, it is instructive to look at the issues involved with antimicrobial advanced technology in connection with chronic wound treatment because the advanced technology applied to chronic wounds generally follows the same pathway. As one of us (MC) was intimately involved with the systematic review of silver-impregnated dressings in 2008, as well as presenting that evidence to a CMS committee for coding of such dressings, this antimicrobial treatment has been chosen as the example.

THE CONCEPT

The use of silver in the form of silver sulfadiazine cream as a topical antimicrobial started in the 1960s and was (and has been) more widely used in burns than wounds, although several RCTs testing the efficacy of the cream in wounds have been published.[94–96] Cationic silver (Ag^+) is highly reactive and interferes at multiple levels with bacterial cell metabolism, including cell wall and membrane permeability, blockage of transport and enzyme systems, alteration of proteins, and binding of microbial RNA and DNA.[97] Because silver binding to protein and other anions does occur, it is advantageous to provide a continual release mechanism of the ionic form, and this was the impetus for the development of silver-impregnated dressings of which several are now on the market.

One major advantage of these kinds of dressings is that resistance to silver is far lower than the resistance to antibiotics that can develop because the silver attacks bacterial metabolism at several levels simultaneously. In addition, systemic toxicity resulting from such dressings is very low and allergies to silver are nonexistent.[97,98]

CLINICAL USE OF SILVER-IMPREGNATED DRESSINGS

Some of the more popular dressings on the market include Contreet® foam (sustained silver release in a foam dressing; Coloplast), Aquacel® (hydrofiber dressing with ionic silver; Convatec), Restore Contact Layer Silver with TRIACT technology® (polyester textile mesh with hydrocolloid, and sustained silver release; Hollister), and Silvercel® (silver hydroalginate, Johnson & Johnson Wound Management). All of these dressings have been tested in RCTs as well as observational trials on patients with a variety of wounds with varying degrees of success.

Past systematic reviews have been critical of the evidence supporting their use. For example, Vermeulen et al.[99] concluded: "Only three trials with a short follow-up duration were found. There is insufficient evidence to recommend the use of silver-containing dressings or topical agents for treatment of infected or contaminated wounds." Chambers, Dumville, and Cullum[100] characterized the studies as providing "inconsistent evidence regarding the effects of silver-based dressings and topical agents on leg ulcer healing. Studies generally provided poor evidence due to a lack of statistical power, poor study designs, and incomplete reporting." In conclusion, they stated: "the current evidence base on the use of these silver-based products on leg ulcers is limited, both in terms of the quantity available and the quality of the evidence." The systematic review of Lo et al.[101] had the broadest scope but also the least rigorous approach in our opinion, and the authors concluded: "silver-releasing dressings show positive effects on infected chronic wounds. The quality of the trials was limited by the potential for bias associated with inadequate concealment, no detailed description of the outcome measurement and no reported intention-to-treat analysis.... The review clearly highlights the need for well-designed, methodological standardized outcome measurement research into the effectiveness of silver-releasing dressings.... This review strengthens the case for the use of silver dressings when managing infected chronic wounds."

Because many of the patients that are being treated with silver-impregnated dressings are over 60 years of age, several wound-care manufacturers thought it was important that such dressings be accepted by CMS. Acceptance (i.e., reimbursement for the use of such dressings rather than reimbursement based upon general dressing codes) is contingent upon receiving a coding and coverage, which is predicated upon evidence that such dressings do indeed to knock down infections, eliminate critically colonized wounds, and thus contribute to healing of the wound. However, while most clinicians and payers accepted that such dressings could achieve the first two goals, controversy has centered on the third goal, i.e., do such dressings influence wound healing, and if so, by how much? It was with this point in mind that MC was asked to conduct an impartial systematic review of the evidence for several wound care manufacturers and communicate that evidence to the CMS subcommittee responsible for approving new codings for medical devices and treatments.

SYSTEMATIC REVIEW AND META-ANALYSIS

The initial review confirmed the previously published findings of the systematic reviews. One of the major issues was the lack of complete reporting as specified by the CONSORT statement for reporting RCTs.[102] In terms of understanding the details of RCTs when crucial information is omitted it is not possible to give a "benefit of the doubt" when quality analysis is conducted, so although the trial might have been properly conducted it is going to be downgraded because information is missing. External and internal (bias and confounding) trial validities are also critical issues to quantify, because problems in these areas suggest that the results may not be valid or cannot be generalized to more general population groups. Our analysis found several problems with these issues in the RCTs.

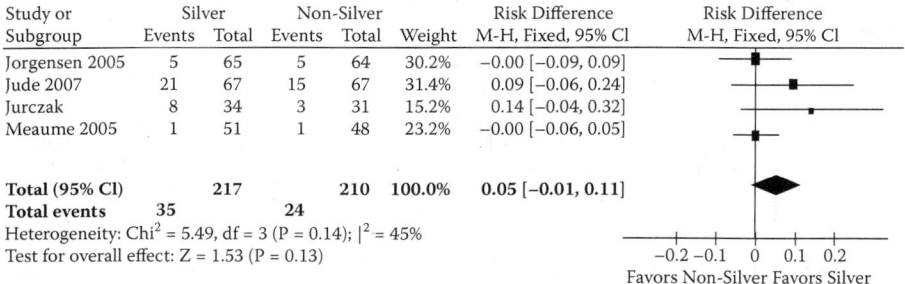

Study or Subgroup	Silver Events	Total	Non-Silver Events	Total	Weight	Risk Difference M-H, Fixed, 95% CI
Jorgensen 2005	5	65	5	64	30.2%	−0.00 [−0.09, 0.09]
Jude 2007	21	67	15	67	31.4%	0.09 [−0.06, 0.24]
Jurczak	8	34	3	31	15.2%	0.14 [−0.04, 0.32]
Meaume 2005	1	51	1	48	23.2%	−0.00 [−0.06, 0.05]
Total (95% CI)		217		210	100.0%	0.05 [−0.01, 0.11]
Total events	35		24			

Heterogeneity: Chi^2 = 5.49, df = 3 (P = 0.14); I^2 = 45%
Test for overall effect: Z = 1.53 (P = 0.13)

FIGURE 13.2   Meta-analysis of complete wound healing outcomes for randomized controlled trials.

Study or Subgroup	Silver Mean	SD	Total	Non-Silver Mean	SD	Total	Weight	Mean Difference IV, Fixed, 95% CI
Jorgensen 2005	34.7	34.6	65	24.4	38.9	64	26.2%	10.30 [−2.41, 23.01]
Jude 2007	58.1	53.1	67	60.5	42.7	67	15.9%	−2.40 [−18.72, 13.92]
Lazareth	36.6	48.8	51	6.2	80.2	48	6.1%	30.40 [4.05, 56.75]
Meaume 2005	23.7	43.6	51	24	41.6	48	15.0%	−0.30 [−17.08, 16.48]
Munter 2006	41.5	46.7	291	27.8	77.5	268	36.8%	13.70 [2.98, 24.42]
Total (95% CI)			525			495	100.0%	9.17 [2.66, 15.67]

Heterogeneity: Chi^2 = 6.37, df = 4 (P = 0.17); I^2 = 37%
Test for overall effect: Z = 2.76 (P = 0.006)

FIGURE 13.3   Meta-analysis of relative wound reduction (%) partial wound healing outcomes for randomized controlled trials.

Another concern focused on outcomes. The trials were conducted over a relatively short period of time varying from 2 to 8 weeks because silver-impregnated dressings are applied to wounds for a relatively short period of time. If the dressings bring the bacterial colonization or infection under control, there is no point to prolonging dressing applications, and in fact much longer times may be detrimental to wound healing. Therefore, there is likely to be very little difference regarding complete wound healing data between control and treatment groups as much longer times (e.g., 16 or 20 weeks) would be needed to more accurately estimate any differences. Indeed a pooling of this data clearly showed little difference (Figure 13.2). On the other hand, when all the partial wound healing outcomes were pooled (relative wound reduction), an unequivocal result was obtained (Figure 13.3) with little heterogeneity.

After extensive discussion of the data, the initial results were presented to a CMS HCPCS (Common Procedure Coding System) coding meeting in April 2008. During the several months that this information was being digested, additional data requested from two study authors had arrived, which enabled a meta-analysis subgroup analysis. The idea behind this meta-analysis was to take the outcomes from the RCTs and divide them into two time periods: 2 to 4 weeks and 8 weeks. It was hypothesized that since the dressings were applied for a short period of time, a substantial difference might be observed between the two groups for partial wound healing rates, but no complete wound healing.

The results confirmed the hypothesis. While no difference between complete wound healing was observed (risk difference for the 2- to 4-week and 8-week groups, 95% CI −0.03 to 0.09 and −0.02 to 0.06; p = 0.33 and 0.31, respectively), in terms of wound size reduction a very different results was obtained (weighted mean difference for the 2- to 4-week and 8-week groups, 95% CI 5.62 to 19.61 and −7.12 to 20.62; p = 0.0004 and 0.34; heterogeneity [chi squared], p = 0.39, p = 0.04; I^2 = 1% and 77%, respectively). The conclusion is that the wound-healing effect lasts at least 4 weeks, but by 8 weeks the effects between the experimental and control groups are starting to equalize. The end result should be a shortening of the average time to heal, although this cannot be estimated without trial data lasting some 16 to 20 weeks.

A more extensive quality analysis was also performed by modifying the approach of Downs and Black,[103] shortening the power subsection and substituting for external validity the scheme outlined in our previous work.[73] In addition, RCTs were graded using both the SIGN methodology[104] and the approach of Atkins et al.[105] based on the quality score ranges and strengths/weaknesses of each study to help minimize the judgment factor that can arise, particularly when assigning an "A" or "B" grade to the evidence. Although the majority of the trials were still graded B, with some graded A, we had more confidence that unknown systemic bias was absent from the group as a whole.

At the time of writing, more meetings are being convened regarding coding and coverage of the dressings with additional data being distributed, and it is hoped that favorable decisions will be reached in 2010. This treatment approach also desperately needs a cost-effectiveness analysis (e.g., cost utility study) to convince users and payers that the dressings are cost-effective, but this is somewhat a catch-22 process because cost utility studies cannot be performed without a complete and specific wound-healing model, which is still lacking.

The take-home message for this AT technology is that despite an appropriate concept, as well as a considerable number of clinical trials, from an EBM point of view it will likely take nearly 10 years from the dates of the phase 1 clinical trials before these dressings are accepted as a viable treatment by payers for chronic wounds because of the limitations of the RCTs involved as well as our understanding of wound healing.

CONCLUSIONS

Many factors are involved in wound healing, yet if basic wound care were successfully applied in all cases we could heal nearly 80% to 85% of venous ulcers, 75% to 80% of pressure ulcers, and perhaps two-thirds of diabetic foot ulcers. Advanced technologies are often not instituted until after the average wound would have healed, i.e., until conservative care has failed for more than 60 days. By initiating advanced technology sooner, one wonders if the outcome of some of these refractory wounds would be even better since the longer wounds persist, the more difficult they are to heal. However, some rational method is needed to ensure that simpler approaches are used first. Developing a consistent approach to the complex armamentarium of advanced technology is a dauntingly difficult task.

Through meticulous evaluation and the application of basic wound care principles we can identify those patients that might be healed by basic wound care alone, those whose chronic wounds might benefit from advanced technology, those patients who need revascularization before their wounds can be healed by any means, and those patients whose vasculature supply is so inadequate that it is unlikely we can do much for them. The problem is that nearly all product trial design research excludes patients with vascular disease, which is the final common denominator for *nonhealing*. In addition, most of the patients that are seen at wound care centers are exactly the patients who are excluded for a variety of reasons (besides vascular disease) from the RCTs that test new wound care products. In essence, our "real-world" patients are poorly represented in these RCTs.

Comorbidities heavily influence the prognosis of wounds. If these comorbidities are not properly controlled and/or treated, no amount of advanced technology will succeed. Far too many trials exclude patients with serious comorbidities, with the inference by some clinicians that if the trials appear to be successful in these carefully selected populations then any patient can be successfully treated. Honest conclusions by trial authors in terms of patients most likely to benefit would provide an up-front education for clinicians. Instead, practitioners must engage in mass "clinical experiments" with their wound center patients to determine whether products tested in RCTs will in fact, provide any benefit to their highly compromised patients. Thus we prove advanced wound care technology in patients who mostly do not really need advanced technology, exclude many of the patients who fill our wound care clinics, some of whom might benefit if we had better data, and spend huge amounts of money on developing products that do not improve the vascular supply. Summarizing, our data suggest that we develop the wrong products for the wrong patients while providing poor care to the rest. If clinicians, policy makers, and third-party payers were better educated regarding the basics of wound care so that the right treatments were applied in the right order for the right situation, and covered when they are demonstrated to provide benefit, we could heal the wounds of many more patients in a timely fashion.

In the words of George Bernard Shaw: "We are made wise not by the recollection of our past, but by the responsibility for our future."

REFERENCES

1. Paré, A. (1571). Journeys in diverse places. In *The Harvard classics*, ed. C. W, Eliot, trans. S. Paget, Vol. 38, Part 2. New York: P.F. Collier & Son, 1909–14; Bartleby.com, 2001. http://www.bartleby.com/38/2/.
2. Glimore, O. J. (1977). 150 years after. A tribute to Joseph Lister. *Ann R Coll Surg Engl* 59:199–204.
3. Singer, M., and P. Glynne. (2005). Treating critical illness: The importance of first doing no harm. *PLoS Med* 2:e167.
4. Maclean A. A., A. M. O'Neill, H. L. Pachter, and M. A. Miglietta. (2006). Devastating consequences of subway accidents: Traumatic amputations. *Am Surg* 72:74–76.

5. Witkowski, J. E., and L. C. Parish. (1982). Histopathology of the decubitus ulcer. *J Am Acad Dermatol* 6:1014–1021.

6. Daniel, R. K., D. Wheatley, and D. Priest. (1985). Pressure sores and paraplegia. *Ann Plast Surg* 15:41–49.

7. Nola, G. T., and L. M. Vistnes. (1980). Differential response of skin and muscle in the experimental production of pressure sores. *Plast Reconstr Surg* 66:728–733.

8. Zulkowski, K., D. Langemo, and M. E. Posthauer. (2005). Coming to consensus on deep tissue injury. *Adv Skin Wound Care* 18:28–29.

9. Black, J., M. M. Baharestani, and J. Cuddigan, et al. (2007). National Pressure Ulcer Advisory Panel's updated pressure ulcer staging system. *Adv Skin Wound Care* 20:269–274.

10. Barczak, C. A., R. I. Barnett, E. J. Childs, and L. M. Boslev. (1997). Fourth national pressure ulcer survey. *Adv Wound Care* 10:18–26.

11. Coleman, E. A., J. M. Martau, M. K. Lin, and A. M. Kramer. (2002). Pressure ulcer prevalence in long-term nursing home residents since the implementation of OBRA '87. Omnibus Budget Reconciliation Act. *J Am Geriatr Soc* 50:728–732.

12. Spector, W. D., and R. H. Fortinsky. (1998). Pressure ulcer prevalence in Ohio nursing homes: Clinical and facility correlates. *J Aging Health* 10:62–80.

13. Lahmann, N. A., R. J. Halfens, and T. Dassen. (2006). Pressure ulcers in German nursing homes and acute care hospitals: Prevalence, frequency, and ulcer characteristics. *Ostomy Wound Manage* 52:20–33.

14. Barrois, B., C. Labalette, and P. Rousseau, et al. (2008). A national prevalence study of pressure ulcers in French hospital inpatients. *J Wound Care* 17:373–376, 378–379.

15. Glass, A. G., J. V. Lacey, Jr, J. D. Carreon, and R. N. Hoover. (2007). Breast cancer incidence, 1980–2006: Combined roles of menopausal hormone therapy, screening mammography, and estrogen receptor status. *J Natl Cancer Inst* 99:1152–1161.

16. Robson, M. C., and J. P. Heggers. (1969). Bacterial quantification of open wounds. *Mil Med* 134:19–24.

17. Schraibman, I. G. (1990). The significance of beta-haemolytic streptococci in chronic leg ulcers. *Ann R Coll Surg Engl* 72:123–124.

18. Gardner, S. E., R. A. Frantz, and B. N. Doebbeling. (2001). The validity of the clinical signs and symptoms used to identify localized chronic wound infection. *Wound Repair Regen* 9:178–186.

19. Nelson, E. A., S. O'Meara, and D. Craig, et al. (2006). A series of systematic reviews to inform a decision analysis for sampling and treating infected diabetic foot ulcers. *Health Technol Assess* 10:1–221.

20. Serena, T., M. C. Robson, D. M. Cooper, and J. Ignatius. (2006). Lack of reliability of clinical/visual assessment of chronic wound infection: The incidence of biopsy-proven infection in venous leg ulcers. *Wounds* 18:197–202.

21. Gardner, S. E., R. A. Frantz, C. L. Saltzman, S. L. Hillis, H. Park, and M. Scherubel. (2006). Diagnostic validity of three swab techniques for identifying chronic wound infection. *Wound Repair Regen* 14:548–557.

22. Levine, N. S., R. B. Lindberg, A. D. Mason, and B. A. Pruitt. (1976). The quantitative swab culture and smear: a quick, simple method for determining the number of viable aerobic bacteria on open wounds. *J Trauma* 16:89–94.

23. Flores, A., A. Kingsley. (2007). Topical antimicrobial dressings: an overview. *Wound Essentials* 2:182–185.

24. Vazquez, J. A. (2006). The emerging problem of infectious diseases: The impact of antimicrobial resistance in wound care. *Wounds* S11:1–9.

25. Waldrop, R. D., C. Prejean, and R. Singleton. (1998). Overuse of parenteral antibiotics for wound care in an urban emergency department. *Am J Emerg Med* 16:343–345.

26. Keast, D. H., and H. Orsted. (1998). The basic principles of wound care. *Ostomy Wound Manage* 44:24–28, 30–31.
27. Yager, D. R., R. A. Kulina, and L. A. Gilman. (2007). Wound fluids: A window into the wound environment? *Int J Low Extrem Wounds* 6:262–272.
28. Burton, C. S. (1991). Practical management of leg ulcers. *South Med J* 84:2–17.
29. Margolis, D. J., J. A. Berlin, and B. L. Strom. (1999). Risk factors associated with the failure of a venous leg ulcer to heal. *Arch Dermatol* 135:920–926.
30. Steed, D. L., D. P. Hill, M. E. Woodske, W. G. Payne, and M. C. Robson. (2006). Wound-healing trajectories as outcome measures of venous stasis ulcer treatment. *Int Wound J* 3:40–47.
31. Margolis, D. J., L. Allen-Taylor, O. Hoffstad, J. A. Berlin. (2005). Healing diabetic neuropathic foot ulcers: Are we getting better? *Diabet Med* 22:172–176.
32. Pound, N., S. Chipchase, K. Treece, F. Game, W. Jeffcoate. (2005). Ulcer-free survival following management of foot ulcers in diabetes. *Diabet Med* 22:1306–1309.
33. Beitz J. M., J. Fey, and D. O'Brien. (1998). Perceived need for education vs. actual knowledge of pressure ulcer care in a hospital nursing staff. *Medsurg Nurs* 7:293–301.
34. Vermeulen, H., D. Ubbink, S. Schreuder, and M. Lubbers. (2006). Inter- and intra-observer (dis)agreement among physicians and nurses as to the choice of dressings in surgical patients with open wounds. *Wounds* 18:286–293.
35. Bux, M., and J. S. Malhi. (1996). Assessing the use of dressings in practice. *J Wound Care* 5:305–308.
36. Graham I. D., M. B. Harrison, M. Shafey, and D. Keast. (2003). Knowledge and attitudes regarding care of leg ulcers. *Can Fam Physician* 49:896–902.
37. Flanagan, M., L. Rotchell, J. Fletcher, and J. Schofield. (2001). Community nurses', home carers' and patients' perceptions of factors affecting venous leg ulcer recurrence and management of services. *J Nurs Manag* 9:153–159.
38. Boulton, A. J. (2004). Pressure and the diabetic foot: clinical science and offloading techniques. *Am J Surg* 187:17S–24S.
39. Bolton, L., L. Corbett L, and L. Bernato, et al. (2006). Development of a content-validated venous ulcer guideline. *Ostomy Wound Manage* 52:32–48.
40. Van Hecke, A., M. Grypdonck, and T. Defloor. (2008). Interventions to enhance patient compliance with leg ulcer treatment: A review of the literature. *J Clin Nurs* 17:29–39.
41. Heinen, M. M., C. van der Vleuten, and M. J. de Rooij, et al. (2007). Physical activity and adherence to compression therapy in patients with venous leg ulcers. *Arch Dermatol* 143:1283–1288.
42. Mudge, E., N. Ivins, W. Simmonds, and P. Price. (2007). Adherence to a 2-layer compression system for chronic venous ulceration. *Br J Nurs* 16:S4,S6,S8,
43. Association for the Advancement of Wound Care. (2006). Summary algorithm for venous ulcer care with annotations of available evidence. Retrieved January 26, 2008 from: http://www.aawconline.org/pdf/Algorithm%20for%20Venous%20Ulcer%20Care.pdf.
44. Royal College of Nursing. (2006). The nursing management of patients with venous leg ulcers. Recommendations. Retrieved January 26, 2009 from: http://www.rcn.org.uk/__data/assets/pdf_file/0003/107940/003020.pdf.
45. Clark, M. (2003). Barriers to the implementation of clinical guidelines. *J Tissue Viability* 13:62–64, 66, 68.
46. Helberg, D., E. Mertens, R. J. Halfens, T. Dassen. (2006). Treatment of pressure ulcers: Results of a study comparing evidence and practice. *Ostomy Wound Manage* 52:60–72.
47. Paquay, L., R. Wouters, T. Defloor, F. Buntinx, R. Debaillie, and L. Geys. (2008). Adherence to pressure ulcer prevention guidelines in home care: A survey of current practice. *J Clin Nurs* 17:627–636.

48. Jones, K. R., K. Fennie, and A. Lenihan. (2007). Evidence-based management of chronic wounds. *Adv Skin Wound Care* 20:591–600.
49. Marchionni, C., and J. Ritchie. (2008). Organizational factors that support the implementation of a nursing best practice guideline. *J Nurs Manag* 16:266–274.
50. Salcido, R. (2002). AHCPR clinical practice guidelines, a decade later. *Adv Skin Wound Care* 15:52,54.
51. Shekelle, P. G., E. Ortiz, and S. Rhodes, et al. Validity of the Agency for Healthcare Research and Quality clinical practice guidelines: How quickly do guidelines become outdated? *JAMA* 286:1461–1467.
52. Levine, J. M., M. Peterson, and F. Savino. (2005). Implementing the new CMS guidelines for wound care: Areas for potential citations explained. *Nurs Homes* 54:110.
53. Hess, C. T. (2008). Developing a wound care compliance program (Parts 1 and 2). *Adv Skin Wound Care* 21:496,544.
54. Fife, C., D. Walker, B. Thomson, and M. Carter. (2007). Limitations of daily living activities in patients with venous stasis ulcers undergoing compression bandaging: Problems with the concept of self-bandaging. *Wounds* 19:255–257.
55. Kluger, J. (2008). *Simplexity: Why simple things become complex* (and how *complex things* can be made *simple*). New York: Hyperion.
56. Cullen, B., L. Morrison, and P. Watt (2000). Mechanism of action of a next generation wound treatment. Paper presented at First Wound Healing Congress, September 12, Melbourne, Australia.
57. Motta, G., G. B. Ratto, and A. De Barbieri, et al. (1983). Can heterologous collagen enhance the granulation tissue growth? An experimental study. *Ital J Surg Sci* 13:101–108.
58. Nagata, H., H. Ueki, and T. Moriguchi. (1985). Fibronectin. Localization in normal human skin, granulation tissue, hypertrophic scar, mature scar, progressive systemic sclerotic skin, and other fibrosing dermatoses. *Arch Dermatol* 121:995–999.
59. Veves, A., P. Sheehan, and H. T. Pham. (2002). A randomized, controlled trial of Promogran (a collagen/oxidized regenerated cellulose dressing) vs standard treatment in the management of diabetic foot ulcers. *Arch Surg* 137:822–827.
60. Ghatnekar, O., M. Willis, and U. Persson. (2002). Cost-effectiveness of treating deep diabetic foot ulcers with Promogran in four European countries. *J Wound Care* 11:70–74.
61. Kakagia, D. D., K. J. Kazakos, and K. C. Xarchas, et al. (2007). Synergistic action of protease-modulating matrix and autologous growth factors in healing of diabetic foot ulcers. A prospective randomized trial. *J Diabetes Complications* 21:387–391.
62. Nisi, G., C. Brandi, L. Grimaldi, M. Calabrò, and C. D'Aniello. (2005). Use of a protease-modulating matrix in the treatment of pressure sores. *Chir Ital* 57:465–468.
63. Vin, F., L. Teot, and S. Meaume. (2002). The healing properties of Promogran in venous leg ulcers. *J Wound Care* 11:335–341.
64. Kakagia, D. D., K. J. Kazakos, and K. C. Xarchas, et al. (2007). Synergistic action of protease-modulating matrix and autologous growth factors in healing of diabetic foot ulcers. A prospective randomized trial. *J Diabetes Complications* 21:387–391.
65. Harding, K. G., H. L. Morris, and G. K. Patel. (2002). Science, medicine and the future: Healing chronic wounds. *BMJ* 324:160–163.
66. Stojadinovic, A., J. W. Carlson, G. S. Schultz, T. A. Davis, and E. A. Elster. (2008). Topical advances in wound care. *Gynecol Oncol* 111:S70–S80.
67. Wieman, T. J., J. M. Smiell, and Y. Su. (1998). Efficacy and safety of a topical gel formulation of recombinant human platelet-derived growth factor-BB (becaplermin) in patients with chronic neuropathic diabetic ulcers. A phase III randomized placebo-controlled double-blind study. *Diabetes Care* 21:822–827.
68. Hunter, J. E., L. Teot, R. Horch, and P. E. Banwell. (2007). Evidence-based medicine: vacuum-assisted closure in wound care management. *Int Wound J* 4:256–269.

69. Thompson, J. T., and M. W. Marks. (2007). Negative pressure wound therapy. *Clin Plast Surg* 34:673–684.
70. Webb, L. X., and H. C. Pape. (2008). Current thought regarding the mechanism of action of negative pressure wound therapy with reticulated open cell foam. *J Orthop Trauma* 22:S135–S137.
71. Carter, M. J. and R. A. Warriner III. (2009). Evidence-based medicine in wound care: time for a new paradigm. *Adv Skin Wound Care* 22:12,14–16.
72. Britton, A., M. McKee, N. Black, K. McPherson, C. Sanderson, and C. Bain. (1998). Threats to applicability of randomised trials: Exclusions and selective participation. *Health Technol Assess* 2:1–124.
73. Carter, M. J., C. E. Fife, D. Walker, and B. Thomson. (2009). Estimating the applicability of wound-care randomized controlled trials to general wound care populations by estimating the percentage of individuals excluded from a typical wound care population in such trials. *Adv Skin Wound Care* 22:316–324.
74. Benson, K., and A. J. Hartz. (2000). A comparison of observational studies and randomized, controlled trials. *N Engl J Med* 342:1878–1886.
75. Concato, J., N. Shah, R. I. Horwitz. (2000). Randomized, controlled trials, observational studies, and the hierarchy of research designs. *N Engl J Med* 342:1887–1892.
76. Ioannidis, J. P., A. B. Haidich, and M. Pappa, et al. (2001). Comparison of evidence of treatment effects in randomized and nonrandomized studies. *JAMA* 286:821–830.
77. McKee, M., A. Britton, N. Black, K. McPherson, C. Sanderson, C. Bain. (1999). Methods in health services research. Interpreting the evidence: choosing between randomised and non-randomised studies. *BMJ* 319:312–315.
78. Shikata, S., T. Nakayama, Y. Noguchi, Y. Taji, and H. Yamagishi. (2006). Comparison of effects in randomized controlled trials with observational studies in digestive surgery. Ann Surg 244:668–676.
79. Dooley, J., G. King, and B. Slade. (1997). Establishment of reference pressure of transcutaneous oxygen for the comparative evaluation of problem wounds. *Undersea Hyperb Med* 24:235–244.
80. Dowd, G. S. E., K. Linge, and G. Bentley. (1983). Measurement of transcutaneous oxygen pressure in normal and ischaemic skin. *J Bone Joint Surg* 65B:79–83.
81. Hauser, C. J., and W. C. Shoemaker. (1983). Use of transcutaneous PO2 regional perfusion index to quantify tissue perfusion in peripheral vascular disease. *Ann Surg* 197:337–343.
82. Wipke-Tevis, D. D., N. A. Stotts, D. A. Williams, E. S. Froelicher, and T. K. Hunt. (2001). Tissue oxygenation, perfusion, and position in patients with venous leg ulcers. Nurs Res 50:24–32.
83. Fife, C. E., D. R. Smart, P. J. Sheffield, H. W. Hopf, G. Hawkins, and D. Clarke. (2009). Transcutaneous oximetry in clinical practice: Consensus statements from an expert panel based on evidence. *Undersea Hyperb Med* 36:43–53.
84. Moffatt, C. J., P. J. Franks, D. C. Doherty, R. Martin, R. Blewett, and F. Ross. (2004). Prevalence of leg ulceration in a London population. *Q J Med* 97:431–437.
85. Nelzén, O., D. Bergqvist, and A. Lindhagen. (1991). Leg ulcer etiology—A cross sectional population study. J Vasc Surg 14:557–564.
86. Ince, P., D. Kendrick, F. Game, and W. Jeffcoate. (2007) The association between baseline characteristics and the outcome of foot lesions in a UK population with diabetes. *Diabet Med* 24:977–981.
87. Prompers, L., M. Huijberts, and J. Apelqvist, et al. (2007). High prevalence of ischaemia, infection and serious comorbidity in patients with diabetic foot disease in Europe. Baseline results from the Eurodiale study. *Diabetologia* 50:18–25.
88. Viswanathan, V. (2007). The diabetic foot: Perspectives from Chennai, South India. *Int J Low Extrem Wounds* 6:34–36.

89. Gray, B. H., M. S. Conte, and M. D. Dake, et al. (2008). Atherosclerotic Peripheral Vascular Disease Symposium II: Lower-extremity revascularization: State of the art. *Circulation* 118:2864–2872.

90. Ubbink, D. T., G. H. Spincemaille, R. S. Reneman, and M. J. Jacobs. (1999). Prediction of imminent amputation in patients with non-reconstructible leg ischemia by means of microcirculatory investigations. *J Vasc Surg* 30:114–121.

91. Arora, S., F. Pomposelli, F. W. LoGerfo, and A. Veves. (2002). Cutaneous microcirculation in the neuropathic diabetic foot improves significantly but not completely after successful lower extremity revascularization. *J Vasc Surg* 35:501–505.

92. Sen, C. K. (2009). Wound healing essentials: Let there be oxygen. *Wound Repair Regen* 17:1–18.

93. Nikol, S., I. Baumgartner, and E. Van Belle, et al. (2008). Therapeutic angiogenesis with intramuscular NV1FGF improves amputation-free survival in patients with critical limb ischemia. *Mol Ther* 16:972–978.

94. Bishop, J. B., L. G. Phillips, and T. A. Mustoe, et al. (1992). A prospective randomized evaluator-blinded trial of two potential wound healing agents for the treatment of venous stasis ulcers. *J Vasc Surg* 16:251–257.

95. Blair, S. D., C. M. Backhouse, D. D. I. Wright, E. Riddle, and C. N. McCollum. (1988). Do dressings influence the healing of chronic venous ulcers? *Phlebology* 3:129–134.

96. Fumal, I., C. Braham, P. Paquet, C. Piérard-Franchimont, and G. E. Piérard. (2002). The beneficial toxicity paradox of antimicrobials in leg ulcer healing impaired by a polymicrobial flora: A proof-of-concept study. *Dermatology* 204:70–74.

97. Leaper, D. J. (2006). Silver dressings: Their role in wound management. *Int Wound J* 3:282–294.

98. Lansdown, A. B., A. Williams, S. Chandler, and S. Benfield. (2005). Silver absorption and antibacterial efficiency of silver dressings. *J Wound Care* 14:155–160.

99. Vermeulen, H., J. M. van Hattem, M. N. Storm-Versloot, and D. T. Ubbink. (2007). Topical silver for treating infected wounds. *Cochrane Database Syst Rev* 1:CD005486.

100. Chambers, H., J. C. Dumville, and N. Cullum. (2007). Silver treatments for leg ulcers: A systematic review. *Wound Repair Regen* 15:165–173.

101. Lo, S. F., M. Hayter, C. J. Chang, W. Y. Hu, and L. L. Lee. (2008). A systematic review of silver-releasing dressings in the management of infected chronic wounds. *J Clin Nurs* 17:1973–1985.

102. Altman, D. G., K. F. Schulz, and D. Moher, et al. (2001). The revised CONSORT statement for reporting randomized trials: Explanation and elaboration. *Ann Intern Med* 134:663–694.

103. Downs, S. H., and N. Black. (1998). The feasibility of creating a checklist for the assessment of the methodological quality both of randomised and non-randomised studies of health care interventions. *J Epidemiol Community Health* 52:377–384.

104. Harbour, R, and J. Miller. (2001). A new system for grading recommendations in evidence based guidelines. *BMJ* 323:334–336.

105. Atkins, D., D. Best, and P. A. Briss, et al. (2004). Grading quality of evidence and strength of recommendations. *BMJ* 328:1490.

Index

Key: page numbers for tables are boldface, and for figures are italic.

A

abrasions, 104
acetic acid, 308–309
    as antimicrobial intervention for wounds, 308–309
    efficacy on biofilms, 309
    mode of action, 309
acetic acid (vinegar), 308
Acinobacter species, 72
acute wounds, 187
    beta-hemolytic *streptococcus*, 197
    complicated, 188
    matrix metalloproteinases (MMPs) in, *256*, *379*
    Pseudomonas aeruginosa, 198
    species diversity in, 196–197
    Staphylococcus aureus, 197
    tissue inhibitors of MMPs (TIMP) in, *256*, *379*
adaptive immunity–wound interactions, 281
adhesins, 335
aerobic bacteria. *see* bacteria, aerobic
African sleeping sickness, 12
Agency for Health Care Policy and Research (AHCPR), 353
Agency for Healthcare Research and Quality (AHRQ), 353
aging, healing of chronic and impaired wounds and, 171
algicides, 13
alginate dressings. *see* wound dressings, alginate
allograft, 137
American Civil War, amputation and, 346
amputation
    American Civil War and, 346
    diabetes and, 221
    mortality rates and, 346
anaerobic bacteria. *see* bacteria, anaerobic
ankle/brachial index (ABI), 122
anthrax, 4
antibiotics, 301–302
    effectiveness of, 14
    mode of action of, **14**
    resistance to, 16, 36
    topical, 301
    treatment of wound infections and, **303**
antibiotics used systematically in treating wound infections, target sites for, **303**

antimicrobial agents
    action mechanism of, 14
    control of microorganisms and, 13
    death of microorganisms following exposure to, 13
    effectiveness of, 14
    intervention for wounds, 295–296
    types of, 295–296
antimicrobial agents in bacteria, target sites of, *298*
antimicrobial interventions for wounds, 293–328
    acetic acid, 308–309
    antibiotics, 301–302
    antimicrobial agents, types of, 295–296
    antiseptics, 303
    bacteriophages, 313–314
    chlorhexidine, 307–308
    essential oils, 316
    Ethylenediamine Tetraacetic Acid (EDTA), 314–315
    honey, 309
    hydrogen peroxide, 311–312
    iodine, 304–305
    lactoferrin, 312
    physical/mechanical methods for reduction/removal of bioburden and biofilms, 299–300
    polyhexamethylene biguanides (PHMB), 306
    quorum-sensing inhibitors, 316
    rationale for using, 297–298
    silver, 305–306
    Xylitol, 312–313
antimicrobial peptides, 280
antimicrobial strategies, conventional, **297**
antiseptics, 13, 303
    target sites for, **302**
Apligraf, 358–359, 362
Aquacel®, 334, 362
Argenta Morykwas technique, 332
Aristotle, 336
arterial ulcers, 120
aspirate/tissue, 205–206

B

B lymphocytes (B cells), 282–283
bacteria, 5–9
    aerobic, 262
    anaerobic, 263

biofilms, 260–261
chemotaxis, 8
components of, **6**
Gram negative, 6
Gram positive, 6
MMPs, 260–261
role of in wound healing, *190*
shapes, 5
sizes, 5
unculturable, 201–202
VBNC (viable but not curable), 208
bacteria, growth of, 9–10
death phase, 9
exponential phase, 9
factors affecting, 11
lag phase, 9
oxygen and, 11
pH and, 11
stationary phase, 9
temperature and, 11
bacteria, structure of, *6*, 6–8
capsule/slime layer, 7
cell wall, 6, 7
flagella, 8
nucleoid, 8
pili and fimbriae, 8
plasmids, 8
ribosomes, 8
bacterial myonecrosis, 100–101
diagnosis, 101
management, 101
bacterial resistance, mechanism of, 15
bacterial skin residents, associated dermatoses
and, **66–67**
bacterial virulence determinants, control of, 335
adhesins, 335
biofilm formation, 335
immuno-evasive measures, 335
invasins, including bacterial (i.e., exogenous)
proteases, 335
pili, 335
quorum sensing, 335
synergy, 335
toxins: endo- and exotoxins, 335
wound dressings and topical treatments in
bioburden control, 335
bacteriocides, 13
bacteriophages, 313–314
as antimicrobial intervention for wounds,
313–314
bacteriophage virus replication within
a bacterial cell (viral infection of a
bacterium), *311*
efficacy on biofilms, 314
mode of action, 313

Bassi, Agostino, 4
beta-hemolytic streptococcus, 197
bioburden, physical/mechanical methods for
reduction/removal of, 299–300
as antimicrobial intervention for wounds,
299–300
biological debridement, 300
debridement, 299
enzymic debridement, 300
biochemical reaction sequence (homostasis,
inflammation, proliferation, maturation),
349
biocide overview, *296*
bioengineered skin products, 355
biofilm infection
characterizing, 237–239
significance of biofilms to wound healing
and, 237–239
biofilm resistance, 43–45
binding/failure of the antimicrobial to
penetrate the biofilm, 44
heterogeneity, 45
induction of a biofilm phenotype, 45
slow growth and the stress response, 44
biofilmology, 1–59
biofilms. *see also* wound biofilms
acetic acid and, 309
bacteria and, 260–261
chronic wounds and, 284–285
detachment and dispersal in wounds, 243
development of, *23*
development of the conditioning film and
substratum effects, 22–23
endocarditis and, 38
evidence of in chronic wounds, 236
evidence of on the lumen of endotracheal
tubes, *41*
examples of where they occur, **21**
factors governing the development of, 32
formation of, 287–288, 335
formed on the inner lumen of an extubated
endotracheal tube, *42*
historical aspects of, 18–20
immune retardation and, 240–242
immune stimulation and, 240–242
mature, *23*
MMPs and, 260–261
occurence of, 20–21
public and medical health consequences of,
36–42
quorum sensing and, 32
significance to wound healing, 233–248
stages in formation of, 22
wound healing immunology and, 284–285
biofilms, physical/mechanical methods for
reduction/removal of, 299–300

as antimicrobial intervention for wounds, 299–300
biological debridement, 300
debridement, 299
enzymic debridement, 300
biofilms, public and medical health and, 36–42
central venous catheters, 40
chronic wounds, 43
cystic fibrosis, 38
dental water units, 37
drinking water, 36
endocarditis, 38
endotracheal tubes (ETTs), 40
hospital and domestic water, 37
indwelling and medical devices, 39
intra-amniotic infection, 39
kidney stones, 37
opthalmic infections, 42
oral infections, 42
osteomyelitis, 38
otitis media, 38
prostatitis, 39
rhinosinusitus, 40–41
urinary tract infections, 39
biofilms, wound healing and, 233, 233–248
biofilm detachment and dispersal in wounds, 243
biology of wound healing and, 233
characterizing biofilm infection, 237–239
development of a wound biofilm, 234
evidence of biofilms in chronic wounds, 236
extracellular polymeric substances (EPS), 235
immune stimulation and retardation by the biofilm, 240–242
biological debridement, 300
biopsy, indications for, **109**
biosurgery, 332
Black Death (bubonic plague), 3
Braden Scale, 117
bTEFAP, 201, 203, 206
bubonic plague, 3
burn care, wound management in, 135
burn center referral criterion, *96*
burn disease, 94
burn injuries, in United States, 90
burn wound management, 135–150
burn wound infection, 135
chemical lesions, 142
deep partial thickness burns (deep second degree), 139
donor sites, 141
first-degree burns, 136
full thickness burns (third degree), 139–140
long-term results, 142–143
mixed partial thickness burns, 139

principles of wound management in burn care, 135
superficial partial thickness burns (superficial second degree), 137–138
burn wounds
infection, 135
physiology of, 93
burns, degree classification
deep partial thickness burns (deep second degree), 139
deep partial thickness (deep second degree), 139
depth of, 92
first degree, 92
fourth degree, 93
full thickness burns (third degree), 139–140
superficial partial thickness burns (superficial second degree), 137–138
burns, size of, 94
rule of nines and, 94
burns, types of
contact burn, 91
electrical burns, 91
flame, 90
flash, 91
radiation, 91
scald, 90
thermal injuries, 90
Buruli ulcers, 105

C

catheter-related bloodstream infections (CRBSIs), 40
cell biology of normal and impaired healing
properties of cytokines and, 166–170
properties of key growth factors and, 166–170
cell interactions during normal healing, 156–161
hemostasis, 156
inflammation, 157
lymphocytes, 161
macrophages, 158
neutrophils, 157
resolution of inflammation, 159–160
cells, healing and, 152–155
resolution of inflammation, 159–160
cellular defects with the chronic wound, 174
Centers for Medicaid Services (CMS), 224, 362
CPCS (Common Procedure Coding System), 365
pressure ulcer guidelines and, 353
central venous catheters, biofilms and, 40
Chamberland, Charles, 5
Charcot foot, 115
Chariker-Jeter technique, 332

chemical injuries, 99
chemical lesions, 142
chemokines, 169
　cytokines and, 169
chemotaxis inhibitory protein of staphylococci
　　(CHIPS), 71, 287
chlorhexidine, 307–308
　as antimicrobial intervention for wounds,
　　307–308
　efficacy on biofilms, 308
　mode of action, 308
　structure of, *308*
　surgical site infections (SSI) and, 308
chronic rhinosinusitis, 228
chronic venous insufficiency (CVI), 361
chronic wound etiology, 173–174
　cell biology of normal and impaired healing,
　　173–174
　diabetic ulcers, 173
　pressure ulcers, 174
　venous leg ulcers, 173
chronic wound infection, stages of microbiology
　　progression, 189–192
　colonization: irreversible adhesion, 191
　colonization: reversible adhesion, 191
　contamination stage, 189–190
　critical colonization, 192
　infection—local and systemic, 192
chronic wounds, 105, **105**, 171–173, 188, 226, 228
　aging and, 171
　beta-hemolytic *Streptococcus*, 197
　biofilms and, 284–285
　cell biology of normal and impaired healing,
　　171–173
　cellular defects and, 174
　evidence of biofilms in, 236
　general factors impacting on healing, 171–173
　menopausal effects and, 172
　nutrition and, 172
　occurence and impact, 105–115
　Pseudomonas aeruginosa, 198
　public and medical health consequences of
　　biofilms and, 43
　species diversity in, 196–197
　Staphylococcus aureus, 197
　in United States, 105
　wound healing immunology and, 284–285
chronic wounds, factors affecting healing, 345–347
　colonization, infections, and wound healing,
　　347
　compliance with basic wound care, 350–354
　iconoclastic view of, 345–372
　pressure ulcers as an example of chronic
　　wounds, 346
　silver-impregnated wound dressings as
　　an example of advanced technology,
　　363–365
　technology ceiling and, 355–362
　treatment of, 348–349
chronic wounds, occurence and impact, 105–115
　diabetic ulcer, 112–115
　venous ulcer, 107–111
chronic wounds, treatment of, 348–349
　basic wound care, 350–351
Church, John, 332
climax community, 31
clinical practice guidelines (CPGs), 351
collagen, 163
collagenases, 250–255
colonization
　infections and, 347
　irreversible adhesion, 191
　reversible adhesion, 191
　wound healing and, 347
community, role of
　community hypothesis and wound infections/
　　healing, 199–200
complement system, 280
conditioning film, *23*
confocal laser scanning microscopy (CLSM), 20
CONSORT statement, 362
contact burn, 91
contamination stage, 189–190
Contreet®, 362
corynebacterium, 69
Costerton, Bill, 20
Courtois, Bernard, 304
critical colonization, 192
Crohn's disease, 228
cystic fibrosis, biofilms and, 38
cytokines
　cell biology of normal and impaired healing,
　　165
　chemokines and, 169
　epidermal growth factor and, 168
　fibroblast growth factor and, 167
　insulin-like growth factor and, 169
　interleukin-1 and, 169
　platelet-derived growth factor and, 166
　properties of, 166–170
　in regulation of healing, 165
　transforming growth factor-β, 167
　tumor necrosis factor-α, 170
　vascular endothelial growth factor and, 168
cytokines, properties of
　cell biology of normal and impaired healing,
　　166–170

D

Darwin, Charles, 20
debridement, 299–300
　biological, 300–301
　enzymic, 300

denaturing gradient gel electrophoresis (DGGE), 202, 203
depth diagnosis, thermal injuries, 93
Dermagraft, 362
dermal healing, normal, *153*
dermatoses, skin bacteria and, **66–67**
diabetes, 285
    amputation and, 221
    in Mexico, 113
    peripheral neuropathy, 221
    in Sub-Saharan Africa, 113
    Type II, 116
    in United States, 112
diabetic foot ulcers (DFUs), 171, 220–221, 350
diabetic neuropathy, 113
diabetic ulcers (DUs), 112–115, 173
dialysis, 39
disease, control of microorganisms and, 17
disinfection, 296
disseminated intravascular coagulation (DIC), 102
DNA, 208, 228–229, 235, 300, 307
donor sites, 139
Doppler study, 122
dressings. *see* wound dressings
drinking water
    biofilms and, 36
    public and medical health and, 36

E

edema, 285
    of the lower limb, *110, 373*
electrical burns. *see* burns, electrical
endocarditis, 228
    biofilms and, 38
endothelial cells, 154
endotracheal tubes (ETTs), biofilms and, 40
enzymes, and proteases, 249
    factors that stimulate MMPs, 260–263
    history and structure of MMPs, 250
    management of MMPs, 263
    overall MMPs in chronic wounds, 259
    regulation of MMPs, 257–259
    types, modes of action, and sources of MMPs, 250–257
    wounds and, 249–270
enzymes and proteases
    factors that stimulate MMPs, 260–263
    history and structure of MMPs, 250
    management of MMPs, 263
    overall MMPs in chronic wounds, 259
    regulation of MMPs, 257–259
    types, modes of action, and sources of MMPs, 250–257
    wounds and, 249, 249–270
enzymic debridement, 300

epidemiology, control of microorganisms and, 16
epidermal growth factor (EGF), 168, 275
epidermis of the skin, components of, **61**
epithelialization, 178
Erlich, Paul, 12
essential oils, as antimicrobial intervention for wounds, 316
Ethylenediamine Tetraacetic Acid (EDTA), 314–315
    as antimicrobial intervention for wounds, 314–315
    efficacy on biofilms, 315
    mode of action, 314–315
extracellular polymeric substances (EPS), 235, 300
    biofilms and, 235

F

FDA. *see* U.S. Food and Drug Administration (FDA)
fibroblast growth factor, 167
fibroblasts, 155
first-degree burns, 136
flame burn, 90
flash burn, 91
Flemming, Alexander, 12, 346
florescent in situ hybridization (FISH), 201
fluid handling properties
    absorption/retention, 337
    lateral wicking, 337
    sequestration, 337
    wound dressings and topical treatments in bioburden control, 337
Food and Drug Administration (FDA). *see* U.S. Food and Drug Administration (FDA)
Fournier's gangrene, 100
Fracastroro, Girolamo, 3
frostbite, 98
full thickness burns (third degree), 139–140
fungal species, 73
    bacterial interactions on the skin, 73
fungicides, 13

G

Galen of Pergamum, 4, 345
gangrene, 100
gelatinases, 256
germ theory, 4
Gibbs free energy, 25
Gram, Christian, 6
Gram staining, 6–7
Gram-negative bacteria, 67
granulation tissue formation, 162–163, 176
Gravitational Platelet Separation System, 353

Group A Streptococcus *(Streptococcus pygogenes)*, 72
growth factors
    cell biology of normal and impaired healing, 165
    cytokines and chemokines, 169
    epidermal growth factor, 168
    fibroblast growth factor, 167
    insulin-like growth factor, 169
    interleukin-1, 169
    platelet-derived growth factor, 166
    properties of, 166–170
    regulation of healing and, 165
    transforming growth factor-β, 167
    tumor necrosis factor-α, 170
    vascular endothelial growth factor, 168

H

healing
    factors affecting stages of and treatments to overcome problems associated with these factors, **356–358**
    principle cells involved in, 152–155
healing, normal and impaired, 152
    cell biology of, 151–186
    cell interactions during normal healing, 156–161
    cellular defects with the chronic wound, 174
    chronic and impaired wounds, 171–173
    chronic wound etiology, 173–174
    cytokines and growth factors in regulation of healing, 165
    epithelialization, 178
    granulation tissue formation, 162–163, 176
    inflammation, 174–175
    modulation of wound cell biology by clinical interventions, 178–179
    principle cells involved in healing, 152–155
    properties of key growth factors and cytokines, 166–170
    reepithelialization, 164
    remodeling, 165
healing, principle cells involved in, 152–155
    endothelial cells, 154
    fibroblasts, 155
    keratinocytes, 155
    macrophages, 154
    neutrophils, 153
    platelets and, 152
Health Insurance Portability and Accountability Act (HIPAA) privacy requirements, 351
healthcare-associated infections (HAIs), 73
hemostasis, 156
Hesse, Walther, 4
Hippocrates, 309

HIV/AIDS, 109
hollowing, 35
honey, 333
    as antimicrobial intervention for wounds, 309–310
    efficacy on biofilms, 310
    history of wound treatment with, 309–310
    mode of action, 310
    as a wound dressing, 333
hormone replacement therapy (HRT), 172
hospital and domestic water, biofilms and, 37
Human Microbiome Project (HMP), 77
human skin
    anatomy and characteristics of, 60–62
    distribution of microbial flora on the skin, 63–64
    protective and defensive mechanisms of, 63–65
human skin, microbial flora and, 59–83
    inhibitory factors, 63
    other factors, 64
hydrocolloids, 331
hydrogen peroxide, 309, 311–312
    as antimicrobial intervention for wounds, 311–312
    efficacy on biofilms, 312
    mode of action, 311
hyperbaric oxygen (HBO), 334
hyperbaric oxygen therapy (HBOT), 355, 362
hypertrophic scarring, 95, 142

I

IgG antibodies, *284*
immune evasion, bacterial, 286
immune retardation, biofilms and, 240–242
immune stimulation, biofilms and, 240–242
immune system
    coordinated responses of innate and adaptive, 284
    innate and adaptive compartments of, *273, 379*
    overview of, 272–273
immune system, adaptive
    components of, 282–283
immune system, innate
    components of, 277–280
    wound healing and, 274–275
immuno-evasive measures, 335
immunology
    adaptive immunity–wound interactions, 281
    bacterial immune evasion, 286
    biofilm formation, 287–288
    biofilms and, 271–292
    chronic wounds, 284–285
    components of the adaptive immune system, 282–283

components of the innate immune system,
     277–280
coordinated responses of the innate and
     adaptive immune systems, 284
focus on *S. aureus* and *P. aeruginosa* immune
     evasion, 286
identification of microbial intruders, 276
the innate immune system and wound
     healing, 274–275
overview of the immune system, 272–273
wound healing and, 271–292
immunosuppression, 204
impaired wounds, 171–173
aging and, 171
cell biology of normal and impaired healing,
     171–173
general factors impacting on healing, 171–173
menopausal effects and, 172
nutrition and, 172
other factors and, 173
indigenous microbiota, distribution of, 65–77
fungal species, 73–74
molecular approaches to the investigation of
     skin microbiota, 68
normal residents of skin in health and
     infection, 69–73
skin flora and infection, 74–75
skin microflora and human immunity, 76–77
indwelling and medical devices, biofilms and, 39
infections, 226–229. *see also* biofilm infections;
     wound infections
causes of, 226
colonization and, 347
control of microorganisms and, 17
intra-amniotic, 39
local and systemic, 192
medical devices and, 228
opthalmic, 42
oral, 42
skin flora and, 74–75
urinary tract, 39
wound healing and, 347
infections, nosocomial, 36
biofilms and, 36
in the United States, 36
inflammation, 157
cell biology of normal and impaired healing,
     174–175
during normal healing, 157
resolution of during normal healing, 159–160
inhalation injuries, 94
Institute for Advanced Wound Care, 109
insulin-like growth factor, 169
Integra®, 140

Intellicure Clinical Documentation and Facility
     Management Software Intellicure
     database, 352
"Intellicure Research Consortium," 351
interleukin-1, 169
intra-amniotic infection, public and medical
     health consequences of biofilms and, 39
invasins, including bacterial (i.e., exogenous)
     proteases, 335
iodine, 304–305
as antimicrobial intervention for wounds,
     304–305
effectiveness on biofilms, 304–305
mode of action, 304–305
ischemia and cellular hypoxia, 361

J

Jenner, Edward, 4
Johnson & Johnson, 355
Johnson & Johnson Wound Management, 362
Joint Commission on Accreditation of Healthcare
     Organizations (JCAHO), 353

K

keloid formation, 98
keratinocytes, 155
kidney stones, biofilms and, 37
Koch, Robert, 4, 226, 229
Koch's postulates, 4
Kremlin technique, 332

L

lactoferrin, 312
as antimicrobial intervention for wounds, 312
efficacy on biofilms, 312
mode of action, 312
Langerhans cells, 61, 63
Laser Doppler flowmetry, 93
leprosy, 221
Levine technique, 205, 348
lipodermatosclerosis, 109, *111*
Lister, Joseph, 4, 303, 346
Lucretius, 3
lymphocytes, normal healing and, 161

M

macrophage-derived products relevant to healing,
     160–161
macrophages, 154, 278
cell interactions during normal healing, 158
involvement in healing, 154
normal healing and, 158

maggots, 332–333
    debridement and, 300–301
Marjolin's ulcer, 143
mast cells, 279
matrix metalloproteinases (MMPs)
    bacteria and, 260–261
    biofilms and, 260–261
    classification of, **251–254**
    collagenases, 250–255
    expression and cellular source of in acute
        wounds, *256, 379*
    expression in the epidermis and dermis
        during wound healing, *255*
    gelatinases, 256
    history and structure of, 250
    management of, 263
    membrane-type metalloproteinases, 257
    mutual activation of, *261*
    stromelysins, 256
    types, modes of action, and sources of,
        250–257
matrix metalloproteinases (MMPs), factors that
        stimulate, 260–263
    aerobic bacteria, 262
    anaerobic bacteria, 263
    bacteria, biofilms, and MMPs, 260–261
matrix metalloproteinases (MMPs), regulation
        of, 257–259
    chronic wounds and, 259
    tissue inhibitors of metalloproteinases
        (TIMPs), 258
    transcriptional level regulation, 257
    zymogen activation, 257
mechanical blisters, 104
medical device infections, 228
melanin, 61
Memorial Hermann Hospital, 350
menopausal effects, healing of chronic and
        impaired wounds and, 172
Merkel cells, 61
metalloproteinases, membrane-type, 257
microbial adhesion, 26–28
    role of pili and fimbriae in adhesion, 27–28
microbial barrier, properties of wound dressings
        and, 336
microbial control, terms used to describe, 13
microbial flora, human skin and, 59, 59–83, 63–64
    distribution of indigenous microbiota, 65–77
    investigations into the normal flora of the
        skin of healthy adults, 65
microbial growth, typical location in a batch
        culture test tube, *12*
microbial intruders, identification of, 276
microbial life cycle, *21*
microbial world, 5
microbially induced corosion (MIC), 20

microbiological specimen handling for laboratory
        culture-based diagnostics, 207
microbiological specimen handling for laboratory
        molecular-based diagnostics, 208
microbiology, 1–59
    history of, 3–4
micrococcaceae, 69
microorganisms
    mean generation times for, **9**
    numbers of and delayed healing, *195*
    sensitivity to available antimicrobial agents in
    types of, *295*
microorganisms, close proximity with the surface
        and, 24–25
    detachment and dispersal of the biofilm, 33–35
    factors governing the development of
        biofilms, 32
    growth and division of microorganisms at the
        colonized surface, 29
    microbial adhesion, 26–28
microorganisms, control of, 12–17
    acquisition of pathogens, 17
    control of disease and infection, 17
    control of nosocomial infections, 17
    death of microorganisms following exposure
        to antimicrobial agents, 13
    effectiveness of the antibiotic or
        antimicrobial, 14
    epidemiology, 16
    mechanism of action of antimicrobials, 14
    mechanism of bacterial resistance, 15
    sources of pathogens, 17
    terms used to describe microbial control, 13
    transmission of antibiotic resistance, 16
microorganisms, growth and division of at the
        colonized surface, 29
    biofilm structure, 31
    extracellular polymeric substances, 30
    gene transfer, 30
    microcolony and biofilm formation, 29
microorganisms, role in wound healing, 193–195
    numbers, 194–195
moist exposed burn ointment (MEBO), 136
moist wound healing (MWH), 330
MRSA (methicillin-resitant *S. aureus*), 71, 196,
        197, 204, 316

N

National Guideline Clearinghouse Web site, 352
National Health System (NHS), 352
National Institute of Clinical Excellence (NICE),
        352
National Institutes of Health, 227
National Institutes of Health Roadmap for
        Medical Research, 77
    Human Microbiome Project (HMP), 77

National Pressure Ulcer Advisory Panel
     (NPUAP), 117–118, 347
natural killer (NK) cells, 277, 279
necrosis, pressure and, 224
necrotizing fasciitis, 100–101
     diagnosis, 101
     management, 101
Needham, John, 3
negative pressure wound therapy (NPWT), 332,
     355
     Argenta Morykwas technique, 332
     Chariker-Jeter technique, 332
     Kremlin technique, 332
neomycin, 12
neutrophils, 153
     in healing, 153, 157
nod-like receptors (NLRs), 156
Noridian, 353
normal healing, cell interactions during, 156–161
     hemostasis, 156
     inflammation, 157
     lymphocytes, 161
     macrophages, 158
Norton Scale, 117
nosocomial infections, control of, 17
nutrition, healing of chronic and impaired
     wounds and, 172

O

ocular infections, biofilms and, 42
opsonization, 280
opthalmic infections, biofilms and, 42
oral infections, biofilms and, 42
osteomyelitis, biofilms and, 38
otitis media, biofilms and, 38

P

Paré, Ambroise, 345–346
Pasteur, Louis, 3, 4, 5
pasteurization, 3
pathogen-associated molecular patterns (PAMP),
     228, 276
pathogenicity, 17
pathogens, 17
penicillin, 12, 346
Penicillium notatum, 12
periodontal diseases, 42
peripheral arterial disease (PAD), **357**, 361
peripheral neuropathy, 221
peripheral vascular disease, evaluation of a
     patient with, **120**
Petri, Richard, 4
phagocytosis, process of, *277, 380*
physical and chemical barriers to entry at the
     surfaces of the body, *274*

physical/mechanical methods for reduction/
     removal of bioburden and biofilms,
     299–300
     as antimicrobial intervention for wounds,
          299–300
     biological debridement, 300
     debridement, 299
     enzymic debridement, 300
pili, 335
planktonic bacteria, growth of in liquid, *9*
plantar hyperkeratosis in an immobile patient,
     241, 378
platelet-derived bioactive molecules involved in
     initiation of healing, **157**
platelet-derived growth factor (PDGF), 152, 166,
     275
     gene-cloned, 355
platelets, healing and, 152
polyhexamethylene biguanides (PHMB), 307
     as antimicrobial intervention for wounds, 306
     efficacy against biofilms, 307
     mode of action, 307
polymerase chain reaction (PCR), 201
pressure ulcers (PU), 116–119, 171, 224
     as an example of chronic wounds, 346
     incidence of in United States, 224
     other pressure ulcers, 119
     predisposing factors for development, **116**
     prevention of, 119
     Stage I, 116–117
     Stage II, 118
     Stage III, 118
     Stage IV, 118
     stages of, **117**
pretibial laceration, 103
Promogran, 353
Propionibacterium spp., 72
prostatitis, biofilms and, 39
proteinases, bacterial
     functions of, *262*
Pseudomonas aeruginosa, 72, 198
     in acute and chronic wounds, 198
     immune evasion, 286
     species diversity, 198
psoriasis, 73
purpura fulminans, 102

Q

quorum sensing, 335
quorum-sensing inhibitors, 316
     as antimicrobial intervention for wounds, 316

R

rabies, 5
radiation burn, 91

radiation necrosis, 91
rate-determining steps (RDS), 348–349
reactive nitrogen intermediates (RNIs), 64
reactive oxygen species (ROS), 84
reepithelialization, 164
Regranex®, 358–359, 362
relative wound reduction (%) partial wound
       healing outcomes for randomized
       controlled trials, meta-analysis of, *365*
remodeling, 165
RescuDerm®, 315
rhinosinusitus, 40–41
Riedl-Pfleiderer process, 311, *311*
RNA, 235
road rash, 104
Roadmap for Medical Research (National
       Institutes of Health), 77
Royal College of Nursing (UK), 352
rules of nines, *94*

S

sanitation, 13
scald, 90
Semmes-Weinstein nylon monofilament test, 114
septicemia, 73
sequestration, 334
Shaw, George Bernard, 365
Sherman, Ron, 332
silver
       as antimicrobial intervention for wounds,
             305–306
       efficacy on biofilms, 306
       mode of action, 306
silver nitrate solutions, 136
silver sulfadiazine (SSD), 305
Silvercel®, 362
silver-impregnated wound dressings
       clinical use of silver-impregnated dressings,
             363
       concept of, 363
       as example of advanced technology, 363–365
       systematic review and meta-analysis, 364–365
skin
       defense mechanisms of, **63**
       overall structure of, *61*
skin, normal microorganisms in health and
       infection, 69–73
       Acinobacter species, 72
       corynebacterium, 69
       Group A Streptococcus *(Streptococcus
             pygogenes)*, 72
       micrococcaceae, 69
       Propionibacterium spp., 72
       Pseudomonas aeruginosa, 72
       Staphylococcus aureus, 71
       Staphylococcus epidermidis, 70

skin donor sites, 100
skin microbiota
       bacteria and, **66–67**
       infection and, 74–75
       molecular approaches to the investigation
             of, 68
skin products, bioengineered, 355
skin tears, 103
slough, 238, 244
       overlapping the wound edge, *239*, *377*
smallpox, 4
species diversity
       in acute and chronic wounds, 196–197
       beta-hemolytic *Streptococcus*, 197
       Pseudomonas aeruginosa, 198
       Staphylococcus aureus, 197
Staphylococcus aureus, 71. see also MRSA
       (methicillin-resitant *S. aureus*); VISA
       (vancomycin-intermediate *S. aureus*);
       VRSA (vancomycin-resistant *S. aureus*)
       immune evasion, 286
       species diversity, 197
Staphylococcus epidermidis
       biofilms and, 94–95
       normal residents of skin in health and
             infection, 69–71
sterilization, 13, 296
Stevens-Johnson syndrome (SJS), 102, 104
streaming, 33
Streptococcus, beta-hemolytic, 197
Streptococcus pygogenes (Group A
       Streptococcus), 72
streptomycin, 12
stromelysins, 256
superficial wound swabs, 205
surgical site infections (SSI), 307
surgical wound management
       general guidelines, 87–99
surgical wounds. *see* wounds, surgical
synergy, 335
systemic inflammatory response syndrome
       (SIRS), 88, 95

T

T lymphocytes (T cells), 282–283
terramycin, 12
tetracycline, 12
Therapeutic Goods Agency, 309
thermal injuries, 90–97
       depth diagnosis, 93
       depth of burns, 92
       first aid and guidelines for referral, 95–96
       long-term results, 97
       physiology of the burn wound, 93
       radiation necrosis, 91

size of the burn, inhalation injury, and burn
    disease, 94
types of burns, 90
tissue, inflammation/infection, moisture, edge
    (TIME), 330
tissue inhibitors of metalloproteinases (TIMPs),
    258
    regulation of MMPs (matrix
        metalloproteinases) and, 258
tissue inhibitors of MMPs (TIMP)
    expression and cellular source of in acute
        wounds, *256, 379*
Tissue Viability Society, 353
toll-like receptors (TLRs), 156
total body surface area (TBSA), 94–95
toxic epidermal necrolysis, 102
toxic epidermal necrolysis (TEN), 102
toxins: endo- and exotoxins, 335
transcriptional level regulation, regulation of
    MMPs (matrix metalloproteinases) and,
    257
transforming growth factor-β, 167, 275
transport mechanisms (laminar flow and
    turbulent flow), 25
tumor necrosis factor-α, 170

U

ulcer, malignant
    masquerading as a venous ulcer, *112, 375*
ulcers, 84
    arterial, 120–121
    Buruli, 105
    diabetic. *see* diabetic ulcers (DUs)
    diabetic (foot). *see* diabetic foot ulcers
        (DFUs)
    Marjolin's ulcer, 143
    venous. *see* venous ulcers (VU)
    venous leg. *see* venous leg ulcers (VLU)
ultraviolet (UV) light, 61
urinary tract infections, biofilms and, 39
U.S. Food and Drug Administration (FDA), 311,
    353, 359
U.S. Surgeon General, 294

V

vaccines, 4, 5
van Leeuwenhoek, Antonie, 19, 42
varicose eczema, *110, 373*
varicose veins, extensive, *111, 374*
vascular endothelial growth factor (VEGF), 152,
    168, 334
VBNC (viable but not curable) bacteria, 208
venous leg ulcers (VLU), 107–111, 171, 173,
    222–223
venous ulceration, *112, 375*

venous ulcers (VU), 107–111, 348–349
ventilator-associated pneumonia (VAP), 40
VersaJet, 138, 140
viricides, 13
VISA (vancomycin-intermediate *S. aureus*), 71
von Willebrand Factor, 152
VRSA (vancomycin-resistant *S. aureus*), 71

W

Waksman, Selman, 12
wound biofilms
    development of, 234. *see also* biofilms
    wound healing and, 234
wound border, bright red as positive sign of host
    control of wound biofilm, *240, 378*
wound care, basic, 350–351
    clinical practice guidelines, 352–354
    compliance with, 350–354
wound care, technology and, 355–362
    advanced wound-healing technologies, 358
    efficacy vs. effectiveness, 359–360
    future of, 362
    high-tech wound healing treatments,
        355–357
    ischemia and cellular hypoxia, 361
wound cell biology
    cell biology of normal and impaired healing,
        178–179
    modulation of by clinical interventions,
        178–179
wound chronicity, generation of, *175, 376*
wound dressings, 330
    alginate, 331
    biological, 138
    biosurgery, 332
    generic classes of, 331–332
    leakage strike-through and, 336
    microbial barrier properties of, 336
    other topical treatments in bioburden control,
        329–344
    the perfect dressing/skin substitute, **138**
    silver-impregnated, 363
    topical treatments in bioburden control, 336
    value of close association with the wound
        bed, 336
wound dressings, generic classes
    alginate dressings, 331
    hydrocolloids, 331
wound dressings, other topical treatments in
    bioburden control and, 329, 330, 335
    close association of dressing with wound
        bed, 336
    control of bacterial virulence determinants,
        335
    dressing and leakage strike-through, 336

fluid handling properties: absorption/
retention, lateral wicking, sequestration,
337
generic dressing classes, 331–332
honey, 333
hyperbaric oxygen, 334
microbial barrier properties of dressings, 336
negative pressure wound therapy, 332
sequestration, 334
wound dressings, silver-impregnated
clinical use of, 363
the concept, 363
as an example of advanced technology,
363–365
systematic review and meta-analysis,
364–365
wound healing
biofilms and, 233
biology of, 233
colonization and, 347
factors delaying, **188**
factors influencing, **106**
four phases of normal healing, **106**
infections and, 347
outcomes for randomized controlled trials,
meta-analysis of, *365*
significance of biofilms and, 233–248
wound healing, biofilms and, 233
biofilm detachment and dispersal in wounds:
clinical significance, 243
biology of wound healing and, 233
characterizing biofilm infection, 237–239
development of a wound biofilm, 234
evidence of biofilms in chronic wounds, 236
extracellular polymeric substances (EPS),
235
immune stimulation and retardation by the
biofilm, 240–242
wound healing immunology, 271–292
adaptive immunity–wound interactions, 281
bacterial immune evasion, 286
biofilm formation, 287–288
biofilms and, 271–292
chronic wounds, 284–285
components of the adaptive immune system,
282–283
components of the innate immune system,
277–280
coordinated responses of the innate and
adaptive immune systems, 284
focus on *S. aureus* and *P. aeruginosa* immune
evasion, 286
identification of microbial intruders, 276
the innate immune system and wound
healing, 274–275
overview of the immune system, 272–273

Wound Healing Society, 105
wound healing treatments, high tech
short history of, 355–357
wound infections, 226
in American military injured in Iraq and
Afghanistan, 73
clinical indicators for, **204**
target sites for antibiotics used systematically
in treating, **303**
target sites for antiobiotics used
systematically in treating, **303**
wound infections/healing, community hypothesis
and, 198–202
the role of a community, 199–200
unculturable bacteria and wounds, 201–202
wound management, in burn care, 135
wound types, 220–225
diabetic foot ulcers (DFUs), 220–221
pressure ulcers (PU), 224
venous leg ulcers (VLU), 222–223
wound characteristics, 225
wound-healing technologies, advanced, 358
wound–microbiology life cycle, *190*
wounds, 83–134
antimicrobial interventions, rationale for
using, 297–298
biofilm detachment and dispersal in, 243
characteristics of, 225
classification and terminology, 84–87
edges of, 226, 239
immune defense of, *275, 380*
unculturable, 201–202
wounds, acute. *see* acute wounds
wounds, antimicrobial interventions for, 293–328
acetic acid, 308–309
antibiotics, 301–302
antimicrobial agents, types of, 295–296
antiseptics, 303
bacteriophages, 313–314
chlorhexidine, 307–308
essential oils, 316
Ethylenediamine Tetraacetic Acid (EDTA),
314–315
honey, 309
hydrogen peroxide, 311–312
iodine, 304–305
lactoferrin, 312
physical/mechanical methods for reduction/
removal of bioburden and biofilms,
299–300
polyhexamethylene biguanides (PHMB), 306
quorum-sensing inhibitors, 316
rationale for using, 297–298
silver, 305–306
Xylitol, 312–313
wounds, chronic. *see* chronic wounds

wounds, classification/terminology, 84–87
    conclusion, 86
    conversion, 86
    differences in physiology, 84
    differences in treatment, 85
    influence of microorganisms, 85
    treatment objectives and outcomes, 86
wounds, enzymes, and proteases and, 249,
        249–270
    factors that stimulate MMPs, 260–263
    history and structure of MMPs, 250
    management of MMPs, 263
    overall MMPs in chronic wounds, 259
    regulation of MMPs, 257–259
    types, modes of action, and sources of
        MMPs, 250–257
wounds, introduction to, 84–134
    arterial ulcers, 120
    chronic wounds: their occurence and impact,
        105–115
    classification/terminology, 84–87
    general guidelines of surgical wound
        management, 87–99
    necrotizing fasciitis and bacterial
        myonecrosis, 100–101
    pressure ulcers, 116–119
    pretibial laceration, 103
    purpura fulminans, 102
    road rash, abrasions, mechanical blisters, 104
    skin donor sites, 100
    skin tears, 103
    toxic epidermal necrolysis and Stevens-
        Johnson syndrome, 102
    zoonoses: bite and scratch wounds, 104
wounds, microbiology of, 187–218
    community hypothesis and wound infections/
        healing, 198–202
    introduction, 187
    microbiological specimen handling for
        laboratory molecular-based diagnostics,
        208

    role of microorganisms in wound healing,
        193–195
    sampling of a wound, 203–207
    species diversity in acute and chronic
        wounds, 196–197
    stages involved in microbiology progression
        of a chronic wound infection, 189–192
wounds, sampling of, 203–207
    aspirate/tissue, 205–206
    Levine technique, 205
    microbiological specimen handling for
        laboratory culture-based diagnostics, 207
    overview, 207
    superficial wound swabs, 205
wounds, surgical, 89–90
    chemical injuries, 99
    frostbite, 98
    general guidelines, 87–99
    lesions with pus, 89
    management of, 87–99
    sutured wounds, 89
    thermal injuries, 90–97
wounds and infections, types of, 219–232
    infection, 226–229
    introduction, 219
    wound types, 220–225

X

xenografts, 138
Xylitol, 312–313
    as antimicrobial intervention for wounds,
        312–313
    efficacy on biofilms, 313
    mode of action, 313

Z

zoonoses: bite and scratch wounds, 104
zymogen activation, 257
    regulation of MMPs and, 257